VOLUME NINETY NINE

ADVANCES IN PARASITOLOGY

VOLUME NINETY NINE

ADVANCES IN PARASITOLOGY

Edited by

D. ROLLINSON

Life Sciences Department
The Natural History Museum,
London, United Kingdom

J.R. STOTHARD

Department of Parasitology
Liverpool School of Tropical Medicine
Liverpool, United Kingdom

Academic Press is an imprint of Elsevier
125 London Wall, London EC2Y 5AS, United Kingdom
The Boulevard, Langford Lane, Kidlington, Oxford OX5 1GB, United Kingdom
50 Hampshire Street, 5th Floor, Cambridge, MA 02139, United States
525 B Street, Suite 1800, San Diego, CA 92101-4495, United States

First edition 2018

ISBN: 978-0-12-815192-1
ISSN: 0065-308X

For information on all Academic Press publications
visit our website at https://www.elsevier.com/books-and-journals

ELSEVIER
Book Aid International
Working together to grow libraries in developing countries
www.elsevier.com • www.bookaid.org

Publisher: Zoe Kruze
Acquisition Editor: Ashlie Jackman
Editorial Project Manager: Ana Claudia A. Garcia
Production Project Manager: Abdulla Sait
Cover Designer: Matthew Limbert

Typeset by SPi Global, India

CONTENTS

Contributors *ix*

1. Parasites of the Giant Panda: A Risk Factor in the Conservation of a Species **1**

Tao Wang, Yue Xie, Youle Zheng, Chengdong Wang, Desheng Li, Anson V. Koehler, and Robin B. Gasser

1. Introduction 2
2. Parasite Records for the Giant Panda 3
3. Conclusions and a Perspective on Future Research 23

Acknowledgement 26
References 26

2. The Evolutionary Biology, Ecology and Epidemiology of Coccidia of Passerine Birds **35**

Alex Knight, John G. Ewen, Patricia Brekke, and Anna W. Santure

1. Introduction 36
2. The Taxonomy and Life Cycle of Coccidia 37
3. Coccidia and Passerine Health 40
4. Epidemiology of Coccidia in Wild Passerines 42
5. Natural Selection 48
6. Sexual Selection 50
7. Future Directions 51
8. Conclusion 54

Acknowledgements 54
References 54

3. Monogenean Parasite Cultures: Current Techniques and Recent Advances **61**

Kate Suzanne Hutson, Alexander Karlis Brazenor, David Brendan Vaughan, and Alejandro Trujillo-González

1. Introduction 62
2. Establishing Monogenean Infections 66
3. Maintaining Monogenean Cultures 72
4. Hyperparasite Cultures 79

5. Viviparous Cultures 80
6. Amphibian Monogenean Cultures 80
7. Troubleshooting and Virulence 82
8. Animal Ethics and Biosecurity 83
9. Further Considerations and Conclusive Comments 84
Acknowledgements 84
References 84

4. Molecular Epidemiology of *Anisakis* and Anisakiasis: An Ecological and Evolutionary Road Map 93

Simonetta Mattiucci, Paolo Cipriani, Arne Levsen, Michela Paoletti, and Giuseppe Nascetti

1. Introduction 94
2. When Can an *Anisakis* Parasite Be Considered as a 'Biological Species'? 96
3. How Many Valid Species Are in the Genus *Anisakis*? 115
4. Reconciling Molecular and Morphological Results 146
5. How Does *Anisakis* spp. Diversity Vary Across Host Species? 157
6. Molecular Epidemiology of *Anisakis* spp. in Fisheries 167
7. Do *Anisakis* spp. Always Occupy the Same Site in Fish? 193
8. Human Anisakiasis: Which Are the *Anisakis* spp. Infective to Humans? 208
9. What Key Questions for Future Research Challenges? 228
Acknowledgements 239
References 240
Further Reading 262

5. Evolution, Systematics, and Biogeography of the Triatominae, Vectors of Chagas Disease 265

Fernando Araujo Monteiro, Christiane Weirauch, Márcio Felix, Cristiano Lazoski, and Fernando Abad-Franch

1. Introduction 266
2. Evolution of the Triatominae: From Predators to Blood Feeders 267
3. Systematics of the Triatominae 274
4. Biogeography of the Triatominae 296
5. Closing Thoughts and Conclusions 323
Acknowledgements 325
References 325

6. Expanding the Vector Control Toolbox for Malaria Elimination: A Systematic Review of the Evidence **345**

Yasmin A. Williams, Lucy S. Tusting, Sophia Hocini, Patricia M. Graves, Gerry F. Killeen, Immo Kleinschmidt, Fredros O. Okumu, Richard G.A. Feachem, Allison Tatarsky, and Roly D. Gosling

1. Introduction 346
2. Methods 348
3. Results 354
4. Discussion 370

Acknowledgements 375
Contributors 375
Conflict of Interest 375
References 375

CONTRIBUTORS

Fernando Abad-Franch
Grupo Triatomíneos, Instituto René Rachou, FIOCRUZ, Belo Horizonte, Brazil

Alexander Karlis Brazenor
Marine Parasitology Laboratory, Centre for Sustainable Tropical Fisheries and Aquaculture and the College of Science and Engineering, James Cook University, Townsville, QLD, Australia

Patricia Brekke
Institute of Zoology, Zoological Society of London, London, United Kingdom

Paolo Cipriani
Sapienza—University of Rome, Rome, Italy; "Umberto I" University Hospital, Rome, Italy; Laboratory affiliated to Istituto Pasteur Italia-Fondazione Cenci Bolognetti; Tuscia University, Viterbo, Italy; Institute of Marine Research, Bergen, Norway

John G. Ewen
Institute of Zoology, Zoological Society of London, London, United Kingdom

Richard G.A. Feachem
Malaria Elimination Initiative, Global Health Group, University of California, San Francisco, San Francisco, CA, United States

Márcio Felix
Laboratório de Biodiversidade Entomológica, Instituto Oswaldo Cruz, FIOCRUZ, Rio de Janeiro, Brazil

Robin B. Gasser
Faculty of Veterinary and Agricultural Sciences, The University of Melbourne, Parkville, VIC, Australia

Roly D. Gosling
Malaria Elimination Initiative, Global Health Group, University of California, San Francisco, San Francisco, CA, United States

Patricia M. Graves
College of Public Health, Medical and Veterinary Sciences and Australian Institute of Tropical Health and Medicine, James Cook University, Cairns, QLD, Australia

Sophia Hocini
Malaria Elimination Initiative, Global Health Group, University of California, San Francisco, San Francisco, CA, United States

Kate Suzanne Hutson
Marine Parasitology Laboratory, Centre for Sustainable Tropical Fisheries and Aquaculture and the College of Science and Engineering, James Cook University, Townsville, QLD, Australia

Gerry F. Killeen
Ifakara Health Institute, Ifakara, Tanzania; Liverpool School of Tropical Medicine, Liverpool, United Kingdom

Immo Kleinschmidt
MRC Tropical Epidemiology Group, London School of Hygiene and Tropical Medicine, London, United Kingdom; School of Pathology, Faculty of Health Sciences, University of Witwatersrand, Johannesburg, South Africa; Elimination 8, Windhoek, Namibia

Alex Knight
School of Biological Sciences, University of Auckland, Auckland, New Zealand

Anson V. Koehler
Faculty of Veterinary and Agricultural Sciences, The University of Melbourne, Parkville, VIC, Australia

Cristiano Lazoski
Instituto de Biologia, Universidade Federal do Rio de Janeiro, Rio de Janeiro, Brazil

Arne Levsen
Institute of Marine Research, Bergen, Norway

Desheng Li
China Conservation and Research Centre for the Giant Panda, Ya'an, Sichuan, China

Simonetta Mattiucci
Sapienza—University of Rome, Rome, Italy; "Umberto I" University Hospital, Rome, Italy; Laboratory affiliated to Istituto Pasteur Italia-Fondazione Cenci Bolognetti

Fernando Araujo Monteiro
Laboratório de Epidemiologia e Sistemática Molecular, Instituto Oswaldo Cruz, FIOCRUZ, Rio de Janeiro, Brazil

Giuseppe Nascetti
Tuscia University, Viterbo, Italy

Fredros O. Okumu
Ifakara Health Institute, Ifakara, Tanzania

Michela Paoletti
Tuscia University, Viterbo, Italy

Anna W. Santure
School of Biological Sciences, University of Auckland, Auckland, New Zealand

Allison Tatarsky
Malaria Elimination Initiative, Global Health Group, University of California, San Francisco, San Francisco, CA, United States

Alejandro Trujillo-González
Marine Parasitology Laboratory, Centre for Sustainable Tropical Fisheries and Aquaculture and the College of Science and Engineering, James Cook University, Townsville, QLD, Australia

Lucy S. Tusting
Big Data Institute, University of Oxford, Oxford, United Kingdom

David Brendan Vaughan
Marine Parasitology Laboratory, Centre for Sustainable Tropical Fisheries and Aquaculture and the College of Science and Engineering, James Cook University, Townsville, QLD, Australia

Chengdong Wang
China Conservation and Research Centre for the Giant Panda, Ya'an, Sichuan, China

Tao Wang
Faculty of Veterinary and Agricultural Sciences, The University of Melbourne, Parkville, VIC, Australia

Christiane Weirauch
University of California, Riverside, Riverside, CA, United States

Yasmin A. Williams
Malaria Elimination Initiative, Global Health Group, University of California, San Francisco, San Francisco, CA, United States

Yue Xie
College of Veterinary Medicine, Sichuan Agricultural University, Chengdu, Sichuan, China; Agricultural Research Service, Beltsville Human Nutrition Research Center, Diet, Genomics and Immunology Laboratory, Beltsville, MD, United States

Youle Zheng
College of Veterinary Medicine, Sichuan Agricultural University, Chengdu, Sichuan, China

CHAPTER ONE

Parasites of the Giant Panda: A Risk Factor in the Conservation of a Species

Tao Wang*[,1], Yue Xie[†,‡], Youle Zheng[†], Chengdong Wang[§], Desheng Li[§], Anson V. Koehler*, Robin B. Gasser*

*Faculty of Veterinary and Agricultural Sciences, The University of Melbourne, Parkville, VIC, Australia
[†]College of Veterinary Medicine, Sichuan Agricultural University, Chengdu, Sichuan, China
[‡]Agricultural Research Service, Beltsville Human Nutrition Research Center, Diet, Genomics and Immunology Laboratory, Beltsville, MD, United States
[§]China Conservation and Research Centre for the Giant Panda, Ya'an, Sichuan, China
[1]Corresponding author: e-mail address: tao.wang1@unimelb.edu.au

Contents

1. Introduction 2
2. Parasite Records for the Giant Panda 3
   2.1 Key Parasites Reported to Cause Clinical Problems 8
   2.2 Protists: An Emerging Issue? 21
3. Conclusions and a Perspective on Future Research 23
Acknowledgement 26
References 26

Abstract

The giant panda, with an estimated population size of 2239 in the world (in 2015), is a global symbol of wildlife conservation that is threatened by habitat loss, poor reproduction and limited resistance to some infectious diseases. Of these factors, some diseases caused by parasites are considered as the foremost threat to its conservation. However, there is surprisingly little published information on the parasites of the giant panda, most of which has been disseminated in the Chinese literature. Herein, we review all peer-reviewed publications (in English or Chinese language) and governmental documents for information on parasites of the giant pandas, with an emphasis on the intestinal nematode *Baylisascaris schroederi* (McIntosh, 1939) as it dominates published literature. The purpose of this chapter is to: (i) review the parasites recorded in the giant panda and describe what is known about their biology; (ii) discuss key aspects of the pathogenesis, diagnosis, treatment and control of key parasites that are reported to cause clinical problems and (iii) conclude by making some suggestions for future research. This chapter shows that we are only just 'scratching the surface' when it comes to parasites and parasitological research of the giant panda. Clearly, there needs to be a concerted research effort to support the conservation of this iconic species.

Advances in Parasitology, Volume 99
ISSN 0065-308X
https://doi.org/10.1016/bs.apar.2017.12.003

1. INTRODUCTION

The giant panda, *Ailuropoda melanoleuca*, is one of the world's most recognised and rarest animals. It is a solitary bear of the subfamily Ailuropodinae in the family Ursidae, whose diet is mainly bamboo (Swaisgood et al., 2006). Giant pandas usually live for 14–20 years in the wild and up to 38 years in captivity (China Conservation and Research Centre for the Giant Panda, CCRCGP, unpublished data; Schaller et al., 1985). From the 16th to the 19th centuries, this panda was distributed widely in Western China (Gansu, Hubei, Hunan, Shaanxi and Sichuan provinces; Zhu and Long, 1983). However, today, this species is restricted to 30–40 populations in six isolated mountain ranges at the eastern edge of the Tibetan plateau in China, i.e., the Minshan, Qionglai, Qinling, Daxiangling, Xiaoxiangling and Liangshan mountains (Fig. 1), with a total estimated population size of 1864 in the wild (The State Forestry Administration of China, 2015).

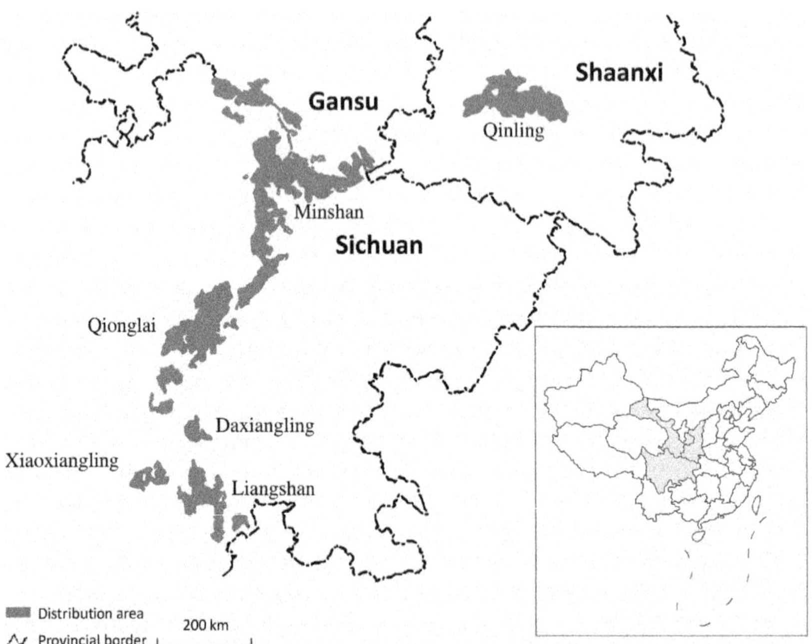

Fig. 1 Distribution of wild giant pandas in six mountain regions (Qinling, Minshan, Qionglai, Liangshan, Daxiangling and Xiaoxiangling) in China. *Adapted from Zhang, L., Wu, Q., Hu, Y., Wu, H., Wei, F., 2015. Major histocompatibility complex alleles associated with parasite susceptibility in wild giant pandas. Heredity (Edinb) 114, 85–93.*

To protect this iconic and threatened animal, more than 375 giant pandas have been raised in captivity in conservation centres and zoos (The State Forestry Administration of China, 2015).

Many factors threaten this endangered species, including habitat loss, degradation and fragmentation, poor reproduction and limited resistance to some infectious diseases (Feng et al., 1985; Wei et al., 2015; cf. Tables 1 and 2). Of these factors, diseases caused by parasites are reported to be a major threat to the conservation of the giant panda. In particular, a disease (baylisascariasis) caused by the ascaridoid nematode *Baylisascaris schroederi* (McIntosh, 1939) is a leading cause of deaths in wild populations (Zhang et al., 2008). Indeed, between 2001 and 2005, visceral larva migrans (VLM) linked to *B. schroederi* infection was reported to be responsible for 50% (12/24) of mortalities in the giant panda (Zhang et al., 2008).

However, surprisingly, there is limited published information on the parasites of the giant panda ($n=91$ peer-reviewed publications and archived governmental reports; 14 January 2017), most of which have been published in the Chinese literature ($n=55$ publications). Unpublished information from zoos, animal parks and breeding centres (sources available from authors) suggests that parasites of the giant panda continue to be a persistent and chronic issue, adversely impacting the health and conservation of this iconic animal. Therefore, we consider it appropriate and timely to review the literature on parasites of the giant panda, with an emphasis on *B. schroederi* because it dominates published literature. The purpose of this chapter is to: (i) review the parasites recorded in the giant panda and describe what is known about their life cycles; (ii) discuss key aspects of the pathogenesis, diagnosis, treatment and control of common parasites and (iii) conclude by making some suggestions for future research.

2. PARASITE RECORDS FOR THE GIANT PANDA

Since *B. schroederi* (originally named *Ascaris schroederi*) (McIntosh, 1939) was found in the small intestine from a giant panda in the New York Zoological Park (now the Bronx Zoo) in 1939, an increasing number of parasites have been identified in this animal species. To date, at least 29 parasite taxa, including 11 endoparasites (5 nematode, 1 trematode and 5 protozoan taxa) and 18 ectoparasites (13 tick, 2 mite, 2 flea taxa and 1 'blow fly') have been recorded in the giant panda (Table 2). However, many of these records are in the Chinese literature (publications or archival documents),

Table 1 Viruses and Bacteria Which Have the Potential to Threaten the Health and Conservation of the Giant Panda

Pathogens	Classification	Location	Signs or Problem	Comments	References
Viruses					
Canine distemper virus	Paramyxoviridae	Nervous, digestive and respiratory systems	Fever and inflammation	The most dangerous virus to the giant pandas	Hvistendahl (2015) and Zhao et al. (2017)
Canine adenovirus	Adenoviridae	Liver and brain			Qin et al. (2011)
Canine coronavirus	Coronaviridae	Stomach and intestine	Mental depression, vomiting, flatulence and watery diarrhoea	There is no comprehensive knowledge on CCV of giant pandas	Mainka et al. (1994)
Canine parvovirus	Parvoviridae	Unknown	Watery diarrhoea and vomiting		Mainka et al. (1994) and Qin et al. (2011)
Rotavirus	Reoviridae	Stomach and small intestine	Depression, anorexia vomiting and diarrhoea		Wang et al. (2008a)
Bacteria					
Clostridium perfringens (*Clostridium welchii*)	Bacillaceae	Intestine	Subclinical	Opportunistic pathogen; sudden death	Pan et al. (2001)
Escherichia coli	Enterobacteriaceae	Intestinal tract and vagina	Diarrhoea, hemorrhagic enterocolitis and intestinal mucosal inflammation	Opportunistic pathogen	Wang et al. (2013b)
Klebsiella pneumoniae	Enterobacteriaceae	Respiratory and intestinal tracts	Lethargy, mental depression, diarrhoea, inappetence, emaciation vomiting, haematochezia and haemorrhagic enterocolitis		Yang et al. (2016)
Proteus mirabilis	Enterobacteriaceae	Urogenital tract	Urinary urogenital infections	Opportunistic pathogen; might cause reproductive problems in females	Wang et al. (2007a)

Table 2 Parasites Recorded in the Giant Panda

Parasite	Classification	Location	First Report	Complications	Comments	References
Helminths						
Ancylostoma ailuropodae sp. nov.	Ancylostomatidae	Small intestine	First reported as *Ancylostoma caninum* in Sichuan province, China (2005). Renamed as *Ancylostoma ailuropodae*, based on morphological study (Xie et al., 2017)	n/a	Validity needs to be confirmed	Lai et al. (1991) and Xie et al. (2017)
Baylisascaris schroederi sp. nov.	Ascarididae	Small intestine	First reported in the Bronx Zoo, USA (1939)	Intestinal obstruction, emaciation, inflammation (enteritis and pneumonia), sometimes death	The most studied parasite of the giant panda	McIntosh (1939) and Sprent (1968)
Lungworm	n/a	n/a	First reported in Sichuan province, China (1991)	n/a	No genus and species names Identification based on egg examination	Lai et al. (1991)
Strongyloides sp.	Strongyloididae	Small intestine	First reported in Sichuan province, China (1991)	n/a	Identification based on egg examination. Parasite name misspelled as '*Storonglata*'	Lai et al. (1991)
Toxascaris seleactis	Ascarididae	Small intestine	First reported in Sichuan province, China (1991)	n/a	Identification based on egg examination	Lai et al. (1991)
Ogmocotyle sikae	Notocotylidae	Small intestine	First reported from an informal publication by He et al. (1987) in Shanxi province, China (without morphological description) First detailed morphological and phylogenetic studies conducted by Song et al. (2016)	n/a	Life cycle and pathogenesis unknown	He et al. (1987) and Song et al. (2016)
Protists						
Cryptosporidium sp.	Cryptosporidiidae	Gastrointestinal tract	First reported with a prevalence of 1.75% (1/57) in Sichuan province, China (2013), with morphological and phylogenetic analysis	No associated clinical signs	Potential novel genotype	Liu et al. (2013)
Cryptosporidium andersoni	Cryptosporidiidae	Gastrointestinal tract	First reported with a prevalence of 15.6% (19/122) and 0.5% (1/200) in captive and wild giant pandas, Sichuan province, China (2015)	No associated clinical signs		Wang et al. (2015)

Continued

Table 2 Parasites Recorded in the Giant Panda—cont'd

Parasite	Classification	Location	First Report	Complications	Comments	References
Enterocytozoon bieneusi	Microsporidia	Small intestine	First reported with a prevalence of 8.70% (4/46) in the northwest of China (2015)	No associated clinical signs	Potential novel genotype	Tian et al. (2015)
Toxoplasma gondii	Sarcocystidae	Liver, spleen, lungs, kidneys and intestines	First reported in a dead captive panda in Henan province, China (2015)	Fatal toxoplasmosis associated with serious respiratory and gastroenteritis signs	Potential novel genotype	Ma et al. (2015)
Sarcocystis sp.	Sarcocystidae	Muscle	Described in two reviews of parasites of the giant panda (Yang, 1998; Zhang et al., 2010); however, original literature is not available from either review	n/a		n/a
Ectoparasites						
Blowfly	Unknown		The gastroenteritis maggots were found in a giant panda (2007), without morphological description	n/a	No genus and species names	Li et al. (2007)
Dermacentor taiwanensis	Ixodidae	Skin surface	Described in two reviews of parasites of the giant panda (Yang, 1998; Zhang et al., 2010); however, original literature is not available from either review	n/a		n/a
Haemaphysalis aponommoides	Ixodidae	Skin surface	First reported in Sichuan province, China (1985)	n/a		Lai et al. (1990)
H. flava	Ixodidae	Skin surface	First reported in Sichuan province, China (1990) without morphological description; first detailed morphological and phylogenetic analysis were conducted by Cheng et al. (2013)	n/a		Cheng et al. (2013) and Lai et al. (1990)
H. hystricis	Ixodidae	Skin surface	Described in two reviews of parasites of the giant panda (Yang, 1998; Zhang et al., 2010); however, original literature is not available from either review	n/a		n/a
H. kitaotai	Ixodidae	Skin surface	Described in two reviews of parasites of the giant panda (Yang, 1998; Zhang et al., 2010); however, original literature is not available from either review	n/a		n/a

H. longicornis	Ixodidae	Skin surface	First reported in Sichuan province, China (1992) with detailed morphological description	n/a		Chen and Shi (1992)
H. megaspinosa	Ixodidae	Skin surface	First reported in Gansu province, China (1987) without morphological description	n/a		Ma (1987)
H. montgomeryi	Ixodidae	Skin surface	First reported in Gansu province, China (1987) without morphological description	n/a		Ma (1987)
H. warburtoni	Ixodidae	Skin surface	First reported in Sichuan province, China (1985) with detailed morphological description	n/a		Wu and Hu (1985a)
H. ailuropodae sp. nov.	Ixodidae	Skin surface	First reported in Shanxi province, China (1998) with detailed morphological description; no associated clinical signs reported	n/a	Validity needs to be confirmed	Yu et al. (1998)
Ixodes acutitarsus	Ixodidae	Skin surface	First reported in Gansu province, China (1987) without morphological description	n/a		Ma (1987)
I. granulatus	Ixodidae	Skin surface	First reported in Gansu province, China (1987) without morphological description	n/a		Ma (1987)
I. ovatux	Ixodidae	Skin surface	First reported in Gansu province, China (1987) without morphological description	n/a		Ma (1987)
Chorioptes panda sp. nov.	Psoroptidae	Skin surface	First reported in Paris Zoo, France (1975)	Mild alopecia, erythema and crusting affecting sleep and appetite		Fain and Leclerc (1975)
Chaetopsylla ailuropodae	Vermipsyllidae	Skin surface	First reported in Sichuan province, China (1991) with detailed morphological description	n/a		Qiu et al. (1991)
C. mikado	Vermipsyllidae	Skin surface	First reported in Sichuan province, China (1990) with detailed morphological description	n/a		Lai et al. (1990)
Demodex ailuropodae	Demodicidae	Hair follicles, sebaceous gland	First reported in Shanghai, China (1985), with detailed morphological description	n/a		Xu et al. (1986)

n/a, no information available.

often with very limited descriptions of morphology and/or other aspects. Beyond these records, reports of clinical cases often involve *B. schroederi*, the mite *Chorioptes panda* and ixodid ticks.

2.1 Key Parasites Reported to Cause Clinical Problems

2.1.1 Baylisascaris schroederi

B. schroederi (Nematoda: Ascaridoidea) is a large parasitic nematode inhabiting the small intestine of giant pandas and belongs to the genus *Baylisascaris*, of which there are 11 species, namely, *B. procyonis* of raccoons, *B. columnaris* of skunks, *B. potosis* of kinkajous, *B. ailuri* of red pandas, *B. transfuga* of bears, *B. melis* of badgers, *B. laevis* of marmots and ground squirrels, *B. devosi* of marten and fishers, *B. tasmaniensis* of Tasmanian devils and quolls and *B. venezuelensis* of spectacled bears (Kazacos, 2008; Pérez Mata et al., 2016; Sprent, 1968; Tokiwa et al., 2014; Tranbenkova and Spiridonov, 2017; Table 3). There is no evidence that *B. schroederi* affects other animal species or humans under natural conditions. Despite its massive health impact on wild giant panda populations in the early 20th century (Qiu and Mainka, 1993; Zhang et al., 2008), *B. schroederi* was first described in 1939 as *Ascaris schroederi* (McIntosh, 1939), before being renamed *B. schroederi* in 1968 (Sprent, 1968). Presently, baylisascariasis is considered the most harmful parasitic disease of the giant panda (Feng et al., 1985; Zhang et al., 2015).

Table 3 Recognized Species of *Baylisascaris*

Species[a]	Primary Definitive Host(s)[b]
Baylisascaris ailuri (Wu et al., 1987)	Red panda
Baylisascaris columnaris (Leidy, 1856)	Skunks
Baylisascaris devosi (Sprent, 1952)	Marten, fisher
Baylisascaris melis (Gedoelst, 1920)	European badger
Baylisascaris potosis (Tokiwa et al., 2014)	Kinkajou
Baylisascaris procyonis (Stefanski and Zarnowski, 1951)	Raccoon
Baylisascaris schroederi (McIntosh, 1939)	Giant panda
Baylisascaris transfuga (Rudolphi, 1819)	Bears
Baylisascaris laevis (Leidy, 1856)	Ground hog, ground squirrels
Baylisascaris tasmaniensis (Sprent, 1970)	Tasmanian devil, quolls
Baylisascaris venezuelensis (Pérez Mata et al., 2016)	Spectacled bear

[a]Adapted from Kazacos, K.R., 2008. *Baylisascaris procyonis* and related species. In: Samuel, W.M., Margo, J.P., Kocan A.A. (Eds.), Parasitic of Wild Mammals. Iowa State University Press, Ames, USA, pp. 301–341.
[b]List of definitive hosts is not extensive due to the numerous species recorded; see Sprent (1968) and Kazacos (2016) for extensive list.

Thus, *B. schroederi* has been the most studied parasite of this animal. Most published studies have investigated important aspects of this parasite's biology, epidemiology, pathogenesis, diagnosis, treatment and control.

2.1.1.1 Life Cycle

Similar to many other ascaridoids, *B. schroederi* has a complex life cycle in a single host, involving larval moults and development in several organ systems. Although not all aspects of this cycle have been proven, based on postmortem findings for giant pandas (CCRCGP, unpublished data) and experimentally infection studies in mice (Li, 1989, 1990a,b, 1993), evidence indicates that giant pandas become infected via the faecal–oral route, and larvae undergo hepatopulmonary and somatic migration prior to establishing as dioecious adults in the small intestine. Currently, there is no evidence of vertical or transmammary transmission of *B. schroederi*.

The current understanding is that, after eggs hatch in the intestine, the infective larvae penetrate the mucosa of the intestine. Subsequently, based on current knowledge, the larvae migrate via the mesenteric and portal blood system to the liver and then the lungs, and eventually return (via the trachea) to the intestinal lumen, where they mature to adults, mate and reproduce. Unembryonated eggs are released from female worms into chyme and faeces into the environment, where they can remain viable in moist soil for many years (Hou et al., 2012; Li, 1988; Yang and Zhang, 2013). Previous studies have shown that eggs of *B. schroederi* survive better than those of some other ascaridoids, such as *Ascaris lumbricoides* and *Ascaris suum*, and are characterised by rapid embryonation and resistance to low temperatures (4–12°C) (Li, 1988; Wu and Hu, 1985b). It usually takes 2–4 weeks for fertile eggs of *B. schroederi* to become infective at 22°C in moist and shaded soil (Wu et al., 1985), whereas 6–10 weeks are required by *Ascaris* eggs under the same conditions (Maung, 1978). Li (1988) found that 43% of *B. schroederi* eggs contained infective larvae within 30 days at −10°C.

2.1.1.2 Epidemiology

Since *Baylisascaris* eggs are highly resistant to environmental pressures (e.g. temperature and desiccation), it is difficult for panda individuals to avoid exposure once the environment is contaminated (Zhang and Wang, 2003). According to the limited, but valuable reports on the prevalence of *B. schroederi* infection (Feng et al., 1985; Lai et al., 1991; Li et al., 2014; Peng et al., 1989; Wang et al., 2001, 2013a; Yang, 1993; Ye, 1989;

Yu et al., 1998; Zhang et al., 2011, 2015; Zhou et al., 2013b), baylisascariasis commonly occurs in wild and captive populations of the giant panda (Loeffler et al., 2006). It is noteworthy that *B. schroederi* infection has been frequently detected in giant pandas upon arrival at international zoos, such as Adelaide Zoo (Australia), Chiang Mai Zoo (Thailand), Edinburgh Zoo (UK), Kebecity Oji Zoo (Japan), Pairi Daiza Zoo (Belgium), San Diego Zoo (USA), Smithsonian National Zoo (USA), Ueno Zoo (Japan) and Zoo Negara (Malaysia) (personal communication, Chengdong Wang, 10 February 2017).

Table 4 shows up-to-date, published prevalence information for *B. schroederi* infection in giant pandas (in English or Chinese language). Although there are questions surrounding the validity of some information in some Chinese publications, due to incomplete descriptions of methodologies and/or result sections, it is reassuring to see that independent studies, applying different methodologies, reveal similar findings, namely (1) that *B. schroederi* infection is commonly found in the wild giant panda populations across all six isolated mountain ranges in China (Fig. 1), (2) that the prevalence of infection in captive giant pandas is high (7%–88%) in most of breeding centres and zoos, (3) that the infection rate has not changed significantly over the last four decades in both captive and wild populations and (4) that there is no significant difference in the prevalence of *B. schroederi* infection among different age groups of giant pandas or between the sexes from previous large-scale surveys ($n=2680$, Lai et al., 1991; $n=336$, Yang, 1993). In this context, it is noteworthy to mention that Zhang et al. (2011) employed a genetic testing technique to link each stool sample to individual wild giant pandas, when estimating the prevalence of *B. schroederi* infection. The use of this noninvasive, genetic approach simultaneously provides the precise number of individual pandas sampled as well as the *B. schroederi* infection status (Zhang et al., 2011).

2.1.1.3 The Disease: Baylisascariasis

Most pathological and clinical features of baylisascariasis in the giant panda are caused by the migration of *B. schroederi* larvae in various tissues and by adult worms in the gastrointestinal tract (Feng et al., 1985; Ye, 1989). When the *Baylisascaris* larvae migrate through the tissues of their host, significant tissue damage relates predominantly to extensive inflammation and (subsequent) scarring in the intestinal wall as well as in the parenchymata of the liver and lungs (Li, 1990a,b; Loeffler et al., 2006). In some extreme cases, in addition to inflammation, entangled or clumped adult worms can

Table 4 Published Prevalence Information for *Baylisascaris schroederi* of the Giant Panda

Year	Location[a]	Population	Technique[b]	Reported Prevalence	References
1974–1986	Minshan and Qionglai	Wild	n/a	100% (50/50)	Ye (1989)
1984	Shanghai Zoo	Captive	Sedimentation–flotation technique	67% (2/3)	Peng et al. (1989)
1985	Minshan and Qionglai	Wild	Necropsy	100% (13/13)	Feng et al. (1985)
1985–1988	Minshan, Qionglai, Daxiangling, Xiaoxiangling and Liangshan	Wild	Sedimentation–flotation technique	56% (1050/2680)	Lai et al. (1991)
1985–1988	Minshan	Wild	Sedimentation–flotation technique	78% (262/336)	Yang (1993)
	Qionglai			83% (160/194)	
	Liangshan			60% (81/135)	
	Daxiangling and Xiaoxiangling			47% (15/32)	
1998	Qinling mountain	Wild	Sedimentation–flotation technique	100% (2/2)	Yu et al. (1998)
2001	Chengdu Zoo	Captive	Sedimentation–flotation technique	7% (1/14)	Wang et al. (2001)
2006–2008	Qinling	Wild	Sedimentation–flotation technique	66% (31/47)	Zhang et al. (2011)
	Minshan			44% (10/23)	
	Qionglai			48% (14/29)	
	Liangshan			57% (8/14)	
	Daxiangling			80% (4/5)	
	Xiaoxiangling			13% (1/8)	
2009–2010	Minshan	Wild	PCR/CE-SSCP	48% (15/31)	Zhang et al. (2012)
2012	Ya'an, CCRCGP	Captive	PCR (*cox*2)	68% (34/50)	Wang et al. (2013a)
2013	Ya'an, CCRCGP	Captive	PCR (12S rRNA)	88% (44/50)	Zhou et al. (2013b)
2014	Ya'an, CCRCGP and Chengdu, CRBGP	Captive	Sedimentation–flotation technique	26% (54/210)	Li et al. (2014)
2014	Minshan, Qionglai, Qinling, Xiaoxiangling and Liangshan	Wild	McMaster method	55% (48/87)	Zhang et al. (2015)

[a]*CCRCGP*, China Conservation and Research Centre for the Giant Panda; *CRBGP*, Chengdu Research Base of Giant Panda.
[b]*12S rRNA*, mitochondrial 12S ribosomal RNA gene; *cox*2, mitochondrial cytochrome *c* oxidase subunit 2 gene; n/a, not available; *PCR/CE-SSCP*, PCR-based capillary electrophoretic single-strand conformation polymorphism analysis.

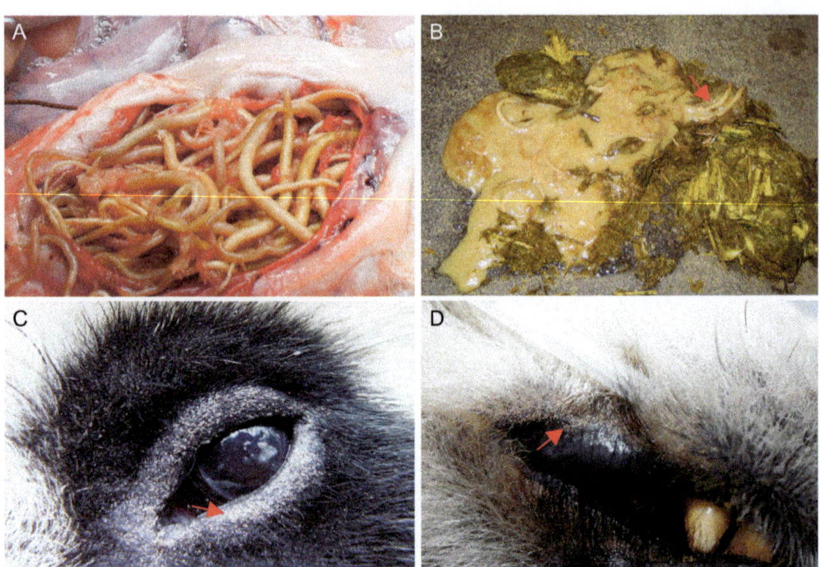

Fig. 2 Selected parasites of the giant panda. (A) The parasitic nematode *Baylisascaris schroederi* in the gastrointestinal tract of the giant panda, causing obstruction. (B) *Baylisascaris schroederi* expelled from an infected giant panda following anthelmintic treatment. (C) The mite *Chorioptes panda* and its typical predilections sites on the eyelid and/or lips (D). *Red arrows* indicating *B. schroederi* (B) or skin affected by *C. panda* (C and D).

lead to mechanical intestinal obstruction, which can be life-threatening (Fig. 2A) (Li, 1990b; Wang et al., 2007b). Moreover, although not yet reported, it is possible that pulmonary injury caused by migrating *Baylisascaris* larvae may enable or exacerbate bacterial infections (cf. Yang and Zhang, 2013). To date, there have been no reports of neural larva migrans associated with *B. schroederi* as seen for *B. procyonis* in raccoons (*Procyon lotor*) in North America (Graeff-Teixeira et al., 2016; Kazacos, 2016).

Morbidity and mortality associated with baylisascariasis are reported to relate directly to the intensity of *B. schroederi* infection (Qiu and Mainka, 1993), whereas individual pandas harbouring small numbers of worms tend to be asymptomatic. In captive giant panda populations, where there is a focus on controlling *B. schroederi*, this nematode rarely causes specific clinical symptoms (CCRCGP, unpublished clinical records), although the migration of larvae through lungs can cause acute symptoms, such as coughing and wheezing, particularly in young cubs (Yang and Zhang, 2013). Nonetheless, it is commonly observed that captive giant pandas, especially juveniles, can pass whole adult worms in the faeces or vomit

(CCRCGP, unpublished data; Loeffler et al., 2006). Nevertheless, *B. schroederi* infection is presently recognised as the biggest threat to free-ranging panda populations (Qiu and Mainka, 1993). Based on the literature (including publications in Chinese scientific journals, governmental reports and websites) between 1971 and 2005, Zhang et al. (2008) studied the causes of death in 789 adult wild giant pandas in natural habitats. These authors concluded that VLM caused by *B. schroederi* appeared to be the most significant threat of the three major factors of wild giant panda mortality during that period, the other two factors being food shortage (i.e. flowering and 'die-off' of bamboo; Reid et al., 1989) and poaching (Li et al., 2003). Surprisingly, baylisascariasis was reported to be responsible for 50% (12/24) of all deaths in free-ranging giant pandas between 2001 and 2005 (Zhang et al., 2008).

2.1.1.4 Diagnosis

Although the presence of adult worms in the faeces or vomit indicates infection in giant pandas, the diagnosis of *B. schroederi* infection is commonly performed using the quantitative McMaster test or a semiquantitative sedimentation–flotation method in most breeding centres and zoos (Loeffler et al., 2006; Zhang and Zhang, 2002). However, due to the large amount of undigested bamboo fibers in giant panda's faeces, *B. schroederi* eggs may be challenging to detect using a microscopy upon routine laboratory examination. In some instances, giant pandas with repeatedly 'negative' faecal test results have been reported to suddenly vomit bundles of worms. Hence, test sensitivity appears to be relatively low, in spite of the high reproductive index of *B. schroederi* (see Yang and Zhang, 2013), suggesting that 'false negative' results might relate to the presence of immature worms in the intestines.

Polymerase chain reaction (PCR)-based techniques can overcome this issue. For instance, some researchers (Wang et al., 2013a; Zhou et al., 2013b) have developed a PCR-based tool to directly amplify parts of the mitochondrial 12S ribosomal RNA or cytochrome *c* oxidase subunit 2 (*cox2*) gene from stool DNA samples. Their results showed that this tool is able to detect genomic DNA amounts that are at least equivalent to that from a single egg of *B. schroederi* and are more sensitive (0.5–1 time) than traditional coproscopic methods. In addition, no 'cross-reactivity' with DNA from other nematodes (i.e. *Ancylostoma caninum*, *B. transfuga* or *B. procyonis*) was found using this molecular diagnostic approach. In addition, other workers (Zhang et al., 2012) employed a combined PCR and capillary electrophoretic-based single-strand conformation polymorphism analysis (PCR-based CE-SSCP) using the mitochondrial gene *cox*2 to screen for *B. schroederi* DNA in stool samples from wild giant pandas from the Minshan mountains. Using this approach, these authors concluded that they were able to establish the prevalence and intensity of *B. schroederi* infection.

Apart from conventional and molecular tests, some progress has been made on developing serological detection methods. For instance, an antibody detection enzyme-linked immunosorbent assay (ELISA) employing a *B. schroederi* glutathione *S*-transferase antigen was established for the detection anti-*B. schroederi* serum antibody (IgG) in experimentally infected mice, with a sensitivity of 79.1% and a specificity of 82.0% (Xie et al., 2015a). However, such an assay has not yet been assessed for the diagnosis of baylisascariasis or *B. schroederi* infection in giant pandas. In the meantime, preliminary experiments are being planned at Ghent University (Belgium) and CCRCGP to assess an ELISA using *A. suum* antigen (Vlaminck et al., 2012) for the serological monitoring of *B. schroederi* infection in pandas.

2.1.1.5 Treatment and Control

The transmission of *B. schroederi* infection within and among captive giant panda populations is dependent on various factors, including the housing system, hygiene, management practices and anthelmintic treatment. However, to accomplish the short-term goal of reducing infection intensity and transmission potential, current control strategies rely mainly on monthly coprological examination (for eggs) and a mass anthelmintic treatment strategy (all individuals, including those with possible false-negative results for *B. schroederi*) (Loeffler et al., 2006; Wu and Hu, 1988). Anthelmintics used in practice include pyrantel pamoate; albendazole, fenbendazole, mebendazole; ivermectin, milbemycin oxime, doramectin and selamectin (CCRCGP, unpublished data; Loeffler et al., 2006)—the dosages of these compounds are indicated in Table 5. Usually, multiple (2–4) treatments are given until an individual panda ceases to expel worms and/or eggs in the faeces (Fig. 2B) (Wu and Hu, 1988; CCRCGP, unpublished clinical records; for limitations of conventional diagnostic methods, see Section 2.1.1.4). However, the efficacies of these anthelmintics at the doses routinely used against *B. schroederi* have not yet been critically assessed in captive giant pandas. Thus, there is a need to evaluate and compare the efficacies of these anthelmintics using a standardised protocol (cf. International Harmonisation of Anthelmintic Efficacy Guidelines; Vercruysse et al., 2001, 2002).

Table 5 Suggested Efficacy of Different Anthelmintic Drugs Tested Against *Baylisascaris schroederi*

Anthelmintics	Formulation	Dosage	Effect[b]	References[c]
Albendazole[a]	Oral	6 mg/kg, once	Negative of eggs in faecal floatation examination after 9 days of treatment	Liu and Yang (1994)
Doramectin[a]	Pour on	0.5 mg/kg, once	n/a	CCRCGP (unpublished data)
Febantel	Oral	20 mg/kg, once	Reported as effective without supporting data	Wu and Hu (1985b)
Fenbendazole[a]	Oral	5 mg/kg, once	n/a	CCRCGP (unpublished data)
Levamisole	Oral	7–8 mg/kg per 2 months, four times	Negative of worm or eggs in faecal examination after four times of treatment	Wu and Hu (1988)
	Oral/in feed	5–10 mg/kg, once	Negative of worm in faecal examination after 4 days of treatment	Ye and Zhang (1981)
	Unknown	8–10 mg/kg per 2 weeks	Reported as effective without supporting data	Qiu (1990)
Mebendazole[a]	Unknown	3.2–6.4 mg/kg	Reported as effective without supporting data	Qiu (1990)
Methylimidazole compound	Unknown	Unknown	Reported as effective without supporting data	Zhang and Zhang (2002)
Milbemycin[a]		0.5 mg/kg, once	n/a	CCRCGP (unpublished data)
Ivermectin[a]	Usually oral	0.2 mg/kg per day, twice	None reduction of egg counts in faecal flotation examination after 10–15 days of treatment (without negative control)	Li et al. (2015)
Pyrantel pamoate[a]	Oral	10 mg/kg per day, twice	80.0% (ointment) reduction of egg counts in faecal floatation examination after 10–15 days of treatment	Li et al. (2015)
Piperazine citrate	Oral	150 mg/kg, once	Significant reduction of worms in faecal examination after 5 days of treatment	Ye and Zhang (1981)
	Unknown	140–160 mg/kg per 2 weeks	Reported as effective without supporting data	Qiu (1990)
Selamectin[a]	Spot-on	6–12 mg/kg, once	n/a	CCRCGP (unpublished data)
Trichlorfon	Unknown	55 mg/kg	Negative of worm in faecal examination after 5 days of treatment; significant side effect was observed	Ye and Zhang (1981)

[a]Anthelmintics used in practice currently.
[b]None of these studies followed International Harmonisation of Anthelmintic Efficacy Guidelines (Vercruysse et al., 2001, 2002); n/a, not available.
[c]*CCRCGP*, China Conservation and Research Centre for the Giant Panda; *CRBGP*, Chengdu Research Base of Giant Panda.

Due to the relatively short activity of anthelmintics in the host and the spread of large numbers of resilient *B. schroederi* eggs from infected animals in the environment, reinfection can occur rapidly (usually within 20–40 days; CCRCGP, unpublished records). Given that resistance to anthelmintics is increasing in a number of nematode species of both animals and humans (e.g. Kaplan, 2004; Vercruysse et al., 2011), it is readily possible that resistance could develop in *Baylisascaris* as anthelmintics are routinely (excessively) administered to giant pandas at varying or imperfect dosages. Given the implementation of reintroduction programmes for captive giant pandas to the wild in recent years (Shan et al., 2014), drug resistance genes carried by *Baylisascaris* in released giant pandas could spread to and through wild populations.

Therefore, the possibility or likelihood that drug resistance in *Baylisascaris* could emerge as a problem has stimulated the search for alternative methods of prevention and control. One possibility could be to develop a vaccine against baylisascariasis. Inspired by research towards developing a vaccine against *A. suum* (see Islam et al., 2005; Matsumoto et al., 2009; Tsuji et al., 2001, 2002, 2003, 2004), considerable attention and research effort have been directed towards a recombinant subunit vaccine against baylisascariasis. It is encouraging to know that pigs can mount some protection (58%) following vaccination with recombinant antigens against challenge infection in a pig-*Ascaris* model (Tsuji et al., 2004), suggesting that vaccination against *B. schroederi* might be feasible. The first vaccination trial of a recombinant antigen, Bs-Ag3 (37 kDa), against *B. schroederi* infection in laboratory mice was reported in 2008 (Wang et al., 2008b). Repeated subcutaneous administration (three times at 2-weekly intervals) of recombinant Bs-Ag3 with adjuvant in mice (BALB/c) resulted in a 63% reduction in the number of larvae collected from the lungs and a significant increase in total IgG in serum, in comparison with nonimmunised control mice. Subsequently, similar results were observed when mice were immunised with other recombinant antigens, Bs-Ag1 (64% reduction in number of lung larvae) and Bs-Ag2 (69% reduction), using same experimental design (He et al., 2009, 2012). However, to date, there has been no investigation of the IgG subclasses in the sera from immunised mice or of the precise mechanism(s) by which protection is achieved against *B. schroederi*.

More recently, aside from these immunogens, more attention has focused on targets that play essential roles in the survival of the parasite. For example, Xie et al. (2013) identified PYP-1, a new homologue of inorganic pyrophosphatases (PPases) (Kajander et al., 2013), which likely plays critical roles in nematode development and moulting (Islam et al., 2003; Ko et al., 2007) and is distributed widely in the body wall, gut epithelium, ovary and uterus of adult female *Baylisascaris*. Two separate vaccination experiments in mice (BALB/c) showed that recombinant PYP-1 induced 69%–71% reductions in the number of liver-stage and lung-stage larvae 7 days following challenge infection (Xie et al., 2013). An investigation of the IgG subclasses in the sera of immunised mice showed that the level of IgG1 was significantly higher than IgG2a, with increased levels of IL-4 and IL-10, indicating a type 2 protective immune response (Xie et al., 2013).

Apart from work directed towards a vaccine against *B. schroederi*, efforts have also been made to understand aspects of the molecular biology and genetics of this parasite. In another study (Zhao et al., 2013), the microRNA profile of *B. schroederi* via high-throughput sequencing and real-time quantitative PCR suggested that chitinases, ovarian or egg development related proteins and ribosomes were the targets of large numbers of microRNAs for the regulation of genes at the posttranscriptional level. This study also highlighted the potential of (at least some of) these microRNAs as intervention targets against baylisascariasis (Zhao et al., 2013).

2.1.1.6 Genetics

Following the publication of the complete mitochondrial genome of *B. schroederi* (see Xie et al., 2011), Li et al. (2012) undertook the first molecular characterisation of this parasite using nuclear ribosomal (r) RNA genes (18S and 28S) and a mitochondrial gene (12S) (Li et al., 2012). Importantly, Li et al. (2012) showed that 18S was not a suitable candidate marker for assessing variation within the genus *Baylisascaris*, but that 28S and 12S sequences were capable of distinguishing *B. schroederi* from *B. ailuri* and *B. transfuga*.

Current and published phylogenetic trees for members of the genus *Baylisascaris* using 28S and mitochondrial *cox*1 gene data sets (Tokiwa et al., 2014) consistently show two main groupings: one containing *B. schroederi*, *B. ailuri* from the red panda and *B. transfuga* from a variety of bear hosts, and the other comprising *Baylisascaris* species from raccoon, skunk and the South American kinkajou (see Table 3). An analysis using recent sequence data (accession no. KY465564) places *B. devosi* from mustelids with *B. potosis* from this kinkajou (Fig. 3). Additionally, a new species of *Baylisascaris*, called *B. venezuelensis*, from the South American spectacled bear (which is closely related to the giant panda; Talbot and Shields, 1996) was described recently (Pérez Mata et al., 2016). Clearly, further sequence data

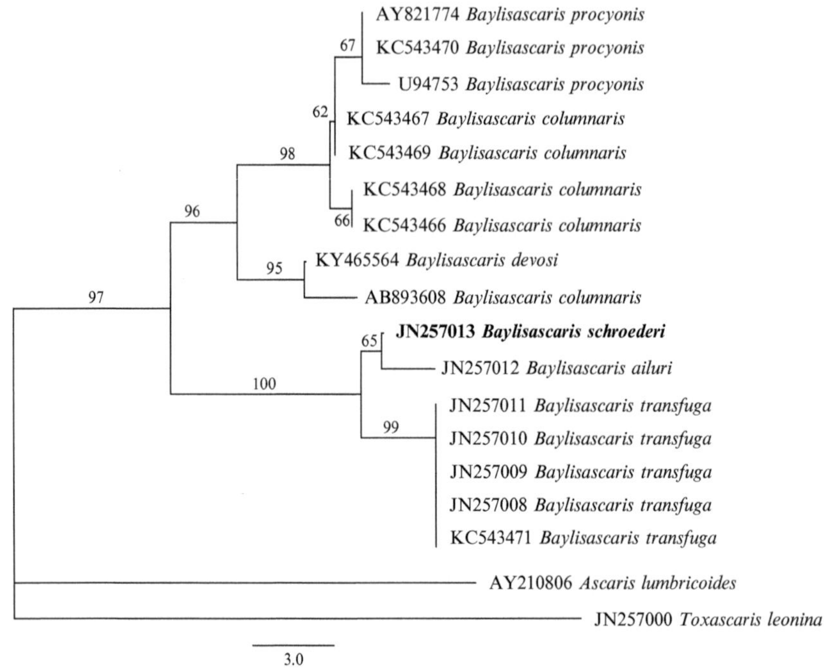

Fig. 3 A phylogenetic tree showing the position of *Baylisascaris schroederi* in relation to all other species of *Baylisascaris* for which nucleotide sequence data are available, using *Ascaris lumbricoides* (from human) and *Toxascaris leonina* as outgroups. This tree is based on the analysis of 28S rRNA gene sequence data using the neighbour joining method. All data were obtained from the GenBank database, and accession numbers precede species names in the phylogenetic tree. *Adapted from fig. 3A from Tokiwa, T., Nakamura, S., Taira, K., Une, Y., 2014. Baylisascaris potosis n. sp., a new ascarid nematode isolated from captive kinkajou, Potos flavus, from the Cooperative Republic of Guyana. Parasitol. Int. 63, 591–596.*

from other members of the genus *Baylisascaris* are needed to gain a better understanding of the phylogeographic history of *B. schroederi*, which likely involves 'host-switching' and 'ecological fitting' (Araujo et al., 2015).

Based on a morphological study by Wan et al. (2003), giant pandas from the Qinling mountain range represent a distinct subspecies (*Ailuropoda melanoleuca qinlingensis*) from *Ailuropoda melanoleuca melanoleuca* and are thought to be genetically distinct from all other populations of pandas (Wei et al., 2012). Although the draft nuclear genome of the giant panda has been published (Li et al., 2010), there is no study describing the use of large nuclear genomic sequence data sets to explore the population genetics of this panda. Several studies (see later) have attempted to search for genetic differences between *B. schroederi* of the two subspecies of giant panda. However, no genetic variation has been found between worms of the two subspecies of giant panda using first and second internal transcribed spacers (ITS-1 and ITS-2 = ITS) of nuclear ribosomal DNA (Lin et al., 2012; Zhao et al., 2012). Other studies attempted to resolve the same issue using various mitochondrial gene markers. First, Zhou et al. (2013a) used the *cytb* gene and concluded that there was a high rate of gene flow among three populations of *B. schroederi* representing the two recognised subspecies. Second, Xie et al. (2014, 2015b) used portions of the mitochondrial *cox*1, *atp*6 and 12S rRNA genes, which displayed very few parsimony informative sites. Again, these authors found no discernable genetic difference in worms among the populations from distinct habitats of the giant panda (i.e. the Minshan, Qionglai and Qinling mountain ranges). A low level of genetic diversity but a high level of gene flow in worms suggested the potential for a rapid spread of drug resistance in *B. schroederi* (see Xie et al., 2014). Third, Zhao et al. (2014) did not find support for the genetic substructuring within *B. schroederi* among samples from the Qinling mountains and one sample from Sichuan province using portions of the mitochondrial genes *cytb*, *cox*3 and *nad*5. Unfortunately, to date, only mitochondrial genes, whose validity as a population genetic markers has been questioned (Galtier et al., 2009), have been used to explore structuring and substructuring in *B. schroederi* populations. Therefore, future studies should be conducted using multiple neutral nuclear genomic markers. A project in China is now underway to sequence, assemble and annotate the nuclear genome of *B. schroederi*, which, if successful, would underpin such a focus.

2.1.2 Chorioptes panda

Chorioptes spp. (Acariformes: Psoroptidae) are skin mites that cause mange in domestic and wild animals. These mites are commonly found in herbivorous hosts, including cattle, sheep, goats, horses, camelids and moose (Yeruham et al., 1999). *Chorioptes* was first found in the ears of captive giant pandas in the Paris Zoo, France and named *Chorioptes panda* (Fain and Leclerc, 1975), and subsequently reported in China in 1986 (Ye, 1986). However, *C. panda* was considered an invalid species by some researchers (Zahler et al., 2001). In ensuing years, other researchers reappraised the morphology of the mite, undertook phylogenetic analyses of mitochondrial *cox*1 and nuclear 18S rRNA gene data sets, and concluded that *C. panda* is a valid species (Bochkov et al., 2014; Hestvik et al., 2007; Wang et al., 2012).

Although no epidemiological data are available, *C. panda* is commonly found in captive populations of the giant panda (Qiu et al., 1984; Wang et al., 2000; Xu and Zhang, 2002; Yang et al., 2001; Zhou et al., 1989), particularly in late spring and early autumn (CCRCGP, unpublished records). However, to date, there is only one published record of *C. panda* infection in wild populations of the giant panda (Wang et al., 2012), although obtaining such information for such populations is a considerable challenge. In addition, wild giant pandas are usually solitary animals, and only occasionally interact with each other, mainly during their short mating season (Schaller et al., 1985). Thus, the probability of direct cross-transmission of this mite appears to be low.

Typically, *C. panda* is found on the eyelids, lips and ears of giant pandas (Fig. 2), causing mild alopecia, erythema and crusting, and affecting the sleep and appetite of giant pandas (Qiu et al., 1984). The control of *Chorioptes* mange relies mainly on chemotherapeutic treatment. Macrocyclic lactones (e.g. ivermectin and selamectin) have been found to be effective when routinely administered on a monthly basis (CCRCGP, unpublished records). Closantel (Wang et al., 2000) and deltamethrin (Xu and Zhang, 2002) have also been proposed to be effective against *C. panda* (see Wang et al., 2000; Xu and Zhang, 2002), but well-controlled experiments are needed to verify the authors' claims.

2.1.3 Ixodidae (Hard Ticks)

Other ectoparasites that can affect the health of the giant panda are (blood-feeding) hard ticks. Since the first description of tick infection by *Haemaphysalis warburtoni* (Wu and Hu, 1985a), an increasing number of hard tick species have been identified on giant pandas in the last two decades (Lai et al., 1990; Ma, 1987; Qiu and Zhu, 1987; Yu et al., 1998). To date, 13 species representing 3 genera of hard ticks have been proposed as recorded (from rescued, sheltered or dead, wild giant pandas). These ticks include members of the genera *Haemaphysalis* (9 species), *Ixodes* (3 species) and *Dermacentor* (1 species) (Table 2). Of these ticks, *Haemaphysalis flava* has been most commonly reported in giant panda populations (Cheng et al., 2013; Ma, 1987; Qiu and Zhu, 1987). Recently, molecular tools have also been employed for the genetic characterisation of ticks from the giant panda. Using mitochondrial and ribosomal DNA markers as well as key morphological characters, *H. flava* was identified to predominate on giant pandas in the Qinling mountain range (Cheng et al., 2013). Although there is no report on mortality caused by such hard ticks, morbidity involving dermatitis and/or weight loss has been recorded in infested giant pandas (CCRCGP, unpublished records). Similar to the treatment of *C. panda*, ivermectin and selamectin are the compounds most commonly used against ticks in breeding centres and zoos (CCRCGP, unpublished clinical records). To date, there are no reports of any associated tick-borne diseases in the giant panda.

2.2 Protists: An Emerging Issue?

Using molecular diagnostic tools, such as PCR, *Cryptosporidium* spp. (Liu et al., 2013; Wang et al., 2015), *Enterocytozoon bieneusi* (see Tian et al., 2015) and *Toxoplasma gondii* (see Ma et al., 2015) have been detected and characterised from giant pandas. During a routine coprological examination (flotation) of 57 faecal samples at CCRCGP (Liu et al., 2013), one sample from an 18-year-old male captive giant panda was test-positive for *Cryptosporidium*, with oocysts being 4–4.6 μm in size. A phylogenetic analysis of partial 18S rRNA (786 bp), 70 kDa heat shock protein (1879 bp) and actin (1044 bp) gene sequence data (GenBank accession nos. JF970610, JN588571 and JN969985) showed that *Cryptosporidium* from the giant panda is genetically similar to that of genotype of *Cryptosporidium* from a black bear (*Ursus americanus*) (98%–99.5%; GenBank accession nos. AF247535, AF247536 and AF382339). Due to its sequence divergence (0.5%–2.0%) from other nucleotide sequences available in the GenBank database, the *Cryptosporidium* taxon from the giant panda was inferred to be a novel genotype, designated *Cryptosporidium* 'giant panda genotype' (Liu et al., 2013). Subsequently, a study of a larger sample size ($n=322$) confirmed the presence of *Cryptosporidium* infection in captive and wild giant panda populations, recording prevalences of 15.6% (19/122) and 0.5% (1/200), respectively (Wang et al., 2015). Interestingly, a genetic analysis of partial 18S rRNA gene sequence data revealed that *Cryptosporidium* from these pandas was more similar genetically (94.1%–99.8%) to *C. andersoni* from cattle than to the *Cryptosporidium* 'giant panda genotype' (84.6%–89.8% similarity) (Liu et al., 2013), even though most test-positive samples ($n=17$) were collected from the same conservation centre (i.e. CCRCGP). This finding raises a question about the spectrum of *Cryptosporidium* genotypes that might infect the giant panda in different environments, geographical localities and times of the year. However, the authors (Wang et al., 2015) suggested that seasonal variation in the distribution of *Cryptosporidium* might be a possible reason for this finding. In addition, these authors reported a higher prevalence of *Cryptosporidium* infection in captive than in wild giant panda populations. Whether this difference relates to a high degree of transmission of *Cryptosporidium* among giant pandas or between other animals and pandas in a captive (high population density) environment remains to be elucidated. Although no clinical signs, such as diarrhoea (often associated with cryptosporidiosis), were observed in either of these two studies (Liu et al., 2013; Wang et al., 2015), systematic investigations should be performed to gain a better understanding of the pathogenicity/virulence of *Cryptosporidium* in the giant panda as a basis for improved prevention or control in the event of clinical outbreaks.

Recently, *E. bieneusi* was identified in captive giant pandas in Shanxi province, with an estimated prevalence of 9% (4/46) (Tian et al., 2015). Subsequent phylogenetic analysis based on the nuclear ITS rDNA sequences of *E. bieneusi* suggested that a novel genotype I-like *E. bieneusi* occurs in giant pandas. Although *E. bieneusi* is known as an emerging and opportunistic enteric pathogen, causing diarrhoea in humans and animals, and progress has been made in our knowledge of the epidemiology of *E. bieneusi*, the transmission routes of this pathogen are still unclear (Matos et al., 2012; Santín, 2015). Given the possibility of transmission via respiratory secretions (Mathis et al., 2005) and the potentially broad host range of this pathogen (Santín and Fayer, 2009), whether this new genotype I-like *E. bieneusi* has a specific host affiliation to the giant panda remains to be explored. Similar to the reports of *Cryptosporidium* infection in giant pandas, no associated clinical signs were observed in *E. bieneusi*-infected animals (Tian et al., 2015).

Meanwhile, other than asymptomatic *Cryptosporidium* and *Enterocytozoon* infections in giant pandas, there is one report of an acute, fatal toxoplasmosis case, characterised by serious respiratory and gastroenteritis symptoms (Ma et al., 2015). Here, a 7-year-old giant panda was found dead at Zhengzhou Zoo, China (Ma et al., 2015). The necropsy findings, and serological results from an immunofluorescence assay and a modified agglutination test as well as PCR results for DNA from tissue biopsy samples (from liver, spleen, lungs, kidneys and intestines) all indicated that the giant panda had died from acute toxoplasmosis. Additional multilocus, nested PCR–RFLP analysis, using 10 genetic markers (*SAG*1, *SAG*2, *SAG*3, *BTYB*, *GRA*6, *c*22-8, *c*29-2, *L*358, *PK*1 and *Apico*; see Dubey et al., 2007), provided evidence that the *T. gondii* isolate causing this panda's death represented an atypical genotype (with reference to strains GT1, PTG, CTG, MAS, TgCgCal, TgCatBr5, TaCatBr40, TgCatBr64 and TgRsCr1). The findings suggested that a *T. gondii*-infected stray cat or rodents in the zoo might have been the source of infection in this fatal case (Ma et al., 2015).

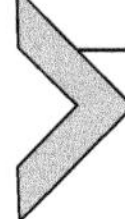

3. CONCLUSIONS AND A PERSPECTIVE ON FUTURE RESEARCH

This chapter has reviewed current information on parasites of the giant panda, but it seems that we have only just 'scratched the surface' when it comes to research of the giant panda. Therefore, in our opinion, there is a need to conduct research in the following areas: (1) detailed genetic comparison of the two subspecies of giant panda using advanced genomic sequencing and analytical approaches employing the genomic resources available for the panda (cf. Li et al., 2010); (2) more morphological and molecular studies of parasites of the giant panda, improved (genetic) classification of known taxa, as well as detailed molecular epidemiological studies to assess the prevalence and distribution of parasites in captive and wild populations; (3) identification and characterisation of any emerging parasites and other pathogens using noninvasive sampling and PCR-based molecular or next-generation sequencing tools (cf. Korhonen et al., 2016); (4) studies to improve our understanding of the fundamental biology and molecular biology of *B. schroederi*; (5) investigations to assess whether anthelmintic resistance is emerging in *B. schroederi* and whether it is spread in the field and (6) work towards improved methods for the prevention or control of baylisascariasis (e.g. vaccination).

Without an accurate catalogue of which parasite species exist in wildlife, and a deep understanding of their life cycles, biology and the diseases that they cause, it is challenging to evaluate their risk to animal health (Colwell et al., 2009). Although numerous parasite taxa ($n=29$) have been detected/recorded in the giant panda (Table 2), many of them, such as *Sarcocystis*, *Strongyloides* and lungworm (see Table 2), have not been described in any detail, and, to our knowledge, no voucher specimens are readily accessible in China. In our opinion, it should be a priority to classify known parasite taxa using international taxonomic rules. In addition, studies are needed to understand the biology and the epidemiology of these parasites, to guide conservation decisions.

Previous observations of *B. schroederi* infection in baby giant pandas (<2 months) indicate the possibility of transplacental transmission (Yang, 1993), but there is no direct evidence of this mode. Thus, it would be interesting to undertake studies to critically assess whether *B. schroederi* can undergo transplacental and/or transmammary transmission. In addition, a major impediment to large-scale epidemiological investigations of giant panda parasites is the difficulty of assessing infection status in free-ranging populations. This limitation seems to have been somewhat overcome by the development of a PCR-based diagnostic approach (see Section 2.1.1.4; Wang et al., 2013a) for the simultaneous genetic 'fingerprinting' of individual pandas (see Section 2.1.1.2; Zhang et al., 2011) and the detection of their parasites in faecal samples, which could be used for field studies, in order to explore the distribution and dynamics of parasitic infections/diseases. Moreover, it would not be surprising to detect new parasite species using PCR-based or high-throughput DNA sequencing technology, considering the recent detection of new genotypes of protists (i.e. *Cryptosporidium*, *Enterocytozoon* and *Toxoplasma*) in the giant panda (Liu et al., 2013; Ma et al., 2015; Tian et al., 2015; Wang et al., 2015), which could not be characterised using conventional parasitological methods.

There is no doubt that baylisascariasis continues to cause serious health problems in the giant panda and will likely remain one of the biggest challenges for the conservation of this animal. Although modern anthelmintics appear to be reasonably effective for the treatment of baylisascariasis, the dissemination of large numbers of eggs into the environment and the resilience of these thick-shelled eggs make this disease/infection challenging to control *B. schroederi* without the implementation of an integrated approach, including management components (pen cleaning protocols and housing infrastructure) and regular monitoring for infection in different age groups of panda. Because of the potential for anthelmintic resistance to develop in *B. schroederi*, as a consequence of routine and excessive use of anthelmintics in captive animals, an integrated approach for baylisascariasis control needs to be explored. Such approaches might include the use of effective disinfectants (against the egg stage) to block transmission, new drugs with different modes of action and/or vaccination. Early work showed that the disinfectant neopredisan (active constituent: chlorocresol or *p*-chloro-*m*-cresol) has a high efficacy (100%) against *A. suum* eggs under laboratory conditions (Mielke and Hiepe, 1998). Whether this chemical has similar efficacy against *B. schroederi* eggs remains to be determined.

On the other hand, lessons learned from previous attempts to develop vaccines against ascaridoid nematodes (Matsumoto et al., 2009; Tsuji et al., 2004) indicate that it is critical to gain knowledge of the immunobiology of the parasite and to ensure that any vaccine candidate consistently induces high-level and long-lasting protective immunity (Geldhof et al., 2007). Also, a deep knowledge of parasite-derived molecules involved in vital developmental processes in *Baylisascaris*, host–parasite interactions and mechanism by which this nematode develops and survives within the host might assist in defining novel candidate vaccine targets. Clearly, developing an anti-*Baylisascaris* vaccine for an endangered animal, such as the giant panda, presents considerable challenges, such that immunogens might need to be selected based on consistent protection and immunorecognition in different laboratory (paratenic) hosts (e.g. mice, rodents and/or rabbits) prior to any well-controlled trials (without experimental challenge infection) in captive panda populations in which *B. schroederi* is known to be highly endemic, to assess whether a reduction in prevalence and intensity of infection is achievable.

In the last three decades, multidrug resistance has emerged across five continents in endoparasites and ectoparasites of animals and, to some extent, those of humans (Blake and Coles, 2007; Coles et al., 1994; Geary, 2005; Kaplan, 2004; McNair, 2015; Srivastava and Misra-Bhattacharya, 2015; Vercruysse et al., 2011). Given that anthelmintics are often used routinely and excessively in a suppressive manner in captive animal populations, there is a probability that drug resistance is emerging (cf. Geary, 2005). If this is the case for *B. schroederi*, what are the implications of this when captive animals are released into the wild? In the first instance, it seems pertinent to assess *B. schroederi* for resistance to commonly used anthelmintics using a faecal egg count reduction testing (FECRT) protocol similar to that recommended by the World Association for the Advancement of Veterinary Parasitology (WAAVP) for livestock animals (Coles et al., 2006). The presence of anthelmintic resistance in *B. schroederi* would make immunoprevention via vaccination even more attractive.

Apart from addressing some of these issues, there is an opportunity to pool expertise from a range of key experts in parasitology, in order to address some of the salient parasite problems and management issues to ensure that the health of captive giant pandas is maximised. By sharing new knowledge and information, this would not only significantly increase the utilisation of available data and clinical information but also enhance research opportunities to ensure the conservation of the giant panda. Although this animal is critically endangered, this review indicates, surprisingly, that parasitological research of this animal is in its infancy. Thus, much can be done to contribute towards the conservation of this iconic species. Although major public and scientific attention has focused on the conservation of the giant panda, with injections of funding from a number of bodies including the Giant Panda Conservation Foundation, it seems that very little funding is spent on the research of infectious (including parasitic) diseases of this species. Therefore, we strongly recommend that more attention be paid to the diseases that are likely to threaten this animal's conservation. We strongly believe that there is little room for error at this time point in history.

ACKNOWLEDGEMENT

The support from the International Cooperation Project of Sichuan Science and Technology Department (Program No. 2014HH0047; T.W.) is gratefully acknowledged.

REFERENCES

Araujo, S.B., Braga, M.P., Brooks, D.R., Agosta, S.J., Hoberg, E.P., von Hartenthal, F.W., Boeger, W.A., 2015. Understanding host-switching by ecological fitting. PLoS One 10, e0139225.

Blake, N., Coles, G., 2007. Flock cull due to anthelmintic-resistant nematodes. Vet. Rec. 161, 36.

Bochkov, A.V., Klimov, P.B., Hestvik, G., Saveljev, A.P., 2014. Integrated Bayesian species delimitation and morphological diagnostics of chorioptic mange mites (Acariformes: Psoroptidae: *Chorioptes*). Parasitol. Res. 113, 2603–2627.

Chen, K.L., Shi, X.Q., 1992. Identification of *Haemaphysalis longicornis* and *H. megaspinosa* in giant panda. Shanghai Commun. Anim. Husb. Vet. Med. 1, 8 (In Chinese).

Cheng, W.Y., Zhao, G.H., Jia, Y.Q., Bian, Q.Q., Du, S.Z., Fang, Y.Q., Qi, M.Z., Yu, S.K., 2013. Characterization of *Haemaphysalis flava* (Acari: Ixodidae) from Qingling subspecies of giant panda (*Ailuropoda melanoleuca* qinlingensis) in Qinling Mountains (Central China) by morphology and molecular markers. PLoS One 8, e69793.

Coles, G.C., Borgsteede, F.H., Geerts, S., 1994. Recommendations for the control of anthelmintic resistant nematodes of farm animals in the EU. Vet. Rec. 134, 205–206.

Coles, G.C., Jackson, F., Pomroy, W.E., Prichard, R.K., von Samson-Himmelstjerna, G., Silvestre, A., Taylor, M.A., Vercruysse, J., 2006. The detection of anthelmintic resistance in nematodes of veterinary importance. Vet. Parasitol. 136, 167–185.

Colwell, D.D., Otranto, D., Stevens, J.R., 2009. Oestrid flies: eradication and extinction versus biodiversity. Trends Parasitol. 25, 500–504.

Dubey, J.P., Sundar, N., Gennari, S.M., Minervino, A.H.H., Farias, N.A.D.R., Ruas, J.L., Dos Santos, T.R.B., Cavalcante, G.T., Kwok, O.C.H., Su, C., 2007. Biologic and genetic comparison of *Toxoplasma gondii* isolates in free-range chickens from the northern Para state and the southern state Rio Grande do Sul, Brazil revealed highly diverse and distinct parasite populations. Vet. Parasitol. 143, 182–188.

Fain, A., Leclerc, M., 1975. A case of mange in a giant panda caused by a new species of *Chorioptes* (Acarina: Psoroptidae). Acarologia 17, 177–182 (In French).

Feng, W.H., Hu, T.Q., Bi, F.Z., Chui, Y.T., He, G.X., Ye, Z.Y., 1985. A study on the endangering causes of giant panda. Anim. World 2, 1–7 (In Chinese).

Galtier, N., Nabholz, B., Glémin, S., Hurst, G.D., 2009. Mitochondrial DNA as a marker of molecular diversity: a reappraisal. Mol. Ecol. 18, 4541–4550.

Geary, T.G., 2005. Ivermectin 20 years on: maturation of a wonder drug. Trends Parasitol. 21, 530–532.

Geldhof, P., De Maere, V., Vercruysse, J., Claerebout, E., 2007. Recombinant expression systems: the obstacle to helminth vaccines? Trends Parasitol. 23, 527–532.

Graeff-Teixeira, C., Morassutti, A.L., Kazacos, K.R., 2016. Update on baylisascariasis, a highly pathogenic zoonotic infection. Clin. Microbiol. Rev. 2016 (29), 375–399.

He, C.D., Liu, S.X., Chen, X.H., 1987. First report of parasites of giant panda from Qinling. Selected Acadmic Papers of Giant Panda Disease Control. China Forestry Press, Beijing, China, pp. 25–29 (In Chinese).

He, G., Wang, T., Yang, G., Fei, Y., Zhang, Z., Wang, C., Yang, Z., Lan, J., Luo, L., Liu, L., 2009. Sequence analysis of Bs-Ag2 gene from *Baylisascaris schroederi* of giant panda and evaluation of the efficacy of a recombinant Bs-Ag2 antigen in mice. Vaccine 27, 3007–3011.

He, G., Chen, S., Wang, T., Yan, Y., Zhang, Z., Li, D., Yu, H., Xie, Y., Wang, C., Gu, X., Wang, S., Peng, X., Yang, G., 2012. Sequence analysis of the Bs-Ag1 gene of *Baylisascaris schroederi* from the giant panda and an evaluation of the efficacy of a recombinant *Baylisascaris schroederi* Bs-Ag1 antigen in mice. DNA Cell Biol. 31, 1174–1181.

Hestvik, G., Zahler-Rinder, M., Gavier-Widen, D., Lindberg, R., Mattsson, R., Morrison, D., Bornstein, S., 2007. A previously unidentified *Chorioptes* species infesting outer ear canals of moose (*Alces alces*): characterization of the mite and the pathology of infestation. Acta Vet. Scand. 49, 21.

Hou, Z.J., He, S.W., Zhang, C.S., Ma, J.Z., 2012. Reseach advances in the knowledge of *Baylisascaris schroederi* of the giant panda. Jiangsu Agr. Sci. 40, 215–217 (In Chinese).

Hvistendahl, M., 2015. Captive pandas succumb to killer virus. Science 347, 700–701.

Islam, M.K., Miyoshi, T., Kasuga-Aoki, H., Isobe, T., Arakawa, T., Matsumoto, Y., Tsuji, N., 2003. Inorganic pyrophosphatase in the roundworm *Ascaris* and its role in the development and molting process of the larval stage parasites. Eur. J. Biochem. 270, 2814–2826.

Islam, M.K., Miyoshi, T., Tsuji, N., 2005. Vaccination with recombinant *Ascaris suum* 24-kilodalton antigen induces a Th1/Th2-mixed type immune response and confers high levels of protection against challenged *Ascaris suum* lung-stage infection in BALB/c mice. Int. J. Parasitol. 35, 1023–1030.

Kajander, T., Kellosalo, J., Goldman, A., 2013. Inorganic pyrophosphatases: one substrate, three mechanisms. FEBS Lett. 587, 1863–1869.

Kaplan, R.M., 2004. Drug resistance in nematodes of veterinary importance: a status report. Trends Parasitol. 20, 477–481.

Kazacos, K.R., 2008. *Baylisascaris procyonis* and related species. In: Samuel, W.M., Margo, J.P., Kocan, A.A. (Eds.), Parasitic of Wild Mammals. Iowa State University Press, Ames, USA, pp. 301–341.

Kazacos, K.R., 2016. *Baylisascaris* Larva Migrans: U.S. Geological Survey Circular 1412. 122 p.

Ko, K.M., Lee, W., Yu, J.R., Ahnn, J., 2007. PYP-1, inorganic pyrophosphatase, is required for larval development and intestinal function in *C. elegans*. FEBS Lett. 581, 5445–5453.

Korhonen, P.K., Young, N.D., Gasser, R.B., 2016. Making sense of genomes of parasitic worms: tackling bioinformatic challenges. Biotechnol. Adv. 34, 663–686.

Lai, C.L., Yang, M.L., Wang, Q., 1990. Ectoparasites of the giant panda. Sichuan J. Zool. 9, 7 (In Chinese).

Lai, C.L., Qiu, X.M., Luo, X.F., Fan, W.A., 1991. The internal parasitic infection in wild giant pandas (*Ailuropoda melanoleuca*). Expl. Nat. 35, 68–71 (In Chinese).

Li, J.H., 1988. Effects of cold on the development of *Baylisascaris schroederi* eggs. Chinese J. Vet. Sci. Technol. 9, 41–42 (In Chinese).

Li, J.H., 1989. The development of *Baylisascaris schroederi* larvae in mice. Chinese J. Vet. Sci. Technol. 8, 24–25 (In Chinese).

Li, J.H., 1990a. Migration, distribution and development of larvae of panda ascarid, *Baylisascaris schroederi*, in mice. Acta Zool. Sin. 36, 236–243 (In Chinese).

Li, J.H., 1990b. Observations on the pathogenicity of *Baylisascaris schroederi* in experimental mice. Chinese J. Zoooses 9, 32–34 (In Chinese).

Li, J.H., 1993. Morphological observations on the larvae of panda Ascarid, *Baylisascaris schroederi*, in experimental host. J. Guiyang Med. Coll. 18, 184–187 (In Chinese).

Li, Y., Guo, Z., Yang, Q., Wang, Y., Niemelä, J., 2003. The implications of poaching for giant panda conservation. Biol. Conserv. 111, 125–136.

Li, D.S., Wang, C.D., Yang, G.Y., 2007. A case report describing the diagnosis and treatment of gastrointestinal myiasis in a giant panda. Progr. Vet. Med. 28, 115–116 (In Chinese).

Li, R., Fan, W., Tian, G., Zhu, H., He, L., Cai, J., Huang, Q., Cai, Q., Li, B., Bai, Y., Zhang, Z., Zhang, Y., Wang, W., Li, J., Wei, F., Li, H., Jian, M., Li, J., Zhang, Z., Nielsen, R., Li, D., Gu, W., Yang, Z., Xuan, Z., Ryder, O.A., Leung, F.C., Zhou, Y., Cao, J., Sun, X., Fu, Y., Fang, X., Guo, X., Wang, B., Hou, R., Shen, F., Mu, B., Ni, P., Lin, R., Qian, W., Wang, G., Yu, C., Nie, W., Wang, J., Wu, Z., Liang, H., Min, J., Wu, Q., Cheng, S., Ruan, J., Wang, M., Shi, Z., Wen, M., Liu, B., Ren, X., Zheng, H., Dong, D., Cook, K., Shan, G., Zhang, H., Kosiol, C., Xie, X., Lu, Z., Zheng, H., Li, Y., Steiner, C.C., Lam, T.T., Lin, S., Zhang, Q., Li, G., Tian, J., Gong, T., Liu, H., Zhang, D., Fang, L., Ye, C., Zhang, J., Hu, W., Xu, A., Ren, Y., Zhang, G., Bruford, M.W., Li, Q., Ma, L., Guo, Y., An, N., Hu, Y., Zheng, Y., Shi, Y., Li, Z., Liu, Q., Chen, Y., Zhao, J., Qu, N., Zhao, S., Tian, F., Wang, X., Wang, H., Xu, L., Liu, X., Vinar, T., Wang, Y., Lam, T.W., Yiu, S.M., Liu, S., Zhang, H., Li, D., Huang, Y., Wang, X., Yang, G., Jiang, Z., Wang, J., Qin, N., Li, L., Li, J., Bolund, L., Kristiansen, K., Wong, G.K., Olson, M., Zhang, X., Li, S., Yang, H., Wang, J., Wang, J., 2010. The sequence and de novo assembly of the giant panda genome. Nature 463, 311–317.

Li, Y., Niu, L., Wang, Q., Zhang, Z., Chen, Z., Gu, X., Xie, Y., Yan, N., Wang, S., Peng, X., Yang, G., 2012. Molecular characterization and phylogenetic analysis of ascarid nematodes from twenty-one species of captive wild mammals based on mitochondrial and nuclear sequences. Parasitology 139, 1329–1338.

Li, D.S., He, Y., Wu, H.L., Wang, C.D., Li, C.W., Lan, J.C., Chen, Z.Q., Xie, Y., Han, H.Y., Yang, G.Y., Wang, C.D., 2014. Prevalence of helminths in captive giant pandas. J. Econ. Anim. 4, 214–216 (In Chinese).

Li, D.S., He, Y., Deng, L.H., Chen, Z.Q., Cheng, Y.X., Xie, Y., Han, H.Y., Yang, G.Y., Wang, C.D., 2015. Comparison of the efficacy of ivermectin and pyrantel pamoate for the control of *Baylisascaris schroederi* infection in the giant panda. Anim. Husb. Vet. Med. 47, 87–90 (In Chinese).

Lin, Q., Li, H.M., Gao, M., Wang, X.Y., Ren, W.X., Cong, M.M., Tan, X.C., Chen, C.X., Yu, S.K., Zhao, G.H., 2012. Characterization of *Baylisascaris schroederi* from Qinling subspecies of giant panda in China by the first internal transcribed spacer (ITS-1) of nuclear ribosomal DNA. Parasitol. Res. 110, 1297–1303.

Liu, W.Z., Yang, G.C., 1994. Dignosis and treatment of baylisascariasis in the giant panda. Henan Anim. Husb. Vet. Med. 15, 48 (In Chinese).

Liu, X., He, T., Zhong, Z., Zhang, H., Wang, R., Dong, H., Wang, C., Li, D., Deng, J., Peng, G., Zhang, L., 2013. A new genotype of *Cryptosporidium* from giant panda (*Ailuropoda melanoleuca*) in China. Parasitol. Int. 62, 454–458.

Loeffler, K., Montali, R.J., Rideout, B.A., 2006. Diseases and pathology of giant pandas. In: Wildt, D.E., Zhang, A., Zhang, H., Janssen, D.L., Ellis, S. (Eds.), Giant Pandas: Biology, Veterinary Medicine and Management. Cambridge University Press, New York, USA, pp. 377–409.

Ma, G.Y., 1987. Records of *Baylisascaris schroederi* and ticks in the giant panda from Wenxian, Gansu province. Sichuan J. Zool. 6, 34 (In Chinese).

Ma, H., Wang, Z., Wang, C., Li, C., Wei, F., Liu, Q., 2015. Fatal *Toxoplasma gondii* infection in the giant panda. Parasite 22, 30.

Mainka, S.A., Qiu, X., He, T., Appel, M.J., 1994. Serologic survey of giant pandas (*Ailuropoda melanoleuca*), and domestic dogs and cats in the Wolong Reserve, China. J. Wildl. Dis. 30, 86–89.

Mathis, A., Weber, R., Deplazes, P., 2005. Zoonotic potential of the microsporidia. Clin. Microbiol. Rev. 18, 423–445.
Matos, O., Lobo, M.L., Xiao, L., 2012. Epidemiology of *Enterocytozoon bieneusi* infection in humans. J. Parasitol. Res. 2012, 981424.
Matsumoto, Y., Suzuki, S., Nozoye, T., Yamakawa, T., Takashima, Y., Arakawa, T., Tsuji, N., Takaiwa, F., Hayashi, Y., 2009. Oral immunogenicity and protective efficacy in mice of transgenic rice plants producing a vaccine candidate antigen (As16) of *Ascaris suum* fused with cholera toxin B subunit. Transgenic Res. 18, 185–192.
Maung, M., 1978. The occurrence of the second moult of *Ascaris lumbricoides* and *Ascaris suum*. Int. J. Parasitol. 8, 371–378.
McIntosh, A., 1939. A new nematode, *Ascaris schroederi*, from a giant panda, *Ailuropoda melanoleuca*. Fortschr. Zool. 24, 355–357.
McNair, C.M., 2015. Ectoparasites of medical and veterinary importance: drug resistance and the need for alternative control methods. J. Pharm. Pharmacol. 67, 351–363.
Mielke, D., Hiepe, T., 1998. The effectiveness of different disinfectants based on *p*-chloro-*m*-cresol against *Ascaris suum* eggs under laboratory conditions. Berl. Munch. Tierarztl. Wochenschr. 111, 291–294 (In German).
Pan, X.W., Yan, Y.X., Lv, J.Y., Ma, Q., Yuan, Y.H., 2001. Studies on the pathogens of sudden death in panda. Chinese J. Zoonoses 17, 53–55 (In Chinese).
Peng, D.W., Huang, S., Wang, J., Wei, F., Xue, G.F., 1989. Investigation of helminths of rare wild animals in the Shanghai Zoo. Chinese J. Zoonoses 5, 62–63 (In Chinese).
Pérez Mata, A., García Pérez, H., José Gauta Parra, J., 2016. Morphological and molecular description of *Baylisascaris venezuelensis*, n. sp. from a natural infection in the South American spectacled bear *Tremarctos ornatus* Cuvier, 1825 in Venezuela. Neotrop. Helminthol. 10, 85–103 (In Spainsh).
Qin, Q., Li, D., Zhang, H., Hou, R., Zhang, Z., Zhang, C., Zhang, J., Wei, F., 2011. Serosurvey of selected viruses in captive giant pandas (*Ailuropoda melanoleuca*) in China. Vet. Microbiol. 142, 199–204.
Qiu, X.M., 1990. The giant panda's treatment and prevention from diseases [sic. Approaches for the treatment and prevention of diseases of the giant panda]. J. Sichuan Teach. Coll. 11, 195–199 (In Chinese).
Qiu, X.M., Mainka, S.A., 1993. Review of mortality of the giant panda (*Ailuropoda melanoleuca*). J. Zoo Wildl. Med. 24, 425–429 (In Chinese).
Qiu, M.H., Zhu, C.J., 1987. Parasites of giant panda and their control. Selected Acadmic Papers of Giant Panda Disease Control. China Forestry Press, Beijing, China, pp. 1–9 (In Chinese).
Qiu, M.H., Zhou, Y.H., Xiang, P.L., 1984. Observations of mites from the giant panda. Sichuan J. Zool. 2, 36–37 (In Chinese).
Qiu, M.H., Wang, D.Q., Li, K.Z., 1991. A new flea of the genus *Chaetopsylla* from giant panda. Sichuan J. Zool. 10, 7–9 (In Chinese).
Reid, D.G., Hu, J., Dong, S., Wang, W., Huang, Y., 1989. Giant panda *Ailuropoda melanoleuca* behaviour and carrying capacity following a bamboo die-off. Biol. Conserv. 49, 85–104.
Santín, M., 2015. Enterocytozoon bieneusi. In: Xiao, L., Ryan, U., Feng, Y. (Eds.), Biology of Foodborne Parasites. CRC Press, Boca Raton, USA, pp. 149–174.
Santín, M., Fayer, R., 2009. *Enterocytozoon bieneusi* genotype nomenclature based on the internal transcribed spacer sequence: a consensus. J. Eukaryot. Microbiol. 56, 34–38.
Schaller, G.B., Hu, J.C., Pan, W.S., Zhu, J., 1985. The Giant Pandas of Wolong. University of Chicago Press, Chicago, USA.
Shan, L., Hu, Y., Zhu, L., Yan, L., Wang, C., Li, D., Jin, X., Zhang, C., Wei, F., 2014. Large-scale genetic survey provides insights into the captive management and reintroduction of giant pandas. Mol. Biol. Evol. 31, 2663–2671.

Song, J.K., Wang, Z.H., Wang, H.B., Peng, X.Q., Wang, X.T., Ren, G.J., Zhao, G.H., 2016. Identification and phylogenetic analysis of *Ogmocotyle* sp. from giant panda in Qinling Mountain. Chinese J. Vet. Sci. 36, 563–566 (In Chinese).

Sprent, J.F., 1968. Notes on *Ascaris* and *Toxascaris*, with a definition of *Baylisascaris* gen.nov. Parasitology 58, 185–198.

Srivastava, M., Misra-Bhattacharya, S., 2015. Overcoming drug resistance for macro parasites. Future Microbiol. 10, 1783–1789.

Swaisgood, R.R., Zhang, G.Q., Zhou, X.P., Zhang, H.M., 2006. The science of behavioural management: creating biologically relevant living environments in captivity. In: Wildt, D.E., Zhang, A., Zhang, H., Janssen, D.L., Ellis, S. (Eds.), Giant Pandas: Biology, Veterinary Medicine and Management. Cambridge University Press, New York, USA, pp. 377–409.

Talbot, S.L., Shields, G.F., 1996. A phylogeny of the bears (Ursidae) inferred from complete sequences of three mitochondrial genes. Mol. Phylogenet. Evol. 5, 567–575.

The State Forestry Administration of China, 2015. The results of the fourth national giant panda survey (Report). (In Chinese)

Tian, G.R., Zhao, G.H., Du, S.Z., Hu, X.F., Wang, H.B., Zhang, L.X., Yu, S.K., 2015. First report of *Enterocytozoon bieneusi* from giant pandas (*Ailuropoda melanoleuca*) and red pandas (*Ailurus fulgens*) in China. Infect. Genet. Evol. 34, 32–35.

Tokiwa, T., Nakamura, S., Taira, K., Une, Y., 2014. *Baylisascaris potosis* n. sp., a new ascarid nematode isolated from captive kinkajou, *Potos flavus*, from the Cooperative Republic of Guyana. Parasitol. Int. 63, 591–596.

Tranbenkova, N.A., Spiridonov, S.E., 2017. Molecular characterization of *Baylisascaris devosi* Sprent, 1952 (Ascaridoidea, Nematoda) from Kamchatka sables. Helminthologia 54, 105–112.

Tsuji, N., Suzuki, K., Kasuga-Aoki, H., Matsumoto, Y., Arakawa, T., Ishiwata, K., Isobe, T., 2001. Intranasal immunization with recombinant *Ascaris suum* 14-kilodalton antigen coupled with cholera toxin B subunit induces protective immunity to *A. suum* infection in mice. Infect. Immun. 69, 7285–7292.

Tsuji, N., Kasuga-Aoki, H., Isobe, T., Arakawa, T., Matsumoto, Y., 2002. Cloning and characterisation of a highly immunoreactive 37 kDa antigen with multi-immunoglobulin domains from the swine roundworm *Ascaris suum*. Int. J. Parasitol. 32, 1739–1746.

Tsuji, N., Suzuki, K., Kasuga-Aoki, H., Isobe, T., Arakawa, T., Matsumoto, Y., 2003. Mice intranasally immunized with a recombinant 16-kilodalton antigen from roundworm *Ascaris* parasites are protected against larval migration of *Ascaris suum*. Infect. Immun. 71, 5314–5323.

Tsuji, N., Miyoshi, T., Islam, M.K., Isobe, T., Yoshihara, S., Arakawa, T., Matsumoto, Y., Yokomizo, Y., 2004. Recombinant *Ascaris* 16-Kilodalton protein-induced protection against *Ascaris suum* larval migration after intranasal vaccination in pigs. J. Infect. Dis. 190, 1812–1820.

Vercruysse, J., Holdsworth, P., Letonja, T., Barth, D., Conder, G., Hamamoto, K., Okano, K., World Organization for Animal Health, 2001. International harmonisation of anthelmintic efficacy guidelines. Vet. Parasitol. 96, 171–193.

Vercruysse, J., Holdsworth, P., Letonja, T., Conder, G., Hamamoto, K., Okano, K., Rehbein, S., Veterinary International Co-operation on Harmonisation Working Group on anthelmintic guidelines, 2002. International harmonisation of anthelmintic efficacy guidelines (Part 2). Vet. Parasitol. 103, 277–297.

Vercruysse, J., Behnke, J.M., Albonico, M., Ame, S.M., Angebault, C., Bethony, J.M., Engels, D., Guillard, B., Nguyen, T.V., Kang, G., Kattula, D., Kotze, A.C., McCarthy, J.S., Mekonnen, Z., Montresor, A., Periago, M.V., Sumo, L., Tchuente, L.A., Dang, T.C., Zeynudin, A., Levecke, B., 2011. Assessment of the anthelmintic efficacy of albendazole in school children in seven countries where soil-transmitted helminths are endemic. PLoS Negl. Trop. Dis. 5, e948.

Vlaminck, J., Nejsum, P., Vangroenweghe, F., Thamsborg, S.M., Vercruysse, J., Geldhof, P., 2012. Evaluation of a serodiagnostic test using *Ascaris suum* haemoglobin for the detection of roundworm infections in pig populations. Vet. Parasitol. 189, 267–273.

Wan, Q.H., Fang, S.G., Wu, H., Fujihara, T., 2003. Genetic differentiation and subspecies development of the giant panda as revealed by DNA fingerprinting. Electrophoresis 24, 1353–1359.

Wang, C.D., Yang, G.Y., Wang, Q., Yu, X.M., Zhao, B., Zhong, S.T., 2000. Experimental by closantel treatment of *Chorioptes panda* in the giant panda. Chinese J. Vet. Sci. Technol. 30, 39 (In Chinese).

Wang, C.D., Yang, G.Y., Wang, Q., Yu, X.M., Zhao, B., Lan, J.C., 2001. Investigation of parasitic infections of rare mammals in Chengdu Zoo. Chinese J. Vet. Sci. 37, 21–22 (In Chinese).

Wang, C.D., Li, D.S., Tan, C.X., Deng, L.H., Huang, Z., Han, H.Y., Zhang, Y., 2007a. *Proteus mirabilis* infection of the reproductive tract of a giant panda. Sichuan J. Zool. 26, 167 (In Chinese).

Wang, C.D., Tang, C.X., Deng, L.H., Li, D.S., 2007b. Case report of rectal prolapse associated with intussusception in a wild giant panda. Chinese J. Vet. Med. 43, 64–65 (In Chinese).

Wang, C.D., Yan, Q.G., Zhang, Z.H., Luo, L., Fan, W.Q., Yang, Z., Lan, J.C., Huang, X.M., Li, M.X., 2008a. Isolation and identification of rotavirus from giant panda cubs. Acta Theriol. Sin. 28, 87–91 (In Chinese).

Wang, T., He, G., Yang, G., Fei, Y., Zhang, Z., Wang, C., Yang, Z., Lan, J., Luo, L., Liu, L., 2008b. Cloning, expression and evaluation of the efficacy of a recombinant *Baylisascaris schroederi* Bs-Ag3 antigen in mice. Vaccine 26, 6919–6924.

Wang, S., Gu, X., Fu, Y., Lai, S., Wang, S., Peng, X., Yang, G., 2012. Molecular taxonomic relationships of *Psoroptes* and *Chorioptes* mites from China based on COI and 18S rDNA gene sequences. Vet. Parasitol. 184, 392–397.

Wang, N., Li, D.S., Zhou, X., Xie, Y., Liang, Y.N., Wang, C.D., Yu, H., Chen, S.J., Yan, Y.B., Gu, X.B., Wang, S.X., Peng, X.R., Yang, G.Y., 2013a. A sensitive and specific PCR assay for the detection of *Baylisascaris schroederi* eggs in giant panda feces. Parasitol. Int. 62, 435–436.

Wang, X., Yan, Q., Xia, X., Zhang, Y., Li, D., Wang, C., Chen, S., Hou, R., 2013b. Serotypes, virulence factors, and antimicrobial susceptibilities of vaginal and fecal isolates of *Escherichia coli* from giant pandas. Appl. Environ. Microbiol. 79, 5146–5150.

Wang, T., Chen, Z., Xie, Y., Hou, R., Wu, Q., Gu, X., Lai, W., Peng, X., Yang, G., 2015. Prevalence and molecular characterization of *Cryptosporidium* in giant panda (*Ailuropoda melanoleuca*) in Sichuan province, China. Parasit. Vectors 8, 344.

Wei, F., Hu, Y., Zhu, L., Bruford, M.W., Zhan, X., Zhang, L., 2012. Black and white and read all over: the past, present and future of giant panda genetics. Mol. Ecol. 21, 5660–5674.

Wei, F., Swaisgood, R., Hu, Y., Nie, Y., Yan, L., Zhang, Z., Qi, D., Zhu, L., 2015. Progress in the ecology and conservation of giant pandas. Conserv. Biol. 29, 1497–1507.

Wu, J., Hu, H.G., 1985a. Giant panda—a new host for *Haemaphysalis warburtoni*. Chinese J. Vet. Sci. 8, 10 (In Chinese).

Wu, J., Hu, H.G., 1985b. Recent advances in parasitic diseases of giant panda. Wildlife 5, 42–43 (In Chinese).

Wu, J., Hu, H.G., 1988. Reports of regular deworming of *Baylisascaris schroederi*. Anim. Husb. Vet. Med. 3, 118–119 (In Chinese).

Wu, J., Jiang, Y.K., He, G.Z., Wu, G.Q., Zhang, D.H., Hu, H.G., 1985. Research of the life cycle of *Baylisascaris schroederi* of the giant panda. Chinese J. Vet. Sci. Technol. 6, 21–23 (In Chinese).

Xie, Y., Zhang, Z., Wang, C., Lan, J., Li, Y., Chen, Z., Fu, Y., Nie, H., Yan, N., Gu, X., Wang, S., Peng, X., Yang, G., 2011. Complete mitochondrial genomes of *Baylisascaris schroederi*, *Baylisascaris ailuri* and *Baylisascaris transfuga* from giant panda, red panda and polar bear. Gene 482, 59–67.

Xie, Y., Chen, S., Yan, Y., Zhang, Z., Li, D., Yu, H., Wang, C., Nong, X., Zhou, X., Gu, X., Wang, S., Peng, X., Yang, G., 2013. Potential of recombinant inorganic pyrophosphatase antigen as a new vaccine candidate against *Baylisascaris schroederi* in mice. Vet. Res. 44, 90.

Xie, Y., Zhou, X., Zhang, Z., Wang, C., Sun, Y., Liu, T., Gu, X., Wang, T., Peng, X., Yang, G., 2014. Absence of genetic structure in *Baylisascaris schroederi* populations, a giant panda parasite, determined by mitochondrial sequencing. Parasit. Vectors 7, 606.

Xie, Y., Zhou, X., Chen, L., Zhang, Z., Wang, C., Gu, X., Wang, T., Peng, X., Yang, G., 2015a. Cloning and characterization of a novel sigma-like glutathione S-transferase from the giant panda parasitic nematode, *Baylisascaris schroederi*. Parasit. Vectors 8, 44.

Xie, Y., Zhou, X., Sun, Y., Gu, X., Yang, G., 2015b. Mitochondrial 12S-based analysis of genetic diversity of *Baylisascaris schroederi* in giant pandas from two mountain ranges in China. Acta Theriol. Sin. 35, 328–335 (In Chinese).

Xie, Y., Hoberg, E.P., Yang, Z., Urban Jr., J.F., Yang, G., 2017. *Ancylostoma ailuropodae* n. sp. (Nematoda: Ancylostomatidae), a new hookworm parasite isolated from wild giant pandas in Southwest China. Parasit. Vectors 10, 277.

Xu, R.H., Zhang, R.Y., 2002. Efficacy of drug treatment against mite infections in the giant panda. Fujian J. Anim. Husb. Vet. Med. 24, 55–56 (In Chinese).

Xu, Y.H., Xie, H.X., Liu, S.L., Zhou, Z.Y., Shi, X.Q., 1986. A new species of the genus *Demodex* (Acariformes: Demodicidae). Acta Zool. Sin. 32, 163–167 (In Chinese).

Yang, X.Y., 1993. Discussion of the relationship between *Baylisascaris* prevalence and habitat of wild giant pandas. Sichuan Forest. Sci. Technol. 14, 70–73 (In Chinese).

Yang, G.Y., 1998. Research advances in the parasite and parasitic disease of the giant panda. Chinese J. Vet. Sci. 18, 206–209 (In Chinese).

Yang, G.Y., Zhang, Z.H., 2013. Wildlife Parasitology. Science Press, Beijing, China, pp. 458–465 (In Chinese).

Yang, W., Xu, G.J., Lan, J.C., 2001. Diagnosis and treatment of skin mite of the giant panda. Sichuan J. Zool. 20, 56–60 (In Chinese).

Yang, X., Yang, J., Wang, H., Li, C., He, Y., Jin, S., Zhang, H., Li, D., Wang, P., Xu, Y., Xu, C., Fan, C., Xu, L., Huang, S., Qu, C., Li, G., 2016. Normal vaginal bacterial Flora of giant pandas (*Ailuropoda melanoleuca*) and the antimicrobial susceptibility patterns of the isolates. J. Zoo Wildl. Med. 47, 374–378.

Ye, Z.Y., 1986. Report of the treatment of itch mites in the giant panda. Chinese J. Vet. Sci. 1, 28–29 (In Chinese).

Ye, Z.Y., 1989. Fifty case reports on diseases and their treament in wild giant pandas. Chinese J. Vet. Sci. 15, 30–31 (In Chinese).

Ye, Z.Y., Zhang, A.J., 1981. Discussion regarding the elimination of *Baylisascaris schroederi*. J. Zool. 2, 48–49 (In Chinese).

Yeruham, I., Rosen, S., Hadani, A., 1999. Chorioptic mange (Acarina: Psoroptidae) in domestic and wild ruminants in Israel. Exp. Appl. Acarol. 23, 861–869.

Yu, S.K., Feng, N., Yuan, W., Pu, P., Li, Y.Z., Ma, Q.Y., 1998. Investigation of parasites of the giant panda, and a record of a new species. J. Chin. Agr. Univ. S2, 117–118 (In Chinese).

Zahler, M., Hendrikx, W.M., Essig, A., Rinder, H., Gothe, R., 2001. Taxonomic reconsideration of the genus *Chorioptes* Gervais and van Beneden, 1859 (Acari: Psoroptidae). Exp. Appl. Acarol. 25, 517–523.

Zhang, H.M., Wang, P.Y., 2003. Breeding Research of Giant Panda. China Forestry Publishing, Beijing, China, pp. 121–174 (In Chinese).

Zhang, H., Zhang, Z.L., 2002. Diagnosis and treatment of giant panda ascarids. Gansu Anim. Vet. Sci. 32, 25–26 (In Chinese).

Zhang, J.S., Daszak, P., Huang, H.L., Yang, G.Y., Kilpatrick, A.M., Zhang, S., 2008. Parasite threat to panda conservation. Ecohealth 5, 6–9.

Zhang, H., Wang, X.H., Fan, W.A., Yuan, M., 2010. Literature review of the giant panda parasitic diseases. Gansu Anim. Vet. Sci. 40, 40–43 (In Chinese).

Zhang, L., Yang, X., Wu, H., Gu, X., Hu, Y., Wei, F., 2011. The parasites of giant pandas: individual-based measurement in wild animals. J. Wildl. Dis. 47, 164–171.

Zhang, W., Yie, S., Yue, B., Zhou, J., An, R., Yang, J., Chen, W., Wang, C., Zhang, L., Shen, F., Yang, G., Hou, R., Zhang, Z., 2012. Determination of *Baylisascaris schroederi* infection in wild giant pandas by an accurate and sensitive PCR/CE-SSCP method. PLoS One 7, e41995.

Zhang, L., Wu, Q., Hu, Y., Wu, H., Wei, F., 2015. Major histocompatibility complex alleles associated with parasite susceptibility in wild giant pandas. Heredity (Edinb) 114, 85–93.

Zhao, G.H., Li, H.M., Ryan, U.M., Cong, M.M., Hu, B., Gao, M., Ren, W.X., Wang, X.Y., Zhang, S.P., Lin, Q., Zhu, X.Q., Yu, S.K., 2012. Phylogenetic study of *Baylisascaris schroederi* isolated from Qinling subspecies of giant panda in China based on combined nuclear 5.8S and the second internal transcribed spacer (ITS-2) ribosomal DNA sequences. Parasitol. Int. 61, 497–500.

Zhao, G.H., Xu, M.J., Zhu, X.Q., 2013. Identification and characterization of microRNAs in *Baylisascaris schroederi* of the giant panda. Parasit. Vectors 6, 216.

Zhao, Z.H., Bian, Q.Q., Ren, W.X., Cheng, W.Y., Jia, Y.Q., Fang, Y.Q., Zhao, G.H., 2014. Genetic variability of *Baylisascaris schroederi* from the Qinling subspecies of the giant panda in China revealed by sequences of three mitochondrial genes. Mitochondrial DNA 25, 212–217.

Zhao, N., Li, M., Luo, J., Wang, S., Liu, S., Wang, S., Lyu, W., Chen, L., Su, W., Ding, H., He, H., 2017. Impacts of canine distemper virus infection on the giant panda population from the perspective of gut microbiota. Sci. Rep. 4, 39954.

Zhou, Y.H., Yang, F.X., Zhu, C.J., Zhao, G.L., 1989. Study on the etiology, treatment and prevention of skin infection of baby panda. J. Chongqing Med. Univ. 14, 134–135 (In Chinese).

Zhou, X., Xie, Y., Zhang, Z.H., Wang, C.D., Sun, Y., Gu, X.B., Wang, S.X., Peng, X.R., Yang, G.Y., 2013a. Analysis of the genetic diversity of the nematode parasite *Baylisascaris schroederi* from wild giant pandas in different mountain ranges in China. Parasit. Vectors 6, 233.

Zhou, X., Yu, H., Wang, N., Xie, Y., Liang, Y.N., Li, D.S., Wang, C.D., Chen, S.J., Yan, Y.B., Gu, X.B., Wang, S.X., Peng, X.R., Yang, G.Y., 2013b. Molecular diagnosis of *Baylisascaris schroederi* infections in giant panda (*Ailuropoda melanoleuca*) feces using PCR. J. Wildl. Dis. 49, 1052–1055.

Zhu, L., Long, Z., 1983. The vicissitudes of the giant panda. Acta Zool. Sin. 29, 90–104 (In Chinese).

CHAPTER TWO

The Evolutionary Biology, Ecology and Epidemiology of Coccidia of Passerine Birds

Alex Knight*,[1], John G. Ewen†, Patricia Brekke†, Anna W. Santure*

*School of Biological Sciences, University of Auckland, Auckland, New Zealand
†Institute of Zoology, Zoological Society of London, London, United Kingdom
[1]Corresponding author: e-mail address: alexknight444@gmail.com

Contents

1. Introduction 36
2. The Taxonomy and Life Cycle of Coccidia 37
3. Coccidia and Passerine Health 40
4. Epidemiology of Coccidia in Wild Passerines 42
   4.1 Diurnal Shedding Patterns 42
   4.2 Age 45
   4.3 Host Migration 46
   4.4 Foraging Ecology 47
5. Natural Selection 48
6. Sexual Selection 50
7. Future Directions 51
8. Conclusion 54
Acknowledgements 54
References 54

Abstract

Coccidia are intracellular parasites of the phylum Apicomplexa that cause a range of pathologies collectively termed coccidiosis. Species of coccidia of commercial importance have been well studied, with the effect of other species on passerine birds receiving increasing attention. In this chapter, we review the literature on coccidia in passerines, with a particular focus on wild populations. The taxonomy and life cycle of passerine coccidia are covered, as is their impact on the health of passerines, their epidemiology and their role in parasite-mediated natural and sexual selection. Coccidia can pose a significant threat to the health of wild passerine populations, and high rates of mortality have been observed in some studies. We examine some of the genetic factors that influence host resistance to coccidia and discuss how these parasites may be important in relation to sexually selected traits. General patterns are beginning to emerge with regard to the epidemiology of the parasites, and the influence of different aspects of the host's ecology on the prevalence and intensity of coccidia is being

Advances in Parasitology, Volume 99
ISSN 0065-308X
https://doi.org/10.1016/bs.apar.2018.01.001

revealed. We examine these, as well exceptions, in addition to the phenomenon of diurnal oocyst shedding that can bias studies if not accounted for. Finally, we discuss potential future directions for research on coccidia in passerines and the importance of understanding parasite ecology in the management of threatened species.

1. INTRODUCTION

Parasitism is one of the most common modes of existence in the animal kingdom, with some estimates suggesting that half of all animal taxa at some point parasitize another during their life cycle (Poulin and Morand, 2000). There is increasing awareness that host–parasite relationships have substantial influence on the ecology of their host species and ecosystems (Hatcher et al., 2012). A common perception of parasites is that they depress the fitness and potentially threaten the persistence of their hosts; while this may be true, parasites can also have beneficial effects on both their hosts and the ecosystem they are in (Hudson et al., 2006; Spencer and Zuk, 2016). Parasites can maintain the genetic diversity of their hosts, act as keystone species and limit the spread of pathogens (Hatcher et al., 2008; Johnson et al., 2013; Weedall and Conway, 2010). Therefore, understanding how parasites interact with their host can be essential in studying the evolutionary ecology of host species and determining how parasites influence community or ecosystem level processes (Paterson and Piertney, 2011).

Moreover, parasite ecology and epidemiology is receiving increasing attention in the context of the conservation of threatened species. Historical examples of species extinctions have demonstrated the ominous potential of infectious parasitic organisms to drive reductions in population size (Warner, 1968; Wyatt et al., 2008). In the wake of global anthropogenic change and the declining population size of many species, the risk of extinction posed by disease-causing agents is likely to increase. Human activity can facilitate parasite transmission in multiple ways, such as by inadvertently creating conditions favourable for the infective life stages of a parasite, or bringing parasites into contact with otherwise isolated host populations (Daszak et al., 2001). Furthermore, wildlife with smaller population sizes is expected to be less genetically diverse (Frankham, 1996). Decreases in genetic diversity have been linked to increases in susceptibility to parasitism and disease (Acevedo-Whitehouse et al., 2003). Disease management for endangered species can be improved upon when informed by ecological and epidemiological knowledge of the pathogen (Haydon et al., 2006).

Coccidia are single-celled, obligate intracellular protozoan parasites of the phylum Apicomplexa. They infect a wide range of taxa including mammals, reptiles, amphibians and birds. In the class Aves, coccidia are ubiquitous parasites, infecting nearly all known orders of birds (Duszynski et al., 2000). Infection can result in a variety of different pathologies, collectively termed coccidiosis. Coccidia have a huge economic impact on livestock and domestic animals, which has resulted in significant efforts being made to understand the biology and epidemiology of the parasites (Chapman, 2014; Williams, 1999). There has been increasing interest in the effect of coccidia on wild animal species to determine how they drive population dynamics, affect host ecology and influence the evolution of their hosts. In this chapter, we review the literature on coccidia that are shed in the faeces of passerines, with the exception of *Cryptosporidium* (no longer considered to be a classical coccidian parasite) and also excluding intestinal sarcocystosis in passerines of the family Corvidae as definitive hosts. Five areas are broadly covered, including the taxonomy and life cycle of passerine coccidia; their impact on the health of passerines; the epidemiology of coccidian infection and the involvement of coccidia in parasite-mediated natural and sexual selection.

2. THE TAXONOMY AND LIFE CYCLE OF COCCIDIA

Coccidia are typically classified as a class or subclass within the phylum Apicomplexa of the protozoa, whereas classifications of orders, suborders and families have been more disputed (reviewed in Tenter et al., 2002). Passerine coccidia that are excreted in faeces are predominantly placed within the genus *Isospora*, but there are also reports of the occurrence in passerines of species belonging to the genera *Eimeria* and *Caryospora* (Berto et al., 2011; Brown et al., 2010; Duszynski et al., 2000). In the past, the generic and species level classifications of coccidia in passerines have been based on their morphology, host species, life cycle patterns and where in the host different stages of the life cycle develop (Berto et al., 2011; Fayer, 1980). The structure of sporulated oocysts, the exogenous stage of the parasite, is frequently used to distinguish coccidian genera; *Isospora* spp. oocysts have two sporocysts, each of which contains four sporozoites; *Eimeria* spp. have four sporocysts with two sporozoites in each and *Caryospora* spp. have one sporocyst and eight sporozoites (Duszynski and Upton, 2001; Duszynski et al., 2000).

The taxonomy of coccidia at fine levels has been criticized because of issues with type specimens (Chapman et al., 2013). For example, often the only specimen available for long-term storage is the exogenous stage of the life cycle, namely, the oocyst (Williams et al., 2010). Endogenous stages that develop in various tissues of the host may be unobtainable because hosts are endangered and legally protected or because coccidia-free hosts are not readily available for controlled experimentation. Tissue stages can be important for determining taxonomic relationships, and, thus, taxonomic classifications may be based on limited data (Williams et al., 2010). Furthermore, assumptions of host specificity, i.e., that different species of coccidia are believed to parasitize different hosts, may not be correct (Chapman et al., 2013). Molecular analysis of coccidia infecting 23 species of passerines did indeed, however, reveal that most passerine species are infected by at least one unique lineage of coccidia (Schrenzel et al., 2005). Yet the same study also revealed that some host species shared coccidian lineages that were much more closely related than their hosts, suggesting recent host switching, or that coccidia can infect multiple hosts. Phylogenetic analyses have been recommended in order to resolve future taxonomic uncertainties; in particular, 18S ribosomal DNA and cytochrome *c* oxidase subunit I gene sequences have proven valuable in making generic and species level distinctions (Ogedengbe et al., 2011, 2015; Schrenzel et al., 2005). Moreover, recent sequencing of complete mitochondrial genomes from coccidian species that infect passerines will provide a valuable resource for further phylogenetic studies (Ogedengbe et al., 2016; Yang et al., 2017).

The life cycle of passerine coccidia has three phases (Fig. 1): merogony (frequently called schizogony), which is asexual multiplication; gamogony (or gametogony), a sexual phase; and sporogony, which results in the formation of sporozoites (Duszynski and Upton, 2001). Outside of their hosts, oocysts undergo sporogony, a meiotic process, which requires oxygen and suitable levels of humidity and temperature (Allen and Fetterer, 2002). During sporogony, sporozoites develop within the oocyst, which is thereafter infective. Once these oocysts have been ingested by the host and are inside the lumen of the gastrointestinal tract, sporozoites emerge from the oocyst and invade epithelial cells (Long, 1993). After the sporozoites have successfully penetrated the gut epithelium, they form meronts (also known as schizonts) and reproduce in an asexual phase called merogony (synonym: schizogony, as has been mentioned earlier), by undergoing mitosis. Merogony results in the formation of merozoites, which exit the host cell to enter new host cells and give rise to another round of merogony. The

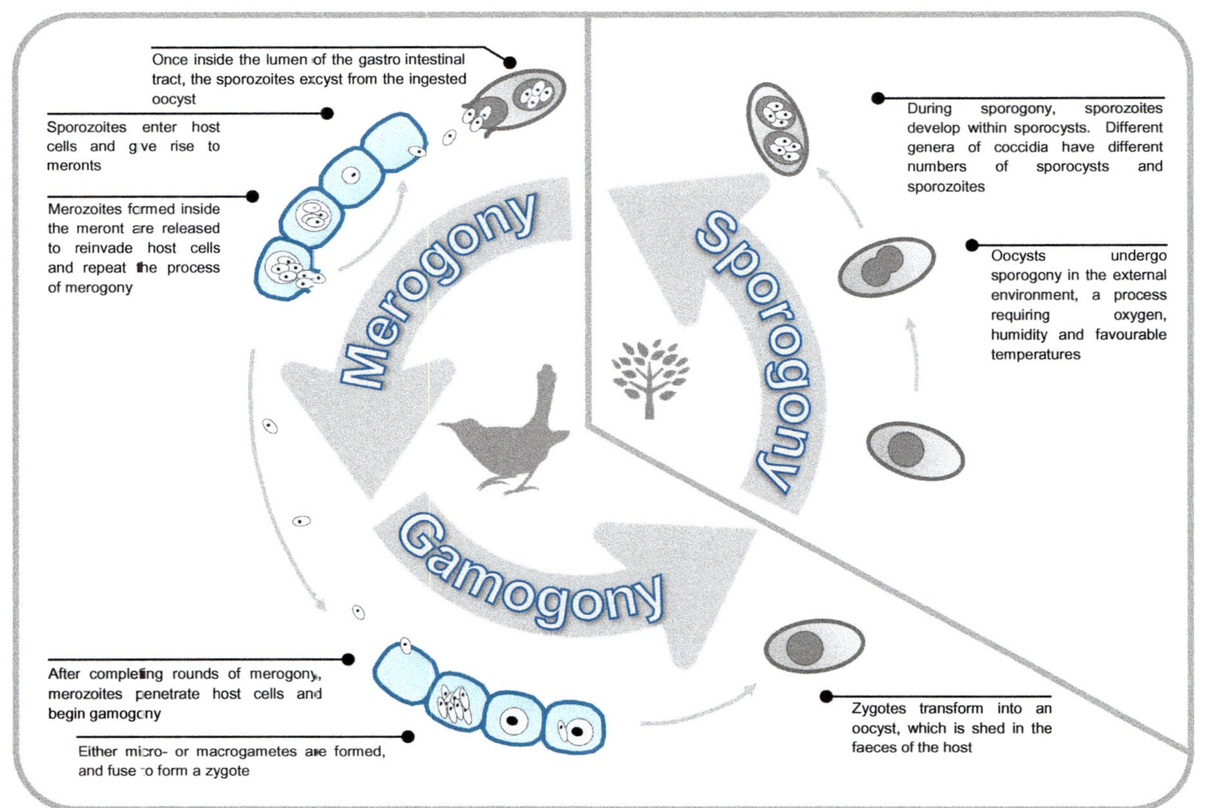

Fig. 1 The life cycle of coccidia.

number of cycles of merogony is thought to be fixed for different species of coccidia (Long, 1993). After the final round of merogony, merozoites enter a host cell and form male and female gamonts, the sexual stages. The final phase results in the production of gametes and fertilization, followed by the formation of a protective wall, creating the oocyst. This is shed in the faeces, thereby completing the life cycle.

Some species of coccidia that infect passerines apparently undergo their life cycle development solely within the gastrointestinal tract of the host (Abd-Al-Aal et al., 2000; Box, 1977). However, Box (1977) also noted an alternative life cycle, with merozoites invading cells of the immune system in canaries (*Serinus canaria*) and entering the viscera of the host. Box (1977) concluded that two species were present in the canary, one with and one without extraintestinal stages. This led to some confusion as to the taxonomic classifications of passerine coccidia that are found outside the intestine, and the genus *Atoxoplasma* has been used by some authors for coccidia found in extraintestinal tissues (Carreno and Barta, 1999; Mohr et al., 2017). Recent work has shown that isosporoid coccidia do proliferate outside of the intestine and that *Atoxoplasma* spp. and *Isospora* spp. found in passerines are in fact one and the same group of organisms (Schrenzel et al., 2005). Extraintestinal stages might be related to specific lineages of coccidia or may also occur in birds of different ages and experiencing different levels of physiological stress (Schrenzel et al., 2005). It appears that merogony is often the only extraintestinal developmental stage, although gamogony has very infrequently been observed to occur outside of the gastrointestinal tract (Baker et al., 1996; Lainson, 1959; Schrenzel et al., 2005).

3. COCCIDIA AND PASSERINE HEALTH

The effects of avian coccidiosis have been studied for well over a century because of its occurrence in domesticated birds (Williams, 2005). Different species of coccidia that infect poultry are known to vary in their pathogenicity, but the development of symptoms is also dependent on the dose of infective oocysts that are transmitted (Long, 1993; Williams, 2005). How different species of coccidia that infect individual passerine hosts vary in their pathogenicity is not as well understood. Box (1977) did note that infection was lethal for four of eight birds when visceral coccidian infection was observed, whereas infection that was limited to the gastrointestinal tract did not result in mortality. She suggested that extraintestinal infection was associated with a specific coccidian species. Box (1977) therefore concluded that different species of coccidia vary in their virulence. High mortality rates have subsequently been noted in wild and captive passerines that have exhibited coccidian infection of the viscera. Isosporan coccidia were identified from a wild Israeli sparrow (*Passer domesticus biblicus*) population that had suffered a large number of deaths; however, many individuals within this population were concurrently infected with another apicomplexan parasite which could have contributed to the increased rate of mortality (Gill and Paperna, 2008).

The captive blue-crowned laughingthrush (*Dryonastes courtoisi*) population at Mulhouse Zoo in France had been unsuccessful at rearing chicks past 1 year of age (Mohr et al., 2017). An investigation revealed that coccidia were present in both the chicks and parents. Postmortems found the main pathological findings to be congested lungs, hepatomegaly and a pale liver (Mohr et al., 2017). Extraintestinal infection of the host by coccidia may therefore be associated with high rates mortality; however, the cause of infection progressing outside the gastrointestinal tract remains unclear.

One of the major symptoms of coccidian infection when it is found within the gastrointestinal tract is believed to be a reduced ability to absorb nutrients (Hõrak et al., 2004; Pap et al., 2011). Autopsies of infected individuals have revealed thickened intestinal walls when sexual stages of coccidia are present, and evidence of ruptured glands in the epithelium, resulting from the release of coccidia (Schoener, 2010; Swayne et al., 1991). It has been asserted in several studies, which have demonstrated an association between coccidian infection and body condition, that malabsorption of nutrients was probably the cause. Experimentally infected house sparrows (*Passer domesticus*) had reduced feather quality, lighter feathers and shorter wings (Pap et al., 2011). A previous study supported their conclusion that this was related to nutrient uptake by demonstrating a link between diet and feather quality (Pap et al., 2008). Although this has not been tested, changes in wing morphology and development have the potential to alter the cost of flight for infected birds (Pap et al., 2011). In greenfinches (*Carduelis chloris*), coccidian infection was also associated with decreases in serum albumin, triglyceride, vitamin E and carotenoid concentrations (Hõrak et al., 2004). These authors proposed that damage to the jejunum and duodenum, where uptake of proteins and lipids occurs, was a possible mechanism for lowering levels of these nutrients. Frequently, infected individuals also display a reduction in body mass, even when food is supplied ad libitum, reinforcing the proposition that infection limits nutrient uptake (Costa and Macedo, 2005; Hõrak et al., 2004).

4. EPIDEMIOLOGY OF COCCIDIA IN WILD PASSERINES

It is a well-recognized phenomenon that not all individuals in a population are equally susceptible to a pathogen, nor are they equally likely to transmit it once infected (Lloyd-Smith et al., 2005). Environmental factors as well as behavioural, demographic, genetic and physiological traits can influence an individual's susceptibility to different parasites and their potential to transmit them (Lloyd-Smith et al., 2005; Woolhouse et al., 1997). Identification of these factors is critical to understanding the distribution, spread and impact of disease-causing agents in a population (Beldomenico and Begon, 2010; Hawley and Altizer, 2011). This information can also be used to target control strategies and improve the management of wild host populations (Perkins et al., 2003).

Determining which individuals are most susceptible to coccidian infection in the wild is often done by estimating the parasite load or infection intensity. To do this, the number of coccidian oocysts in a bird's faeces is often counted and parasite load is calculated as the number of oocysts per gram or droplet of faeces. Faecal oocyst counts have been verified as a reliable measure of infection intensity of coccidia in at least two passerine species (Dolnik, 2006; Filipiak et al., 2009). In addition to parasite load, the prevalence of coccidia has been reported in numerous wild passerine species. However, it is often estimated incidentally in conjunction with other hypotheses or observations (Gill and Paperna, 2008). Thus, many of the recorded estimates of prevalence in passerines have relied on small sample sizes, and may not reflect the true prevalence in the avian population. Furthermore, many studies do not explicitly explain sampling procedures that are designed to eliminate sources of bias when estimating prevalence (Lachish et al., 2012). We consider, below, some of the underlying causes of heterogeneity in susceptibility to coccidia and the prevalence thereof in a population (summarized in Fig. 2), and we discuss the diurnal shedding of coccidia, which can bias studies on infection intensity if not accounted for.

4.1 Diurnal Shedding Patterns

For intestinal coccidia to be transmitted from one host to the next, they have to be shed in the faeces of their current host. Moreover, the shedding of oocysts by passerine birds has a strong circadian pattern that appears to be a ubiquitous phenomenon. To our knowledge, in all passerines tested to

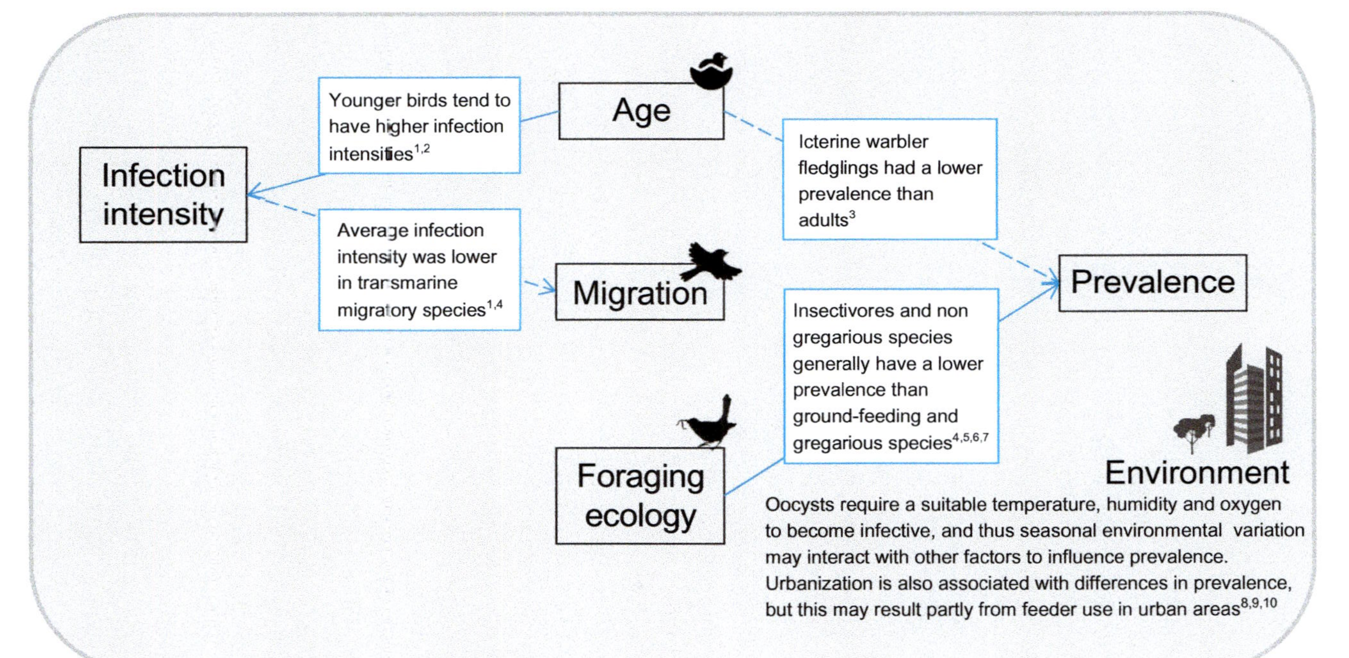

Fig. 2 Diagram showing interactions between aspects of host ecology and the prevalence and intensity of coccidian infection. *Arrows* indicate associations. *Dashed arrows* indicate association, but possible confounding factors were also identified, or the study was contrary to others in which no effect was found. References for the text are in *superscript* and are as follows: 1, Dolnik (2002); 2, López et al. (2007); 3, Svobodová and Cibulková (1995); 4, Zinke et al. (2004); 5, Bandelj et al. (2015); 6, McQuistion (2000); 7, Dolnik et al. (2010); 8, Allen and Fetterer (2002); 9, Delgado and French (2015); 10, Giraudeau et al. (2014).

date, oocyst shedding peaks in the afternoon or evening. In our review of the literature, evidence of a circadian rhythm was found for English house sparrows (*P. domesticus*; Boughton, 1933), cliff swallows (*Petrochelidon pyrrhonota*; Stabler and Kitzmiller, 1972), canaries (*Serinus canarius*; Box, 1977), chaffinches (*Fringilla coelebs*; Gryczyńska et al., 1999), house finches (*Carpodacus mexicanus*; Brawner and Hill, 1999), dark-eyed juncos (*Junco hyemalis*; Hudman et al., 2000), greenfinches (*C. chloris*; Brown et al., 2001), starlings (*Sturnus vulgaris*; Dolnik, 2002), scarlet grosbeaks (*Carpodacus erythrinus*; Dolnik, 2002), reed warblers (*Acrocephalus scirpaceus*; Dolnik, 2002), willow warblers (*Phylloscopus trochilus*; Dolnik, 2002), blackbirds (*Turdus merula*; Misof, 2004), serins (*Serinus serinus*; López et al., 2007), garden warblers (*Sylvia borin*; López et al., 2007), regent honeyeaters (*Xanthomyza phrygia*; Morin-Adeline et al., 2011) and green-winged saltators (*Saltator similis*; Coelho et al., 2013). In some instances, the difference in oocyst shedding can vary from detection of no oocysts whatsoever shed in the morning to several hundreds of thousands in the afternoon (Brawner and Hill, 1999). Acknowledging the variation in circadian rhythms of oocyst shedding is important, otherwise studies that do not account for this temporal variation may draw spurious conclusions.

Multiple hypotheses have been suggested to explain the diurnal patterns of coccidian oocyst shedding. The synchronization of oocyst discharge may be necessary to increase the concentration of oocysts within faeces to levels that are sufficient to ensure transmission to the next host, if successful transmission is dosage dependent (Dolnik et al., 2011). Furthermore, oocyst shedding may be timed to enhance transmission by increasing the number of sporulated oocysts so as to coincide with a higher density of hosts during periods of active feeding activity, as is suggested to occur in the case of the blackbird (*T. merula;* Martinaud et al., 2009). A peak in afternoon shedding may also be an adaptation to increase oocyst survival under external environmental conditions. Strong support for the validity of this last-mentioned theory is that oocysts experimentally exposed to sunlight had a significantly lower survival rate than those left in shade; 1 hour of sunlight decreased oocyst survival by approximately 50% (Martinaud et al., 2009). These authors found that oocysts were more susceptible to UVB than UVA radiation (Martinaud et al., 2009). Their findings are also congruent with the observation that diurnal periodicity in oocyst discharge persists in hosts inhabiting regions at high latitudes (Dolnik et al., 2011). Despite long hours of daylight, UV radiation levels still fluctuate greatly throughout the day, peaking around midday and declining in the afternoon

(Dolnik et al., 2011). Coupled with an absence of diurnal patterns in the activity of snow buntings, these authors reasoned that circadian oocyst shedding was most likely an adaptation to survive conditions outside the host. Martinaud et al. (2009) came to the same conclusion but, in addition, stressed that the number of viable oocysts present in the subsequent morning as a result of afternoon shedding was also likely to be the result of adaptive timing to coincide with the feeding activity of blackbirds.

4.2 Age

It has been well established that parasites have aggregated distributions whereby high parasite loads are frequently limited to a small number of host individuals (Poulin, 2007; Shaw and Dobson, 1995). The age of the host may be an important determinant of heterogeneities in infection, leading to aggregated distributions (Wilson et al., 2002). The association between age and different measures of infection may follow a number of patterns. Parasite loads and prevalence might increase linearly with age or level off, reaching an asymptote at a particular age. Alternatively, the relationship could be convex if parasite load or prevalence decrease during middle age but rises again later in life, for example, as a consequence of senescence (Wilson et al., 2002). Infection may also be restricted to young individuals if strong and lasting immunity is acquired after initial infection. Mechanisms that reduce the likelihood of infection later in life may also include age-dependent mortality or age-related changes in behaviour and physiology.

Investigations into these relationships in passerines suggest that in general, coccidian load but not prevalence may be related to age, although there are exceptions. Several studies did not detect any significant difference in the prevalence of coccidia among passerine birds of different ages (Dolnik, 2002; Hudman et al., 2000; Zinke et al., 2004). The exception to these study results was the finding that in icterine warblers (*Hippolais icterina*), prevalence was higher in adults compared to nestlings, as well as in older nestlings compared to younger nestlings (Svobodová and Cibulková, 1995). There is evidence to suggest that parasite load does appear to be greater in younger birds (Dolnik, 2002; López et al., 2007). The latter authors found that yearling serins (*S. serinus*) had greater parasite loads than adult birds. Similarly, in a study of five passerine species, parasite load was consistently higher in younger birds in contrast to older birds, although the difference was only statistically significant for three of five species (Dolnik, 2002). To elucidate the efficacy of the immune system in birds of different ages, Dolnik (2002) reinfected individuals from different age groups and monitored their response. It was found that all ages were susceptible to reinfection, indicating a lack of strong acquired immunity, but 2-month-old birds had consistently higher parasite loads following reinfection. There was, however, variation within all age groups between those who succumbed to reinfection and those who survived, highlighting that there was variation for resistance to infection among individuals in the experiment (Dolnik, 2002).

4.3 Host Migration

Host migration can play an important and complex role in wildlife epidemiology (Altizer et al., 2011). Migration can allow individuals to escape conditions that favour parasite transmission, thus reducing the risk of infection (Folstad et al., 1991). Conversely, migration may result in increased prevalence if hosts move between locations favourable for parasite transmission, such as when environmental conditions in migratory destinations are conducive to the development of infective stages, or the density of host populations is increased at breeding grounds (Brooker et al., 2006; Zinke et al., 2004). Parasite-mediated selection may be greater among migratory birds if parasitism has significant costs to its hosts. In this situation, selection would be stronger on parasitized individuals if they die during migration, or arrive later and have reduced access to resources such as breeding sites (Altizer et al., 2011). If infection alters the normal development of the host, such as by affecting musculature, migration may not be physiologically possible or mortality might be increased during migration (Pap et al., 2008).

Whether the prevalence of coccidia in migratory passerines differs from that in nonmigratory species has not been conclusively established. Prevalence data might be the result of multiple interacting factors confounding research results, or reflect the absence of any effect(s). López et al. (2013) found that intestinal parasite diversity had a negative impact on the arrival time of garden warblers (*S. borin*). Whereas the most common intestinal parasites identified were coccidia of the genus *Isospora*, it was parasite richness, not the influence of a single parasite species, that had a significant effect (López et al., 2013). Migration times were shown not to be associated with prevalence or intensity of infection of coccidia at two sites in Europe (Dolnik, 2002). The average severity of infection was lower in migratory species on an island in contrast to the mainland, leading the author to suggest that selection may be stronger among infected individuals having transmarine migratory routes (Dolnik, 2002). One study did find higher prevalence of intestinal parasites among migratory passerines, but not significantly so (Bandelj et al., 2015). The same study also concluded that feeding style was likely to be a more important determinant of prevalence than migration. Zinke et al. (2004) examined differences in parasite load among four migratory passerines. The species with the shortest and longest migration routes had the highest and lowest parasite loads, respectively. The authors also noted that this was likely to be confounded by the feeding strategies of the species, as the shortest distance migrant was a ground feeder and potentially had the highest exposure to coccidian oocysts (Zinke et al., 2004).

4.4 Foraging Ecology

It is highly likely that the foraging ecology of hosts is a cause of substantial variation in coccidian prevalence and load among species because of the faecal–oral transmission pathway. As noted previously, several studies cited the feeding habits of hosts as a confounding factor when researching the effect of migration on coccidian infection (Bandelj et al., 2015; Zinke et al., 2004). Insectivorous birds have been found to have a lower prevalence of coccidia (Bandelj et al., 2015; Dolnik et al., 2010). Unsurprisingly, ground-feeding birds tend to have a higher prevalence in comparison to other feeding styles, as the amount of faecal material in the environment is likely to be greatest on the ground (Dolnik et al., 2010; McQuistion, 2000; Zinke et al., 2004). Gregarious feeders also tend to have higher prevalences and intensities of infection than solitary feeders (Dolnik et al., 2010; McQuistion, 2000). However, somewhat counterintuitively, Dolnik et al. (2010) found that the prevalence of coccidia in gleaners, species that collect food from above-ground vegetation, was lower when they foraged in flocks. This was explained by the hypothesis that flocking gleaners spend less time in or return less frequently to feeding patches thereby avoiding oocysts after they had become infective (Dolnik et al., 2010).

Studies on the effect of urbanization on coccidian prevalence offer further support that diet can be an important factor. A comparison of the prevalence in two infected bird species across an urbanization gradient showed that prevalence was affected in only one of them (Delgado and French, 2015). There was no prevalence change in an insectivorous passerine despite the degree of urbanization, while a granivore that also uses bird feeders exhibited increased prevalence in urban areas (Delgado and French, 2015). The authors could not reach a direct conclusion regarding the cause of the change. Similarly, Giraudeau et al. (2014) reported that the prevalence of coccidian infection in male house finches was higher in urbanized areas. The authors tested whether increased levels of stress in urbanized areas might be a corollary for increased levels of infection due to individuals being immunocompromised but found no such relationship. They suggested that a higher density of birds resulting from flocking, potentially at places such as artificial feeders, might drive increases in prevalence (Giraudeau et al., 2014).

5. NATURAL SELECTION

Parasites have the potential to impose a significant selective pressure on their host, driving host population dynamics, shaping the evolution of their genomes and helping to maintain the genetic diversity of their host. For example, under the Red Queen hypothesis, rare alleles can be advantageous in the host when parasites are adapted to invade common host genotypes (Decaestecker et al., 2007). Host evolution in response to selection by parasites requires genetic variation that influences resistance to parasites. Genetic diversity at loci associated with the immune system is an obvious target for identifying variation for resistance. Genes that code for major histocompatibility complex (MHC) molecules are a frequent candidate of such investigations and allelic diversity within this region has been linked to parasite resistance (Meyer-Lucht and Sommer, 2009; Paterson et al., 1998). Toll-like receptors (TLR) are another component of the immune system that are involved in the cellular recognition of foreign organisms and are receiving increasing attention in the study of immunology of wildlife diseases (Villaseñor-Cardoso and Ortega, 2011). Signatures of positive selection, consistent with that expected under parasite-mediated selection, have been found in avian TLR loci (Alcaide and Edwards, 2011; Grueber et al., 2014). Furthermore, individuals and populations that are genetically depauperate through mechanisms such as inbreeding may be more susceptible to parasitism and less likely to survive infection (Acevedo-Whitehouse et al., 2003).

Individual variation in resistance to coccidial infection has been shown in passerines (Dolnik, 2002; Hõrak et al., 2006). The latter authors superinfected wild-caught greenfinches (*C. chloris*) that were already infected with coccidia, with the purpose of testing whether individuals with low preexperimental loads could better cope with reinfection than those with higher preexperimental loads. Indeed, individuals with lower preexperimental parasite loads shed fewer oocysts after inoculation than those with higher loads, suggesting that natural parasite loads may reflect an individual's ability to resist infection (Hõrak et al., 2006). Also, Sild et al. (2014) found that resistance to coccidia was associated with corticosterone content, a stress-related hormone, in the feathers. Contrary to popular notions that stressed individuals are more susceptible to infections, higher levels of corticosterone were positively associated with resistance to coccidia. In an experiment designed to investigate age-related patterns in susceptibility to infection, Dolnik (2002) similarly found individual differences in susceptibility within age groups. Birds of different ages were experimentally infected with coccidia; within age groups, different individuals varied in their ability to cope with infection. Some birds succumbed, suffering weight loss despite food being provided ad libitum, while others endured infection and became heavier. Neither study, however, attempted to identify whether there was a genetic component to this variation in resistance.

As far as we are aware, genomic regions that contribute to coccidian resistance have not been identified in any passerine, but in other avian species there is evidence that both MHC and TLR loci contribute, in addition to other regions. Worley et al. (2010) examined the MHC region in a captive population of the red junglefowl (*Gallus gallus*) following an outbreak of coccidiosis. Mortality among the population was high during the outbreak, with 83 of 98 birds dying within 250 days after hatching. Seven MHC haplotypes were identified, one of which significantly decreased the survival ability of birds if the individual was homozygous for this haplotype (Worley et al., 2010). Another genetic determinant of resistance to coccidia has been demonstrated in poultry to be linked to microsatellite markers on chromosome one (Kim et al., 2006; Zhu et al., 2003). Further analysis of this region highlighted an association with single-nucleotide polymorphisms in the *zyxin* gene, which encodes a protein involved in cell surface adhesion (Hong et al., 2009). Thus, this gene might play an important role in strengthening the epithelial barrier to coccidian infection. The poultry industry has also been responsible for beginning to identify TLRs that may influence susceptibility to infection by coccidia (Zhang et al., 2012). Ten TLRs have been identified in birds (Brownlie and Allan, 2011). Multiple TLRs are expressed during coccidian infection in poultry, but which ones specifically may depend on the number of oocysts that are ingested (Sumners et al., 2011; Zhang et al., 2012). Whether polymorphisms in TLR genes are important to the outcome of coccidian infection has yet to be shown, but certain haplotypes of TLR loci have been shown to confer greater overall survival in passerines (Grueber et al., 2013).

6. SEXUAL SELECTION

Hamilton and Zuk (1982) proposed a good-genes model of sexual selection whereby ornamental traits may convey information regarding an individual's ability to resist infection by parasites. In their model, Hamilton and Zuk (1982) suggested that animals may choose mates with genes that confer disease resistance, and are able to identify individuals that carry these genes by the expression of ornamental traits. Such traits that depend on the vitality of their owner could be expressed more elaborately in individuals that do not succumb to disease or parasitism as easily as other members, thereby conveying information on the genetic quality of that individual with respect to its ability to resist infection. Since the publication of their hypothesis, it has received much attention in studies on parasite-mediated sexual selection (Møller, 1990). Carotenoid-based ornamental traits, such as bright plumage, became an attractive mechanism for investigating this relationship (Balenger and Zuk, 2014; Lozano, 1994). Carotenoids cannot be synthesized by birds and must be harvested from food sources. Thus, it was suggested that bright plumages convey information about a bird's ability to forage and, therefore, about its fitness (Møller et al., 2000). Furthermore, carotenoid-rich diets appear to positively influence immune function, further reinforcing the idea that carotenoid-based ornaments are indicators of an individual's health (Bendich, 1989; Møller et al., 2000; Saks et al., 2003). However, support for the Hamilton–Zuk hypothesis has been equivocal and does not appear to be generalizable across all host–parasite systems (Hamilton and Poulin, 1997). Yet, despite this, it still remains an enticing and intuitive hypothesis which attracts investigation (Balenger and Zuk, 2014).

Coccidian infection in passerines presents a potentially fruitful system for studying the association between parasites and sexually selected traits. Many passerines have bright, carotenoid-based patches of plumage which influence mate choice by females (Gray, 1996; Hill et al., 2002). One of the major symptoms of infection with coccidia is disruption of the uptake of nutrients; thus, infection with coccidia may affect both an individual's foraging ability and its efficiency in absorbing carotenoids. Indeed, one study showed that carotenoid content in the feathers of infected individuals was reduced by 52% (Hõrak et al., 2004). Greenfinches (*C. chloris*) and house finches (*C. mexicanus*) that were experimentally infected with coccidia both had altered hues in their carotenoid-based plumage (Brawner et al., 2000;

Hõrak et al., 2004). Moreover, variation in plumage pigmentation is associated with a host's immune response to a novel antigen (Saks et al., 2003). Carotenoids appear to influence immune responses to coccidian infection. However, the mechanism by which they do so is complicated. One study showed that carotenoid supplementation slowed the rate of replication of *Isospora* in blackbirds (*T. merula*) (Baeta et al., 2008). Greenfinches (*C. chloris*) suppressed coccidian infection better when their diet was supplemented with carotenoids, but only when additionally challenged with phytohaemagglutinin (PHA) (Sepp et al., 2011). The authors suggested that challenge with PHA primed the immune system to respond to coccidian infection but was dependent on sufficient carotenoids being available.

7. FUTURE DIRECTIONS

Despite having been extensively researched in commercially associated bird species for many years, it is during the past two decades that studies on the effects of coccidia in wild passerines have begun to gain momentum. It is clear that coccidia can have a significant effect on the health of passerines and in some instances may be the cause of substantial mortality (Gill and Paperna, 2008; Hõrak et al., 2004). The full dynamics of how coccidian infection drives the evolution of wild bird populations has yet to be determined; for example, whether or not nonlethal infection decreases the reproductive success of individuals in a population. Genes associated with resistance to coccidian infection have been identified in poultry and junglefowl and are potential candidate genes for investigating the genetic effects of resistance to coccidia by passerines. However, the genes involved in resistance to particular parasites can vary among different host species (Wilfert and Schmid-Hempel, 2008). Similarly, the genetic determinants of parasite fitness, such as genes that code for proteins on the surface of the invasive stages of coccidia, will be crucial to understanding the coevolutionary mechanisms via which hosts and coccidia interact (Chapman et al., 2013). Different modes of selection on genes that confer resistance in hosts, or virulence in parasites, such as overdominance or frequency-dependent selection, are likely to impact the evolutionary dynamics of host and parasite coevolution (Worley et al., 2010). Understanding such selection pressures will enable predictions of the impact that perturbations, such as changes in population size and subsequent loss of genetic diversity, may have on rates of infection within a population, and the long-term adaptive potential of hosts and parasites.

The risk that parasitic taxa pose to the longevity of their host populations is becoming increasingly pertinent in the conservation of threatened species (Altizer et al., 2003). This is true for both wild and captive populations. In captivity, hosts may be more susceptible to parasites because they are stressed and immunocompromised, or transmission rates may be artificially increased (Adkesson et al., 2005). Threatened species that have small population sizes are potentially less genetically diverse and experience greater levels of inbreeding (Frankham, 1996; Frankham et al., 2002). This, in turn, further decreases individual genetic diversity and increases the risk of extinction through inbreeding depression (O'Grady et al., 2006). One study highlighted the role of how both specific genomic regions and inbreeding may interact to affect the outcome of parasitism on the fitness of individuals (Acevedo-Whitehouse et al., 2006). Inbreeding in California sea lions was associated with hookworm lesions, whereas the development of anaemia following infection was attributable to genetic diversity at a particular locus (Acevedo-Whitehouse et al., 2006). Genetic rescue, the artificial manipulation of gene-flow typically by the translocation of individuals with the aim of increasing the genetic diversity of a population, has become an increasingly frequent and effective management tool to reduce the effects of inbreeding depression in wild animal populations (Frankham, 2015; Whiteley et al., 2015). However, translocating individuals to induce gene flow may act as a Trojan horse if the individuals being transferred are carrying pathogens. Indeed, one disease risk analysis applied to Eurasian Cranes (*Grus grus*) considered coccidia to be the pathogen with the highest associated risk during translocations (Sainsbury and Vaughan-Higgins, 2012). Furthermore, translocated birds may suffer increased mortality during transfer as a result of increased stress or aviary conditions that permit increased transmission. Indeed, coccidiosis in cirl buntings (*Emberiza cirlus*) was responsible for the death of 4 of 17 birds that were part of a trial translocation. The discovery led to improved conditions and lower mortality during the principal translocation (McGill et al., 2010). Given the ubiquity of coccidia among passerine hosts, studies that demonstrate the capacity of threatened populations or translocated individuals to tolerate parasitic taxa are warranted. In addition, it is not clear whether management actions, such as the provision of supplementary feeders to support a threatened population, may increase infection rates. Supplementary feeders can increase the density of animals in an area, which might then act as a reservoir for coccidia, thus potentially increasing the prevalence of the parasite (Murray et al., 2016). Related to this, further research is needed to address whether feeders can act as receptacles for coccidian oocysts, and/or increased encounter rates between host individuals, and thereby lead to increased transmission.

Further investigation into the taxonomy of passerine coccidia is warranted to determine the diversity, geographical range and specificity of these parasites. The introduction of novel strains or species of parasite to a host population can pose significant risks as host switching can be associated with dramatic increases in virulence (Longdon et al., 2015; Woolhouse et al., 2005). Recent molecular analyses have revealed that many coccidia are likely to have cospeciated with their hosts, promoting host specificity among passerine coccidia (Schrenzel et al., 2005). Host specificity is not absolute; however, there is evidence that some coccidia infect multiple species or that recent host switching has occurred (Schrenzel et al., 2005). Host switching has been shown to be more frequent among closely related taxa at fine phylogenetic scales (Santiago-Alarcon et al., 2014). Thus, phylogenetic studies that are able to resolve taxonomic uncertainties can also provide valuable information in regard to tracing the source of host-switching events in the face of epidemics and evaluate the risks from sympatric host species.

The epidemiology of coccidia is also dependent on various aspects of the host ecology. Foraging ecology is emerging as one of the major influences on transmission, yet many ethological traits of the host, such as sociality, personality and nesting behaviour, have yet to be investigated. Furthermore, the relative importance of behavioural and environmental factors on the likelihood of infection, in contrast to the genetic influences, has not been researched in wild passerines (Hawley and Altizer, 2011). This aside, coccidian oocysts require humidity and a particular temperature range as well as oxygen level for sporulation. Annual patterns in the variation of coccidian load are apparent in wild populations of house sparrows (*P. domesticus*; Pap et al., 2015). Accordingly, host ecology may interact with environmental conditions to increase the incidence and prevalence of infection at particular times of the year when conditions permit (Allen and Fetterer, 2002). We are unaware of any coccidian study concerning the interaction between host ecology and environmental drivers of prevalence and parasite load. Moreover, one of the most fascinating adaptive traits of parasites is their ability to change host behaviour. Parasite-induced alterations in host behaviour are likely to be adaptations to increase the probability of transmission. These interactions are therefore relevant to the epidemiology of coccidian infection (Vyas et al., 2007). For threatened species, identifying how parasites and host ecology, behaviour and the environment interact to influence the transmission of coccidia will help to tailor management actions more efficiently and help to control the parasite.

8. CONCLUSION

Coccidia are ubiquitous parasites of passerine birds and have the potential to affect both individuals and populations adversely. Little is currently known about underlying genetic influences on susceptibility to coccidian infection in passerines specifically, but work in chickens and junglefowl has paved the road ahead for future research. In respect of the handful of species examined so far, the age of birds appears to be associated with the intensity of infection, younger birds tending to have higher parasite loads than older birds. High infection intensity may increase selection on migratory passerines but factors such as the foraging ecology of the host have prevented firm conclusions from being drawn. Indeed, foraging ecology appears to be one of the major underlying causes of interspecies variation in coccidian prevalence. The prevalence in insectivorous avian species is frequently lower than in ground-feeding species. Given their widespread occurrence and selective potential, further studies of coccidia in wild and especially threatened populations of passerines are highly warranted.

ACKNOWLEDGEMENTS

We would like to thank the reviewer and editor for their very helpful comments, which have improved the manuscript considerably. A.K. is supported by a University of Auckland Doctoral Scholarship.

REFERENCES

Abd-Al-Aal, Z., Ramadan, N.F., Al-Hoot, A., 2000. Life-cycle of *Isospora mehlhornii* sp. nov. (Apicomplexa: Eimeriidae), parasite of the Egyptian swallow *Hirundo rubicola savignii*. Parasitol. Res. 86, 270–278.

Acevedo-Whitehouse, K., Gulland, F., Greig, D., Amos, W., 2003. Inbreeding: disease susceptibility in California sea lions. Nature 422, 35.

Acevedo-Whitehouse, K., Spraker, T.R., Lyons, E., Melin, S.R., Gulland, F., Delong, R.L., Amos, W., 2006. Contrasting effects of heterozygosity on survival and hookworm resistance in California sea lion pups. Mol. Ecol. 15, 1973–1982.

Adkesson, M.J., Zdziarski, J.M., Little, S.E., 2005. Atoxoplasmosis in tangers. J. Zoo Wildl. Med. 36, 265–272.

Alcaide, M., Edwards, S.V., 2011. Molecular evolution of the toll-like receptor multigene family in birds. Mol. Biol. Evol. 28, 1703–1715.

Allen, P.C., Fetterer, R.H., 2002. Recent advances in biology and immunobiology of *Eimeria* species and in diagnosis and control of infection with these coccidian parasites of poultry. Clin. Microbiol. Rev. 15, 58–65.

Altizer, S., Harvell, D., Friedle, E., 2003. Rapid evolutionary dynamics and disease threats to biodiversity. Trends Ecol. Evol. 18, 589–596.
Altizer, S., Bartel, R., Han, B.A., 2011. Animal migration and infectious disease risk. Science 331, 296–302.
Baeta, R., Faivre, B., Motreuil, S., Gaillard, M., Moreau, J., 2008. Carotenoid trade-off between parasitic resistance and sexual display: an experimental study in the blackbird (*Turdus merula*). Proc. R. Soc. Lond. B 275, 427–434.
Baker, D.G., Speer, C.A., Yamaguchi, A., Griffey, S.M., Dubey, J.P., 1996. An unusual coccidian parasite causing pneumonia in a northern cardinal (*Cardinalis cardinalis*). J. Wildl. Dis. 32, 130–132.
Balenger, S.L., Zuk, M., 2014. Testing the Hamilton-Zuk hypothesis: past, present, and future. Integr. Comp. Biol. 54, 601–613.
Bandelj, P., Blagus, R., Trilar, T., Vengust, M., Rataj, A.V., 2015. Influence of phylogeny, migration and type of diet on the presence of intestinal parasites in the faeces of European passerine birds (Passeriformes). Wildl. Biol. 21, 227–233.
Beldomenico, P.M., Begon, M., 2010. Disease spread, susceptibility and infection intensity: vicious circles? Trends Ecol. Evol. 25, 21–27.
Bendich, A., 1989. Carotenoids and the immune response. J. Nutr. 119, 112–115.
Berto, B.P., Flausino, W., McIntosh, D., Teixeira-Filho, W.L., Lopes, C.W.G., 2011. Coccidia of new world passerine birds (Aves: Passeriformes): a review of *Eimeria* Schneider, 1875 and *Isospora* Schneider, 1881 (Apicomplexa: Eimeriidae). Syst. Parasitol. 80, 159–204.
Boughton, D.C., 1933. Diurnal gametic periodicity in avian *Isospora*. Am. J. Hyg. 18, 161–184.
Box, E.D., 1977. Life cycles of two *Isospora* species in the canary, *Serinus canarius* Linnaeus. J. Protozool. 24, 57–67.
Brawner III, W.R., Hill, G.E., 1999. Temporal variation in shedding of coccidial oocysts: implications for sexual-selection studies. Can. J. Zool. 77, 347–350.
Brawner III, W.R., Hill, G.E., Sundermann, C.A., 2000. Effects of coccidial and mycoplasmal infections on carotenoid-based plumage pigmentation in male house finches. Auk 117, 952–963.
Brooker, S., Clements, A.C.A., Bundy, D.A.P., 2006. Global epidemiology, ecology and control of soil-transmitted helminth infections. Adv. Parasitol. 62, 221–261.
Brown, M., Ball, S., Holman, D., 2001. The periodicity of isosporan oocyst discharge in the greenfinch (*Carduelis chloris*). J. Nat. Hist. 35, 945–948.
Brown, M.A., Ball, S.J., Snow, K.R., 2010. Coccidian parasites of British wild birds. J. Nat. Hist. 44, 2669–2691.
Brownlie, R., Allan, B., 2011. Avian toll-like receptors. Cell Tissue Res. 343, 121–130.
Carreno, R.A., Barta, J.R., 1999. An eimeriid origin of isosporoid coccidia with Stieda bodies as shown by phylogenetic analysis of small subunit ribosomal RNA gene sequences. J. Parasitol. 85, 77–83.
Chapman, H.D., 2014. Milestones in avian coccidiosis research: a review. Poult. Sci. 93, 501–511.
Chapman, H.D., Barta, J.R., Blake, D., Gruber, A., Jenkins, M., Smith, N.C., Suo, X., Tomley, F.M., 2013. A selective review of advances in coccidiosis research. Adv. Parasitol. 83, 93–171.
Coelho, C.D., Berto, B.P., Neves, D.M., de Oliveira, V.M., Flausino, W., Lopes, C.W.G., 2013. Oocyst shedding by green-winged-saltator (*Saltator similis*) in the diagnostic of coccidiosis and *Isospora similisi* n. sp. (Apicomplexa: Eimeriidae). Rev. Bras. Parasitol. Vet. 22, 64–70.
Costa, F.J., Macedo, R.H., 2005. Coccidian oocyst parasitism in the blue-black grassquit: influence on secondary sex ornaments and body condition. Anim. Behav. 70, 1401–1409.

Daszak, P., Cunningham, A.A., Hyatt, A.D., 2001. Anthropogenic environmental change and the emergence of infectious diseases in wildlife. Acta Trop. 78, 103–116.

Decaestecker, E., Gaba, S., Raeymaekers, J.A.M., Stoks, R., Van Kerckhoven, L., Ebert, D., De Meester, L., 2007. Host–parasite 'Red Queen' dynamics archived in pond sediment. Nature 450, 870–873.

Delgado, V.C.A., French, K., 2015. Differential influence of urbanisation on coccidian infection in two passerine birds. Parasitol. Res. 114, 2231–2235.

Dolnik, O., 2002. Some Aspects of the Biology and Host-Parasite Interactions of Isospora spp. (Protozoa: Coccidiida) of Passerine Birds. PhD thesis, Universität Oldenburg, Germany.

Dolnik, O., 2006. The relative stability of chronic *Isospora sylvianthina* (Protozoa: Apicomplexa) infection in blackcaps (*Sylvia atricapilla*): evaluation of a simplified method of estimating isosporan infection intensity in passerine birds. Parasitol. Res. 100, 155–160.

Dolnik, O.V., Dolnik, V.R., Bairlein, F., 2010. The effect of host foraging ecology on the prevalence and intensity of coccidian infection in wild passerine birds. Ardea 98, 97–103.

Dolnik, O.V., Metzger, B.J., Loonen, M.J.J.E., 2011. Keeping the clock set under the midnight sun: diurnal periodicity and synchrony of avian *Isospora* parasites cycle in the High Arctic. Parasitology 138, 1077–1081.

Duszynski, D.W., Upton, S.J., 2001. Enteric protozoans: *Cyclospora, Eimeria, Isospora*, and *Cryptosporidium* spp. In: Pybus, M.J., Kocan, A.A. (Eds.), Parasitic Diseases of Wild Mammals, second ed. Iowa State University Press, Ames, pp. 416–459.

Duszynski, D.W., Couch, L., Upton, S.J., 2000. The Coccidia of the World. http://www.k-state.edu/parasitology/worldcoccidia/index.html.

Fayer, R., 1980. Epidemiology of protozoan infections: the coccidia. Vet. Parasitol. 6, 75–103.

Filipiak, L., Mathieu, F., Moreau, J., 2009. Caution on the assessment of intestinal parasitic load in studying parasite-mediated sexual selection: the case of blackbirds coccidiosis. Int. J. Parasitol. 39, 741–746.

Folstad, I., Nilssen, A.C., Halvorsen, O., Andersen, J., 1991. Parasite avoidance: the cause of post-calving migrations in *Rangifer*? Can. J. Zool. 69, 2423–2429.

Frankham, R., 1996. Relationship of genetic variation to population size in wildlife. Conserv. Biol. 10, 1500–1508.

Frankham, R., 2015. Genetic rescue of small inbred populations: meta-analysis reveals large and consistent benefits of gene flow. Mol. Ecol. 24, 2610–2618.

Frankham, R., Briscoe, D.A., Ballou, J.D., 2002. Introduction to Conservation Genetics, first ed. Cambridge University Press, Cambridge.

Gill, H., Paperna, I., 2008. Proliferative visceral *Isospora* (atoxoplasmosis) with morbid impact on the Israeli sparrow *Passer domesticus biblicus* Hartert, 1904. Parasitol. Res. 103, 493–499.

Giraudeau, M., Mousel, M., Earl, S., McGraw, K., 2014. Parasites in the city: degree of urbanization predicts poxvirus and coccidian infections in house finches (*Haemorhous mexicanus*). PLoS One 9, e86747.

Gray, D.A., 1996. Carotenoids and sexual dichromatism in North American passerine birds. Am. Nat. 148, 453–480.

Grueber, C.E., Wallis, G.P., Jamieson, I.G., 2013. Genetic drift outweighs natural selection at toll-like receptor (TLR) immunity loci in a re-introduced population of a threatened species. Mol. Ecol. 22, 4470–4482.

Grueber, C.E., Wallis, G.P., Jamieson, I.G., 2014. Episodic positive selection in the evolution of avian toll-like receptor innate immunity genes. PLoS One 9, e89632.

Gryczyńska, A., Dolnik, O., Mazgajski, T.D., 1999. Parasites of chaffinch (*Fringilla coelebs*) population. Part 1. Coccidia (Protozoa, Apicomplexa). Wiad. Parazytol. 45, 495–500.

Hamilton, W.J., Poulin, R., 1997. The Hamilton and Zuk hypothesis revisited: a meta-analytical approach. Behaviour 134, 299–320.
Hamilton, W.D., Zuk, M., 1982. Heritable true fitness and bright birds: a role for parasites? Science 218, 384–387.
Hatcher, M.J., Dick, J.T.A., Dunn, A.M., 2008. A keystone effect for parasites in intraguild predation? Biol. Lett. 4, 534–537.
Hatcher, M.J., Dick, J.T.A., Dunn, A.M., 2012. Diverse effects of parasites in ecosystems: linking interdependent processes. Front. Ecol. Environ. 10, 186–194.
Hawley, D.M., Altizer, S.M., 2011. Disease ecology meets ecological immunology: understanding the links between organismal immunity and infection dynamics in natural populations. Funct. Ecol. 25, 48–60.
Haydon, D.T., Randall, D.A., Matthews, L., Knobel, D.L., Tallents, L.A., Gravenor, M.B., Williams, S.D., Pollinger, J.P., Cleaveland, S., Woolhouse, M.E.J., Sillero-Zubiri, C., Marino, J., Macdonald, D.W., Laurenson, M.K., 2006. Low-coverage vaccination strategies for the conservation of endangered species. Nature 443, 692–695.
Hill, G.E., Inouye, C.Y., Montgomerie, R., 2002. Dietary carotenoids predict plumage coloration in wild house finches. Proc. R. Soc. Lond. B 269, 1119–1124.
Hong, Y.H., Kim, E.S., Lillehoj, H.S., Lillehoj, E.P., Song, K.D., 2009. Association of resistance to avian coccidiosis with single nucleotide polymorphisms in the *zyxin* gene. Poult. Sci. 88, 511–518.
Hõrak, P., Saks, L., Karu, U., Ots, I., Surai, P.F., McGraw, K.J., 2004. How coccidian parasites affect health and appearance of greenfinches. J. Anim. Ecol. 73, 935–947.
Hõrak, P., Saks, L., Karu, U., Ots, I., 2006. Host resistance and parasite virulence in greenfinch coccidiosis. J. Evol. Biol. 19, 277–288.
Hudman, S.P., Ketterson, E.D., Nolan Jr., V., 2000. Effects of time of sampling on oocyst detection and effects of age and experimentally elevated testosterone on prevalence of coccidia in male dark-eyed juncos. Auk 117, 1048–1051.
Hudson, P.J., Dobson, A.P., Lafferty, K.D., 2006. Is a healthy ecosystem one that is rich in parasites? Trends Ecol. Evol. 21, 381–385.
Johnson, P.T.J., Preston, D.L., Hoverman, J.T., LaFonte, B.E., 2013. Host and parasite diversity jointly control disease risk in complex communities. Proc. Natl. Acad. Sci. U.S.A. 110, 16916–16921.
Kim, E.S., Hong, Y.H., Min, W., Lillehoj, H.S., 2006. Fine-mapping of coccidia-resistant quantitative trait loci in chickens. Poult. Sci. 85, 2028–2030.
Lachish, S., Gopalaswamy, A.M., Knowles, S.C.L., Sheldon, B.C., 2012. Site-occupancy modelling as a novel framework for assessing test sensitivity and estimating wildlife disease prevalence from imperfect diagnostic tests. Methods Ecol. Evol. 3, 339–348.
Lainson, R., 1959. *Atoxoplasma* Garnham,1950, as a synonym for *Lankesterella* Labbé, 1899. Its life cycle in the English sparrow (*Passer domesticus domesticus*, Linn.). J. Protozool. 6, 360–371.
Lloyd-Smith, J.O., Schreiber, S.J., Kopp, P.E., Getz, W.M., 2005. Superspreading and the effect of individual variation on disease emergence. Nature 438, 355–359.
Long, P.L., 1993. Chapter 1—Avian coccidosis. In: Kreier, J.P. (Ed.), Parasitic Protozoa, second ed. In: vol. 4. Academic Press, San Diego, pp. 1–88.
Longdon, B., Hadfield, J.D., Day, J.P., Smith, S.C.L., McGonigle, J.E., Cogni, R., Cao, C., Jiggins, F.M., 2015. The causes and consequences of changes in virulence following pathogen host shifts. PLoS Pathog. 11, e1004728.
López, G., Figuerola, J., Soriguer, R., 2007. Time of day, age and feeding habits influence coccidian oocyst shedding in wild passerines. Int. J. Parasitol. 37, 559–564.
López, G., Muñoz, J., Soriguer, R., Figuerola, J., 2013. Increased endoparasite infection in late-arriving individuals of a trans-Saharan passerine migrant bird. PLoS One 8, e61236.
Lozano, G.A., 1994. Carotenoids, parasites, and sexual selection. Oikos 70, 309–311.

Martinaud, G., Billaudelle, M., Moreau, J., 2009. Circadian variation in shedding of the oocysts of *Isospora turdi* (Apicomplexa) in blackbirds (*Turdus merula*): an adaptative trait against desiccation and ultraviolet radiation. Int. J. Parasitol. 39, 735–739.

McGill, I., Feltrer, Y., Jeffs, C., Sayers, G., Marshall, R.N., Peirce, M.A., Stidworthy, M.F., Pocknell, A., Sainsbury, A.W., 2010. Isosporoid coccidiosis in translocated cirl buntings (*Emberiza cirlus*). Vet. Rec. 167, 656–660.

McQuistion, T.E., 2000. The prevalence of coccidian parasites in passerine birds from South America. Trans. Ill. State Acad. Sci. 93, 221–227.

Meyer-Lucht, Y., Sommer, S., 2009. Number of MHC alleles is related to parasite loads in natural populations of yellow necked mice, *Apodemus flavicollis*. Evol. Ecol. Res. 11, 1085–1097.

Misof, K., 2004. Diurnal cycle of *Isospora* spp. oocyst shedding in Eurasian blackbirds (*Turdus merula*). Can. J. Zool. 82, 764–768.

Mohr, F., Betson, M., Quintard, B., 2017. Investigation of the presence of *Atoxoplasma* spp. in blue-crowned laughingthrush (*Dryonastes courtoisi*) adults and neonates. J. Zoo Wildl. Med. 48, 1–6.

Møller, A.P., 1990. Parasites and sexual selection: current status of the Hamilton and Zuk hypothesis. J. Evol. Biol. 3, 319–328.

Møller, A.P., Biard, C., Blount, J.D., Houston, D.C., Ninni, P., Saino, N., Surai, P.F., 2000. Carotenoid-dependent signals: indicators of foraging efficiency, immunocompetence or detoxification ability? Avian Poult. Biol. Rev. 11, 137–159.

Morin-Adeline, V., Vogelnest, L., Dhand, N.K., Shiels, M., Angus, W., Šlapeta, J., 2011. Afternoon shedding of a new species of *Isospora* (Apicomplexa) in the endangered regent honeyeater (*Xanthomyza phrygia*). Parasitology 138, 713–724.

Murray, M.H., Becker, D.J., Hall, R.J., Hernandez, S.M., 2016. Wildlife health and supplemental feeding: a review and management recommendations. Biol. Conserv. 204, 163–174.

O'Grady, J.J., Brook, B.W., Reed, D.H., Ballou, J.D., Tonkyn, D.W., Frankham, R., 2006. Realistic levels of inbreeding depression strongly affect extinction risk in wild populations. Biol. Conserv. 133, 42–51.

Ogedengbe, J.D., Hanner, R.H., Barta, J.R., 2011. DNA barcoding identifies *Eimeria* species and contributes to the phylogenetics of coccidian parasites (Eimeriorina, Apicomplexa, Alveolata). Int. J. Parasitol. 41, 843–850.

Ogedengbe, J.D., Ogedengbe, M.E., Hafeez, M.A., Barta, J.R., 2015. Molecular phylogenetics of eimeriid coccidia (Eimeriidae, Eimeriorina, Apicomplexa, Alveolata): a preliminary multi-gene and multi-genome approach. Parasitol. Res. 114, 4149–4160.

Ogedengbe, M.E., Brash, M., Barta, J.R., 2016. The complete mitochondrial genome sequence of an *Isospora* sp. (Eimeriidae, Eucoccidiorida, Coccidiasina, Apicomplexa) causing systemic coccidiosis in domestic canaries (*Serinus canaria* Linn.). Mitochondrial DNA Part A 27, 3315–3317.

Pap, P.L., Vágási, C.I., Czirják, G.Á., Barta, Z., 2008. Diet quality affects postnuptial molting and feather quality of the house sparrow (*Passer domesticus*): interaction with humoral immune function? Can. J. Zool. 86, 834–842.

Pap, P.L., Vágási, C.I., Czirják, G.Á., Titilincu, A., Pintea, A., Osváth, G., Fülöp, A., Barta, Z., 2011. The effect of coccidians on the condition and immune profile of molting house sparrows (*Passer domesticus*). Auk 128, 330–339.

Pap, P.L., Pătraş, L., Osváth, G., Buehler, D.M., Versteegh, M.A., Sesarman, A., Banciu, M., Vágási, C.I., 2015. Seasonal patterns and relationships among coccidian infestations, measures of oxidative physiology, and immune function in free-living house sparrows over an annual cycle. Physiol. Biochem. Zool. 88, 395–405.

Paterson, S., Piertney, S.B., 2011. Frontiers in host–parasite ecology and evolution. Mol. Ecol. 20, 869–871.

Paterson, S., Wilson, K., Pemberton, J.M., 1998. Major histocompatibility complex variation associated with juvenile survival and parasite resistance in a large unmanaged ungulate population (*Ovis aries* L.). Proc. Natl. Acad. Sci. U.S.A. 95, 3714–3719.

Perkins, S.E., Cattadori, I.M., Tagliapietra, V., Rizzoli, A.P., Hudson, P.J., 2003. Empirical evidence for key hosts in persistence of a tick-borne disease. Int. J. Parasitol. 33, 909–917.

Poulin, R., 2007. Are there general laws in parasite ecology? Parasitology 134, 763–776.

Poulin, R., Morand, S., 2000. The diversity of parasites. Q. Rev. Biol. 75, 277–293.

Sainsbury, A.W., Vaughan-Higgins, R.J., 2012. Analyzing disease risks associated with translocations. Conserv. Biol. 26, 442–452.

Saks, L., Ots, I., Hõrak, P., 2003. Carotenoid-based plumage coloration of male greenfinches reflects health and immunocompetence. Oecologia 134, 301–307.

Santiago-Alarcon, D., Rodríguez-Ferraro, A., Parker, P.G., Ricklefs, R.E., 2014. Different meal, same flavor: cospeciation and host switching of haemosporidian parasites in some non-passerine birds. Parasit. Vectors 7, 286.

Schoener, E.R., 2010. Gastrointestinal Parasites in Endemic, Native, and Introduced New Zealand Passerines with a Special Focus on Coccidia. Master's degree thesis, Massey University, Palmerston North, New Zealand.

Schrenzel, M.D., Maalouf, G.A., Gaffney, P.M., Tokarz, D., Keener, L.L., McClure, D., Griffey, S., McAloose, D., Rideout, B.A., 2005. Molecular characterization of isosporoid coccidia (*Isospora* and *Atoxoplasma* spp.) in passerine birds. J. Parasitol. 91, 635–647.

Sepp, T., Karu, U., Sild, E., Männiste, M., Hõrak, P., 2011. Effects of carotenoids, immune activation and immune suppression on the intensity of chronic coccidiosis in greenfinches. Exp. Parasitol. 127, 651–657.

Shaw, D.J., Dobson, A.P., 1995. Patterns of macroparasite abundance and aggregation in wildlife populations: a quantitative review. Parasitology 111, S111–S133.

Sild, E., Meitern, R., Männiste, M., Karu, U., Hõrak, P., 2014. High feather corticosterone indicates better coccidian infection resistance in greenfinches. Gen. Comp. Endocrinol. 204, 203–210.

Spencer, H.G., Zuk, M., 2016. For host's sake: the pluses of parasite preservation. Trends Ecol. Evol. 31, 341–343.

Stabler, R.M., Kitzmiller, N.J., 1972. *Isospora petrochelidon* sp. n. (Protozoa: Eimeriidae) from the cliff swallow, *Petrochelidon pyrrhonota*. J. Protozool. 19, 248–251.

Sumners, L.H., Miska, K.B., Jenkins, M.C., Fetterer, R.H., Cox, C.M., Kim, S., Dalloul, R.A., 2011. Expression of toll-like receptors and antimicrobial peptides during *Eimeria praecox* infection in chickens. Exp. Parasitol. 127, 714–718.

Svobodová, M., Cibulková, M., 1995. *Isospora* sp. (Apicomplexa: Eimeriidae) in icterine warbler (*Hippolais icterina*, Passeriformes: Sylviidae): the possibility of parents to nestlings transmission. Acta Protozool. 34, 233–235.

Swayne, D.E., Getzy, D., Slemons, R.D., Bocetti, C., Kramer, L., 1991. Coccidiosis as a cause of transmural lymphocytic enteritis and mortality in captive Nashville warblers (*Vermivora ruficapilla*). J. Wildl. Dis. 27, 615–620.

Tenter, A.M., Barta, J.R., Beveridge, I., Duszynski, D.W., Mehlhorn, H., Morrison, D.A., Thompson, R.C.A., Conrad, P.A., 2002. The conceptual basis for a new classification of the coccidia. Int. J. Parasitol. 32, 595–616.

Villaseñor-Cardoso, M.I., Ortega, E., 2011. Polymorphisms of innate immunity receptors in infection by parasites. Parasite Immunol. 33, 643–653.

Vyas, A., Kim, S.K., Giacomini, N., Boothroyd, J.C., Sapolsky, R.M., 2007. Behavioral changes induced by *Toxoplasma* infection of rodents are highly specific to aversion of cat odors. Proc. Natl. Acad. Sci. U.S.A. 104, 6442–6447.

Warner, R.E., 1968. The role of introduced diseases in the extinction of the endemic Hawaiian avifauna. Condor 70, 101–120.

Weedall, G.D., Conway, D.J., 2010. Detecting signatures of balancing selection to identify targets of anti-parasite immunity. Trends Parasitol. 26, 363–369.

Whiteley, A.R., Fitzpatrick, S.W., Funk, W.C., Tallmon, D.A., 2015. Genetic rescue to the rescue. Trends Ecol. Evol. 30, 42–49.

Wilfert, L., Schmid-Hempel, P., 2008. The genetic architecture of susceptibility to parasites. BMC Evol. Biol. 8, 187.

Williams, R.B., 1999. A compartmentalised model for the estimation of the cost of coccidiosis to the world's chicken production industry. Int. J. Parasitol. 29, 1209–1229.

Williams, R.B., 2005. Intercurrent coccidiosis and necrotic enteritis of chickens: rational, integrated disease management by maintenance of gut integrity. Avian Pathol. 34, 159–180.

Williams, R.B., Thebo, P., Marshall, R.N., Marshall, J.A., 2010. Coccidian oocysts as type-specimens: long-term storage in aqueous potassium dichromate solution preserves DNA. Syst. Parasitol. 76, 69–76.

Wilson, K., Bjørnstad, O.N., Dobson, A.P., Merler, S., Poglayen, G., Randolph, S.E., Read, A.F., Skorping, A., Hudson, P.J., Rizzoli, A., Grenfell, B.T., Heesterbeek, H., Dobson, A.P., 2002. Heterogeneities in macroparasite infections: patterns and processes. In: The Ecology of Wildlife Diseases. Oxford University Press, Oxford, pp. 6–44.

Woolhouse, M.E.J., Dye, C., Etard, J.F., Smith, T., Charlwood, J.D., Garnett, G.P., Hagan, P., Hii, J.L.K., Ndhlovu, P.D., Quinnell, R.J., Watts, C.H., Chandiwana, S.K., Anderson, R.M., 1997. Heterogeneities in the transmission of infectious agents: implications for the design of control programs. Proc. Natl. Acad. Sci. U.S.A. 94, 338–342.

Woolhouse, M.E.J., Haydon, D.T., Antia, R., 2005. Emerging pathogens: the epidemiology and evolution of species jumps. Trends Ecol. Evol. 20, 238–244.

Worley, K., Collet, J., Spurgin, L.G., Cornwallis, C., Pizzari, T., Richardson, D.S., 2010. MHC heterozygosity and survival in red junglefowl. Mol. Ecol. 19, 3064–3075.

Wyatt, K.B., Campos, P.F., Gilbert, M.T.P., Kolokotronis, S.-O., Hynes, W.H., DeSalle, R., Daszak, P., MacPhee, R.D.E., Greenwood, A.D., 2008. Historical mammal extinction on Christmas Island (Indian Ocean) correlates with introduced infectious disease. PLoS One 3, e3602.

Yang, R., Brice, B., Oskam, C., Zhang, Y., Brigg, F., Berryman, D., Ryan, U., 2017. Characterization of two complete *Isospora* mitochondrial genomes from passerine birds: *Isospora serinuse* in a domestic canary and *Isospora manorinae* in a yellow-throated miner. Vet. Parasitol. 237, 137–142.

Zhang, L., Liu, R., Ma, L., Wang, Y., Pan, B., Cai, J., Wang, M., 2012. *Eimeria tenella:* expression profiling of toll-like receptors and associated cytokines in the cecum of infected day-old and three-week old SPF chickens. Exp. Parasitol. 130, 442–448.

Zhu, J.J., Lillehoj, H.S., Allen, P.C., Van Tassell, C.P., Sonstegard, T.S., Cheng, H.H., Pollock, D., Sadjadi, M., Min, W., Emara, M.G., 2003. Mapping quantitative trait loci associated with resistance to coccidiosis and growth. Poult. Sci. 82, 9–16.

Zinke, A., Schnebel, B., Dierschke, V., Ryll, M., 2004. Prevalence and intensity of excretion of coccidial oocysts in migrating passerines on Helgoland. J. Ornithol. 145, 74–78.

CHAPTER THREE

Monogenean Parasite Cultures: Current Techniques and Recent Advances

Kate Suzanne Hutson[1], Alexander Karlis Brazenor, David Brendan Vaughan, Alejandro Trujillo-González

Marine Parasitology Laboratory, Centre for Sustainable Tropical Fisheries and Aquaculture and the College of Science and Engineering, James Cook University, Townsville, QLD, Australia

[1]Corresponding author: e-mail address: kate.hutson@jcu.edu.au

Contents

1. Introduction 62
2. Establishing Monogenean Infections 66
   2.1 Collection of Infected Hosts 67
   2.2 Cohabitation of Susceptible Hosts With Infected Stock 68
   2.3 Cohabitation of Susceptible Hosts With Monogenean Eggs and Oncomiracidia 69
   2.4 Transfer of Adult Parasites to New Host Individuals 71
3. Maintaining Monogenean Cultures 72
   3.1 Egg Production 73
   3.2 Egg Collection 74
   3.3 Egg Maintenance and Hatching 76
   3.4 Oncomiracidia Collection and Infection Success 77
   3.5 Time to Sexual Maturity and Assessment of Adult Infection Intensity 78
4. Hyperparasite Cultures 79
5. Viviparous Cultures 80
6. Amphibian Monogenean Cultures 80
7. Troubleshooting and Virulence 82
8. Animal Ethics and Biosecurity 83
9. Further Considerations and Conclusive Comments 84

Acknowledgements 84

References 84

Abstract

Global expansion in fish production and trade of aquatic ornamental species requires advances in aquatic animal health management. Aquatic parasite cultures permit diverse research opportunities to understand parasite–host dynamics and are essential to validate the efficacy of treatments that could reduce infections in captive

Advances in Parasitology, Volume 99
ISSN 0065-308X
https://doi.org/10.1016/bs.apar.2018.01.002

populations. Monogeneans are important pathogenic parasites of captured captive fishes and exhibit a single-host life cycle, which makes them amenable to in vivo culture. Continuous cultures of oviparous monogenean parasites provide a valuable resource of eggs, oncomiracidia (larvae) and adult parasites for use in varied ecological and applied scientific research. For example, the parasite–host dynamics of *Entobdella soleae* (van Beneden and Hesse, 1864) and its fish host, *Solea solea* (Linnaeus, 1758), is one of the most well-documented of all monogeneans following meticulous, dedicated study. *Polystoma* spp. cultures provide an intriguing model for examining evolution in monogeneans because they exhibit two alternative phenotypes depending on the age of infection of amphibians. Furthermore, assessments of the ecological, pathological and immunological effects of fish parasites in aquaculture have been achieved through cultures of *Gyrodactylus* von Nordmann, 1832 spp., *Benedenia seriolae* (Yamaguti, 1934), *Neobenedenia* Yamaguti, 1963 spp. and *Zeuxapta seriolae* (Meserve, 1938). This review critically examines methods to establish and maintain in vivo monogenean monocultures on finfish, elasmobranchs and amphibians. Four separate approaches to establish cultures are scrutinised including the collection of live infected hosts, cohabiting recipient hosts with infected stock, cohabiting hosts with parasite eggs or oncomiracidia (larvae) and direct transfer of live adult parasites onto new fish hosts. Specific parasite species' biology and behaviour permits predictive collection of parasite life stages to effectively maintain a continuous culture, while environmental parameters can be altered to manipulate parasite generation time. Parasite virulence and biosecurity are vital components of a well-managed culture to ensure appropriate animal welfare and uncontaminated surrounding environments. Contemporary approaches and techniques are reviewed to ensure optimised monogenean cultures, which ultimately can be used to further our understanding of aquatic parasitology and identify mechanisms to limit infestations in public aquaria, ornamental trade and intensive aquaculture.

1. INTRODUCTION

Parasite cultivation techniques constitute a substantial component of the current approach for the study of parasites that infect humans and production animals. Cultures are valuable for the identification and production of parasite management strategies including species susceptibility trials, vaccine development and efficiency, antigen production for obtaining serological reagents, detection of drug resistance, screening of potential therapeutic agents, treatment strategies, epidemiology and host–parasite interactions (e.g. Ahmed, 2014; Matile and Pink, 1990). Aquatic parasite cultures, or life cycle closures in a laboratory setting, have been developed for a diversity of species that cause chronic infections on or in their hosts, including protozoans (e.g. Bastos Gomes et al., 2016; Crosbie et al., 2012; Gauthier and Vasta, 1993; Overath et al., 1998; Woo and Li, 1990), myxozoans (e.g. Morris and Adams, 2006), platyhelminths (e.g. Barber, 2013; Barber and Scharsack, 2010; Cable, 2011; Coustau and Yoshino, 2000; Schelkle et al., 2011; Stewart et al., 2017), leeches (e.g. Kua et al., 2010) and crustaceans (e.g. Hamre et al., 2009; Stewart et al., 2017). Nevertheless, the complexity of aquatic parasite life cycles and various culture conditions required for different life stages (e.g. water quality, nutrients, temperature, incubation, lighting) can make continuous parasite cultivation extremely challenging. Successful in vitro and in vivo culture of aquatic parasites is critical for food security as it permits extensive research that can identify tangible solutions for enhanced aquatic animal production.

Monogeneans comprise a class of parasitic platyhelminths that are ubiquitous in aquatic environments and have gained notoriety as harmful parasites of farmed and ornamental fishes. They can also infect cephalopods, amphibians, aquatic chelonian reptiles and hippopotamus. Viviparous species, such as *Gyrodactylus salaris* (Malmberg, 1957), have been associated with catastrophic losses of wild and farmed salmonids (Johnsen and Jensen, 1986), while oviparous species, such as *Neobenedenia* spp., cause mortalities in numerous families of farmed and ornamental fishes (Bullard et al., 2000; Whittington, 2004). Monogeneans have a direct, single-host, water-based life cycle. Monogeneans are hermaphrodites and, with the exception of the gyrodactylids, are protandrous. The advantage of hermaphroditism in parasite cultures is that encounters between any two individuals always have potential for insemination, which may be unilateral or bilateral (mutual) (Kearn and Whittington, 2015). Viviparous species produce live offspring that may directly infect the same or neighbouring hosts, while oviparous species, with some exceptions (e.g. *Udonella* Johnston, 1835 spp.; examples in Whittington and Kearn, 2011), shed eggs freely into water. The egg casing is physically strong, but a detachable lid or operculum permits escape of the infective larva (oncomiracidium). Monogenean species produce either ciliated or nonciliated larvae (Whittington et al., 2000a). Ciliated larvae are free-swimming, but their energy reserves are limited and they need to find and infect a suitable specific host before these reserves are depleted. Transmission for nonciliated larvae transmission is typically passive, although some species can crawl. Until optimal in vitro culture methods are identified for monogeneans, research laboratories will continue to require limited numbers of susceptible host individuals to maintain in vivo cultures.

Monogenean culture systems represent a substantial resource for advanced knowledge on species biology and management in aquaculture (see examples in Table 1). The biology of oviparous *Entobdella soleae*

Table 1 Examples of Cultured Monogenean–Host Models on Fish, Elasmobranchs and Amphibians

Monogenean Species	Environment	Host	Reference
Fish			
Benedenia seriolae (Yamaguti, 1934)	Marine	*Seriola quinqueradiata* Temminck and Schlegel, 1845	Ernst et al. (2005)
		Seriola lalandi Valenciennes, 1833	Tubbs et al. (2005)
Entobdella soleae (van Beneden and Hesse, 1864)	Marine	*Solea solea* (Linnaeus, 1758)	Kearn (1967, 1970, 1973, 1974, 1975, 1984, 1988)
Euryhaliotrema Kritsky and Boeger, 2002 sp.	Marine	*Lutjanus guttatus* (Steindachner, 1869)	Fajer-Ávila et al. (2007)
Gyrodactylus bullatarudis Turnbull, 1956	Freshwater	*Poecilia reticulata* Peters, 1859	Scott and Anderson (1984), Richards and Chubb (1996) and Schelkle et al. (2011)
Gyrodactylus salaris Malmberg, 1957	Freshwater	*Salmo salar* Linnaeus, 1758	Bakke et al. (2007)
Gyrodactylus von Nordmann, 1832 spp.	Freshwater	*Gasterosteus aculeatus* Linnaeus, 1758	Stewart et al. (2017)
Gyrodactylus turnbulli Harris, 1986	Freshwater	*Poecilia reticulata* Peters, 1859	Richards and Chubb (1996), Bakke et al. (2007) and Schelkle et al. (2011, 2013)
Haliotrema Johnston and Tiegs, 1922 sp.	Marine	*Lutjanus guttatus* (Steindachner, 1869)	Fajer-Ávila et al. (2007)
Lepidotrema bidyana Murray, 1931	Freshwater	*Bidyanus bidyanus* (Mitchell, 1838)	Forwood et al. (2013)
Neobenedenia girellae (Hargis, 1955)	Marine	*Verasper variegatus* (Temminck and Schlegel, 1846)	Hirazawa et al. (2004)
		Lates calcarifer (Bloch, 1790)	This study

Table 1 Examples of Cultured Monogenean–Host Models on Fish, Elasmobranchs and Amphibians—cont'd

Monogenean Species	Environment	Host	Reference
Neobenedenia Yamaguti, 1963 sp.	Marine	*Seriola lalandi* Valenciennes, 1833	Trasviña-Moreno et al. (2017)
Sparicotyle chrysophrii (Van Beneden and Hesse, 1863)	Marine	*Sparus aurata* Linnaeus, 1758	Saitjà-Bobadilla and Alvarez-Pellitero (2009)
Zeuxapta seriolae (Meserve, 1938)	Marine	*Seriola dumerili* Risso	Montero et al. (2004)
		Seriola lalandi Valenciennes, 1833	Tubbs et al. (2005) and Mooney et al. (2006)
Elasmobranchs			
Branchotenthes octohamatus Glennon et al., 2005	Marine	*Trygonorrhina fasciata* Müller and Henle, 1841	Glennon et al. (2006a, 2007, 2008)
Calicotyle australis Johnston, 1934	Marine	*Trygonorrhina fasciata* Müller and Henle, 1841	Glennon et al. (2006a, b, 2008)
Pseudoleptobothrium aptychotremae Young, 1967	Marine	*Trygonorrhina fasciata* Müller and Henle, 1841	Glennon et al. (2006a, c, 2008)
Amphibians			
Polystoma australis Kok and van Wyk, 1986	Freshwater	*Kassina senegalensis* (Dumeril and Bibron, 1841); *Kassina wealii* Boulenger, 1882	Kok and Du Preez (1987)
Polystoma gallieni Price, 1938	Freshwater	*Hyla meridionalis* Boettger, 1874	Badets et al. (2009, 2010, 2013)

(van Beneden and Hesse, 1864) and its relationship with its host, *Solea solea* (Linnaeus, 1758), has been meticulously documented following dedicated study on specimens from infected soles maintained in aquaria (Kearn, 1967, 1970, 1973, 1974, 1975, 1984, 1988). Similarly, three monogenean species cultivated on captive populations of the southern fiddler ray, *Trygonorrhina fasciata* Müller and Henle, 1841, facilitated a series of comprehensive biological studies on parasite taxonomy, comparative phylogeography, egg hatching strategies, larval behaviour, infection dynamics and development (Glennon et al., 2006a,b,c, 2007, 2008). Assessments of the ecological, pathological and immunological effects of *Gyrodactylus* spp. have largely used cultures on model fish including guppy *Poecilia reticulata* Peters, 1859 (see Cable, 2011 for review) and stickleback *Gasterosteus aculeatus* Linnaeus, 1758 (see Barber, 2013 for review) because salmonid fry are sensitive to stressors. We have maintained an ethics-approved, continuous in vivo culture of a pathogenic fish monogenean, *Neobenedenia* sp. (=*Neobenedenia girellae*; see Brazenor et al., 2017) on barramundi, *Lates calcarifer* (Bloch, 1790), for 7 years (since 2011), which has supported a diverse research programme on histopathology, reproductive biology, infection dynamics and treatment efficacy (e.g. Dinh Hoai and Hutson, 2014; Hutson et al., 2012; Militz and Hutson, 2015; Militz et al., 2013, 2014; Trujillo-González et al., 2014, 2015). Notably, research using this culture enabled the development of a fish health management plan for the aquaculture industry, designed to precisely time treatments to break the parasite's life cycle (Brazenor and Hutson, 2015).

The aim of this review is to provide the first evaluation of contemporary methods used to establish and maintain continuous in vivo cultures of monogeneans infecting finfish, elasmobranchs and amphibians. Methods that enable precision in initial host parasite intensity for propagation are emphasised. The intricacies of culture techniques for oviparous monogeneans infecting fishes are discussed in detail, as this reproductive strategy is well represented in monogeneans. Techniques associated with viviparous monogeneans that infect fish and oviparous monogeneans that infect amphibians are also examined.

2. ESTABLISHING MONOGENEAN INFECTIONS

Establishing a monogenean infection requires appropriate infrastructure to ensure that adequate water quality (dechlorinated UV-treated freshwater or filtered UV-treated seawater) can be maintained for the host and the parasite. Closed, recirculating aquarium systems provide optimum control over water parameters, and reduce the likelihood of contamination from and to other captive animals or wild ecosystems. Monogenean infections can be established by collecting live infected hosts (wild or farmed), cohabiting recipient hosts with infected stock, cohabiting hosts with parasite eggs or oncomiracidia, or translocating live adult parasites onto new hosts.

Monogenean species are considered among the most host-specific parasites (Bakke et al., 2002; Whittington et al., 2000b), and a specific host species is usually required to maintain a continuous culture. Appropriate permits, permission and/or animal ethics should be obtained prior to host collection and welfare of the animals should be considered following infection (see Section 8).

2.1 Collection of Infected Hosts

Monogenean cultures are commonly established by collecting infected host organisms. Live wild or farmed animals can be transported to the laboratory and housed in closed aquaria where the water is monitored for shed parasite eggs to confirm infection (see Section 3.2). Lethal sampling of representative individuals is sometimes performed to confirm adult monogenean infection (e.g. Fajer-Ávila et al., 2007; Forwood et al., 2013; Sitjà-Bobadilla and Alvarez-Pellitero, 2009); however, this is not desirable if the host species is rare, challenging to collect, or if infection prevalence is low. Collecting infected hosts is advantageous because obtaining specific parasite species' eggs or oncomiracidia from the environment can be challenging. Glennon et al. (2006a) collected wild southern fiddler rays, *T. fasciata*, in South Australia and maintained them live in aquaria to collect eggs laid by *Calicotyle australis* Johnston, 1934 in vivo. In this study, a piece of plastic fly-wire was secured in the tank to promote a continuous and heavy parasite infection by trapping monogenean eggs (see Fig. 1). Entangled eggs remained in tanks where they developed and hatched, thus exposing rays to large numbers of oncomiracidia. Similarly, Ernst et al. (2005) collected farmed Japanese amberjack *Seriola quinqueradiata* Temminck and Schlegel, 1845, infected with *Benedenia seriolae* (Yamaguti, 1934) and increased infections by providing a substrate (nylon mesh) for the eggs to entangle onto.

The introduction of infected hosts to the laboratory presents the possibility for the invasion of other, potentially undesired, pathogens and parasites. It is not feasible for captured animals to be held in quarantine for extended periods or treated on arrival when attempting to establish a monogenean culture as the parasites may be lost, especially if they occur in low abundance. Accordingly, all introduced animals should be held in separate, isolated systems. This enables them to be monitored for monogenean eggs shed into the water and any undesired progressive disease simultaneously. Any viable monogenean eggs or oncomiracidia collected can then be reintroduced to the host following appropriate quarantine and disease

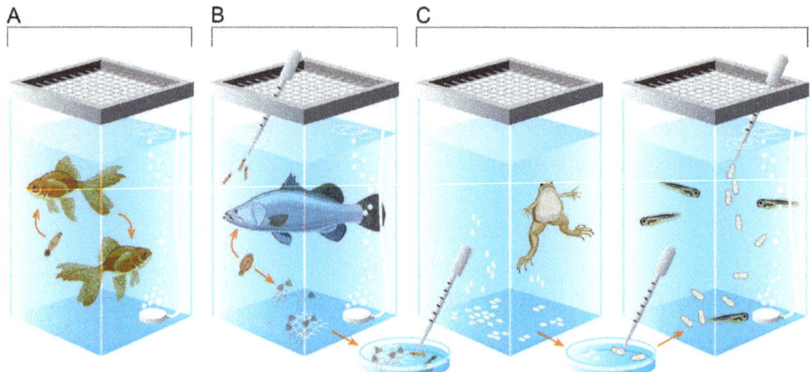

Fig. 1 Generalised experimental set up for maintaining (A) viviparous monogeneans infecting fish; (B) oviparous monogeneans infecting fish and (C) oviparous monogeneans infecting amphibians. (A) Fish infected with viviparous monogeneans are held in tanks with high-quality, aerated water with daily examination of host behaviour and clinical signs of disease; (B) Step 1: Fish infected with oviparous monogeneans are held in tanks with high-quality, aerated water and an egg collector; Step 2: Eggs are collected, cleaned and maintained in Petri dishes with daily monitoring of egg development and water exchanges; Step 3: Vigorously swimming oncomiracidia are aspirated with a pipette and introduced into tanks. (C) Step 1: Frogs are held in dechlorinated, high-quality freshwater which is monitored for monogenean eggs released in the urine; Step 2: Eggs are cleaned and maintained in Petri dishes with fresh water with daily monitoring of egg development and water exchanges; Step 3: Vigorously swimming oncomiracidia are aspirated with a pipette and introduced into tanks containing susceptible tadpoles.

treatment (see Section 2.3). Undesired agents may be difficult to eradicate, or alternatively they may be desirable if the objective is to maintain a polyculture. Indeed, relatively few scientific studies synchronously manipulate multiple species infections on an individual host species, despite this being a more accurate reflection of parasite assemblages in wild and farmed hosts (Sharp et al., 2004). Infection by more than one species of monogenean may be able to be distinguished through unique egg morphology (Glennon et al., 2006a), but some species may require morphological or molecular characterisation. Indeed, molecular analyses of replicate parasite isolates from the founder population should occur to support the composition of a monoculture as opposed to a number of cryptic species.

2.2 Cohabitation of Susceptible Hosts With Infected Stock

Cohabitation of naïve or susceptible hosts with infected stock can be an effective method to establish an in vivo culture. One advantage of this method is that host animals may be supplied from captive populations where the disease history of the animals is known. However, it does not permit precise control of infection levels and may result in varied parasite infection intensities, varied parasite ages and potentially compromise the welfare of some individuals. Two examples of successful cohabitation with infected stock include infection by the polyopisthocotylean *Sparicotyle chrysophrii* (Van Beneden and Hesse, 1863) on gilthead seabream, *Sparus aurata* Linnaeus, 1758 (see Sitjà-Bobadilla and Alvarez-Pellitero, 2009) and the monopisthocotylean *Lepidotrema bidyana* Murray, 1931 infecting silver perch, *Bidyanus bidyanus* (Mitchell, 1838) (see Forwood et al., 2013).

2.3 Cohabitation of Susceptible Hosts With Monogenean Eggs and Oncomiracidia

Cohabitation of susceptible hosts with eggs or oncomiracidia permits finer control over host infection. Embryonated eggs may be collected from the wild by water filtration, collection of aquatic plants and/or objects that eggs may entangle on, or screening wild hosts (see Section 2.1). Some fish monogenean eggs can be obtained from farm environments where they entangle on structures (Shirakashi and Hirano, 2015). Monogenean eggs must not be exposed to air during collection as they can rapidly desiccate (Chen et al., 2010; Ernst et al., 2005). Eggs should be examined prior to cohabitation to ensure that they are embryonated (Hutson et al., 2012; Fig. 2). Egg development should be monitored daily to ensure development of eyespots, hooks and the movement of live oncomiracidia inside eggs prior to hatching (Tubbs et al., 2005; Fig. 2). Eggs that fully develop should be counted prior to cohabitation, to ensure the most accurate indication of hatching success. Eggs exhibit a preweakened operculum, which may detach from the egg even in the event of unsuccessful development and thus empty eggs with open opercula are not necessarily an accurate indication of hatching success. While eggs of some species will hatch spontaneously, some species require specific cues to ensure hatching (see Section 3.3). Cohabitation of susceptible hosts with monogenean eggs is a successful approach to establish an in vivo culture as shown by Sitjà-Bobadilla and Alvarez-Pellitero (2009) who infected gilthead seabream, *S. aurata*, with *Sparicotyle chrysophrii* by introducing containers with parasite eggs into fish tanks.

Oncomiracidia are challenging to capture in the wild because they are fragile and microscopic. It is plausible that positively phototactic species could be captured in light traps although this has never been experimentally tested. Some species' oncomiracidia are almost entirely transparent and make

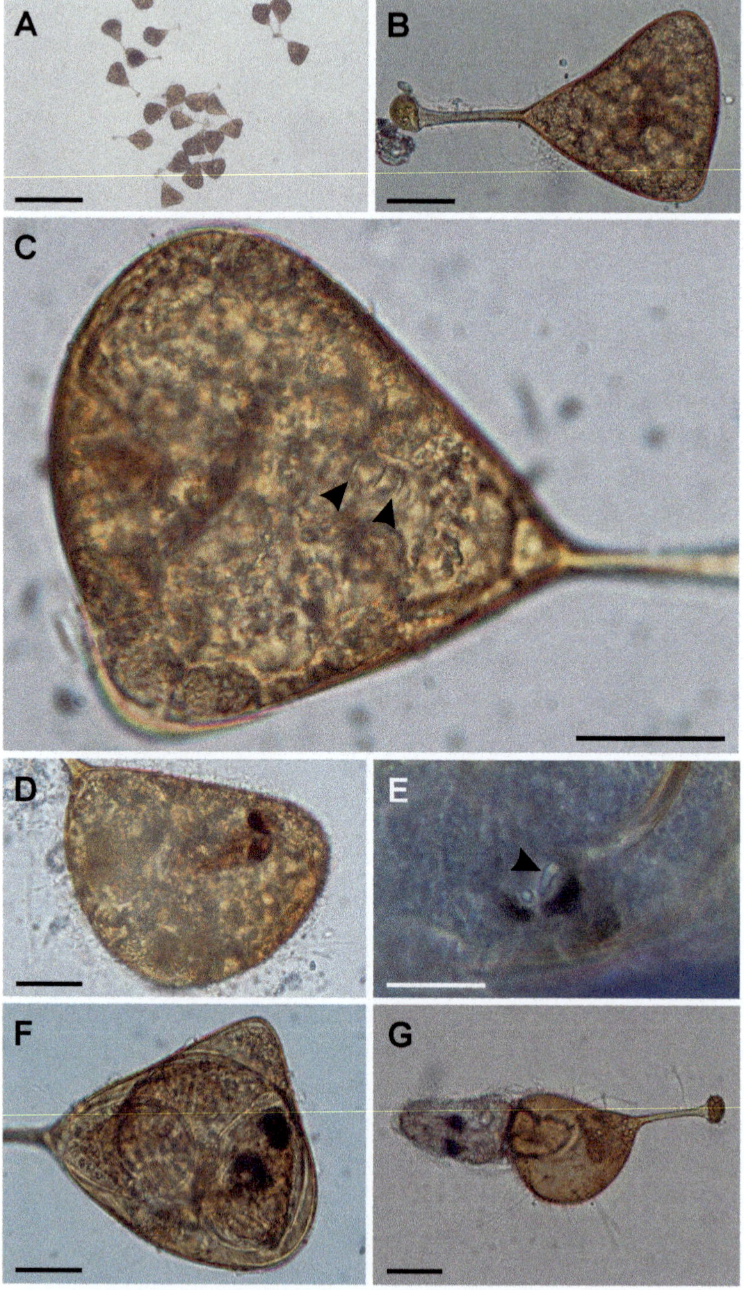

Fig. 2 Egg development of *Dendromonocotyle ukuthena* eggs at 23°C. (A, B) Day 0; (C) Marginal hooklet development (*arrows*), Day 3; (D, E) Eyespot development with crystalline lens (*arrow*); (F) Embryonation, Day 9; (G) Hatch, Day 10. Scale bars: A = 200 μm; B–G = 30 μm.

fast swimming movements, making accurate counts difficult. To overcome this, replicate seawater samples can be taken from agitated water to estimate numbers of oncomiracidia (Hirayama et al., 2009; Hirazawa et al., 2010). Alternatively, individual oncomiracidia can be counted accurately by gently aspirating individuals from egg incubation dishes using a pipette immediately following hatching (Militz et al., 2013; Rubio-Godoy and Tinsley, 2002). Cohabitation of susceptible hosts with oncomiracidia is a well-established and successful infection method (Brazenor and Hutson, 2015; Hirazawa et al., 2010, 2016). For example, Fajer-Ávila et al. (2007) used recently hatched oncomiracidia of *Haliotrema* Johnston and Tiegs, 1922 sp. and *Euryhaliotrema* Kritsky and Boeger, 2002 sp. to infect red snapper, *Lutjanus guttatus* (Steindachner, 1869). Following infection of *L. calcarifer* with *N. girellae* oncomiracidia, Trujillo-González et al. (2015) noted a decline in the infection intensity over time (i.e. between 4 and 16 days post-infection). While the reasons for this remain unclear, it is important to note that infection success and subsequent retention on fish is in a state of flux.

2.4 Transfer of Adult Parasites to New Host Individuals

Transfer of live adult parasites to new host individuals permits ultimate control of parasite infection intensity. It is also a valuable method to reduce infection intensities in a culture, in the event an individual host fish becomes overburdened with infection, yet the parasites are required alive. Stewart et al. (2017) recommend this technique for infection of sticklebacks, *G. aculeatus*, with *Gyrodactylus* spp. They suggest anaesthetising a donor and recipient fish in 0.02% MS222 and allowing *Gyrodactylus* worms to cross from one fish to another by overlapping the stickleback caudal fins. Alternatively, parasites may be gently dislodged using insect pins (Buchmann and Bresciani, 1997), and a known number of parasites brought into close contact with a recipient fish. We have used a similar method to transfer live adult *N. girellae* to other susceptible fish individuals. We initially sedate donor and recipient fish in shallow trays containing seawater. A fine paintbrush or blunt-edged blade is placed underneath the haptor (attachment organ) of a live parasite under stereomicroscopy and the parasite is immediately transferred, ventral side down, onto a recipient fish. Successful transfer is confirmed by observing reattachment and immediate behaviour of the parasite on the sedated recipient fish under a stereomicroscope. The success of the transfer can also be quantified in subsequent days by filtering nonrenewed aquarium water of the recipient fish for dead or detached parasites. An alternative method involves euthanizing donor fish with 2-Phenoxyethanol which effectively anaesthetises the adult parasites. 2-Phenoxyethanol at 1.2 mL/L for 5–10 min can be used to humanely euthanize fishes (see Vaughan et al., 2008a), while the monogeneans can be dislodged easily by gently shaking the donor fish or by carefully brushing them off with a soft paintbrush. Monogeneans are revived in fresh untreated water and attach to new hosts using the method mentioned above.

2-Phenoxyethanol has been used successfully at 1.2 mL/L for 5–10 min to collect both oviparous and viviparous species representing the Monopisthocotylea, and various wild hexabothriids (Polyopisthocotylea) (D.B.V., previously unpublished observation). The effect of 2-Phenoxyethanol on the nervous system of fishes requires further investigation, although it is thought that it has a reversible suppressive effect on the *N*-methyl D-aspartate receptors in the nervous system (Zahl et al., 2009). These receptors are also present in the nervous system of platyhelminths (Ribeiro et al., 2005). A recent study by Grano-Maldonado and Palaiokostas (2015) determined that 2-Phenoxyethanol did not affect the transmission ability of gyrodactylids. However, different monogeneans may be affected at different anaesthetic concentrations, for example, *Haliotrema* sp. was susceptible to the anaesthetic effect of 2-Phenoxyethanol at only 0.3 mL/L (Pironet and Jones, 2000). Sharp et al. (2004) found negligible effect of the fish anaesthetic AQUI-S (8.5 ppm) to remove monogeneans infecting yellowtail kingfish, *Seriola lalandi* Valenciennes, 1833, with only $1.8 \pm 1.34\%$ and $0.1 \pm 0.1\%$ of *B. seriolae* (Monopisthocotylea) and *Z. seriolae* (Meserve, 1938) (Polyopisthocotylea) removed, respectively. The effects of other fish anaesthetics on monogeneans, and their use for collecting monogeneans for cultivation, remain to be experimentally evaluated. The removal of adult monogeneans from their hosts alive may also be useful for the in vitro collection of eggs for embryonation studies and for initiating cultures (see Section 3.2).

3. MAINTAINING MONOGENEAN CULTURES

Once a monogenean infection is established, it is necessary to provide suitable conditions to ensure that the infection can be maintained continuously without overburdening host individuals and to ensure that robust parasites can be reliably collected. Intimate knowledge of parasite species' biology is a considerable advantage to maintain a successful in vivo monogenean culture. Specific species' biology governs the most appropriate times to collect eggs and oncomiracidia for continued maintenance of the infection, while environmental parameters impact generation time, and can be altered to suit the requirements of particular experiment time frames.

3.1 Egg Production

Predictable egg production in a culture can occur when monogenean species exhibit precise egg production outputs over their life span and/or egg-laying rhythms. Knowledge of parasite fecundity over time can allow estimates of expected egg production or when the majority of eggs will be laid, while identification of egg-laying rhythms indicates the best time of day to collect newly laid eggs. For example, fecundity of *N. girellae* peaks a few days following sexual maturity with egg production gradually decreasing with age, and although adult parasites release eggs continuously, more are laid in periods of darkness (Dinh Hoai and Hutson, 2014).

Monogeneans that store eggs temporarily in utero may release them at a specific time in a 24-h period (Whittington and Kearn, 1988). For example, *Z. seriolae* infecting yellowtail kingfish, *S. lalandi*, stores its eggs in utero, prior to releasing ~72% of eggs in the first 3h following sunset (i.e. 1800–2100; Mooney et al., 2006). Moreover, *Z. seriolae* appears to accumulate eggs in utero over a 24-h period (Mooney et al., 2006). While there have been reports of egg-laying rhythms in freshwater and marine gill monogeneans (Macdonald and Jones, 1978; Mooney et al., 2006, 2008) and marine skin monogeneans (Dinh Hoai and Hutson, 2014), not all species exhibit rhythms (e.g. *Discocotyle sagittata* (Leuckart, 1842), see Gannicott and Tinsley, 1997; *B. seriolae*, see Mooney et al., 2008). Furthermore, some species may exhibit different regional differences (e.g. Dinh Hoai and Hutson, 2014, Australia compared to Hirano et al., 2015, Japan) so it is important to determine the presence of rhythms for individual cultures.

To determine monogenean fecundity over time and potential egg-laying rhythms, infected fish can be isolated and the water monitored periodically for eggs over the parasite's life span (Dinh Hoai and Hutson, 2014). Individual, infected hosts are typically housed in isolated, aerated but nonrenewed aquaria. Tanks covered with a clear plastic lid prevent the possible escape of the experimental animal or spillage of water and eggs from the tanks while allowing the tanks to be illuminated. The water in which each animal is maintained is filtered through an appropriately sized mesh sieve and the residue examined microscopically in a Petri dish in solution for eggs laid by parasites in vivo. Air stones and hoses should also be carefully examined for entangled eggs (Mooney et al., 2006).

Temperature is the single most important abiotic factor affecting monogenean egg laying, egg hatching and development, yet remarkably few studies have examined the effect of temperature on parasite fecundity in vivo. Hirazawa et al. (2010) used nylon mesh to catch monogenean eggs to analyse in vivo egg production at various temperatures, but there was high variability in the number of eggs that entangled which prohibited accurate counts. Gannicott and Tinsley (1998a) showed that egg production of *D. sagittata* infecting rainbow trout *Onchorhynchus mykiss* (Walbaum, 1792) was temperature dependent, increasing from a mean of 1.5 eggs per worm per day (e/w/d) at 5°C to 12.0 e/w/d at 18°C. However, while egg production is faster at warmer temperatures, total egg production over the life span of the parasite may not necessarily increase (A.K.B., unpublished data).

3.2 Egg Collection

Regular egg collection is crucial to seed subsequent parasite generations and to ensure that hosts do not become overburdened from reinfection. Oviparous monogeneans may deposit eggs singly, retain eggs in utero prior to deposition on long filamentous strands, or actively attach eggs to their host (e.g. Whittington, 1990). Some monogenean species' eggs shed into the water readily become entangled on fine netting suspended in tanks (Kearn et al., 1992; Fig. 1). Individual eggs or egg masses may also entangle with each other and any surface of the tank including filters, biofoulants, surface scum and even artificial tags on the host. Eggs may also entangle on the gill rakers and opercula of fish. Care must be taken to prevent loss of eggs in the filtration system of recirculating systems. Entangled egg masses can also become trapped in the surface tension of aquaria, which may be vulnerable to predation by host fish (A.K.B., previously unpublished observation). Occasionally eggs and egg masses at the surface may be forced out of the water by the bubbles of air stones; however, the proximity to the surface usually keeps them hydrated and they can be resuspended during water changes or through agitation of the surface. Suspension of fine mesh in tanks with infected fish is useful as the mesh can be cut to a suitable size for subsequent experiments (e.g. Kearn et al., 1992) and limits the number of large, entangled egg masses in the aquaria. Eggs which lack filaments can be concentrated in containers by a gentle rotating action through centripetal force (Theunissen et al., 2014).

Egg collectors, or structures that entangle eggs, can be positioned inside an aquarium to collect eggs in vivo from infected hosts held in tanks. Egg collectors typically comprise cotton threads (Fajer-Ávila et al., 2007), a suspended piece of nylon net, or plastic fly-wire glued to the outflow of the tank (Hirazawa et al., 2010; Kearn et al., 1992). Manual water filtration can also be used to recover eggs. The dimensions of the eggs should be measured so that a suitable filter mesh size can be used to recover all eggs from the tank. For example, Mooney et al. (2006) measured eggs of *Z. seriolae* infecting *S. lalandi* that were approximately 117 μm long and 56 μm wide (joined together by a filament to form a continuous egg string containing up to several hundred eggs). Filtration using a mesh size of 63 μm removed all eggs from seawater that had contained infected fish (Mooney et al., 2006). Marchiori et al. (2013) devised an egg-collecting chamber which used water flow to funnel eggs of *Aphanoblastella mastigatus* (Suriano, 1986) into an inspection apparatus covered by a 41-μm plankton mesh to recover eggs approximately 97 μm long and 47 μm wide. Alternatively, small fish may simply be maintained in small bags of nylon mesh to catch shed monogenean eggs (Buchmann, 1988).

Eggs can also be collected in vitro, following careful detachment of live adult parasites from the host (e.g. Chen et al., 2010; Chisholm and Whittington, 2000; Ellis and Watanabe, 1993; Kritsky and Stephens, 2001). For example, Lackenby et al. (2007) used fine forceps to carefully remove approximately 400 live, adult *B. seriolae* (Yamaguti, 1934) Meserve, 1938 from the skin of several infected farmed *S. lalandi*. Parasites proceeded to lay eggs into filtered seawater over the next 5 h, during which time approximately 15,300 viable eggs (~7.65 eggs/parasite/h) were laid in vitro (Lackenby et al., 2007). Similar work by Vaughan et al. (2008b) for *Dendromonocotyle ukuthena* Vaughan, Chisholm, and Christison, 2008 was performed to identify the sequence of features of the developing oncomiracidium in the egg (see Fig. 2). While some monogenean parasites retain their eggs temporarily in utero, Repullés-Albelda et al. (2012) observed that live gravid specimens of *Sparicotyle chrysophrii* (Van Beneden and Hesse, 1863), removed from their host, *Sparus aurata* L., released their eggs when energetically shaken. Eggs collected in vitro have various uses to maintain monogenean cultures; however, in vitro egg laying is not recommended for assessment of adult fecundity or egg-laying rhythms because starvation of specimens, detached from their host, leads to a progressive decline in egg production rates and quality (Whittington, 1997).

3.3 Egg Maintenance and Hatching

Collection of eggs on a daily basis from the culture will avoid mixing older eggs with recently laid eggs. It also ensures that egg embryonation period and hatching follows a predictable schedule. Following collection, eggs need to be incubated in appropriate conditions to ensure hatching success. The authors routinely spread densely entangled *N. girellae* eggs using fine-tipped forceps in clean seawater to promote dissolved oxygen exchange across the egg casing to developing embryos. Freshwater and marine monogenean eggs can be washed in distilled or dechlorinated freshwater on mesh (Rubio-Godoy and Tinsley, 2002) to remove extraneous organic matter, microorganisms and other contaminants (e.g. bacteria, fish mucus, faeces). Although monogenean eggs are robust and resistant to numerous chemical treatments and temporary rapid changes in salinity, they are susceptible to desiccation and should be maintained in solution at all times (Ernst et al., 2005). Eggs can be incubated in Petri dishes, glass cavity blocks or plastic well plates in natural light or a controlled light:dark regime (Hutson et al., 2012; Repullés-Albelda et al., 2012). The desired temperature can be achieved through air conditioning, water bath or temperature-controlled cabinet. Water changes for each egg batch should be performed daily to maintain constant salinity, to provide adequate dissolved oxygen and to remove potential waste products by microorganisms (Tubbs et al., 2005).

During daily water changes, eggs can be monitored under a stereomicroscope for evidence of embryonation. Embryonated, developing eggs exhibit a granular appearance under light microscopy with eyespots appearing (Fig. 2; Hutson et al., 2012). Larvae can be observed writhing within the eggshell when hatching is imminent (Tubbs et al., 2005). Oncomiracidia exit the egg through the preweakened operculum (Fig. 2; Kearn, 1986). Various factors including light periodicity (e.g. Kearn, 1973), variation in light intensity (e.g. Kearn, 1982), chemicals in host mucus, tissues and urea (e.g. Kearn, 1986; Kearn and Macdonald, 1976) and mechanical disturbance (e.g. Whittington and Kearn, 1988) have been shown to elicit egg hatching in monogeneans and may be required to promote successful hatching and continuous culture.

Temperature is critical for hatching success. Suboptimal temperatures can render eggs unviable, arrest hatching or impact the egg embryonation period. At winter temperatures, eggs of *D. sagittata* slow down or arrest their development and hatch when temperatures rise in spring (Gannicott and Tinsley, 1998a). The embryonation period of most monogeneans typically decreases as temperature increases (Brazenor and Hutson, 2015; Gannicott and Tinsley, 1998a), while few eggs are viable in extremely warm water (<5% hatching success of tropical *N. girellae* eggs at 34°C; see Brazenor and Hutson, 2015). A delay in egg hatching, which can be achieved by incubating eggs in cooler conditions, may be desirable in a laboratory culture to ensure a standby supply of eggs in the event of an undesirable or unexpected impact on the laboratory culture (e.g. power failure or contamination).

3.4 Oncomiracidia Collection and Infection Success

Hatching rhythms have been recorded for numerous monogenean species, which enables predictive collection of oncomiracidia when they have just hatched and have their highest energy reserves (Whittington and Kearn, 2011). Obtaining *N. girellae* oncomiracidia is predictable with the eggs hatching over 2 days in constant conditions (Hutson et al., 2012) and the majority (~81%) hatching within the first 3 h of light (Dinh Hoai and Hutson, 2014). Similarly, Rubio-Godoy and Tinsley (2002) placed *D. sagittata* egg batches in an incubator in total darkness for 1 h to enable mass collection of recently emerged infective stages, because 94% of eggs hatch within the first hour after the onset of darkness (Gannicott and Tinsley, 1997). The positive photoresponse of some monogenean species' oncomiracidia (Glennon et al., 2006a; Hoshina, 1968) could be used to aid collection by inducing the aggregation of larvae in the water, while others have concentrated oncomiracidia by decantation (Tsutsumi et al., 2003). When quantifying oncomiracidia it is important to keep in mind that they can become trapped at the air–water interface and die (Kearn et al., 1992; Tubbs et al., 2005; Whittington and Kearn, 1988). To avoid this, Ernst and Whittington (1996) incubated eggs in wells completely full of water and covered with a coverslip to exclude air; however, care must be taken to ensure parasites are not lost when the coverslip is removed.

Collecting recently hatched oncomiracidia is important because they are typically short-lived; their life span is measured in hours. Gannicott and Tinsley (1998b) found survival of *D. sagittata* larvae to be age and temperature dependent, with parasites exhibiting a longer life span at cooler temperatures. Interestingly, larval survival may not necessarily be directly comparable with infectivity. Lowenberger and Rau (1994) demonstrated that less than 20% of *Plagiorchis elegans* Lühe, 1899, trematode cercariae are infective when they first emerge; however, greater than 75% are able to infect a host 4–6 h postemergence. Similarly, Trujillo-González et al.

(2015) found that in laboratory conditions 20% of *N. girellae* oncomiracidia had infected host fish, *L. calcarifer*, within 15 min, while peak infection success (93%) was reached at 48 h. It has been suggested that some larvae are unable to infect their hosts in the first part of their free-swimming lives as an adaptation to facilitate dispersal in the environment.

Light periodicity, temperature and salinity can also impact infection success. Shirakashi et al. (2015) found that *N. girellae* only infected host fish (*Seriola* sp.) during the day, with almost no infection at night, while cool, hypersaline water provided optimal conditions for *N. girellae* infection success of *L. calcarifer* (Brazenor and Hutson, 2015). To facilitate infection success, many researchers discontinue water flow and/or aeration to tanks for short time periods (Dinh Hoai and Hutson, 2014; Hirayama et al., 2009; Hirazawa et al., 2010; Tsutsumi et al., 2003), but until recently, there was no empirical evidence to support increased infection success with this practice. Shirakashi et al. (2015) showed that *N. girellae* successfully infected fish in low, moderate and high water currents in a circulating water channel (20, 25 and 50 cm/s, respectively), demonstrating that reduced water flow does not necessarily improve infection success.

3.5 Time to Sexual Maturity and Assessment of Adult Infection Intensity

Monogenean generation time can be manipulated using temperature; hence cultures maintained in controlled conditions can be managed to suit experimental requirements and schedules. Sexual maturity of *B. seriolae* is strongly influenced by water temperature, taking longer to attain at cooler temperatures (maturity attained at 41, 24, 16 and 14 days postinfection at 14, 18, 22 and 26 ± 0.5°C, respectively; Lackenby et al., 2007). Similarly, tropical *N. girellae* reach sexual maturity rapidly in warm seawater (6 days between 26 and 32°C) but can take twice as long in cool seawater (12 days at 22°C; Brazenor and Hutson, 2015). Information on the time taken to sexual maturity is not only useful for predicting egg production in monogenean cultures, but is critical to generate strategic treatment plans for captive fishes to ensure that maturing parasites are eradicated before they can recontaminate environments (Brazenor and Hutson, 2015).

Following infection, it can be challenging to determine adult parasite infection intensity without killing the parasites and/or the host because a majority of fish monogeneans are cryptic in nature or deeply penetrate the gill epithelium, while other monogeneans are endoparasites (e.g. polystomes). This can make it challenging to confirm whether infection has been successful until the water can be screened for eggs. Monogeneans that infect the skin of small fish could potentially be observed live using fluorescent staining techniques (Glennon et al., 2007; Trujillo-González et al., 2015), by placing anaesthetised host individuals under a fluorescence microscope. Alternatively, Schelkle et al. (2013) observed guppies *Poecilia reticulata* Peters, 1859, infected with *Gyrodactylus turnbulli* Harris, 1986 and counted the number of attached parasites using a low-power stereomicroscope (Nikon C-DSLS) and fibre optic illumination.

When parasites are not required live, salinity bathing, praziquantel, formalin and hydrogen peroxide are common methods used to remove parasites from live fish (Thoney and Hargis, 1991). Salinity bathing (i.e. dechlorinated freshwater bathing of marine fish and saltwater bathing of freshwater fish) is less harmful to fish hosts compared to more traditional antiparasitic treatments, is cost effective and readily available. Ectoparasites are more severely and rapidly affected by disruption in osmoregulation compared to their fish hosts due to their increased surface area to volume ratio. Praziquantel is also frequently used to detach monopisthocotyleans and polyopisthocotyleans, with some reports showing up to 100% efficacy, depending on its application (Sharp et al., 2004; Vaughan and Chisholm, 2010; Vaughan et al., 2016). Hayward et al. (2007) found that praziquantel bath treatment of mulloway, *Argyrosomus japonicus* (Temminck and Schlegel, 1843), at an initial concentration (not given) that removed the polyopisthocotylean *Sciaenacotyle sciaenicola* (Murray, 1932) did not remove the monopisthocotylean *Calceostoma glandulosum* Johnston and Tiegs, 1922. Only when the concentration of praziquantel was increased, did *C. glandulosum* drop off gills and was able to be collected in bath sediments (Hayward et al., 2007). The ability of monogeneans to penetrate deep into the gills, or under the scales of fish (Trujillo-González et al., 2015) suggests that praziquantel used orally and in combination with host anaesthesia may provide superior results (Vaughan and Chisholm, 2010), and is a useful noninvasive method for monogenean prospecting (Vaughan et al., 2016). Although praziquantel may not immediately kill monogeneans, it causes vacuolisation of the tegument and finally the disruption of the apical tegument layer (Becker et al., 1980).

4. HYPERPARASITE CULTURES

Udonella spp. are hyperparasites that use caligid copepod ectoparasites (sea lice) of fishes as a means of transport from one host to another.

Hyperparasitic monogeneans may prove challenging to culture unless a coculture of infected sea lice can be maintained. *Udonella* spp. attach their eggs to the carapace and egg strings of copepods (Byrnes, 1986), which hatch into juvenile worms, not larvae (Schell, 1972), and these juveniles then transfer onto other copepods during direct contact, or copepod copulation.

5. VIVIPAROUS CULTURES

Viviparous monogeneans can be maintained in continuous culture for several years. Strains of *G. turnbulli* have been maintained experimentally in culture for more than 10 years (Bakke et al., 2007; Schelkle et al., 2011; Table 1) and *G. salaris* for more than 5 years (Bakke et al., 2007; Table 1). Their short generation times and simple culture requirements enable the entire trajectory of infection to be monitored on a single host. *Gyrodactylus* species are well known for their retention of fully grown daughters in utero, until they themselves contain developing embryos. *Gyrodactylus* spp. can transfer between hosts at any stage of the life cycle, without a specific transmission stage (Bakke et al., 2007; Fig. 1). Interestingly, El-Naggar et al. (2004) reported swimming in *Gyrodactylus rysavyi* Ergens, 1973, where parasites exhibited coordinated unidirectional wriggling movements; however, this has not been observed in other gyrodactylids.

6. AMPHIBIAN MONOGENEAN CULTURES

Monogenean parasites of amphibians include polyopisthocotylean genera in the Polystomatidae Gamble, 1896, one of the most diverse families of the Monogenea, whose members also infest Australian lungfish, freshwater turtles and the African hippopotamus. Amphibians can live in terrestrial environments which preclude transmission of monogeneans. Thus, polystomes that infect amphibians typically store eggs in utero with egg laying and transmission in the wild restricted to the period spent by the host in water. Polystomes are oviparous, with some species exhibiting ovoviviparity: a mode of reproduction in which embryos develop inside the eggs and remain in the parent body until release, when the eggs can hatch almost immediately (Tinsley, 1983). Adult polystomes typically attach to the bladder of amphibians using paired haptoral suckers and extract blood from blood vessels in the bladder wall.

A successful amphibian polystome culture requires the supply of adult host amphibians (i.e. frogs/toads) and tadpoles. Adult amphibians and frogspawn are typically collected from the wild to establish cultures. Raising tadpoles from frogspawn also enables a source of parasite-free animals for experimental purposes. Frogspawn should be incubated at optimum temperatures in aerated water and resulting tadpoles reared in the same conditions with water renewed periodically (Badets et al., 2009, 2013). To identify adult amphibians infected with bladder parasites, they can be isolated in tanks or containers containing small amounts of water that is observed daily under a dissection microscope for eggs released with the urine flow (Badets et al., 2013; Theunissen et al., 2014; Fig. 1). Eggs are incubated in clean, dechlorinated freshwater and, when fully formed, oncomiracidia can be seen moving within eggs. Theunissen et al. (2014) found that *Protopolystoma xenopodis* Price, 1943 eggs rapidly hatch following brief exposure to direct sunlight. Active swimming oncomiracidia can be cohabited with susceptible tadpoles and, with some exceptions (see below), will remain attached to the gills until host metamorphosis. Kok and Du Preez (1987) found they could examine the location and numbers of parasites present in sedated *Natalobatrachus bonebergi* Hewitt and Methuen, 1913 tadpoles through in vivo microscopic examination of the gills. A minimal amount of pigment laid down in the ventral body wall of *N. bonebergi* enabled examination at any time without killing the hosts and the destiny of individual parasites could be followed up to metamorphosis (Kok and Du Preez, 1987).

Egg production in *Polystoma* Zeder, 1800 and *Metapolystoma* Yamaguti, 1963 coincides with the brief spawning period spent in water by otherwise terrestrial anuran hosts. *Polystoma nearcticum* (Paul, 1935) exhibits remarkable reproductive synchrony with its tree frog host, *Hyla versicolor* LeConte, 1825, and becomes reproductively active only during the short period of host sexual activity at spawning (Armstrong et al., 1997). Reproduction is typically short lived in *Polystoma* and *Metapolystoma* with ~90% of the total annual egg production taking place in 4 days (Tinsley, 1983) and some of the fastest egg production observed in monogeneans (e.g. *Polystoma integerrimum* (Frölich, 1791) may produce up to 2500 eggs in 24h or ~2 eggs/min; Combes, 1972).

Parasites in *Polystoma* and *Metapolystoma* exhibit two alternative phenotypes depending on the age of infection of amphibians, providing an intriguing model for examining life cycle evolution in monogeneans. The 'branchial' phenotype includes parasites that attach to young tadpoles and exhibit accelerated development while still in the branchial chamber, prior to death at the time of metamorphosis of the tadpoles. The 'bladder' phenotype includes oncomiracidia that migrate to the urinary bladder, where they reach sexual maturity 3 years later, and release eggs that are flushed out with the host urine. These two phenotypes can be produced in vivo as long as there is an adequate source of recently hatched tadpoles. Badets et al. (2009) showed that the branchial phenotype is typically achieved by exposing <7-day-old *Hyla meridionalis* Boettger, 1874 tadpoles to *Polystoma gallieni* Price, 1938 oncomiracidia and the bladder phenotype is exhibited in tadpoles infected at >14 days old.

7. TROUBLESHOOTING AND VIRULENCE

A continuous oviparous monogenean culture relies on regular supply of eggs and successful infection of host individuals. Most importantly, the infection should be in a controlled environment. Problems need to be identified rapidly to ensure the culture is not lost. Daily monitoring of water quality parameters (most importantly temperature), visual monitoring of host behaviour and regular collection and assessment of egg and oncomiracidia quality will identify potential problems with the culture. Low supply can result from multiple scenarios including low infection success, parasite age, parasite–host compatibility, host immune response or suboptimal environmental conditions. Recently hatched, vigorous oncomiracidia should be used to seed the culture and care taken not to damage them. Parasite age can be monitored by appropriate labelling of cultures, as egg production typically decreases towards the end of the parasite's life span (Dinh Hoai and Hutson, 2014).

Low supply of parasite specimens could also be a consequence of reduced parasite virulence. In some monogenean parasite species the relative positions of the male and female reproductive organs could permit self-insemination (Kearn and Whittington, 2015), which may limit the genetic heterogeneity of monogenean cultures and reduce virulence. Indeed, some monogeneans can produce viable eggs in isolation (Dinh Hoai and Hutson, 2014; Gannicott and Tinsley, 1998a). However, the propensity for monogeneans to engage in self-insemination as opposed to unilateral or bilateral insemination is unclear. Interestingly, Dinh Hoai and Hutson (2014) found that reproductively isolated *N. girellae* individuals could produce viable consecutive second and third isolated generations with no significant reduction in infection success. Others have shown that prolonged in vitro cultures of parasitic amoebae can negatively affect virulence (Bridle et al., 2015;

Morrison et al., 2005). Experimental manipulations show that higher genetic heterogeneity of the cestode *Schistocephalus solidus* (Müller, 1776) in its copepod host, *Macrocyclops albidus* (Jurine, 1820), leads to higher infection success and faster development (Christen et al., 2002; Wedekind and Rüetschi, 2000). Reduced genetic heterogeneity could be a concern for the longevity of a culture as well as the application of experimental results to real-world environments. Conversely, isogenic parasite lines can be valuable for experiments where there is a need to reduce genetic variation (e.g. Brazenor et al., 2017; see specific methods for isogenic *Gyrodactylus* spp. lines in Stewart et al., 2017).

8. ANIMAL ETHICS AND BIOSECURITY

To maintain appropriate ethical standards and welfare of hosts, infection intensities should be maintained at low levels and with respect to host–parasite dynamics. Indeed, some monogenean species exhibit high site specificity, which may cause considerable host pathology even in low parasite intensities. For example, *N. girellae* predominantly infects the eyes of cobia *Rachycentron canadum* (Linnaeus, 1766), causing considerable damage to the cornea (Ogawa et al., 2006). Smaller sized hosts may also be more susceptible to disease compared to larger hosts because of increased parasite density. Cultures that maintain low infection intensities on a few host individuals are capable of producing sufficient quantities of eggs to maintain consecutive parasite generations as well as a surplus for use in replicated experiments. For example, a culture consisting of four individual fish infected with 10 *N. girellae* each can theoretically produce ~132,000 eggs over 17 days (Dinh Hoai and Hutson, 2014).

Monogenean cultures maintained in secure facilities minimise the risk of contamination to other captive animals or wild ecosystems. To prevent host animals escaping, animal housing facilities should be contained. Egg production potential can be high, even in small cultures. Excess embryonated eggs should be appropriately disposed of in view of the fact that they exhibit remarkable resilience to environmental conditions (Ernst et al., 2005). Egg disposal is most easily achieved through desiccation followed by disposal in biological waste. Eggs disposed of down the drain and into storm water systems could potentially contaminate wild environments. Biosecurity procedures within the laboratory must be maintained to avoid cross-infection between the culture and animals held in reserve or experimental tanks.

Researchers that do not have extensive culture facilities or the resources required for long-term parasite propagation should consider research exchanges to laboratories that maintain established infections, in preference to the movement of live cultures. Researchers should contact their appropriate regulatory authority in their country, state or territory and be audited to determine the suitability of their laboratory for receiving live specimens. The World Organisation for Animal Health lists a single monogenean species, *G. salaris*, as a notifiable parasitic infection (OIE, 2018).

9. FURTHER CONSIDERATIONS AND CONCLUSIVE COMMENTS

Culture techniques for maintaining monogenean monocultures are increasingly relevant given recent global expansion in farmed fish and the global trade of aquatic ornamental species. Moreover, parasite cultures permit further exploration of broader research questions pertaining to evolutionary biology, ecology and impacts of climate change on host–parasite dynamics. Aquatic parasite cultures provide sufficient numbers of organisms for appropriate replication for applied (e.g. Brazenor and Hutson, 2015) and ecological investigations (e.g. Barber and Scharsack, 2010). Future advances for aquatic parasite cultures, in particular monogeneans, should examine the feasibility and validity of in vivo cultures that would negate reliance on, and maintenance of, host organisms (e.g. Crosbie et al., 2012).

ACKNOWLEDGEMENTS

This review was supported through a James Cook University Special Studies Program (sabbatical) and Development Grant awarded to K.S.H. We thank Eden Cartwright (Bud Design Studio) for graphic art. We express our gratitude to all members of the Marine Parasitology Laboratory, past and present, who have assisted with the maintenance of the monogenean culture.

REFERENCES

Ahmed, N.H., 2014. Cultivation of parasites. Trop. Parasitol. 4, 80–89. https://doi.org/10.4103/2229-5070.138534.

Armstrong, E.P., Halton, D.W., Tinsley, R.C., Cable, J., Johnston, R.N., Johnston, C.F., Shaw, C., 1997. Immunocytochemical evidence for the involvement of an FMRF amide-related peptide in egg production in the flatworm parasite *Polystoma nearcticum*. J. Comp. Neurol. 377, 41–48.

Badets, M., Boissier, J., Brémond, P., Verneau, O., 2009. *Polystoma gallieni*: experimental evidence for chemical cues for developmental plasticity. Exp. Parasitol. 121, 163–166.

Badets, M., Morrison, C., Verneau, O., 2010. Alternative parasite development in transmission strategies: how time flies! J. Evol. Biol. 23, 2151–2162.

Badets, M., Du Preez, L., Verneau, O., 2013. Alternative development in *Polystoma gallieni* (Platyhelminthes, Monogenea) and life cycle evolution. Exp. Parasitol. 135, 283–286.

Bakke, T.A., Harris, P.D., Cable, J., 2002. Host specificity dynamics: observations on gyrodactylid monogeneans. Int. J. Parasitol. 32, 281–308.

Bakke, T.A., Cable, J., Harris, P.D., 2007. The biology of gyrodactylid monogeneans: the Russian-doll killers. Adv. Parasitol. 64, 161–376.

Barber, I., 2013. Sticklebacks as model hosts in ecological and evolutionary parasitology. Trends Parasitol. 29, 556–566. https://doi.org/10.1016/j.pt.2013.09.004.

Barber, I., Scharsack, J.P., 2010. The three-spined stickleback—*Schistocephalus solidus* system: an experimental model for investigating host-parasite interactions in fish. Parasitology 137, 411–424. https://doi.org/10.1017/S0031182009991466.

Bastos Gomes, G., Miller, T.L., Jerry, D.R., Hutson, K.S., 2016. Impacts and current status of parasitic *Chilodonella* spp. (Phyllopharyngea: Chilodonellidae) in freshwater fish aquaculture. J. Fish Dis. 40, 703–715. https://doi.org/10.1111/jfd.12523.

Becker, B., Mehlhorn, H., Andrews, P., Thomas, H., Eckert, E., 1980. Light and electron microscopic studies on the effect of praziquantel on *Schistosoma mansoni*, *Dicrocoelium dendriticum*, and *Fasciola hepatica* (Trematoda) *in vitro*. Z. Parasitenkd. 63, 113–128.

Brazenor, A.K., Hutson, K.S., 2015. Effects of temperature and salinity on the life cycle of *Neobenedenia* sp. (Monogenea: Capsalidae) infecting farmed barramundi (*Lates calcarifer*). Parasitol. Res. 114, 1875–1886. https://doi.org/10.1007/s00436-015-4375-5.

Brazenor, A.K., Saunders, R.J., Miller, T.L., Hutson, K.S., 2017. Morphological variation in the cosmopolitan fish parasite *Neobenedenia girellae* (Capsalidae: Monogenea). Int. J. Parasitol. 48, 125–134. https://doi.org/10.1016/j.ijpara.2017.07.009.

Bridle, A.R., Davenport, D.L., Crosbie, P.B., Polinski, M., Nowak, B.F., 2015. *Neoparamoeba perurans* loses virulence during clonal culture. Int. J. Parasitol. 45, 575–578. https://doi.org/10.1016/j.ijpara.2015.04.005.

Buchmann, K., 1988. Temperature-dependent reproduction and survival of *Pseudodactylogyrus bini* (Monogenea) on the European eel (*Anguilla anguilla*). Parasitol. Res. 75, 162–164.

Buchmann, K., Bresciani, J., 1997. Microenvironment of *Gyrodactylus derjavini* on rainbow trout *Oncorhynchus mykiss*: association between mucous cell density in skin and site selection. Parasitol. Res. 84, 17–24.

Bullard, S.A., Benz, G.W., Overstreet, R.M., Williams, E.H., Hemdal, J., 2000. Six new host records and an updated list of wild hosts for *Neobenedenia melleni* (MacCallum) (Monogenea: Capsalidae). Comp. Parasitol. 67, 190–196.

Byrnes, T., 1986. Five species of Monogenea from Australian bream, *Acanthopagrus* spp. Aust. J. Zool. 34, 65–86.

Cable, J., 2011. Poeciliid parasites. In: Evans, J.P., Pilastro, A., Schlupp, I. (Eds.), Ecology & Evolution of Poeciliid Fishes. Chicago University Press, Chicago, pp. 82–94.

Chen, H.-G., Chen, H.-Y., Wang, C.-S., Chen, S.-N., Shih, H.-H., 2010. Effects of various treatments on egg hatching of *Dendromonocotyle pipinna* (Monogenea: Monocotylidae) infecting the blotched fantail ray, *Taeniurops meyeni*, in Taiwan. Vet. Parasitol. 171, 229–237.

Chisholm, L.A., Whittington, I.D., 2000. Egg hatching in 3 species of monocotylid monogenean parasites from the shovelnose ray *Rhinobatos typus* at Heron Island, Australia. Parasitology 121, 303–313.

Christen, M., Kurtz, J., Milinski, M., 2002. Outcrossing increases infection success and competitive ability: experimental evidence from a hermaphrodite parasite. Evolution 56, 2243–2251. https://doi.org/10.1554/0014-3820.

Combes, C., 1972. Ecologie des Polystomatidae (Monogenea): facteurs influencant le volume et la rythme de la ponte. Int. J. Parasitol. 2, 233–238.

Coustau, C., Yoshino, T.P., 2000. Flukes without snails: advances in the *in vitro* cultivation of intramolluscan stages of trematodes. Exp. Parasitol. 99, 62–66.

Crosbie, P.B., Bridle, A.R., Cadoret, K., Nowak, B.F., 2012. *In vitro* cultured *Neoparamoeba perurans* causes amoebic gill disease in Atlantic salmon and fulfils Koch's postulates. Int. J. Parasitol. 42, 511–515. https://doi.org/10.1016/j.ijpara.2012.04.002.

Dinh Hoai, T., Hutson, K.S., 2014. Reproductive strategies of the insidious fish ectoparasite, *Neobenedenia* sp. (Capsalidae: Monogenea). PLoS One 9 (9), e108801. https://doi.org/10.1371/journal.pone.0108801.

Ellis, E.P., Watanabe, W.O., 1993. The effects of hyposalinity on eggs, juveniles and adults of the marine monogenean, *Neobenedenia melleni*: treatment of ecto-parasitosis in seawater-cultured tilapia. Aquaculture 117, 15–27.

El-Naggar, M.M., El-Naggar, A.A., Kearn, G.C., 2004. Swimming in *Gyrodactylus rysavyi* (Monogenea, Gyrodactylidae) from the Nile catfish, *Clarias gariepinus*. Acta Parasitol. 49, 102–107.

Ernst, I., Whittington, I.D., 1996. Hatching rhythms in the capsalid monogeneans *Benedenia lutjani* from the skin and *B. rohdei* from the gills of *Lutjanus carponotatus* at Heron Island, Queensland, Australia. Int. J. Parasitol. 26, 1191–1204.

Ernst, I., Whittington, I.D., Corneillie, S., Talbot, C., 2005. Effects of temperature, salinity, desiccation and chemical treatments on egg embryonation and hatching success of *Benedenia seriolae* (Monogenea: Capsalidae), a parasite of farmed *Seriola* spp. J. Fish Dis. 28, 157–164.

Fajer-Ávila, E.J., Velásquez-Medina, S.P., Betancourt-Lozano, M., 2007. Effectiveness of treatments against eggs, and adults of *Haliotrema* sp. and *Euryhaliotrema* sp. (Monogenea: Ancyrocephalinae) infecting red snapper, *Lutjanus guttatus*. Aquaculture 264, 66–72.

Forwood, J.M., Harris, J.O., Deveney, M.R., 2013. Efficacy of bath and orally administered praziquantel and fenbendazole against *Lepidotrema bidyana* Murray, a monogenean parasite of silver perch, *Bidyanus bidyanus* (Mitchell). J. Fish Dis. 36, 939–947.

Gannicott, A.M., Tinsley, R.C., 1997. Egg hatching in the monogenean gill parasite *Discocotyle sagittata* from the rainbow trout (*Oncorhynchus mykiss*). Parasitology 114, 569–579.

Gannicott, A.M., Tinsley, R.C., 1998a. Environmental effects on transmission of *Discocotyle sagittata* (Monogenea): egg production and development. Parasitology 117, 499–504.

Gannicott, A.M., Tinsley, R.C., 1998b. Larval survival characteristics and behavioural of the gill monogenean *Discocotyle sagittata*. Parasitology 117, 491–498.

Gauthier, J.D., Vasta, G.R., 1993. Continuous *in vitro* culture of the Eastern oyster parasite *Perkinsus marinus*. J. Invertebr. Pathol. 62, 321–323.

Glennon, V., Chisholm, L.A., Whittington, I.D., 2006a. Three unrelated species, 3 sites, same host–monogenean parasites of the southern fiddler ray, *Trygonorrhina fasciata*, in South Australia: egg hatching strategies and larval behaviour. Parasitology 133, 55–66.

Glennon, V., Chisholm, L.A., Whittington, I.D., 2006b. A redescription of *Calicotyle australis* Johnston, 1934 (Monogenea: Monocotylidae) from the type-host *Trygonorrhina fasciata* (Rhinobatidae) off Adelaide, South Australia, including descriptions of live and silver stained larvae. Syst. Parasitol. 63, 29–40. https://doi.org/10.1007/s11230-005-5501-z.

Glennon, V., Chisholm, L.A., Whittington, I.D., 2006c. *Pseudoleptobothrium aptychotremae* Young, 1967 (Monogenea, Microbothriidae) redescribed from a new host, *Trygonorrhina fasciata* (Rhinobatidae) in South Australia with a description of the larva and post-larval development. Acta Parasitol. 51, 40–46.

Glennon, V., Chisholm, L.A., Whittington, I.D., 2007. Experimental infections, using a fluorescent marker, of two elasmobranch species by unciliated larvae of *Branchotenthes octohamatus* (Monogenea: Hexabothriidae): invasion route, host specificity and post-larval development. Parasitology 134, 1243–1252.

Glennon, V., Perkins, E.M., Chisholm, L.A., Whittington, I.D., 2008. Comparative phylogeography reveals host generalists, specialists and cryptic diversity: hexabothriid, microbothriid and monocotylid monogeneans from rhinobatid rays in southern Australia. Int. J. Parasitol. 38, 1599–1612.

Grano-Maldonado, M.I., Palaiokostas, C., 2015. Does the anaesthetic influence behavioural transmission of the monogenean *Gyrodactylus gasterostei* Gläser, 1974 off the host? Helminthologia 52, 144–147.

Hamre, L.A., Glover, K.A., Nilsen, F., 2009. Establishment and characterisation of salmon louse (*Lepeophtheirus salmonis* (Krøyer 1837)) laboratory strains. Parasitol. Int. 58, 451–460. https://doi.org/10.1016/j.parint.2009.08.009.

Hayward, C.J., Bott, N.J., Itoh, N., Makoto, I., Okihiro, M., Nowak, B.F., 2007. Three species of parasites emerging on the gills of mulloway, *Argyrosomus japonicus* (Temminck and Schlegel, 1843), cultured in Australia. Aquaculture 265, 27–40.

Hirano, C., Ishimaru, K., Shirakashi, S., 2015. Egg laying and hatching rhythms of the skin fluke *Neobenedenia girellae*. Fish Pathol. 50, 23–28.

Hirayama, T., Kawano, F., Hirazawa, N., 2009. Effect of *Neobenedenia girellae* (Monogenea) infection on host amberjack *Seriola dumerili* (Carangidae). Aquaculture 288, 159–165. https://doi.org/10.1016/j.aquaculture.2008.11.038.

Hirazawa, N., Mitsuboshi, T., Hirata, T., Shirasu, K., 2004. Susceptibility of spotted halibut *Verasper variegatus* (Pleuronectidae) to infection by the monogenean *Neobenedenia girellae* (Capsalidae) and oral therapy trials using praziquantel. Aquaculture 238, 83–95.

Hirazawa, N., Takano, R., Hagiwara, H., Noguchi, M., Narita, M., 2010. The influence of different water temperatures on *Neobenedenia girellae* (Monogenea) infection, parasite growth, egg production and emerging second generation on amberjack *Seriola dumerili* (Carangidae) and the histopathological effect of this parasite on fish skin. Aquaculture 299, 2–7. https://doi.org/10.1016/j.aquaculture.2009.11.025.

Hirazawa, N., Tsubone, S., Takano, R., 2016. Anthelmintic effects of 75 ppm hydrogen peroxide treatment on the monogeneans *Benedenia seriolae*, *Neobenedenia girellae*, and *Zeuxapta japonica* infecting the skin and gills of greater amberjack *Seriola dumerili*. Aquaculture 450, 244–249.

Hoshina, T., 1968. On the monogenetic trematode, *Benedenia seriolae*, parasitic on yellowtail, *Seriola quinqueradiata*. Bull. Off. Int. Epizoot. 69, 1179–1191.

Hutson, K.S., Mata, L., Paul, N.A., de Nys, R., 2012. Seaweed extracts as a natural control against the monogenean ectoparasite, *Neobenedenia* sp., infecting farmed barramundi (*Lates calcarifer*). Int. J. Parasitol. 42, 1135–1141. https://doi.org/10.1016/j.ijpara.2012.09.007.

Johnsen, B.O., Jensen, A.J., 1986. Infestations of Atlantic salmon, *Salmo salar*, by *Gyrodactylus salaris* in Norwegian rivers. J. Fish Biol. 29, 233–241.

Kearn, G.C., 1967. Experiments on host-finding and host-specificity in the monogenean skin parasite *Entobdella soleae*. Parasitology 57, 585–605.

Kearn, G.C., 1970. The production, transfer and assimilation of spermatophores by *Entobdella soleae*, a monogenean skin parasite of the common sole. Parasitology 60, 301–311. https://doi.org/10.1017/S0031182000078136.

Kearn, G.C., 1973. An endogenous circadian hatching rhythm in the monogenean skin parasite *Entobdella soleae*, and its relationship to the activity rhythm of the host (*Solea solea*). Parasitology 66, 101–122.

Kearn, G.C., 1974. The effects of fish skin mucus on hatching in the monogenean parasite *Entobdella solea* from the skin of the common sole, *Solea solea*. Parasitology 68, 173–188.

Kearn, G.C., 1975. The mode of hatching of the monogenean *Entobdella soleae*, a skin parasite of the common sole (*Solea solea*). Parasitology 71, 419–431.

Kearn, G.C., 1982. Rapid hatching induced by light intensity reduction in the monogenean *Entobdella diadema*. J. Parasitol. 68, 171–172.

Kearn, G.C., 1984. The migration of the monogenean *Entobdella soleae* on the surface of its host, *Solea solea*. Int. J. Parasitol. 14, 63–69. https://doi.org/10.1016/0020-7519(84)90013-4.

Kearn, G.C., 1986. The eggs of monogeneans. Adv. Parasitol. 25, 175–273. https://doi.org/10.1016/S0065-308X(08)60344-9.

Kearn, G.C., 1988. The monogenean skin parasite *Entobdella soleae*: movement of adults and juveniles from host to host (*Solea solea*). Int. J. Parasitol. 18, 313–319. https://doi.org/10.1016/0020-7519(88)90139-7.

Kearn, G.C., Macdonald, S., 1976. The chemical nature of host hatching factors in the monogenean skin parasites *Entobdella soleae* and *Acanthocotyle lobianchi*. Int. J. Parasitol. 6, 457–466. https://doi.org/10.1016/0020-7519(76)90082-5.

Kearn, G.C., Whittington, I.D., 2015. Sperm transfer in monogenean (platyhelminth) parasites. Acta Parasitol. 60, 567–600. https://doi.org/10.1515/ap-2015-0082.

Kearn, G.C., Ogawa, K., Maeno, Y., 1992. Hatching patterns of the monogenean parasites *Benedenia seriolae* and *Heteraxine heterocerca* from the skin and gills, respectively, of the same host fish, *Seriola quinqueradiata*. Zool. Sci. 9, 451–455.

Kok, D.J., Du Preez, L.H., 1987. *Polystoma australis* (Monogenea): life cycle studies in experimental and natural infections of normal and substitute hosts. J. Zool. 212, 235–243. https://doi.org/10.1111/j.1469-7998.1987.tb05986.x.

Kritsky, D.C., Stephens, F., 2001. *Haliotrema abaddon* n. sp. (Monogenoidea: Dactylogyridae) from the gills of wild and maricultured west Australian dhufish *Glaucosoma hebraicum* (Teleostei: Glaucosomatidae), in Australia. J. Parasitol. 87, 749–754.

Kua, B.C., Azmi, M.A., Hamid, N.K.A., 2010. Life cycle of the marine leech (*Zeylanicobdella arugamensis*) isolated from sea bass (*Lates calcarifer*) under laboratory conditions. Aquaculture 302, 153–157.

Lackenby, J.A., Chambers, C.B., Ernst, I., Whittington, I.D., 2007. Effect of water temperature on reproductive development of *Benedenia seriolae* (Monogenea:Capsalidae) from *Seriola lalandi* in Australia. Dis. Aquat. Organ. 74, 235–242.

Lowenberger, C., Rau, M., 1994. *Plagiorchis elegans*: emergence, longevity and infectivity of cercariae, and host behavioural modifications during cercarial emergence. Parasitology 109, 65–72.

Macdonald, S., Jones, A., 1978. Egg-laying and hatching rhythms in the monogenean *Diplozoon homoion* gracile from the southern barbel (*Barbus meridionalis*). J. Helminthol. 52, 23–28.

Marchiori, N., Tancredo, K., Roumbedakis, K., Gonçalves, E.L., Pereira, J., Martins, M.L., 2013. New technique for collecting eggs from monogenean parasites. Exp. Parasitol. 134, 138–140. https://doi.org/10.1016/j.exppara.2013.02.011.

Matile, H., Pink, J.R.L., 1990. *Plasmodium falciparum* malaria parasite cultures and their use in immunology. In: Lefkovits, I., Pernis, B. (Eds.), In: Immunological Methods, vol. IV. Academic Press, San Deigo, CA, pp. 221–234.

Militz, T.A., Hutson, K.S., 2015. Beyond symbiosis: cleaner shrimp clean up in culture. PLoS One 10 (2), e0117723. https://doi.org/10.1371/journal.pone.0117723.

Militz, T.A., Southgate, P.C., Carton, A.G., Hutson, K.S., 2013. Dietary supplementation of garlic (*Allium sativum*) to prevent monogenean infection in aquaculture. Aquaculture 408–409, 95–99. https://doi.org/10.1016/j.aquaculture.2013.05.027.

Militz, T.A., Southgate, P.C., Carton, A.G., Hutson, K.S., 2014. Efficacy of garlic (*Allium sativum*) extract applied as a therapeutic immersion treatment for *Neobenedenia* sp. management in aquaculture. J. Fish Dis. 37, 451–461. https://doi.org/10.1111/jfd.12129.

Montero, F.E., Crespo, S., Padrós, F., de la Gándara, F., García, A., Raga, J.A., 2004. Effects of the gill parasite *Zeuxapta seriolae* (Monogenea: Heteraxinidae) on the amberjack *Seriola dumerili* Risso (Teleostei: Carangidae). Aquaculture 232, 153–163. https://doi.org/10.1016/S0044-8486(03)00536-2.

Mooney, A.J., Ernst, I., Whittington, I.D., 2006. An egg-laying rhythm in *Zeuxapta seriolae* (Monogenea: Heteraxinidae), a gill parasite of yellowtail kingfish (*Seriola lalandi*). Aquaculture 253, 10–16.

Mooney, A.J., Ernst, I., Whittington, I.D., 2008. Egg-laying patterns and *in vivo* egg production in the monogenean parasites *Heteraxine heterocerca* and *Benedenia seriolae* from Japanese yellowtail *Seriola quinqueradiata*. Parasitology 135, 1295–1302.

Morris, D.J., Adams, A., 2006. Transmission of *Tetracapsuloides bryosalmonae* (Myxozoa: Malacosporea), the causative organism of salmonid proliferative kidney disease, to the freshwater bryozoan *Fredericella sultana*. Parasitology 133, 701–709.

Morrison, R.N., Crosbie, P.B., Cook, M.T., Adams, M.B., Nowak, B.F., 2005. Cultured gill-derived *Neoparamoeba pemaquidensis* fails to elicit amoebic gill disease (AGD) in Atlantic salmon *Salmo salar*. Dis. Aquat. Organ. 66, 135–144.

Ogawa, K., Miyamoto, J., Wang, H.C., Lo, C.F., Kou, G.H., 2006. *Neobenedenia girellae* (Monogenea) infection of cultured cobia *Rachycentron canadum* in Taiwan. Fish Pathol. 41, 51–56.

OIE, 2018. World Organisation for Animal Health. OIE-Listed Diseases, Infections and Infestations in Force in 2018. http://www.oie.int/en/animal-health-in-the-world/oie-listed-diseases-2018/ (accessed 13 January 2018).

Overath, P., Ruoff, J., Stierhof, Y.D., Haag, J., Tichy, H., Dyková, I., Lom, J., 1998. Cultivation of bloodstream forms of *Trypanosoma carassii*, a common parasite of freshwater fish. Parasitol. Res. 84, 343–347. https://doi.org/10.1007/s004360050408.

Pironet, F.N., Jones, J.B., 2000. Treatments for ectoparasites and diseases in captive Western Australian dhufish. Aquac. Int. 8, 349–361. https://doi.org/10.1023/A:1009257011431.

Repullés-Albelda, A., Holzer, A.S., Raga, J.A., Montero, F.E., 2012. Oncomiracidial development, survival and swimming behaviour of the monogenean *Sparicotyle chrysophrii* (Van Beneden and Hesse, 1863). Aquaculture 338–341, 47–55. https://doi.org/10.1016/j.aquaculture.2012.02.003.

Ribeiro, P., El-Shehabi, F., Patocka, N., 2005. Classical transmitters and their receptors in flatworms. Parasitology 10, S19–S40.

Richards, G.R., Chubb, J.C., 1996. Host response to initial and challenge infections, following treatment, of *Gyrodactylus bullatarudis* and *G. turnbulli* (Monogenea) on the guppy (*Poecilia reticulata*). Parasitol. Res. 82, 242–247.

Rubio-Godoy, M., Tinsley, R.C., 2002. Trickle and single infection with *Discocotyle sagittata* (Monogenea:Polyopisthocotylea): effect of exposure mode on parasite abundance and development. Folia Parasitol. 49, 269–278.

Schelkle, B., Doetjes, R., Cable, J., 2011. The salt myth revealed: treatment of gyrodactylid infections on ornamental guppies, *Poecilia reticulata*. Aquaculture 311, 74–77. https://doi.org/10.1016/j.aquaculture.2010.11.036.

Schelkle, B., Snellgrove, D., Cable, J., 2013. *In vitro* and *in vivo* efficacy of garlic compounds against *Gyrodactylus turnbulli* infecting the guppy (*Poecilia reticulata*). Vet. Parasitol. 198, 96–101.

Schell, S.C., 1972. The early development of *Udonella caligorum* Jonhston, 1835 (Trematoda: Monogenea). J. Parasitol. 58, 1119–1121.

Scott, M., Anderson, R.M., 1984. The population dynamics of *Gyrodactylus bullatarudis* (Monogenea) within laboratory populations of the fish host *Poecilia reticulata*. Parasitology 89, 159–194.

Sharp, N.J., Diggles, B.K., Poortenaar, C.W., Willis, T.J., 2004. Efficacy of Aqui-S, formalin and praziquantel against the monogeneans, *Benedenia seriolae* and *Zeuxapta seriolae*, infecting yellowtail kingfish *Seriola lalandi lalandi* in New Zealand. Aquaculture 236, 67–83.

Shirakashi, S., Hirano, C., 2015. Accumulation and distribution of skin fluke *Neobenedenia girellae* eggs on a culture cage. Aquaculture 299, 2–7.

Shirakashi, S., Hirano, C., Ogawa, K., 2015. In: Reproduction and infectious biology of capsalid monogenean *Neobenedenia girellae*. International Symposium for Fish Parasites, Valencia, Spain, p. 174. Abstract P-152.

Sitjà-Bobadilla, A., Alvarez-Pellitero, P., 2009. Experimental transmission of *Sparicotyle chrysophrii* (Monogenea: Polyopisthocotylea) to gilthead seabream (*Sparus aurata*) and histopathology of the infection. Folia Parasitol. 56, 143–151.

Stewart, A., Jackson, J., Barber, I., Eizaguirre, C., Paterson, R., van West, P., Williams, C., Cable, J., 2017. Hook, line and infection: a guide to culturing parasites, establishing infections and assessing immune responses in the three-spined stickleback. Adv. Parasitol. 98, 39–109.

Theunissen, M., Tiedt, L., Du Preez, L.H., 2014. The morphology and attachment of *Protopolystoma xenopodis* (Monogenea: Polystomatidae) infecting the African clawed frog *Xenopus laevis*. Parasite 21, 20.

Thoney, D.A., Hargis, W.J., 1991. Monogenea (Platyhelminthes) as hazards for fish in confinement. Annu. Rev. Fish Dis. 1, 133–153.

Tinsley, R.C., 1983. Ovoviviparity in platyhelminth life-cycles. Parasitology 86, 161–196.

Trasviña-Moreno, A.G., Ascencio, F., Angulo-Valadez, C., Hutson, K.S., Avilés-Quevedo, A., Inohuye-Rivera, R.B., Pérez-Urbiola, J.C., 2017. Plant extracts as a natural treatment against the fish ectoparasite *Neobenedenia* sp. (Monogenea: Capsalidae). J. Helminthol. https://doi.org/10.1017/S0022149X17001122 (Epub ahead of print).

Trujillo-González, A., Johnson, L.K., Constantinoiu, C.C., Hutson, K.S., 2014. Histopathology associated with haptor attachment of the ectoparasitic monogenean *Neobenedenia* sp. (Capsalidae) to barramundi *Lates calcarifer* (Bloch). J. Fish Dis. 38, 1063–1067. https://doi.org/10.1111/jfd.12320.

Trujillo-González, A., Constantinoiu, C.C., Rowe, R., Hutson, K.S., 2015. Tracking transparent monogenean parasites on fish from infection to sexual development. Int. J. Parasitol. Parasites Wildl. 4, 316–322. https://doi.org/10.1016/j.ijppaw.2015.06.002.

Tsutsumi, N., Yoshinaga, T., Kamaishi, T., Nakayasu, C., Ogawa, K., 2003. Effects of water temperature on the development of the monogenean *Neoheterobothrium hirame* on Japanese flounder *Paralichthys olivaceus*. Fish Pathol. 38, 41–47.

Tubbs, L.A., Poortenaar, C.W., Sewell, M.A., Diggles, B.K., 2005. Effects of temperature on fecundity *in vitro*, egg hatching and reproductive development of *Benedenia seriolae* and *Zeuxapta seriolae* (Monogenea) parasitic on yellowtail kingfish *Seriola lalandi*. Int. J. Parasitol. 35, 315–327. https://doi.org/10.1016/j.ijpara.2004.11.008.

Vaughan, D.B., Chisholm, L.A., 2010. *Heterocotyle tokoloshei* sp. nov. (Monogenea, Monocotylidae) from the gills of *Dasyatis brevicaudata* (Dasyatidae) kept in captivity at Two Oceans Aquarium, Cape Town, South Africa: description and notes on treatment. Acta Parasitol. 55, 108–114.

Vaughan, D.B., Penning, M.R., Christison, K.W., 2008a. 2-Phenoxyethanol as anaesthetic in removing and relocating 102 species of fishes representing 30 families from Sea World to uShaka Marine World, South Africa. Onderstepoort J. Vet. Res. 75, 189–198.

Vaughan, D.B., Chisholm, L.A., Christison, K., 2008b. Overview of South African *Dendromonocotyle* (Monogenea: Monocotylidae), with descriptions of 2 new species from stingrays (Dasyatidae) kept in public aquaria. Zootaxa 1826, 26–44.

Vaughan, D.B., Chisholm, L.A., Hansen, H., 2016. *Electrocotyle whittingtoni* n. gen., n. sp. (Monogenea: Monocotylidae: Heterocotylinae) from the gills of a captive onefin electric ray, *Narke capensis* (Narkidae) at Two Oceans Aquarium, Cape Town, South Africa. Parasitol. Res. 115, 3575–3584.

Wedekind, C., Rüetschi, A., 2000. Parasite heterogeneity affects infection success and the occurrence of within-host competition: an experimental study with a cestode. Evol. Ecol. Res. 2, 1031–1043.

Whittington, I.D., 1990. The egg bundles of the monogenean *Dionchus remorae* and their attachment to the gills of the remora, *Echeneis naucrates*. Int. J. Parasitol. 20, 45–49.

Whittington, I.D., 1997. Reproduction and host-location among the parasitic Platyhelminthes. Int. J. Parasitol. 27, 705–714. https://doi.org/10.1016/S0020-7519(97)00012-X.

Whittington, I.D., 2004. The Capsalidae (Monogenea: Monopisthocotylea): a review of diversity, classification and phylogeny with a note about species complexes. Folia Parasitol. 51, 109–122.

Whittington, I.D., Kearn, G.C., 1988. Rapid hatching of mechanically-disturbed eggs of the monogenean gill parasite *Diclidophora luscae*, with observations on sedimentation of egg bundles. Int. J. Parasitol. 18, 847–852.

Whittington, I.D., Kearn, G.C., 2011. Hatching strategies in monogenean (Platyhelminth) parasites that facilitate host infection. Integr. Comp. Biol. 51, 91–99.

Whittington, I.D., Chisholm, L.A., Rohde, K., 2000a. The larvae of Monogenea (Platyhelminthes). Adv. Parasitol. 44, 139–232.

Whittington, I.D., Cribb, B.W., Hamwood, T.E., Halliday, J.A., 2000b. Host-specificity of monogenean (platyhelminth) parasites: a role for anterior adhesive areas? Int. J. Parasitol. 30, 305–320.

Woo, P.T.K., Li, S., 1990. *In vitro* attenuation of *Cryptobia salmositica* and its use as a live vaccine against cryptobiosis in *Oncorhynchus mykiss*. J. Parasitol. 76, 752–755.

Zahl, I.H., Kiessling, A., Samuelsen, O.B., Hansen, M.K., 2009. Anaesthesia of Atlantic cod (*Gadus morhua*)—effect of pre-anaesthetic sedation, and importance of body weight, temperature and stress. Aquaculture 295, 52–59.

CHAPTER FOUR

Molecular Epidemiology of *Anisakis* and Anisakiasis: An Ecological and Evolutionary Road Map

Simonetta Mattiucci*[,1], Paolo Cipriani*[,†,‡], Arne Levsen[‡], Michela Paoletti[†], Giuseppe Nascetti[†]
*Sapienza—University of Rome, Rome, Italy; "Umberto I" University Hospital, Rome, Italy; Laboratory affiliated to Istituto Pasteur Italia-Fondazione Cenci Bolognetti
[†]Tuscia University, Viterbo, Italy
[‡]Institute of Marine Research, Bergen, Norway
[1]Corresponding author e-mail address: simonetta.mattiucci@uniroma1.it

Contents

1. Introduction 94
   1.1 Basic Biology of *Anisakis* spp. 95
2. When Can an *Anisakis* Parasite Be Considered as a 'Biological Species'? 96
   2.1 What Types of Markers Resolve the Identification of *Anisakis* spp.? 100
   2.2 What Types of Markers Are Used in the Study of Hybridization and Introgression Between Cryptic Species of *Anisakis*? 111
3. How Many Valid Species Are in the Genus *Anisakis*? 115
   3.1 *Anisakis* Species Included in Clade 1 128
   3.2 *Anisakis* spp. Included in Clade 2 138
   3.3 *Anisakis* Species Included in Clade 3 140
   3.4 *A. typica* (Diesing, 1860) 143
   3.5 *Anisakis* sp. 1 145
4. Reconciling Molecular and Morphological Results 146
   4.1 Diagnostic Morphological Features at Clade Level 147
   4.2 Diagnostic Morphological Features Between 'Cryptic' Species of the Genus *Anisakis* 150
   4.3 Larval Morphological Features in the Species of Genus *Anisakis* 154
5. How Does *Anisakis* spp. Diversity Vary Across Host Species? 157
   5.1 Host Preference vs Definitive Hosts 157
   5.2 Host Preference vs Intermediate/Paratenic Hosts 165
6. Molecular Epidemiology of *Anisakis* spp. in Fisheries 167
   6.1 What Drivers Shape the Distribution of *Anisakis* spp. in Wild Fisheries? 167
   6.2 *Anisakis* spp. in Farmed Fish 190
7. Do *Anisakis* spp. Always Occupy the Same Site in Fish? 193
   7.1 Detection of *Anisakis* spp. Larvae in Fishery Products 193

Advances in Parasitology, Volume 99
ISSN 0065-308X
https://doi.org/10.1016/bs.apar.2017.12.001

7.2 Site of Infection by *Anisakis* spp. Larvae in Fish 198
7.3 *Intra Vitam* and *Post Mortem* Larval Migration 205
8. Human Anisakiasis: Which Are the *Anisakis* spp. Infective to Humans? 208
8.1 The Zoonotic Role of *Anisakis* spp. 209
8.2 Molecular Epidemiology of Human Anisakiasis 210
8.3 Clinical Manifestation of Human Anisakiasis Caused by *A. simplex* (s.s.) and *A. pegreffii* 220
8.4 Diagnosis of *Anisakis* spp. in Humans 224
9. What Key Questions for Future Research Challenges? 228
9.1 What Molecular/Genetic Tools to Use in Future Studies of *Anisakis*? 228
9.2 What Approaches for Future Analysis of Distribution and Epidemiology? 231
Acknowledgements 239
References 240
Further Reading 262

Abstract

This review addresses the biodiversity, biology, distribution, ecology, epidemiology, and consumer health significance of the so far known species of *Anisakis*, both in their natural hosts and in human accidental host populations, worldwide. These key aspects of the *Anisakis* species' biology are highlighted, since we consider them as main driving forces behind which most of the research in this field has been carried out over the past decade. From a public health perspective, the human disease caused by *Anisakis* species (anisakiasis) appears to be considerably underreported and underestimated in many countries or regions around the globe. Indeed, when considering the importance of marine fish species as part of the everyday diet in many coastal communities around the globe, there still exist significant knowledge gaps as to local epidemiological and ecological drivers of the transmission of *Anisakis* spp. to humans. We further identify some key knowledge gaps related to *Anisakis* species epidemiology in both natural and accidental hosts, to be filled in light of new 'omic' technologies yet to be fully developed. Moreover, we suggest that future *Anisakis* research takes a 'holistic' approach by integrating genetic, ecological, immunobiological, and environmental factors, thus allowing proper assessment of the epidemiology of *Anisakis* spp. in their natural hosts, in human populations, and in the marine ecosystem, in both space and time.

1. INTRODUCTION

A review is like a journey; it requires a map, a direction, and a destination. Moving from the overviews offered by Mattiucci and Nascetti (2008) and Audicana and Kennedy (2008), and a vast literature, here we will draw a road map through the last 10 years of research on the anisakid nematodes of the genus *Anisakis*, and on the fish-borne zoonotic disease (anisakiasis) that these parasites are able to provoke in humans. We will outline key findings of basic and applied value within this research arena, emphasizing their possible significance for parasitological research as a whole, and scanning the horizon to identify 'big questions' and research priorities for the near future.

Travelling throughout the enormous literature on the diversity, evolution, ecology, and epidemiology of *Anisakis* spp. and anisakiasis, has not been easy for the authors. To make this task easier for the readers, we decided to draw the road map using key questions that have dominated recent research efforts in this field. Those are: When can a given *Anisakis* parasite be considered as a 'biological species'? What are the suitable molecular tools to discover and confirm the existence of biological species in the genus *Anisakis*? How many *Anisakis* species are there and how do we identify them? Are morphological features sufficient for the differentiation of *Anisakis* spp., at least at their adult stage? How does *Anisakis* spp. diversity vary across host species and geographical areas? What are the driving factors in shaping the actual distribution and epidemiology of *Anisakis* spp.? What are the main predictors in the infection by larval *Anisakis* spp. in wild fisheries and marine aquaculture? Have *Anisakis* spp. larvae a different localization in fish host? Are all the *Anisakis* spp. infective to humans? What is the human pathogenic role of *Anisakis* spp., and which species are involved? Do climate change and anthropogenic impacts play a role in the infection dynamics of these parasites?

After a short summary of key features of *Anisakis* spp. biology and epidemiology, in the following sections we will present recent research findings useful to answer these questions, as well as the research frontiers that these findings have opened.

1.1 Basic Biology of *Anisakis* spp.

Anisakid nematodes of the genus *Anisakis* Dujardin, 1845 are ascaridoid nematodes colonizing the digestive system of marine vertebrates. These parasites have an indirect life cycle which involves various hosts at different levels across the food webs. Marine mammals (mainly cetaceans) serve as definitive hosts, fish and squids are intermediate or paratenic hosts, while planktonic or semiplanktonic crustaceans act as first intermediate hosts. Eggs are released with the definitive hosts' faeces and lead eventually to free swimming unsheathed larvae in the marine environment. The larvae undergo one or two moults before being ingested by invertebrates, mostly small crustaceans. These intermediate hosts are then eaten by a wide variety of transport or paratenic hosts, including cephalopods and fishes. At this stage, many different life cycle patterns may be observed, mostly depending on the species of *Anisakis*. The fish may be directly eaten by the definitive host when the third-stage larva (L3) moults to L4 and then matures to the adult form of the parasite. Alternatively, the fish may be eaten by another fish: the larvae reencapsulate in this new host. This phenomenon may occur several times, possibly inducing accumulation of *Anisakis* spp. larvae in predatory fish, along the trophic web (Münster et al., 2015; Zuo et al., 2016). Moreover, the behaviour of the larvae may be different depending on several factors, some still unknown: in some cases, most parasites will remain in the visceral cavity of the fish or within the visceral organs, whereas in other cases, parasites can migrate to the musculature of the fish (Cipriani et al., 2016; Karl et al., 2011; Levsen and Lunestad, 2010; Levsen et al., 2017a; Roepstorff et al., 1993).

Humans may act as accidental host acquiring infection through consumptions of raw, smoked, marinated salted, or undercooked fish and squids infected by larval stages. Worldwide, there is an increasing recognition that the fish-borne zoonosis, known as anisakiasis, is an important emerging human disease (Buchmann and Mehrdana, 2017; Chai et al., 2005). Risk factors for the transmission of the zoonotic species include consumption of locally harvested wild fish and traditional home-made seafood preparations. Preferred fish species vary among countries and local communities, being consumed as fresh, partially cooked, fermented, or dried preparations. Some of these preparations can be a potential risk to human health and have been linked to cases of human anisakiasis in different countries worldwide.

There is also an increasing interest in prioritizing fish-borne zoonoses not only in terms of their impact on human health, but also in predicting the effect of climate and marine ecosystem changes on the ecology and epidemiology of *Anisakis* species involved, and any consequences on their behaviour in their natural hosts (both definitive and intermediate/paratenic ones) (Jenkins et al., 2013; Mattiucci and Nascetti, 2008; Zarlenga et al., 2014).

2. WHEN CAN AN *ANISAKIS* PARASITE BE CONSIDERED AS A 'BIOLOGICAL SPECIES'?

... without gene flow, it is inevitable that there will be speciation ...

(Wolpoff, 1989)

The precise identification of parasites of the genus *Anisakis* is essential for understanding their distribution and epidemiology. Any biodiversity assessment of anisakid nematodes is incomplete if inferred only from morphology, although morphological analysis and older (historical) considerations in some cases show a considerable degree of congruence with modern molecular-based taxonomic assessments (Mattiucci et al., 2014a; see also Section 4). In this group of nematodes, the taxonomic classification based on morphological characters is impeded by a strong tendency to adaptive convergence. In fact, selective pressure can result in identical phenotypes, being adapted to a similar host environment/habitat. This phenomenon is called *convergent evolution*. Generally, most descriptions of parasite species conformed with that regarded as the 'morphological, or typological, species concept'. But because speciation is not always accompanied by corresponding morphological change, the actual number of biological species is likely to be greater than the current tally of nominal species, most of which are delineated on morphological grounds.

Deciding what species are, and by which method they may be identified (*species delimitation*), are the prerequisites for characterizing biodiversity. The correct identification of parasite species of the genus *Anisakis* is thus at the core of our understanding of their biodiversity, at both species and gene levels. Yet deciding what species are, and how they should be recognized in nature, have been problematic for biologists. However, some general rules of the '*New Taxonomy Approach*' (see Wheeler, 2008) can be applied in the case of anisakid nematodes.

With regard to the anisakid nematodes, sexually reproducing diecious organisms (or having separate sexes), the biological species concept (BSC) (Mayr, 1963) can be tested under the following definition: '... the species are systems of populations: the gene exchange between these systems is limited or prevented by a reproductive isolating mechanisms (RIM), i.e., by a combination of prezygotic and postzygotic barriers'. In other words, species are characterized 'groups of actually or potentially interbreeding natural populations which are reproductively isolated from other such groups' (Mayr, 1963).

The BSC has regained importance by the use of nuclear genetic markers, which, in the case of nematodes of genus *Anisakis*, has enabled us to demonstrate the existence of reproductive isolation between various sympatric and allopatric populations. The concept was first supported by the application of allozyme markers in certain morphospecies within the genus *Anisakis*, such as in the case of the nominal species *Anisakis simplex* (s.l.) (Mattiucci et al., 2014a, b; Nascetti et al., 1986) or in the morphospecies *Anisakis physeteris* (Mattiucci et al., 2001, 2005). The discovery of cryptic and sibling species represents one the most challenging aspects in marine parasitological studies. In accordance with Nadler and Pèrez-Ponce de León (2011), the detection of cryptic biodiversity in a nominal species, should start with observations of: (1) considerable variability in morphological characters in a nominal species and (2) broad host range, often including species of different ecology and belonging to different families. Further steps would include demonstrating that a given nominal species is indeed a complex of several 'cryptic' and/or 'sibling' species (i.e. species morphologically similar, but reproductively isolated). According to Nadler and Pèrez-Ponce de León (2011), 'sibling' species implies a close phylogenetic relationship, technically meaning 'sister taxa', whereas, the term 'cryptic' does not directly address recency of common ancestry. Thus, those authors, in order to avoid misunderstanding, suggest to use the term of cryptic species as terminology to be used, '... whether or not newly discovered cryptic species share a recent most common ancestor, because an incomplete species sampling may obscure true "sister species" relationships ...' (Nadler and Pèrez-Ponce de León, 2011).

Thus, detecting and defining as 'cryptic *valid species*' a taxon among anisakid nematodes of the genus *Anisakis*, via the study of the genetic variation over a large geographical range and several hosts, have challenged parasitologists for the last decades. For instance, reproductive isolation and absence of gene flow between sympatric and allopatric sibling or cryptic species within nominal species of the genus *Anisakis* were demonstrated by allozyme loci, based on the Hardy–Weinberg equilibrium expectations and estimates of population genetic structure. These applications can be used to test the hypothesis of panmixia for individuals sampled from a sympatric distribution, potentially revealing reproductively isolated biological species. The statistically significant deviation from the H–W equilibrium at a certain number of nuclear loci may indicate the existence of a genetic heterogeneity within a population, which could be explained with the existence of different gene pools, likely corresponding to different species, having no gene flow between them. As a consequence, those loci can be also considered as 'diagnostic loci' (Box 1), allowing differentiation between the discovered species. The corollarium of testing the BSC, based on nontree methods, is the high number of loci and individuals to be tested over a large number of both definitive and intermediate-paratenic hosts, in order to support the existence of biological species (Pérez-Ponce de León and Nadler, 2010). Based on such nontree methods, the specific status of the species *A. simplex* (s.s.), *Anisakis pegreffii*, and *Anisakis berlandi* was established, and

BOX 1 Glossary

Diagnostic markers: Generally, in taxonomy, a marker is said to be 'diagnostic' if it presents fixed different states in different taxa. As a consequence, a locus/genetic marker can be designated as 'diagnostic', only when fixed alleles or fixed nucleotide site differences between two taxa are present. In other words, those fixed differences are not shared between the two taxa, thus indicating that there is not gene flow between them.

Ancestral polymorphism: Presence of alleles (or haplotypes) at a polymorphic nuclear or mitochondrial gene, which have been not fixed as alternative between two taxa during the speciation process. As a consequence, shared alleles (or haplotypes) could be found in both the two species.

Hybridization: Mating between individuals from genetically distinct species that produces offspring, which represents hybrid of first (F1) generation.

Introgression: Incorporation of alleles from a species, into the gene pool of a second divergent species, through hybridization and backcrossing of F1 hybrids with individuals from one of the two parental species. The resultant offspring hybrid categories are 'backcrosses' individuals.

it could be also routinely tested, among sympatric and allopatric populations of these taxa (Mattiucci et al., 2014a).

A supporting method for determining the existence of biological species of *Anisakis* is based on the evidence of their historical lineage independence. Under the phylogenetic species concept (PSC), 'species are monophyletic and speciation results from cladogenesis' (Avise, 1994). A disadvantage in the use of only PSC in the parasite species delimitation, would be the finding of an intraspecific genetic lineage within a species—for instance, from different geographical areas or hosts—which could confound and lead to a misinterpretation in considering those as historically independent species lineages. To overcome this, the concordance principle (CP) (Avise and Ball, 1990) states that the results achieved by sequencing several gene loci should be acquired, and evidence of concordant phylogenetic partitions among multiple genes should be evaluated (Avise, 1994). This is the reason why BSC and PSP concepts should be applied and a multigene molecular approach adopted, when delimiting species of anisakid nematodes.

This, in turn, provides a basis to study patterns and processes in the evolution and ecology of these taxa, including biogeography, host preference and coevolution, and allows an assessment of their biodiversity patterns at a global scale, in space and time.

2.1 What Types of Markers Resolve the Identification of *Anisakis* spp.?

'No more time to stay 'single' in the detection of *Anisakis* spp.'

The power of resolution of molecular/genetic methodologies in detecting anisakid species revolutionized the taxonomy of these species during the last 20 years (Mattiucci and Nascetti, 2008). However, when exploring the molecular markers so far available in *Anisakis* parasites, and based on the existing literature, an important first step should be the choice of the marker set most appropriate to the specific study aims: (1) disentangling cryptic species; (2) establishing phylogenetic relationships between related taxa; (3) assessing hybridization and introgression phenomena between related species; (4) providing an intraspecific population analysis; (5) estimating their genetic diversity; and (6) providing suitable markers for species identification at any stage of their life cycle.

Here, we discuss the most commonly utilized genetic markers in studying several aspects of the molecular systematics and taxonomy of *Anisakis* spp. at both interspecific and intraspecific level, showing some advantages and disadvantages in their use.

2.1.1 Genotyping Approach From Multilocus Allozyme Electrophoresis (MAE)

The BSC was well supported by the application of multilocus allozyme electrophoresis (MAE) for several anisakid species (Mattiucci and Nascetti, 2008). Initial attempts to apply population genetics to the study of genetic variation among large samples of *Anisakis* collected from different intermediate/paratenic and definitive hosts employed the use of MAE (19–24 enzyme loci). Reproductive isolation and absence of gene flow were demonstrated by allozymes between sympatric and allopatric cryptic species, establishing their specific status (Andrews and Chilton, 1999; Nadler and Pèrez-Ponce de León, 2011). Polymorphic allozyme loci showing deviation from the Hardy–Weinberg equilibrium (i.e. deficiency of heterozygosity) for sympatric population samples, or indirect estimating of the gene flow, are results that can falsify the null hypothesis of the existence of a 'single panmictic species'. Based on this inference, MAE tools revealed the existence of high genetic heterogeneity within certain morphotypes, such as those within *Anisakis*. Over the past 2 decades, it allowed disclosure of a number of cryptic species within several nominal species of anisakid nematodes included in the genus *Anisakis*. On the other hand, these markers have proven to be suitable in disclosing several cryptic species within other groups of nematodes (for instance, see Chilton et al., 2016 and references therein).

Indeed, after the first applications of this methodology, the known diversity of species belonging to *Anisakis* quickly increased. After the detection of several cryptic species (Mattiucci et al., 1997, 2014a; Nascetti et al., 1993), this methodology led to the discovery and description of several new species (Mattiucci et al., 2005, 2009; Paggi et al., 1998), and also supported the validity of species of previously synonymized taxa (Mattiucci et al., 1986, 2001, 2002). Allozyme markers in *Anisakis* spp. have allowed us to: (a) genetically characterize different species of anisakid nematodes, (b) estimate their genetic differentiation, (c) identify their larval stages which lack diagnostic morphological characters, (d) disclose hybridization events between cryptic species, and (e) estimate genetic variability of *Anisakis* spp. populations (Mattiucci et al., 2009, 2014a, 2015a, 2016, 2017a; Mattiucci and Nascetti, 2008). Thus, based on allozyme diagnostic loci for different anisakid taxa, the easy and rapid identification of large numbers of individuals can be achieved. The methodology is particularly valuable for identifying larval individuals collected from intermediate/paratenic hosts, where different *Anisakis* species occur in mixed infections. Since numerous allozyme analyses have so far been applied to thousands of individuals, they have contributed greatly to our knowledge about the population genetic structure of *Anisakis* spp. in fish and marine mammals from all over the oceanic waters (Mattiucci et al., 2009, 2014a, 2015a, 2017a).

However, some disadvantages exist in the use of these markers. MAE data do not allow construction of a highly supported phylogenetic tree among related taxa, especially between those showing a high level of genetic differentiation; data from MAE are better analysed in the distance-based tree method (Mattiucci and Nascetti, 2008). Finally, and although standard methodologies are well established for these parasites, the interpretation of the MAE results requires expertise. MAE also requires fresh or frozen material, which is not always easily obtained for these parasites, especially at their adult stage.

2.1.2 Nuclear Genotyping Approach From Internal Transcribed Spacers of rDNA (ITS rDNA)

Anisakid individuals recognized at the species level by allozymes have been used to develop DNA-based approaches for species identification, including the RFLPs-PCR DNA (D'Amelio et al., 2000) of nuclear ITS region of rDNA (Table 1), and the direct sequencing of the same gene (Nadler et al., 2005). The polymorphisms inferred by RFLPs-PCR at the

Table 1 Primers at Those Molecular/Genetic Markers (mtDNA, rDNA, nDNA Genes) so far Disclosed in the Species of Genus *Anisakis*

Gene/ Locus	Primer Name	*Anisakis* Species	Primer Sequence 5′–3′	References
COI	JB3 (F)	All *Anisakis* spp.	TTTTTTGGGCATCCTGAGGTTTAT	Hu et al. (2001)
	JB4.5 (R)	All *Anisakis* spp.	TAAAGAAAGAACATAATGAAAATG	Hu et al. (2001)
	AnCO1-F (F)	All *Anisakis* spp.	ATTTGGTCTTTGATCTGGTATGG	Cross et al. (2007)
	AnCO1-R (R)	All *Anisakis* spp.	TGGCAGAAATAACATCCAAACTAG	Cross et al. (2007)
COII	211 (F)	All *Anisakis* spp.	TTTTCTAGTTATATAGATTGRTTYAT	Nadler and Hudspeth (2000); Valentini et al. (2006)
	210 (R)	All *Anisakis* spp.	CACCAACTCTTAAAATTATC	Nadler and Hudspeth (2000); Valentini et al. (2006)
	RTpegF	*A. pegreffii*	CTTTTGGAGGTTGATAATCG	Paoletti et al. (2017)
	RTpegR	*A. pegreffii*	CCCACAAATCTCTGAACATT	Paoletti et al. (2017)
	pegHyPr[a]		CTTGGGCTTTGCCTAGGATGTC	Paoletti et al. (2017)
	RTsimF	*A. simplex* (s.s.)	CTTTAATTTTGGTTGCTCAGAT	Paoletti et al. (2017)
	RTsimR	*A. simplex* (s.s.)	CGATTATCAACCTCCAAAAG	Paoletti et al. (2017)
	simHyPr[a]		ATGACCAGTGACTTTCACAGTCAAAT[b]	Paoletti et al. (2017)
ITS	NC5 (F)	All *Anisakis* spp.	GTAGGTGAACCTGCGGAAGGATCATT	Zhu et al. (2000); Abollo et al. (2003)
	NC2 (R)	All *Anisakis* spp.	TTAGTTTCTTCCTCC GCT	Zhu et al. (2000); Abollo et al. (2003)

rrnS	MH3 (F)	All *Anisakis* spp.	TTGTTCCAGAATAATC GGCTAGACTT	Hu et al. (2001); Abollo et al. (2003)
	MH4 (R)	All *Anisakis* spp.	TCTACTTTACTACAACTTACTCC	Hu et al. (2001); Abollo et al. (2003)
EF1 α-1	EF-F (F)	*A. pegreffii* and *A. simplex* (s.s.)	TCCTCAAGCGTTGTTATCTGTT	Mattiucci et al. (2016)
	EF-R (R)	*A. pegreffii* and *A. simplex* (s.s.)	AGTTTTGCCACTAGCGGTTCC	Mattiucci et al. (2016)
AnisL4	ASIM_scaffold000000148 (F)	*A. pegreffii* and *A. simplex* (s.s.)	ACAAGGAATGCTCGTCGATATGA	Mladineo et al. (2017a)
	ASIM_scaffold000000148 (R)	*A. pegreffii* and *A. simplex* (s.s.)	TCACTCTTGTCCAGCACTGC	Mladineo et al. (2017a)
AnisL7	ASIM_scaffold00000029 (F)	*A. pegreffii* and *A. simplex* (s.s.)	TGATTGCGATTACTATTATTGTCAGT	Mladineo et al. (2017a)
	ASIM_scaffold00000029 (R)	*A. pegreffii* and *A. simplex* (s.s.)	GGTGGTGTCGTGACAGTGAT	Mladineo et al. (2017a)
AnisL10	ASIM_scaffold000001215 (F)	*A. pegreffii* and *A. simplex* (s.s.)	CAGACACGGTGGTTTTGTGAC	Mladineo et al. (2017a)
	ASIM_scaffold000001215 (R)	*A. pegreffii* and *A. simplex* (s.s.)	TGGCAAGCCAAAATGTCGTT	Mladineo et al. (2017a)
AnisL13	ASIM_scaffold000003311 (F)	*A. pegreffii* and *A. simplex* (s.s.)	TGCCTTAAGGATCTCCGTGG	Mladineo et al. (2017a)
	ASIM_scaffold000003311 (R)	*A. pegreffii* and *A. simplex* (s.s.)	GCCGACTTTTGAAAGAGGTCA	Mladineo et al. (2017a)
AnisL14	ASIM_scaffold000003617 (F)	*A. pegreffii* and *A. simplex* (s.s.)	ACAGCAGCAATACGGCAGAT	Mladineo et al. (2017a)
	ASIM_scaffold000003617 (R)	*A. pegreffii* and *A. simplex* (s.s.)	CACCGAAGTGCCCTTATCGA	Mladineo et al. (2017a)
AnisL17	ASIM_scaffold000000639 (F)	*A. pegreffii* and *A. simplex* (s.s.)	GTGAAGAGCGATCTAGCCGG	Mladineo et al. (2017a)
	ASIM_scaffold000000639 (R)	*A. pegreffii* and *A. simplex* (s.s.)	ATGGTGTCGTAGTTGCACGT	Mladineo et al. (2017a)
AnisL18	ASIM_scaffold000011514 (F)	*A. pegreffii* and *A. simplex* (s.s.)	TCGTTATGGTGGTGACAATCGT	Mladineo et al. (2017a)

Continued

Table 1 Primers at Those Molecular/Genetic Markers (mtDNA, rDNA, nDNA Genes) so far Disclosed in the Species of Genus *Anisakis*—cont'd

Gene/Locus	Primer Name	*Anisakis* Species	Primer Sequence 5′–3′	References
	ASIM_scaffold000011514 (R)	*A. pegreffii* and *A. simplex* (s.s.)	TCGCATCGTCATTACCAGCA	Mladineo et al. (2017a)
AnisL21	ASIM_scaffold00019081 (F)	*A. pegreffii* and *A. simplex* (s.s.)	ACCGTGAACATCTTATCGCA	Mladineo et al. (2017a)
	ASIM_scaffold00019081 (R)	*A. pegreffii* and *A. simplex* (s.s.)	ACATTGCCTTTTGAAAGCGCA	Mladineo et al. (2017a)
AnisL22	ASIM_scaffold00024391 (F)	*A. pegreffii* and *A. simplex* (s.s.)	TCAGAGGTGCGTAACGTTCA	Mladineo et al. (2017a)
	ASIM_scaffold00024391 (R)	*A. pegreffii* and *A. simplex* (s.s.)	GCCACATCCATCCAAAAGAGC	Mladineo et al. (2017a)
AnisL25	ASIM_scaffold000003429 (F)	*A. pegreffii* and *A. simplex* (s.s.)	GGTGCTGGAGACGATAACGA	Mladineo et al. (2017a)
	ASIM_scaffold000003429 (R)	*A. pegreffii* and *A. simplex* (s.s.)	CGAACCGTAAAATGACATCGACA	Mladineo et al. (2017a)
AnisL29	ASIM_scaffold00014654 (F)	*A. pegreffii* and *A. simplex* (s.s.)	TGCAGTGAATGAAGAATGCAGA	Mladineo et al. (2017a)
	ASIM_scaffold00014654 (R)	*A. pegreffii* and *A. simplex* (s.s.)	AGACATGTTTTCGCTGTTGACA	Mladineo et al. (2017a)

[a] Species-specific hydrolysis probes.
[b] The probe for *A. simplex* (s.s.) was designed on the complement strand.

ITS region of DNA have provided a set of nuclear markers for the identification of species of the genus *Anisakis* (D'Amelio et al., 2000).

The internal transcribed spacers (ITS) is part of the rDNA array and lies between the SSU rDNA- and the LSU rDNA-coding regions. It is divided into ITS-1 and ITS-2 and separated by the gene coding for the 5.8S rDNA. While in the past it has been assumed that ITS had no function, it is later proposed that ITS-1, for instance, plays a role in ribosome biogenesis (Lalev and Nazar, 1998). While 5.8S rDNA is considered as 'slowly evolving', the ITS is less conserved, with the ITS-1 more variable than ITS-2 (Van Herwerden et al., 1999). This is likely due to the fact that generally in the ITS-2 there are not variable repeat units, which are, on the contrary, present in the ITS-1.

According to D'Amelio et al. (2000), using three restriction endonucleases (*HhaI*, *HinfI*, and *TaqI*) based on RFLPS of the ITS region of rDNA, it was possible to identify six different species of *Anisakis* among those previously characterized by allozymes (i.e. *A. pegreffii*, *A. simplex* (s.s.), *A. berlandi*, *Anisakis typica*, *Anisakis ziphidarum*, and *A. physeteris*). Subsequently, restriction patterns were provided also for *Anisakis brevispiculata* (see Mattiucci et al., 2002), *Anisakis paggiae* (see Mattiucci et al., 2005), and *Anisakis nascettii* (see Mattiucci et al., 2009), by the same RFLPs and endonucleases (D'Amelio et al., 2000). Thus, according to those restriction patterns, they were considered to be of diagnostic value (see Box 1) among the different species. Indeed, the same authors did not detect any variation at the intraspecific level; however, those nuclear markers at that locus (i.e. the ITS region of rDNA) were first established on a very low number of specimens tested (i.e. between 1 and 13 specimens by each species) (D'Amelio et al., 2000).

The direct sequencing of the same locus (ITS region of the rDNA) was also performed on those species. Some fixed nucleotide positions in different species of *Anisakis* spp. have also been proposed (Nadler et al., 2005; Pontes et al., 2005). For instance, fixed differences at two nucleotide positions of the ITS-1 (i.e. 278 and 294) region have been considered as diagnostic between *A. pegreffii* and *A. simplex* (s.s.) (Nadler et al., 2005; Pontes et al., 2005). In addition, the same sequence analysis was used to infer phylogenetic relationships between related species of *Anisakis* (Nadler et al., 2005), suggesting genes as able to support the existence of different species of *Anisakis*—previously discovered by allozymes—based on the evolutionary lineage concept PSC, in a combined phylogenetic analysis including also other mitochondrial loci (Mattiucci et al., 2014a).

However, generally speaking, several limitations have been identified in studies based on ITS markers. In other helminth species

(Criscione et al., 2007), it has been suggested that the ITS region of rDNA (rDNA complex in general) is a multicopy gene, which undergoes concerted evolution phenomena (Elder Jr and Turner, 1995; Ganley and Kobayashi, 2007) which violates the assumption of Hardy–Weinberg. Therefore, genotypes inferred from PCR-RFLP analysis of ITS rDNA cannot be used for testing H–W assumption, when describing 'biological species'. It was also demonstrated that single molecular marker based on the ITS region of rDNA, when used exclusively, are not always able to recognize 'pure' specimens belonging to the two cryptic species *A. pegreffii* and *A. simplex* (s.s) and, even less, to successfully detect their hybrid categories (Mattiucci et al., 2016). It was suggested that the two nucleotide positions found in the ITS-1 of rDNA (i.e. 278 showing C in *A. pegreffii* and T in *A. simplex* (s.s.), and 294 showing C in *A. pegreffii* and T in *A. simplex* (s.s.)), are not actually fixed diagnostic markers (Mattiucci et al., 2016) (Box 1). As a consequence, the nucleotide position 294, detected at the PCR-RFLPs analysis by *HinfI* in the ITS1 rDNA, cannot be retained as a 100% diagnostic marker between the two species—as it has been conventionally stated (D'Amelio et al., 2000)—just because there is a shared polymorphism between the two *Anisakis* species, at those nucleotide positions 278 and 294 of the ITS-1 region. This polymorphism could have been misinterpreted as the occurrence of 'putative hybrids' between the two cryptic species (Abollo et al., 2003), when not supported by other diagnostic nuclear markers on the same specimens (Mattiucci et al., 2016). The shared polymorphism found by a multimarker approach in both *A. pegreffii* and *A. simplex* (s.s.), occurring at those positions of the ITS-1 region of rDNA, could be likely the outcome of the incomplete lineage sorting of a shared ancestral variation (polymorphism) at that gene, or the result of a historical introgression at that nuclear marker, after their recontact under sympatric conditions (Mattiucci et al., 2016) (see also Section 2.2).

2.1.3 Nuclear Genotyping From EF1 α-1 nDNA

The exploration of molecular markers inferred from other nuclear loci, also based on the existent literature, is still limited in *Anisakis* spp. We anticipate, however, that the new technologies, such as next-generation sequencing, will improve the current lack of other genetic markers for recognizing species of *Anisakis*.

On the other hand, in more recent years, primers for direct sequencing of the elongation factor (EF1 α-1 nDNA) nuclear gene, in individuals belonging to the two species, *A. pegreffii* and *A. simplex* (s.s.), previously identified by allozymes, were designed (Mattiucci et al., 2016) (Table 1 and Fig. 1).

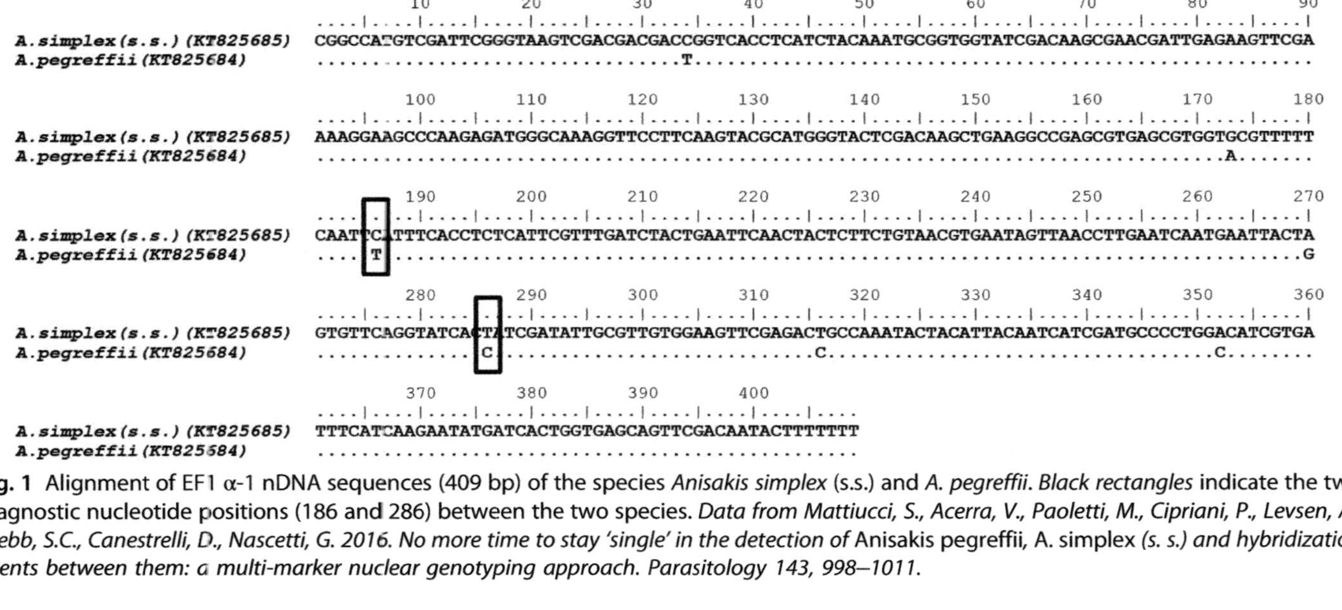

Fig. 1 Alignment of EF1 α-1 nDNA sequences (409 bp) of the species *Anisakis simplex* (s.s.) and *A. pegreffii*. *Black rectangles* indicate the two diagnostic nucleotide positions (186 and 286) between the two species. *Data from Mattiucci, S., Acerra, V., Paoletti, M., Cipriani, P., Levsen, A., Webb, S.C., Canestrelli, D., Nascetti, G. 2016. No more time to stay 'single' in the detection of* Anisakis pegreffii, A. simplex *(s. s.) and hybridization events between them: a multi-marker nuclear genotyping approach. Parasitology 143, 998–1011.*

A fragment of 409 bp in length of the EF1 α-1 nDNA region was obtained. It revealed the presence of two diagnostic nucleotide sites between *A. pegreffii* and *A. simplex* (s.s.). These positions are: 186, showing a T in *A. pegreffii*, whereas it was C in the parental taxon *A. simplex* (s.s.); and position 286, showing a C in *A. pegreffii* but always a T in *A. simplex* (s.s.). The alignment of EF1 α-1 nDNA sequences are reported in Fig. 1, showing the fixed nucleotide sites which are able to discriminate the two species of the genus *Anisakis*. These diagnostic positions were validated, so far, over a large number ($N=689$) of specimens of those *Anisakis*, from both sympatric and allopatric populations (Mattiucci et al., 2016).

2.1.4 Nuclear Genotyping From Microsatellites

Microsatellites are, as the allozymes, codominant markers with repetitive sequences of DNA, which are widely dispersed along and among the chromosomes (Avise, 1994). Because of their mutation rate (10^{-2}, 10^{-6} mutations per generation), they are mostly used for fine-scale genetic population studies, including assessing genetic structure at the intraspecific level.

Recently, Mladineo et al. (2017a) found a panel of microsatellites for *A. simplex* s.s. and *A. pegreffii*. Although they were highly polymorphic, these microsatellite loci did not allow to clearly distinct between the two species; only three specimens for *A. simplex* (s.s.), and a single population of *A. pegreffii* from the Adriatic Sea were analysed by those authors.

Furthermore, a panel of seven DNA microsatellite loci have been developed and investigated in several populations and hundreds specimens of *A. pegreffii, A. simplex* (s.s.) and *A. berlandi* (Mattiucci et al., unpublished data). These markers have been selected and characterized in populations of the two species collected from both sympatric and allopatric areas. The large representative sample of the two species so far analysed has shown that these nuclear loci are sufficiently robust and polymorphic to be valuable population markers for these parasites at the intraspecific level. In addition, allele frequencies gathered at those loci at the intraspecific level allow the estimation of gene flow between geographically distant populations of those species, and to calculate the genetic differentiation, and variability at the intrapopulation level (Mattiucci et al., unpublished data). Also the combined use of those seven DNA microsatellite loci allow the identification of the *A. simplex* (s.s.), *A. pegreffii* and *A. berlandi* in a multimarkers genotyping approach (Mattiucci et al., unpublished data). Furthermore, genotyping of these nematodes based on microsatellites offers a way to address questions concerning population genetic epidemiology and molecular ecology.

2.1.5 Mitochondrial DNA Loci

Applying PCR DNA molecularly derived methodologies on the systematics of anisakid nematodes has gained a wider application. Thus, a multigene approach including both nuclear and mitochondrial markers, together with allozymes, permits robust identification of the species so far included in the genus *Anisakis*. Among the mtDNA loci, the most widely used is the mitochondrial cytochrome oxidase II gene (mtDNA *cox2*) (Mattiucci et al., 2014a; Mattiucci and Nascetti, 2006; Valentini et al., 2006) and the small subunit of rRNA (*rrnS*) in the mitochondrial genome (Mattiucci et al., 2014a; Nadler et al., 2005). However, primers for the mitochondrial cytochrome oxidase I gene (mtDNA *cox1*) have been also developed (Cross et al., 2007; Kijewska et al., 2009) and used in supporting data gathered from other nuclear and mitochondrial markers in the identification of some species of *Anisakis* (Table 1).

Mitochondrial genes, due to their general lack of recombination, relatively rapid rate of substitution coupled with the smaller effective population size, and maternal inheritance, provided data to be used in both phylogenetic (Cavallero et al., 2011; Mattiucci et al., 2014a; Nadler et al., 2005; Valentini et al., 2006) and phylogeographic studies (Baldwin et al., 2011; Blažeković et al., 2015; Cross et al., 2007; Kijewska et al., 2009) of this group of parasites.

Mitochondrial DNA genes were very useful for supporting the hypothesis for the existence of cryptic species within nominal species and to establish phylogenetic relationships among closely related taxa of *Anisakis*. In addition, combined analysis of all sequence data inferred from different mtDNA (*cox2* and *rrnS)* and nuclear ribosomal DNA loci (ITS region) warrants delimitation of *Anisakis* species, and it provides the best estimate of the clade support for delimitation and relationships among the species. Indeed, reciprocal monophyly recovered from multiple independent (not simply more genes) loci analysis, was more strongly supported with respect to the inference from a single locus (Mattiucci et al., 2014a) (see also Fig. 2).

Another mitochondrial marker inferred from RFLP of *HinfI* of mtDNA *cox1* locus was provided by Umehara et al. (2006): the restriction fragments obtained at that locus were diagnostic between *A. simplex* (s.s.) and *A. pegreffii* (Umehara et al., 2006).

Mitochondrial DNA genes proved very powerful not only in phylogenetic analysis of *Anisakis* spp. as operational taxonomic units, but also as applied for elucidating population genetic structure at the intraspecific level. For instance, mitochondrial mtDNA *cox2* haplotype distribution and analysis of molecular variance revealed panmictic distribution of their haplotype in three species of *Anisakis*, i.e., *A. simplex* (s.s.), *A. pegreffii*, and *A. berlandi*,

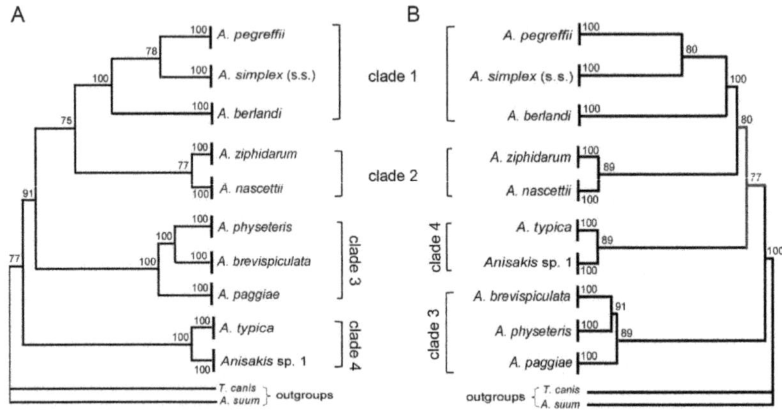

Fig. 2 (A) Maximum parsimony (MP) bootstrap consensus tree, inferred by PAUP*4.0 (bootstrap method with heuristic search) (Swofford, 2003) based on the combined mitochondrial (*cox2* and *rrnS* genes) and nuclear (ITS rDNA) sequence of *Anisakis* spp. (B) MP bootstrap consensus tree inferred by PAUP*4.0 (bootstrap method with heuristic search) (Swofford, 2003) based on the combined mitochondrial *cox2* and *rrnS* genes. The sequence datasets were first determined to be combinable by using the partition homogeneity test (Farris et al., 1994), as executed in PAUP* (Swofford, 2003). Bootstrap values (≥70), reported at the nodes, were obtained over 1000 pseudoreplicates. *Toxocara canis* and *Ascaris suum* were used as outgroups. *Data from Mattiucci, S., Cipriani, P., Webb, S.C., Paoletti, M., Marcer, F., Bellisario, B., Gibson, D.I., Nascetti, G. 2014a. Genetic and morphological approaches distinguishing the three sibling species of the* Anisakis simplex *species complex, with a species designation as* Anisakis berlandi *n. sp. for* A. simplex *sp. C (Nematoda: Anisakidae). J. Parasitol. 15, 12–15; unpublished.*

identified from *Sardinops sagax* (Baldwin et al., 2011) along the Pacific American coast. Similarly, haplotype distributions of mtDNA cox2 were studied in definitive and intermediate hosts of *A. pegreffii* from the Adriatic Sea (Mladineo et al., 2012), as well as in *A. simplex* (s.s.) from *Merluccius merluccius* from NE Atlantic waters (Ceballos-Mendiola et al., 2010). More recently, differential distribution of mtDNA *cox2* haplotype distributions and genetic differentiation, at intraspecific level, has been studied to detect genetic substructuring in populations of *Anisakis* spp. collected in various fish populations from different fishing grounds (Baldwin et al., 2011; Klapper et al., 2015). A genetic substructuring at the mtDNA *cox2* sequences analysis of *A. simplex* (s.s.) was obtained from herring in different areas, with the population from the Norwegian Sea being the most differentiated one, and with North Sea and Baltic Sea populations being most similar. The population genetic structure of *A. simplex* (s.s.) was in accordance with the herring population genetic structure throughout the host's geographical range in the NE Atlantic. Results suggest that mtDNA *cox2* is a suitable genetic marker for

A. simplex (s.s.) population genetic structure analysis and a valuable tool to elucidate the herring stock structure in the NE Atlantic Ocean (Mattiucci et al., 2017b).

2.2 What Types of Markers Are Used in the Study of Hybridization and Introgression Between Cryptic Species of *Anisakis*?

'... disentangling the genetic architecture of hybridization and/or introgression requires a multi-locus approach ...'

It has been frequently reported that sympatric populations of closely related parasite species, including parasitic nematodes, might interbreed (Agatsuma et al., 2000; Anderson, 2001; Criscione et al., 2007; Detwiler and Criscione, 2010; Dunams-Morel et al., 2012; Steinauer et al., 2008). Hybridization, through interspecific crossing between closely related species, could have major evolutionary consequences for species and populations, by either promoting or preventing divergence, depending on the viability and reproductive abilities of the hybrids in producing backcrosses with parental species. Adaptive traits could be also acquired through hybridization, resulting in increased or lower fitness in hybrid individuals (Criscione et al., 2007; Gilabert and Wasmuth, 2013). Hybridization and introgression could lead to some phenotypic changes of pathogens, including parasites, such as the invasion of new hosts, new geographical areas, and sites of infection (Criscione et al., 2007; Detwiler and Criscione, 2010).

Therefore, the accurate detection and characterization of hybridization are important both in basic and applied biology, and not only in free-living species (Canestrelli et al., 2016), but also in parasitic animals. Again, molecular/genetic markers, especially the nuclear ones, are making these analyses accessible via a large number of tested individuals. However, hybridization, introgression, and the retention of ancestral polymorphism in closely related parasitic taxa, could generate patterns of genetic variation, which complicate their disclosure. In this unclear situation, choosing inappropriate genetic/molecular markers, and/or inferring the detection of hybridization from the analysis of a single locus of a genetic markers could lead to misleading results.

The two cryptic species, *A. simplex* (s.s.) and *A. pegreffii*, were found to differ in their geographical distribution (Mattiucci et al., 1997, 2014a; Nascetti et al., 1986) (see also Section 3; Fig. 3). However, their ranges overlap in some oceanic basins, such as the NE Atlantic Ocean (Spanish-Portuguese coast) and the

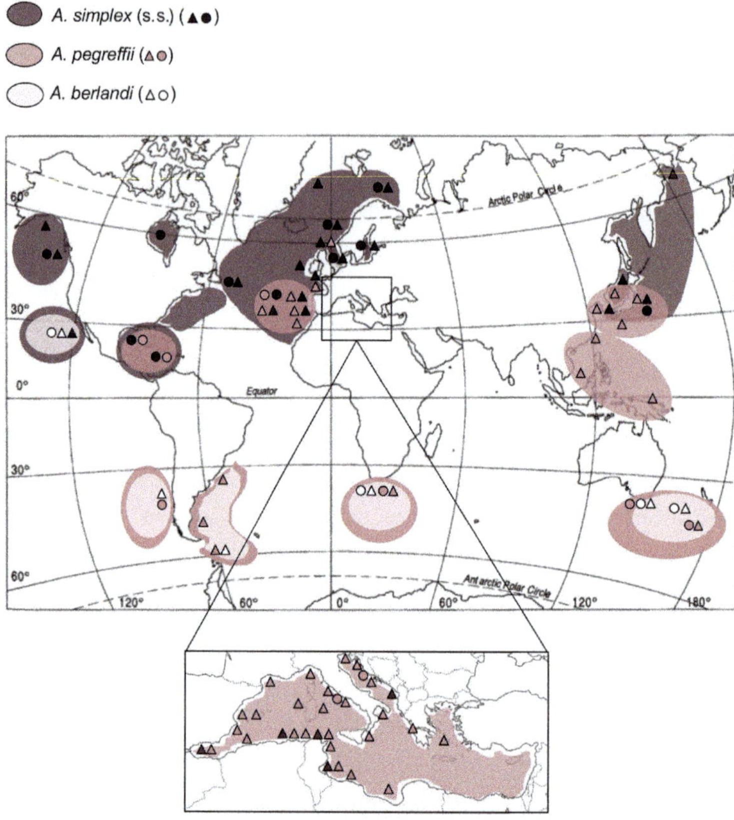

Fig. 3 World map showing range of distribution of *Anisakis simplex* (s.s.), *A. pegreffii*, and *A. berlandi*. The *symbols* are related to the sampling localities of their definitive (*circle*) and intermediate/paratenic (*triangle*) hosts, so far identified by molecular markers.

Western Pacific Ocean (East China Sea and Sea of Japan) (Abollo et al., 2001; Chou et al., 2011; Lee et al., 2009; Mattiucci et al., 1997, 2014a; Mattiucci and Nascetti, 2008; Pontes et al., 2005; Quiazon et al., 2011; Suzuki et al., 2010). In those areas, coinfection by *Anisakis* species pairs can occur in both pelagic and demersal fish hosts, as well as in oceanic dolphin species, and whales (Abollo et al., 2001; Farjallah et al., 2008b; Mattiucci et al., 1997, 2004, 2014a; Mattiucci and Nascetti, 2008; Pontes et al., 2005; Umehara et al., 2006).

The occurrence of specimens, sometimes indicated as 'putative hybrids', 'recombinant genotypes' or just 'hybrids', between the two species *A. simplex* (s.s.) and *A. pegreffii* has been largely reported by using PCR-RFLPs of the ITS region of rDNA and sequence analysis of the same gene (Abollo et al., 2003). These specimens were either larval stages collected from various fish species, or adults in cetaceans, from both allopatric (Cavallero et al., 2012, 2014; Chaligiannis et al., 2012; Costa et al., 2016; Farjallah et al., 2008a; Meloni et al., 2011; Pekmezci et al., 2014) and sympatric areas of the two species (Abollo et al., 2003; Chou et al., 2011; Du et al., 2010; Kong et al., 2015; Lee et al., 2009; Martín-Sánchez et al., 2005; Molina-Fernández et al., 2015; Quiazon et al., 2011; Suzuki et al., 2010; Umehara et al., 2006). The above studies used a single nuclear marker (the ITS region of rDNA), and consequently it remains unknown whether or not the observed shared polymorphism between the two taxa was caused by incomplete lineage sorting, or historical introgression.

A Bayesian clustering approach (STRUCTURE) of genotypes obtained from a multimarker nuclear diagnostic loci analysis of hundreds specimens of *Anisakis* spp. collected from allopatric and sympatric areas of the two species (*A. pegreffii* and *A. simplex* (s.s.)) was recently carried out (Mattiucci et al., 2016). The analysis allowed to recognize that current hybridization (F1 hybrid generation) occurs in sympatric areas (Iberian Atlantic coast, an Alboran Sea). Those hybrid nematodes were indeed heterozygous for all of the diagnostic allozyme loci, and for the diagnostic nucleotide positions observed in the EF1 α-1 nDNA gene (Mattiucci et al., 2016). However, no introgressive hybridization was detected in the study, supported by evidence that no backcrossing with the two parental species—which represents subsequent generations of hybrids—was observed. Nuclear loci chosen for analysis carried out by STRUCTURE in this multilocus approach were allozyme loci (biparentally inherited codominant markers) exhibiting fixed allelic differences between the two species, and sequences analysis of the EF1 α-1 nDNA, also showing fixed nucleotide sites in the two parental species. In the same paper by Mattiucci et al. (2016), it has been demonstrated that if the identification of the same specimens had been inferred by means of a single molecular marker—i.e., the analysis of ITS region of rDNA—a larger number of 'hybrids' would be misidentified not only in sympatric, but also in allopatric areas (including strictly allopatric areas, i.e. New Zealand waters and the North Sea).

Those findings suggest that, even if current hybridization could occur between *A. pegreffii* and *A. simplex* (s.s.) in some sympatric areas (i.e. the NE Atlantic waters off the Spanish-Portuguese coast and the Alboran Sea), the resulting F1 offspring appears to have a reduced fitness. It is likely that either the F1 offspring is infertile, even if it reaches the adult stage, or that selection disadvantages the survival of hybrid offspring, or that some post-mating reproductive isolating mechanisms do not permit backcrossing of F1 hybrids with the parental types. As a consequence, the two species maintain their identity. The study also demonstrated that current hybridization outside the sympatric areas of the two cryptic species does not occur. Indeed, in the eastern (Adriatic Sea) and western (Tyrrhenian Sea) Mediterranean, as well as in strict allopatric areas (New Zealand coast), no F1 hybrids were detected by the multilocus approach (Mattiucci et al., 2016).

The frequency of current hybridization in sympatric areas appears to be about 5% in Spanish-Atlantic waters (Mattiucci et al., 2016), while it was around 3% in the Japanese Sea area (Umehara et al., 2006). In those sympatric geographical areas the two parasite species can be often found in mixed infections in the same individual cetacean host (Mattiucci et al. 2014; Umehara et al., 2006). In those condition, considering that the number of mature females of these anisakids is generally higher than that of mature males in their definitive hosts (Mattiucci et al., 2016), it could be that when the intensity of the infection by one of these two species is low, mating events between the two species could occur, resulting in a first generation (F1) of larval hybrids. Mattiucci et al. (2016) and Cipriani et al. (2015) reported the occurrence of F1 hybrids (≈5%) identified from the sympatric areas studied. From data obtained by the sequence analysis of the mtDNA *cox2* region on these hybrid specimens, 10 out of 11 resulted from the mating of female *A. simplex* (s.s.) with male of *A. pegreffii*, causing a maternal inheritance of the mtDNA of *A. simplex* (s.s.), while one exhibited a *A. pegreffii* matrilineage (Mattiucci et al., 2016). Similarly, in the Japanese Sea, the analysis of the mtDNA *cox1* analysis has shown that, among the four hybrids detected, three had a maternal inheritance of the mtDNA of *A. simplex* (s.s.), and one specimen showed a *A. pegreffii* matrilineage. On the other hand, it has been suggested that there is a tendency for hybridization to take place preferentially between parental species differing greatly in abundance. This hypothesis suggests that the absence of conspecific pairing partners and mating stimuli for females of rarer species may be important factors in increasing the likelihood of interspecific current hybridization (Avise and Saunders, 1984). Indeed, in those sympatric areas, the prevalence and abundance levels of *A. pegreffii* in both fish and cetacean hosts have been found at lower levels with respect to those observed for the species *A. simplex* (s.s.) (Mattiucci and Nascetti, 2008; Umehara et al., 2006) (see also Section 3). Furthermore, a prerequisite for current hybridization is represented by the occurrence in sympatry of the adult mature specimens of the two parental species, as well as the ecological conditions favourable to the development of first parasite stages, otherwise this phenomenon could not physically takes place.

Some recombinant genotypes at the ITS rDNA locus between *A. simplex* (s.s.) and *A. pegreffii* were found in allopatric area of the Adriatic Sea (Mladineo et al., 2017a, b). The same authors argued that the nucleo-mitochondrial mosaicism found by sequence analysis of the ITS region of rDNA and the mtDNA *cox2* loci from the Adriatic Sea, could be the product of ancestral polymorphism and consequent remote genetic introgression (Mladineo et al., 2017b).

So far, the finding of F1 hybrids between *A. simplex* (s.s.) and *A. pegreffii* appears not to be associated with a particular epidemiological or ecological pathway. All of the F1 larvae identified in Mattiucci et al. (2016) were collected from nine specimens of a demersal fish, the European hake *M. merluccius*, caught in waters where the parasites occur in sympatry.

Further research, using a larger number of correctly identified hybrid individuals, might highlight different scenarios in some phenotypic traits, such as, a differential capacity of invading host tissues between the F1 hybrids and the parental species. In other helminths, such as a species of the digenean genus *Microphallus*, the infectivity of F1 hybrid parasites was lower with respect to that exhibited by the parental species (Dybdahl et al., 2008). The authors suggested that the lower fitness of the hybrids was due to outbreeding depression.

Finally, the study of hybridization and introgression has taken advantage of a Bayesian clustering method to assess the identification of pure samples of parental species and samples of mixed ancestry. Such tools, provided by the STRUCTURE and NEWHYBRIDS software, are advantageous because the analyses are able to indicate both contemporary hybridization and events of past hybridization, detecting hybrid categories going back to two or three generations (Gilabert and Wasmuth, 2013).

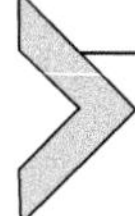

3. HOW MANY VALID SPECIES ARE IN THE GENUS *ANISAKIS*?

'... 47 years after the Davey's revision (Davey, 1971)'

To date, nine species belonging to the genus *Anisakis* have been identified worldwide. All those species were characterized by distinct diagnostic genetic markers, possess distinct gene pools, and are reproductively isolated. The existence of nine species as distinct phylogenetic units was also demonstrated by various concatenated phylogenetic analyses, as inferred from nuclear and mitochondrial genes (Cavallero et al., 2011; Mattiucci et al., 2014a; Mattiucci and Nascetti, 2008; Valentini et al., 2006). According to these phylogenetic analyses, four distinct clades within the genus *Anisakis* clearly exist (Fig. 2).

Below we review the *Anisakis* species, which are so far reported worldwide. Sampling localities and intermediate/paratenic and definitive hosts for each species are given in Tables 2–5.

Table 2 Definitive and Intermediate/Paratenic Hosts so far Recognized, by Molecular/Genetic Markers, of *Anisakis simplex* (s.s.), *A. pegreffii*, and *A. berlandi*

		***A. simplex* (s.s.)**	***A. pegreffii***	***A. berlandi***
Definitive Hosts				
Cetaceans				
Delphinoidea				
Delphinidae	*Cephalorhynchus hectori*	—	NZ	—
	Delphinus delphis	IC, NO	AU, IC	—
	Globicephala melas	IC, NW	CHI, IC, NZ	CHI, NZ, SA
	Globicephala macrorhynchus	CS, GM	—	—
	Grampus griseus	—	EM, ZN	—
	Lagenorhynchus albirostris	NW, BS, NO	—	—
	Lagenodelphis hosei	—	CS, GM	—
	Lissodelphis borealis	—	—	CAL
	Orcinus orca	NEP	—	—
	Pseudorca crassidens	CAN	—	CAN
Monodontidae	*Delphinapterus leucas*	HB, NWA	—	—
Phocoenidae	*Phocoena phocoena*	BA, BS, CAN, IC, NO, NW	IC	—
	Stenella coeruleoalba	EM, IC	EM, IC	—
	Steno bredanensis	CS, GM	—	—
	Tursiops truncatus	EM, IC	AU, CM, EM, IC, SA	—

Mysticeti				
Balenopteridae	*Balaenoptera acutorostrata*	BS, NW, WNPO	—	—
Neobalenidae	*Caperea marginata*	—	SA	—
Physeteroidea				
Kogiidae	*Kogia breviceps*	CS, GM, IC, NO	—	—
Physeteridae	*Physeter macrocephalus*	NO	—	—
Ziphiidae				
Ziphiidae	*Mesoplodon layardii*	—	—	SAU
	Ziphius cavirostris	—	EM	—
Pinnipeds				
Phocidae	*Mirounga angustirostris*	—	—	CAL
	Mirounga leonina	—	—	SSI
Intermediate/Paratenic Hosts				
Cephalopods				
Ommastrephidae	*Illex coindetii*	IC	WM	—
	Nototodarus sloanii	—	NZ	—
	Todaropsis angolenis	—	SA	SA
	Todaropsis eblanae	IC	IC, SA	—
	Todarodes sagittatus	IC	NAM	—
Sepiidae	*Sepia officinalis*	IC	—	—

Continued

Table 2 Definitive and Intermediate/Paratenic Hosts so far Recognized, by Molecular/Genetic Markers, of *Anisakis simplex* (s.s.), *A. pegreffii*, and *A. berlandi*—cont'd

		***A. simplex* (s.s.)**	***A. pegreffii***	***A. berlandi***
Fish				
Anoplopomatidae	*Anoplopoma fimbria*	NEP	—	CAL
Belonidae	*Belone belone*	IC	IC	—
Bothidae	*Arnoglossus imperialis*	—	PC	—
	Arnoglossus laterna	PC	—	—
Bramidae	*Brama brama*	—	SA	—
Carangidae	*Decapterus maruadsi*	—	YS	—
	Pseudocaranx dentex	—	AU	—
	Seriola dumerili	—	CHS	—
	Seriola lalandi	—	YS	—
	Trachurus capensis	—	SA	—
	Trachurus mediterraneus	—	CM, WM	—
	Trachurus picturatus	AZ	AZ, WM	—
	Trachurus trachurus	BS, EM, IC, NO, NW, WM	CM, EM, IC, MA, NAM, NZ, WM	—
Centracanthidae	*Spicara smaris*	—	EM	—
Citharidae	*Citharus linguatula*	PC	PC	—
Clupeidae	*Alosa alosa*	IC	—	—
	Alosa fallax	IC	IC	—
	Clupanodon punctatus	—	YS	—

	Clupea harengus	BA, BS, EC, NEP, NO, NW	—	—
	Clupea pallasi	—	YS	—
	Sardina pilchardus	—	EM, WM	—
	Sardinops sagax	CAL, CAN	—	—
	Sardinella zunasi	—	YS	—
Congridae	*Conger myriaster*	—	CHS, JA, YS	—
	Conger conger	IC	CM	—
Coryphaenidae	*Coryphaena hippurus*	EM	YS	—
Cottidae	*Myoxocephalus scorpius*	GS	—	—
Dussumieriidae	*Etrumeus whiteheadi*	—	SA	SA
Emmelicththydae	*Emmelicththys nitidus nitidus*	—	SA	—
Engraulidae	*Engraulis encrasicolus*	—	CM, EM, WM	—
	Engraulis japonicus	—	YS	—
Exocoetidae	*Cheilopogon agoo*	—	YS	—
Gadidae	*Boreogadus saida*	IC	—	—
	Gadus chalcogramma	BE, NEP, JA	JA	—
	Gadus macrocephalus	JA	CHS, YS	—
	Gadus morhua	BA, BS, EC, NO	—	—
	Melanogrammus aeglefinus	BS, NO	—	—

Continued

Table 2 Definitive and Intermediate/Paratenic Hosts so far Recognized, by Molecular/Genetic Markers, of *Anisakis simplex* (s.s.), *A. pegreffii*, and *A. berlandi*—cont'd

		***A. simplex* (s.s.)**	***A. pegreffii***	***A. berlandi***
	Micromesistius poutassou	AS, BS, IC, NW, PC	AS, CM, EM, IC, NAM, WM	—
	Pollachius virens	NO	—	—
	Trisopterus luscus	IC	—	—
Gempylidae	*Thyrsites atun*	—	—	SA
Hexagrammidae	*Hexagrammos agrammus*	—	YS	—
	Hexagrammos otakii	—	YS	—
	Pleurogrammus azonus	NO, JA	—	—
Lateolabracidae	*Lateolabrax japonicus*	—	YS	—
Lotidae	*Molva dypterygia*	IC	—	—
	Brosme brosme	BS, NW	—	—
Lophiidae	*Lophius budegassa*	GSB, PC	PC	—
	Lophius litulon	—	YS	—
	Lophius piscatorius	GSB, IC, PC	EM, NAM, PC, WM	—
	Lophius vomerinus	—	SA	—
Macrouridae	*Macrourus berglax*	GS	—	—
Merluccidae	*Macruronus novazelandiae*	—	—	NZ
	Merluccius capensis	—	SA	—

	Merluccius hubbsi	—	FA	—
	Merluccius merluccius	AS, GSB, IC, NAM, NW, NWA, PC	AS, CM, EM, NAM, NW, MA, PC, WM	—
Myctophidae	*Electrona carlsbergi*	—	—	SSI
	Gymnoscopelus nicholsi	—	SSI	SSI
	Myctophum punctatum	CA	CA	—
Monacanthidae	*Thamnaconus modestus*	—	YS	—
Moridae	*Pseudophycis bachus*	—	NZ	NZ
Moronidae	*Dicentrarchus labrax*	IC	—	—
Muraenesocidae	*Muraenesox cinereus*	—	YS	—
Muraenidae	*Muraena helena*	—	NAM	—
Ophidiidae	*Genypterus capensis*	—	SA	—
Oreosomatidae	*Allocyttus niger*	—	—	NZ
	Pseudocyttus maculatus	—	—	CHI, NZ
Osmeridae	*Hypomesus japonicus*	JA	—	—
Paralichthyidae	*Pseudorhombus cinnamoneus*	—	YS	—
Pholidae	*Pholis nebulosa*	—	YS	—
Phycidae	*Phycis blennoides*	—	NAM	—
	Pyicis phycis	—	NAM	—
Pinguipedidae	*Parapercis colias*	—	NZ	NZ
Platycephalidae	*Cociella crocodilus*	—	YS	—
	Platycephalus richardsoni	—	AU	—

Continued

Table 2 Definitive and Intermediate/Paratenic Hosts so far Recognized, by Molecular/Genetic Markers, of *Anisakis simplex* (s.s.), *A. pegreffii*, and *A. berlandi*—cont'd

		***A. simplex* (s.s.)**	***A. pegreffii***	***A. berlandi***
Pleuronectidae	*Cleisthenes herzensteini*	—	CHS, YS	—
	Hippoglossoides dubius	—	JA	—
	Hippoglossus hippoglossus	BE	—	—
	Pleuronectes platessa	NO, EC	—	—
	Pseudopleuronectes yokohamae	—	YS	—
Pomacentridae	*Lutjanus erythropterus*	—	YS	—
Rachycentridae	*Rachycentron canadum*	—	YS	—
Salmonidae	*Oncorhynchus gorbuscha*	AK	—	—
	Oncorhynchus keta	AK, JA	—	—
	Oncorhynchus nerka	AK	—	—
	Salmo salar	NEA, NWA	—	—
Schophtalmidae	*Lepidorhombus boscii*	GSB, IC	IC, NAM	—
Sciaenidae	*Collichthys niveatus*	—	YS	—
	Larimichthys polyactis	—	CHS, YS	—
	Nibea albiflora	—	YS	—
	Pennahia argentata	—	YS	—
Scomberesocidae	*Scomberesox saurus*	NWA	—	—

Scombridae	*Auxis thazard*	—	YS	—
	Euthynnus alletteratus	EM	—	—
	Scomber australasicus	TW, JA	JA	—
	Scomber colias	AS	—	—
	Scomber japonicus	AZ, JA	AS, AZ, CHS, EM, MD, NAM, JA, WM	—
	Scomberomorus niphonius	—	CHS, YS	—
	Scomber scombrus	AS, BS, EC, EM, FIC, IC, NAM, NO, NW	AS, CM, EM, IC, NAM, WM	—
	Thunnus thynnus	EM, JA	BR, CM, NAM	—
Scorpaenidae	*Hoplosebastes armatus*	—	YS	—
	Scorpaena scrofa	IC	IC	—
Sebastidae	*Helicolenus dactylopterus*	—	CM, SA	—
	Sebastolobus alascanus	—	—	CAL
	Sebastodes fuscescens	—	YS	—
Sillaginidae	*Sillago sihama*	—	YS	—
Soleidae	*Dicologlossa cuneata*	—	PC	—
	Solea senegalensis	PC	—	—
Sparidae	*Diplodus annularis*	—	EM	—
	Pagellus bogaraveo	PC	PC	—
	Spondyliosoma cantharus	IC	—	—

Continued

Table 2 Definitive and Intermediate/Paratenic Hosts so far Recognized, by Molecular/Genetic Markers, of *Anisakis simplex* (s.s.), *A. pegreffii*, and *A. berlandi*—cont'd

		***A. simplex* (s.s.)**	***A. pegreffii***	***A. berlandi***
Sternoptychidae	*Maurolicus muelleri*	BS, NO, NW	—	—
Stromateidae	*Pampus argenteus*	—	YS	—
Synanceiidae	*Inimicus japonicus*	—	YS	—
Synodontidae	*Saurida elongata*	—	YS	—
Tetraodontidae	*Takifugu niphobles*	—	YS	—
	Takifugu poecilonotus	—	JA	—
Trachichthyidae	*Hoplostethus atlanticus*	—	—	CHI
	Hoplostethus mediterraneus	—	CM	—
Trachinidae	*Echiichtys vipera*	—	NAM	—
	Trachinus draco	—	NAM	—
Trichiuridae	*Aphanopus carbo*	—	MD	—
	Lepidopus caudatus	—	CM, SA, WM	—
	Trichiurus lepturus	CHS	CHS, JA, NAM, YS	—
Trichodontidae	*Arctoscopus japonicus*	—	JA	—
Triglidae	*Chelidonichthys kumu*	—	YS	—
	Eutrigla gurnardus	IC, NO	—	—
	Mullus barbatus	AS, EM	—	—
	Mullus surmuletus	—	WM	—
Xiphiidae	*Xiphias gladius*	—	CM, NAM	—

Zeidae	*Zeus faber*	—	AS, WM	—
	Zeus japonicus	—	YS	—
Zoarcidae	*Zoarces elongatus*	—	YS	—

Sampling locality codes: AS: Aegean Sea; AK: Alaska coast; AU: Australian coast; AZ: Azores Islands; BA: Baltic Sea; BE: Bering Sea; BR: Brazilian Atlantic coast; BS: Barents Sea; CA: Central Atlantic Ocean; CAL: Californian coast, CAN: Canadian coast; CHI: Chilean coast; CHS: China Sea; CM: Central Mediterranean Sea; CS: Caribbean Sea; EC: English Channel; EM: East Mediterranean Sea; FA: Falkland Islands; FIC: Faroe Islands; GM: Gulf of Mexico; GS: Greenland Sea; GSB: Great Sole Bank; HB: Hudson Bay; IC: Iberian Atlantic coast; JA: Japan Sea; MA: Mauritanian coast; MD: Madeira Island; NAM: North African Mediterranean Sea; NEA: North-East Atlantic; NEP: North-East Pacific; NO: North Sea; NW: Norwegian Sea; NWA: North West Atlantic Ocean; NZ: New Zealand coast; PC: Portuguese coast; SA: South African coast; SAU: South East Australian coast; SSI: Southern Shetland coast; TW: Taiwanese coast; WM: West Mediterranean Sea; WNPO: Western North Pacific Ocean; YS: Yellow Sea. Host families are given by alphabetical order.

Data from Mattiucci, S., Nascetti, G., 2008. Advances and trends in the molecular systematics of *Anisakis nematodes*, with implications for their evolutionary ecology and host-parasite co-evolutionary processes. Adv. Parasitol. 66, 47–148 and references therein; Colón-Llavina, M.M., Mignucci-Giannoni, A.A., Mattiucci, S., Paoletti, M., Nascetti, G., Williams, E.H., 2009. Additional records of metazoan parasites from Caribbean marine mammals, including genetically identified anisakid nematodes. Parasitol. Res. 105, 1239; Lee, M.H., Cheon, D., Choi, C. 2009. Molecular genotyping of *Anisakis* species from Korean sea fish by polymerase chain reaction–restriction fragment length polymorphism (PCR-RFLP). Food Control 20, 623–626; Suzuki, J., Murata, R., Hosaka, M., Araki, J. 2010. Risk factors for human Anisakis infection and association between the geographic origins of Scomber japonicus and anisakid nematodes. Int. J. Food Microbiol. 137, 88–93; Du, C., Zhang, L., Shi, M., Ming, Z., Hu, M., Gasser, R.B. 2010. Elucidating the identity of *Anisakis* larvae from a broad range of marine fishes from the Yellow Sea, China, using a combined electrophoretic-sequencing approach. *Electrophoresis* 31, 654–658; Klimpel, S., Busch, M.W., Kuhn, T., Rohde, A., Palm, H.W. 2010. The *Anisakis simplex* complex off the South Shetland Islands (Antarctica): endemic populations versus introduction through migratory hosts. Mar. Ecol. Prog. Ser. 403, 1–11; Murphy, T.M., Berzano, M., O'keeffe, S.M., Cotter, D.M., McEvoy, S.E., Thomas, K.A., PòMaoiléidigh, N., Whelan, K.F. 2010. Anisakid larvae in Atlantic salmon (*Salmo salar* L.) grilse and post-smolts: molecular identification and histopathology. J. Parasitol. 96, 77–82; Baldwin, R.E., Rew, M.B., Johansson, M.L., Banks, M.A., Jacobson, K.C. 2011. Population structure of three species of *Anisakis* nematodes recovered from Pacific sardines (*Sardinops sagax*) distributed throughout the California current system. J. Parasitol. 97, 545–554; Cavallero, S., Nadler, S.A., Paggi, L., Barros, N.B., D'Amelio, S. 2011. Molecular characterization and phylogeny of anisakid nematodes from cetaceans from southeastern Atlantic coasts of USA, Gulf of Mexico and Caribbean Sea. Parasitol. Res. 108, 781–792; Chou, Y.Y., Wang, C.S., Chen, H.G., Chen, H.Y., Chen, S.N., Shih, H.H. 2011. Parasitism between *Anisakis simplex* (Nematoda: Anisakidae) third-stage larvae and the spotted mackerel *Scomber australasicus* with regard to the application of stock identification. Vet. Parasitol. 177, 324–331; Quiazon, K.M., Yoshinaga, T., Ogawa, K., 2011. Distribution of *Anisakis* species larvae from fishes of the Japanese waters. Parasitol. Int. 60, 223–226; Karl, H., Baumann, F., Ostermeyer, U., Kuhn, T., Klimpel, S. 2011. *Anisakis simplex* (s. s.) larvae in wild Alaska salmon: no indication of post-mortem migration from viscera into flesh. Dis. Aquat. Org. 94, 201–209; Klimpel, S., Kuhn, T., Busch, M.W., Karl, H., Palm, H.W. 2011. Deep-water life cycle of *Anisakis paggiae* (Nematoda: Anisakidae) in the Irminger Sea indicates kogiids distribution in the north Atlantic waters. Polar Biol. 34, 899–906; Meloni, M., Angelucci, G., Merella, P., Siddi, F., Deiana, C., Orrù, G., Salati, F. 2011. Molecular characterization of *Anisakis* larvae from fish caught off Sardinia. J. Parasitol. 97, 908–914; Chaligiannis, I., Lalle, M., Pozio, E., Sotiraki, S. 2012. Anisakidae infection in fish of the Aegean Sea. Vet. Parasitol. 184, 362–366; Hermida, M., Mota, R., Pacheco, C.C., Santos, C.L. Cruz, C., Saraia, A., Tamagnini, P. 2012. Infection levels and diversity of anisakid nematodes in blackspot seabream, *Plagellus bogaraveo*, from Portuguese waters. Parasitol. Res. 110, 1919–1928; Shamsi, S., Gasser, R., Beveridge, I. 2012. Genetic characterisation and taxonomy of species of *Anisakis* (Nematoda: Anisakidae) parasitic in Australian marine mammals. Invertebr. Syst. 26, 204–212; Jabbar, A., Fong, R.W., Kok, K.X., Lopata, A.L., Gasser, R.B., Beveridge, I. 2013. Molecular characterization of anisakid nematode larvae from 13 species of fish from Western Australia. Int. J. Food Microbiol. 161, 247–253; Kuhn, T., Hailer, F., Palm, H.W., Klimpel, S. 2013. Global assessment of molecularly identified *Anisakis* Dujardin, 1845 (Nematoda: Anisakidae) in their teleost

Continued

Table 2 Definitive and Intermediate/Paratenic Hosts so far Recognized, by Molecular/Genetic Markers, of *Anisakis simplex* (s.s.), *A. pegreffii*, and *A. berlandi*—cont'd

intermediate hosts. Folia Parasitol. 60, 123–134; Bak, T.J., Jeon, C.H., Kim, J.H. 2014. Occurrence of anisakid nematode larvae in chub mackerel (*Scomber japonicus*) caught off Korea. Int. J. Food Microbiol. 191, 149–156; Levsen, A., Karl, H. 2014. *Anisakis simplex* (sl) in Grey gurnard (*Eutrigla gurnardus*) from the North Sea: food safety considerations in relation to fishing ground and distribution in the flesh. Food Control 36, 15–19; Mattiucci, S., Cipriani, P., Webb, S.C., Paoletti, M., Marcer, F., Bellisario, B., Gibson, D.I., Nascetti, G. 2014a. Genetic and morphological approaches distinguishing the three sibling species of the *Anisakis simplex* species complex, with a species designation as *Anisakis berlandi* n. sp. for *A. simplex* sp. C (Nematoda: Anisakidae). J. Parasitol. 15, 12–15; Mehrdana, F., Bahlool, Q.Z., Skov, J., M-arana, M.H., Sindberg, D., Mundeling, M., Overgaard, B. C., Korbut, R., Strøm, S.B., Kania, P.W., Buchmann, K. 2014. Occurrence of zoonotic nematodes *Pseudoterranova decipiens*, *Contracaecum osculatum* and *Anisakis simplex* in cod (*Gadus morhua*) from the Baltic Sea. Vet. Parasitol. 205, 581–587; Pekmezci, G.Z., Onuk, E.E., Bolukbas, C.S., Yardimci, B., Gurler, A.T., Acici, M., Umur, S. 2014. Molecular identification of *Anisakis* species (Nematoda: Anisakidae) from marine fishes collected in Turkish waters. Vet. Parasitol. 201, 82–94; Skrzypczak, M., Rokicki, J., Pawliczka, I., Najda, K., Dzido, J. 2014. Anisakids of seals found on the southern coast of Baltic Sea. Acta Parasitol. 59, 165–172; Bao, M., Roura, A., Mota, M., Pascual, S. 2015. Macroparasites of allis shad (*Alosa alosa*) and twaite shad (*Alosa fallax*) of the Western Iberian Peninsula Rivers: ecological, phylogenetic and zoonotic insight. Parasitol. Res. 114, 3721–3739; Blažeković, K., Pleić, I.L., Đuras, M., Gomerčić, T., Mladineo, I. 2015. Three *Anisakis* spp. isolated from toothed whales stranded along the eastern Adriatic Sea coast. Int. J. Parasitol. 45, 17–31; Chen, H.Y., Shih, H.H. 2015. Occurrence and prevalence of fish-borne Anisakis larvae in the spotted mackerel Scomber australasicus from Taiwanese waters. Acta Trop. 145, 61–67; Cho, J., Lim, H., Jung, B.K., Shin, E.H., Cha, J.Y. 2015. *Anisakis pegreffii* larvae in Sea Eels (*Astroconger myriaster*) from the South Sea, Republic of Korea. Korean J. Parasitol. 53, 49–353; Kong, Q., Fan, L., Zhang, J., Akao, N., Dong, K., Lou, D., Ding, J., Tong, Q., Zheng, B., Chen, R., Ohta, N., Lu, S. 2015. Molecular identification of *Anisakis* and *Hysterothylacium* larvae in marine fishes from the East China Sea and the Pacific coast of central Japan. Int. J. Food Microbiol. 199, 1–7; Münster, J., Klimpel, S., Fock, H.O., MacKenzie, K., Kuhn, T. 2015. Parasites as biological tags to track an ontogenetic shift in the feeding behaviour of *Gadus morhua* off West and East Greenland. Parasitol. Res. 114, 2723–273; Najda, K., Simard, M., Osewska, J., Dziekońska-Rynko, J., Dzido, J., Rokicki, J. 2015. Anisakidae in beluga whales *Delphinapterus leucas* from Hudson Bay and Hudson Strait. Dis. Aquat. Org. 115, 9–14; Costa, A., Cammilleri, G., Graci, S., Buscemi, M.D., Vazzana, M., Principato, D., Giangrosso, G., Ferrantelli, V. 2016. Survey on the presence of *A. simplex* (s.s.) and *A. pegreffii* hybrid forms in Central-Western Mediterranean Sea. Parasitol. Int. 65, 696–701; Gómez-Mateos, M., Valero, A., Morales-Yuste, M., Martín-Sánchez, J. 2016. Molecular epidemiology and risk factors for *Anisakis simplex* sl infection in blue whiting (*Micromesistius poutassou*) in a confluence zone of the Atlantic and Mediterranean: differences between *A. simplex* ss and *A. pegreffii*. Int. J. Food Microbiol. 232, 111–116; Picó-Durán, G., Pulleiro-Potel, L., Abollo, E., Pascual, S., Muñoz, P. 2016. Molecular identification of *Anisakis* and *Hysterothylacium* larvae in commercial cephalopods from the Spanish Mediterranean coast. Vet. Parasitol. 220, 47–53; Mladineo, I., Trumbić, Ž., Radonić, I., Vrbatović, A., Hrabar, J., Bušelić, I., 2017a. *Anisakis simplex* complex: ecological significance of recombinant genotypes in an allopatric area of the Adriatic Sea inferred by genome-derived simple sequence repeats. Int. J. Parasitol. 47, 215–223 doi:10.1016/j.ijpara.2016.11.003; Li, L., Zhao, J.Y., Chen, H.X., Ju, H.D., An, M., Xu, Z., Zhang, L.P. 2017. Survey for the presence of ascaridoid larvae in the cinnamon flounder *Pseudorhombus cinnamoneus* (Temminck & Schlegel) (Pleuronectiformes: Paralichthyidae). Int. J. Food Microbiol. 241, 108–116; Mattiucci, S., Paoletti, M., Cipriani, P., Webb, S.C., Timi, J.T., Nascetti, G. 2017a. Inventorying biodiversity of *Anisakid* nematodes from the Austral region: a hotspot of genetic diversity? In: Biodiversity and Evolution of Parasitic Life in the Southern Ocean, Springer International Publishing, pp. 109–140; and data not published.

Table 3 Definitive Hosts so far Recognized, by Molecular/Genetic Markers, of *Anisakis nascetti* and *A. ziphidarum*

		A. nascettii	***A. ziphidarum***
Definitive Hosts			
Cetaceans			
Physeteroidea			
	Kogia breviceps	—	NZ
Ziphiidae			
Ziphiidae	*Mesoplodon bowdoini*	NZ	NZ
	Mesoplodon densirostris	—	SA
	Mesoplodon europaeus	BR	CS, FL, GM
	Mesoplodon grayi	NZ, SA	SA
	Mesoplodon layardii	NZ	SA
	Mesoplodon mirus	NZ, SA	SA
	Mesoplodon sp.	—	CHI
	Ziphius cavirostris	—	CHI, CM, CS, GM, SA
Intermediate/Paratenic Hosts			
Cephalopods			
Onychoteuthidae	*Moroteuthis ingens*	MI, NZ	—
Fish			
Carangidae	*Trachurus trachurus*	AZ	—
Merluccidae	*Merluccius merluccius*	IC, MC	EM, IC, MA, MC
Oreosomatidae	*Allocyttus niger*	—	NZ
Scombridae	*Scomber japonicus*	MC	AZ, MC
	Scomber scombrus	MC	MC
Sparidae	*Pagellus bogaraveo*	—	PC
Trachichthyidae	*Hoplostethus cadenati*	—	MC
Trichiuridae	*Aphanopus carbo*	AB, AZ, MC	MD
	Lepidopus caudatus	—	AB
Xiphiidae	*Xiphias gladius*	—	CA

Continued

Table 3 Definitive Hosts so far Recognized, by Molecular/Genetic Markers, of *Anisakis nascetti* and *A. ziphidarum*—cont'd

Sampling locality codes: AB: Alboran Sea; AZ: Azores Islands; BR: Brazilian Atlantic coast; CA: Central Atlantic Ocean; CHI: Chilean coast; CM: Central Mediterranean Sea; CS: Caribbean Sea; EM: East Mediterranean Sea; FL: Floridian coast; GM: Gulf of Mexico; IC: Iberian Atlantic coast; MA: Mauritanian coast; MC: Morocco Atlantic coast; MD: Madeira Island; MI: Macquarie Island; NAA: North African Atlantic Ocean; NZ: New Zealand coast; PC: Philippine Coast; SA: South African coast; SEA: South East Atlantic Ocean. Host families are given by alphabetical order.

Data from Mattiucci, S., Nascetti, G., 2008. Advances and trends in the molecular systematics of *Anisakis nematodes*, with implications for their evolutionary ecology and host-parasite co-evolutionary processes. Adv. Parasitol. 66, 47–148 and references therein; Colón-Llavina, M.M., Mignucci-Giannoni, A.A., Mattiucci, S., Paoletti, M., Nascetti, G., Williams, E.H., 2009. Additional records of metazoan parasites from Caribbean marine mammals, including genetically identified anisakid nematodes. Parasitol. Res. 105, 1239; Kijewska, A., Dzido, J., Rokicki, J. 2009. Mitochondrial DNA of *Anisakis simplex* ss as a potential tool for differentiating populations. J. Parasitol. 95, 1364–1370; Mattiucci, S., Paoletti, M., Webb, S.C. 2009. *Anisakis nascettii* n. sp. (Nematoda: Anisakidae) from beaked whales of the southern hemisphere: morphological description, genetic relationships between congeners and ecological data. Syst. Parasitol. 74, 199–217; Cavallero, S., Nadler, S.A., Paggi, L., Barros, N.B., D'Amelio, S. 2011. Molecular characterization and phylogeny of anisakid nematodes from cetaceans from southeastern Atlantic coasts of USA, Gulf of Mexico and Caribbean Sea. Parasitol. Res. 108, 781–792; Kuhn, T., Hailer, F., Palm, H.W., Klimpel, S. 2013. Global assessment of molecularly identified *Anisakis* Dujardin, 1845 (Nematoda: Anisakidae) in their teleost intermediate hosts. Folia Parasitol. 60, 123–134; Quiazon, K.M., Santos, M.D., Yoshinaga, T. 2013. *Anisakis* species (Nematoda: Anisakidae) of Dwarf Sperm Whale *Kogia simus* (Owen, 1866) stranded off the Pacific coast of southern Philippine archipelago. Vet. Parasitol. 197, 221–230; Di Azevedo, M.I.N., Knoff, M., Carvalho, V.L., Mello, W.N., Torres, E.J.L., Gomes, D.C., Iñiguez, A.M. 2015. Morphological and genetic identification of *Anisakis paggiae* (Nematoda: Anisakidae) in dwarf sperm whale *Kogia sima* from Brazilian waters. Dis. Aquat. Org. 113, 103–111; Di Azevedo, M.I.N., Carvalho, V.L., Iñiguez, A.M. 2016. First record of the anisakid nematode *Anisakis nascettii* in the Gervais' beaked whale *Mesoplodon europaeus* from Brazil. J. Helminthol. 90, 48–53; Mattiucci, S., Acerra, V., Paoletti, M., Cipriani, P., Levsen, A., Webb, S.C., Canestrelli, D., Nascetti, G. 2016. No more time to stay 'single' in the detection of *Anisakis pegreffii*, *A. simplex* (s. s.) and hybridization events between them: a multi-marker nuclear genotyping approach. Parasitology 143, 998–1011; Mattiucci, data not published.

3.1 *Anisakis* Species Included in Clade 1

Clade 1 is formed by the so-called *A. simplex* (s.l.) complex; it includes *A. berlandi*, *A. pegreffii*, and *A. simplex* (s.s.) (Fig. 2), with the support of a bootstrap value of 100 at the MP inference, also indicating that *A. simplex* (s.s.) and *A. pegreffii* are sister taxa (Mattiucci et al., 2014a). This clade is supported by concatenated inference, obtained from both MP analysis based on combined mitochondrial sequences dataset and (Fig. 2A) and by combining both nuclear and mitochondrial sequence data (Mattiucci et al., 2014a). They gave identical topologies in supporting, the existence of the three species *A. simplex* (s.s.), *A. pegreffii*, and *A. berlandi* as distinct phylogenetic lineages.

3.1.1 A. simplex *(s.s.) (See Nascetti et al., 1986)*

A. simplex (s.s.) distribution ranges approximately from 35°N to the Arctic Seas, circumpolarly (Fig. 3). Actually, *A simplex* (s.s.) is the Arctic Boreal species of the

Table 4 Definitive and Intermediate/Paratenic Hosts so far Recognized, by Molecular/Genetic Markers, of *Anisakis brevispiculata*, *A. paggiae*, *A. physeteris*, and *Anisakis* sp. 2

		A. brevispiculata	***A. paggiae***	***A. physeteris***	***Anisakis* sp. 2**
Definitive Hosts					
Cetaceans					
Physeteroidea					
Physeteridae	*Physeter macrocephalus*	—	—	CS, CM, EM, GM, IC, NO	CA
Kogiidae	*Kogia breviceps*	CS, FL, GM, IC, PHC, SA	CS, FL, GM, PHC, SA	BR, CS, GM	—
	Kogia sima	CS, GM, PHC	CS, BR, GM, PHC, SA	CS, GM	—
Ziphiidae					
Ziphiidae	*Ziphius cavirostris*	—	—	EM	—
Intermediate/Paratenic Hosts					
Cephalopods					
Histioteuthidae	*Histioteuthis bonnellii*	—	—	EM	—
Ommastrephidae	*Dosidicus gigas*	—	—	EM	—
	Illex coindetii	—	—	WM	—
	Todarodes sagittatus	—	—	CM	—
Onychoteuthidae	*Ancistroteuthis lichtensteinii*	—	—	EM	—
Fish					
Anoplogastridae	*Anoplogaster cornuta*	—	IS	—	—
Berycidae	*Beryx splendens*	JA	JA	JA	—

Continued

Table 4 Definitive and Intermediate/Paratenic Hosts so far Recognized, by Molecular/Genetic Markers, of *Anisakis brevispiculata*, *A. paggiae*, *A. physeteris*, and *Anisakis* sp. 2—cont'd

		A. brevispiculata	***A. paggiae***	***A. physeteris***	***Anisakis* sp. 2**
Carangidae	*Trachurus trachurus*	—	—	CM, WM	—
Gadidae	*Gadus chalcogramma*		JA	—	—
	Micromesistius poutassou	—	—	WM	—
Lophiidae	*Lophius piscatorius*	—	—	WM	—
Merluccidae	*Merluccius merluccius*	MA	GC	CM, EM, IC, MA, NAM, WM	—
Phycidae	*Phycis phycis*	—	—	NAM	—
	Phycis blennoides	—	—	NAM	—
Scombridae	*Auxis thazard*	—	—	BR	—
	Scomber australasicus	TW	TW	TW	—
	Scomber japonicus	—	—	CHS, MD	—
	Scomber scombrus	—	—	NAM	—
Synaphobranchidae	*Synaphobranchus kaupii*	—	—	SA	—
Sparidae	*Pagellus bogaraveo*	—	—	PHC	—
Trichiuridae	*Aphanopus carbo*	AZ	AZ, MD	AZ	—
	Lepidopus caudatus	—	—	WM	—
Xiphiidae	*Xiphias gladius*	AZ, CSA, NWA, TEA	AZ, CSA, NEA, TEA	AZ, CM, CSA, EM, IC, TEA	TEA

Sampling locality codes: AZ: Azores Islands; BR: Brazilian Atlantic coast; CA: Central Atlantic Ocean; CEA: Central East Atlantic Ocean; CM: Central Mediterranean Sea; CS: Caribbean Sea; CSA: central South Atlantic Ocean; EM: East Mediterranean Sea; FL: Floridian coast; GC: Galician coast; GM: Gulf of Mexico; IC: Iberian Atlantic coast; IS: Irminger Sea; JA: Japan Sea; MA: Mauritanian coast; MD: Madeira Island; NAM: North African Mediterranean Sea; NWA: North West Atlantic Ocean; PHC: Philippine coast; SA: South African coast; SAU: South East Australian coast; TEA: Tropical Equator Atlantic Ocean; TW: Taiwanese coast; WM: West Mediterranean Sea. Host families are given by alphabetical order.

Data from Mattiucci, S., Nascetti, G., 2008. Advances and trends in the molecular systematics of *Anisakis nematodes*, with implications for their evolutionary ecology and host-parasite co-evolutionary processes. Adv. Parasitol. 66, 47–148 and references therein; Iñiguez, A.M., Santos, C.P., Vicente, A.C.P. 2009. Genetic characterization of *Anisakis typica* and *Anisakis physeteris* from marine mammals and fish from the Atlantic Ocean off Brazil. Vet. Parasitol. 165, 350–356; Colón-Llavina, M.M., Mignucci-Giannoni, A.A., Mattiucci, S., Paoletti, M., Nascetti, G., Williams, E.H., 2009. Additional records of metazoan parasites from Caribbean marine mammals, including genetically identified anisakid nematodes. Parasitol. Res. 105, 1239; Luque, J.L., Muniz-Pereira, L.C., Siciliano, S., Siqueira, L.R., Oliveira, M.S., Vieira, F.M. 2010. Checklist of helminth parasites of cetaceans from Brazil. Zootaxa 2548, 57–68; Cavallero, S., Nadler, S.A., Paggi, L., Barros, N.B., D'Amelio, S. 2011. Molecular characterization and phylogeny of anisakid nematodes from cetaceans from southeastern Atlantic coasts of USA, Gulf of Mexico and Caribbean Sea. Parasitol. Res. 108, 781–792; Garcia, A., Mattiucci, S., Damiano, S., Santos, M.N., Nascetti, G. 2011. Metazoan parasites of swordfish, *Xiphias gladius* (Pisces: Xiphiidae) from the Atlantic Ocean: implications for host stock identification. ICES J. Mar. Sci. 68, 175–182; Klimpel, S., Kuhn, T., Busch, M.W., Karl, H., Palm, H.W. 2011. Deep-water life cycle of *Anisakis paggiae* (Nematoda: Anisakidae) in the Irminger Sea indicates kogiids distribution in the north Atlantic waters. Polar Biol. 34, 899–906; Meloni, M., Angelucci, G., Merella, P., Siddi, R., Deiana, C., Orrù, G., Salati, F. 2011. Molecular characterization of *Anisakis* larvae from fish caught off Sardinia. J. Parasitol. 97, 908–914; Murata, R., Suzuki, J., Sadamasu, K., Kai, A. 2011. Morphological and molecular characterization of *Anisakis* larvae (Nematoda: Anisakidae) in *Beryx splendens* from Japanese waters. Parasitol. Int. 60, 193–198; Hermida, M., Mota, R., Pacheco, C.C., Santos, C.L. Cruz, C., Saraiva, A., Tamagnini, P. 2012. Infection levels and diversity of anisakid nematodes in blackspot seabream, *Plagellus bogaraveo*, from Portuguese waters. Parasitol. Res. 110, 1919–1928; Kuhn, T., Hailer, F., Palm, H.W., Klimpel, S. 2013. Global assessment of molecularly identified *Anisakis* Dujardin, 1845 (Nematoda: Anisakidae) in their teleost intermediate hosts. Folia Parasitol. 60, 123–134; Mattiucci, S., Garcia, A., Cipriani, P., Santos, M.N., Nascetti, G., Cimmaruta, R. 2014b. Metazoan parasite infection in the swordfish, *Xiphias gladius*, from the Mediterranean Sea and comparison with Atlantic populations: implications for its stock characterization. Parasite 21, 35; and data not published; Di Azevedo, M.I.N., Knoff, M., Carvalho, V.L., Mello, W.N., Torres, E.J.L., Gomes, D.C., Iñiguez, A.M. 2015. Morphological and genetic identification of *Anisakis paggiae* (Nematoda: Anisakidae) in dwarf sperm whale *Kogia sima* from Brazilian waters. Dis. Aquat. Org. 113, 103–111; Blažeković, K., Pleić, I.L., Đuras, M., Gomerčić, T., Mladineo, I. 2015. Three *Anisakis* spp. isolated from toothed whales stranded along the eastern Adriatic Sea coast. Int. J. Parasitol. 45, 17–31; Chen, H.Y., Shih, H.H. 2015. Occurrence and prevalence of fish-borne Anisakis larvae in the spotted mackerel Scomber australasicus from Taiwanese waters. Acta Trop. 145, 61–67; Picó-Durán, G., Pulleiro-Potel, L., Abollo, E., Pascual, S., Muñoz, P. 2016. Molecular identification of *Anisakis* and *Hysterothylacium* larvae in commercial cephalopods from the Spanish Mediterranean coast. Vet. Parasitol. 220, 47–53; Quiazon, K.M.A. 2016. *Anisakis* Dujardin, 1845 infection (Nematoda: Anisakidae) in Pygmy Sperm Whale *Kogia breviceps* Blainville, 1838 from west Pacific region off the coast of Philippine archipelago. Parasitol. Res. 115, 3663–3668.

Table 5 Definitive and Intermediate/Paratenic Hosts so far Recognized, by Molecular/Genetic Markers, of *Anisakis typica* and *Anisakis* sp. 1

		A. typica	***Anisakis* sp. 1**
Definitive Hosts			
Cetaceans			
Delphinoidea			
Delphinidae	*Globicephala melas*	CS	—
	Globicephala macrorhynchus	CS, FL, GM	—
	Lagenodelphis hosei	CS, GM	—
	Peponocephala electra	BR	—
	Sotalia fluviatilis	BR	—
	Sotalia guianensis	BR	—
	Stenella attenuata	CS, FL, GM	—
	Stenella coeruleoalba	CS, EM, GM	—
	Stenella frontalis	CS	—
	Stenella longirostris	CS, GM	—
	Steno bredanensis	BR, CS, GM	—
	Tursiops aduncus	IO	—
	Tursiops truncatus	CS, FL, GM	—
Physeteroidea			
Kogiidae	*Kogia breviceps*	CS, PHC	—
	Kogia sima	PHC	—
Intermediate/Paratenic Hosts			
Cephalopods			
Ommastrephidae	*Todarodes pacificus*	JA	—
Fish			
Caesionidae	*Caesio cuning*	—	INC
Carangidae	*Decapterus macarellus*	PNG	—
	Selar crumenophthalmus	CHS, HW, PNG, PY	—
	Trachurus picturatus	AZ, MD	—
	Trachurus trachurus	EM	—

Table 5 Definitive and Intermediate/Paratenic Hosts so far Recognized, by Molecular/Genetic Markers, of *Anisakis typica* and *Anisakis* sp. 1—cont'd

		A. typica	***Anisakis* sp. 1**
Carcharhinidae	*Carcharhinus brevipinna*	NC	
Congridae	*Conger myriaster*	JA	—
Coryphaenidae	*Coryphaena hippurus*	IO	—
Gadidae	*Micromesistius poutassou*	AS	—
Gerreidae	*Gerres oblongus*	PNG	—
Lutjanidae	*Pinjalo lewisi*	PNG	—
	Pinjalo pinjalo	PNG	—
Merlucciidae	*Merluccius merluccius*	EM, MA, NAM	—
Nemipteridae	*Nemipterus bathybius*	CHS	MAL
	Nemipterus virgatus	CHS	—
Phycidae	*Phycis blennoides*	NAM	—
	Phycis phycis	NAM	—
Pleuronectidae	*Platichthys flesus*	PHC	—
Scombridae	*Auxis rochei*	—	INC
	Auxis thazard	BR, YS	INC
	Katsuwonus pelamis	HW, PY	INC
	Euthynnus affinis	IO	—
	Rastrelliger kanagurta	TC	—
	Sarda orientalis	IO	—
	Scomber australasicus	TW	—
	Scomber japonicus	AS, AZ, MD, NAM	—
	Scomber scombrus	NAM	—
	Scomberomorus commerson	IO, PG	—
	Scomberomorus maculatus	PNG	—
	Thunnus albacares	PNG	—
	Thunnus thynnus	BR	—
Sparidae	*Pagellus bogaraveo*	PC	—
Trichiuridae	*Lepturacanthus savala*	—	INC
	Trichiurus lepturus	BR, CHS, JA	INC
Triglidae	*Chelidonichthys kumu*	YS	—
Xiphiidae	*Xiphias gladius*	CA	—

Continued

Table 5 Definitive and Intermediate/Paratenic Hosts so far Recognized, by Molecular/Genetic Markers, of *Anisakis typica* and *Anisakis* sp. 1—cont'd

Sampling locality codes: AS: Aegean Sea; AZ: Azores Islands; BR: Brazilian Atlantic coast; CHS: China Sea; CS: Caribbean Sea; EM: East Mediterranean Sea; FL: Floridian coast; GM: Gulf of Mexico; HW: Hawaiian coast; INC: Indonesian coast; IO: Indian Ocean; JA: Japan Sea; MA: Mauritanian coast; MAL: Malaysian coast; MD: Madeira Island; NAM: North African Mediterranean Sea; NC: New Caledonia; PC: Portuguese coast; PG: Persian Gulf; PHC: Philippine coast; PNG: Papua New Guinea; PY: Polynesian Islands; TC: Thailandian coast; TW: Taiwanese coast. Host families are given by alphabetical order.

Data from Mattiucci, S., Nascetti, G., 2008. Advances and trends in the molecular systematics of *Anisakis nematodes*, with implications for their evolutionary ecology and host-parasite co-evolutionary processes. Adv. Parasitol. 66, 47–148 and references therein; Palm H.W., Damriyasa I.M., Linda Oka I.B.M. 2008. Molecular genotyping of *Anisakis* Dujardin, 1845 (Nematoda: Ascaridoidea: Anisakidae) larvae from marine fish of Balinese and Javanese waters, Indonesia. Helminthologia 45, 3–12; Colón-Llavina, M.M., Mignucci-Giannoni, A.A., Mattiucci, S., Paoletti, M., Nascetti, G., Williams, E.H., 2009. Additional records of metazoan parasites from Caribbean marine mammals, including genetically identified anisakid nematodes. Parasitol. Res. 105, 1239; Iñiguez, A.M., Santos, C.P., Vicente, A.C.P. 2009. Genetic characterization of *Anisakis typica* and *Anisakis physeteris* from marine mammals and fish from the Atlantic Ocean off Brazil. Vet. Parasitol. 165, 350–356; Lee, M.H., Cheon, D., Choi, C. 2009. Molecular genotyping of *Anisakis* species from Korean sea fish by polymerase chain reaction–restriction fragment length polymorphism (PCR-RFLP). Food Control 20, 623–626; Umehara, A., Kawakami, Y., Ooi, H.K., Uchida, A., Ohmae, H., Sugiyama, H. 2010. Molecular identification of *Anisakis* type I larvae isolated from hairtail fish off the coasts of Taiwan and Japan. Int. J. Food Microbiol., 143, 161–165; Cavallero, S., Nadler, S.A., Paggi, L., Barros, N.B., D'Amelio, S. 2011. Molecular characterization and phylogeny of anisakid nematodes from cetaceans from southeastern Atlantic coasts of USA, Gulf of Mexico and Caribbean Sea. Parasitol. Res. 108, 781–792; Iñiguez, A.M., Carvalho, V.L., Motta, M.R.A., Pinheiro, D.C.S.N., Vicente, A.C.P. 2011. Genetic analysis of *Anisakis typica* (Nematoda: Anisakidae) from cetaceans of the northeast coast of Brazil: new data on its definitive hosts. Vet. Parasitol. 178, 293–299; Klimpel, S., Kuhn, T., Busch, M.W., Karl, H., Palm, H.W. 2011. Deep-water life cycle of *Anisakis paggiae* (Nematoda: Anisakidae) in the Irminger Sea indicates kogiids distribution in the north Atlantic waters. Polar Biol. 34, 899–906; Borges, J.N., Cunha, L.F.G., Santos, H.L.C., Monteiro-Neto, C., Santos, C.P. 2012. Morphological and molecular diagnosis of anisakid nematode larvae from cutlassfish (*Trichiurus lepturus*) off the coast of Rio de Janeiro, Brazil. PLoS ONE 7, e40447; Koinari, M., Karl, S., Elliot, A., Ryan, U., Lymbery, A.J. 2013. Identification of *Anisakis* species (Nematoda: Anisakidae) in marine fish hosts from Papua New Guinea. Vet. Parasitol. 193, 126–133; Kuhn, T., Hailer, F., Palm, H.W., Klimpel, S. 2013. Global assessment of molecularly identified *Anisakis* Dujardin, 1845 (Nematoda: Anisakidae) in their teleost intermediate hosts. Folia Parasitol. 60, 123–134; Quiazon, K.M., Santos, M.D., Yoshinaga, T. 2013. *Anisakis* species (Nematoda: Anisakidae) of Dwarf Sperm Whale *Kogia simus* (Owen, 1866) stranded off the Pacific coast of southern Philippine archipelago. Vet. Parasitol. 197, 221–230; Anshary, H., Sriwulan, Freeman, M.A., Ogawa, K. 2014. Occurrence and molecular identification of Anisakid Dujardin, 1984 from marine fish in Southern Makassar Strait, Indonesia. Korean J. Parasitol. 1, 9–19; Kleinertz, S., Hermosilla, C., Ziltener, A., Kreicker, S., Hirzmann, J., Abdel-Ghaffar, F., Taubert, A. 2014. Gastrointestinal parasites of free-living Indo-Pacific bottlenose dolphins (*Tursiops aduncus*) in the Northern Red Sea, Egypt. Parasitol. Res. 113, 1405–1415; Pekmezci, G.Z., Onuk, E.E., Bolukbas, C.S., Yardimci, B., Gurler, A.T., Acici, M., Umur, S. 2014. Molecular identification of *Anisakis* species (Nematoda: Anisakidae) from marine fishes collected in Turkish waters. Vet. Parasitol. 201, 82–94; Shamsi, S. 2014. Recent advances in our knowledge of Australian anisakid nematodes. Int. J. Parasitol. Parasites Wildl. 3, 178–187; Shamsi, S., Ghadam, M., Suthar, J., Mousavi, H.E., Soltani, M., Mirzargar, S. 2016. Occurrence of ascaridoid nematodes in selected edible fish from the Persian Gulf and description of Hysterothylacium larval type XV and *Hysterothylacium persicum* n. sp. (Nematoda: Raphidascarididae). Int. J. Food Microbiol. 236, 65–73; Shamsi, S., Briand, M.J., Justine, J.L. 2017. Occurrence of *Anisakis* (Nematoda: Anisakidae) larvae in unusual hosts in Southern hemisphere. Parasitol. Int. in press; Cho, J., Lim, H., Jung, B.K., Shin, E.H., Cha, J.Y. 2015. *Anisakis pegreffii* larvae in Sea Eels (*Astroconger myriaster*) from the South Sea, Republic of Korea. Korean J. Parasitol. 53, 49–353; Chen, H.Y., Shih, H.H. 2015. Occurrence and prevalence of fish-borne Anisakis larvae in the spotted mackerel Scomber australasicus from Taiwanese waters. Acta Trop. 145, 61–67; Kong, Q., Fan, L., Zhang, J., Akao, N., Dong, K., Lou, D., Ding, J., Tong, Q., Zheng, B., Chen, R., Ohta, N., Lu, S. 2015. Molecular identification of *Anisakis* and *Hysterothylacium* larvae in marine fishes from the East China Sea and the Pacific coast of central Japan. Int. J. Food Microbiol. 199, 1–7; Quiazon, K.M.A. 2016. *Anisakis* Dujardin, 1845 infection (Nematoda: Anisakidae) in Pygmy Sperm Whale *Kogia breviceps* Blainville, 1838 from west Pacific region off the coast of Philippine archipelago. Parasitol. Res. 115, 3663–3668; Mattiucci et al., data not published.

genus *Anisakis*, a parasite, at the adult stage, of 12 species of dolphins, porpoises, and whales (see also Section 4) and, at the third stage, of 50 pelagic, bentho-pelagic and demersal teleost fish and 4 squid species (Table 2).

Its life cycle occurs in subarctic and temperate waters of the Northern hemisphere. It has been recorded in both the Western and Eastern Atlantic and Pacific Oceans. Its southern limit of distribution in the North-Eastern Atlantic waters seems to be the Spanish-Portuguese Atlantic coast. The species occurs at rare prevalence of infection in some fish species from the Morocco-Mauritania Atlantic coast (Farjallah et al., 2008a). A rare infection in the pygmy sperm whale, *Kogia breviceps* and few larval L4 specimens in the ziphiid *Ziphius cavirostris* from the Caribbean Sea were identified (Cavallero et al., 2011). There are no data about the occurrence of this species in latitudes southern of 35°N, thus suggesting that this geographical record of the species could be related to the acquired infection, by those cetacean species, in northern areas of the Atlantic Ocean, or throughout migratory intermediate hosts (Cavallero et al., 2011).

The species has not been identified, at the larval stage, in the Mediterranean Sea, except from the Alboran Sea, which is a basin actually considered as part of the Atlantic oceanographic waters rather than the Mediterranean Sea (Mattiucci et al., 2014a, b; Tintore et al., 1988). Spot occurrence below that latitude is reported in some fish species, such in *Scomber scombrus* (Levsen et al., 2017a), or bluefin tuna *Thunnus thynnus* fished in the eastern part of the Mediterranean Sea (Blažeković et al., 2015; Mladineo et al., 2017a). In the same basin, the species has been (*N* specimens = 3) recognized at the adult stage, by sequences analyses of the ITS region of rDNA from *Tursiops truncatus* and *Stenella coeruleoalba* (Blažeković et al., 2015; Mladineo et al., 2017a). However, the occurrence of *A. simplex* (s.s.) at the southern limit of its geographical distribution and in the eastern part of the Mediterranean Sea, would be related to the large migratory route of those intermediate and definitive host species in which it was observed, which likely acquired the infection in other oceanographic areas, in which the life cycles of the two species occur, such as the Iberian Atlantic coast. In the North Pacific Ocean, its southern limit of distribution seems to be represented, in the western area, by the Pacific coast of Japan, while in the eastern part, along the North American coast (see also Section 6).

3.1.2 A. pegreffii *Campana-Rouget and Biocca, 1955*

Previously indicated as *A. simplex* A (see Nascetti et al., 1986), *A. pegreffii* is the dominant species of *Anisakis* in the Mediterranean Sea, being widespread in all the fish species. Indeed, it is presently the most important anisakid nematode in several pelagic and demersal fish from Mediterranean waters (Cavallero et al., 2012; Cipriani et al., 2014, Cipriani et al., 2016, 2017a, b; Farjallah et al., 2008a; Mattiucci et al., 1997, 2014a, b; Meloni et al., 2011) (Table 2). In the Atlantic waters, the northerly limit of its geographical range is represented by the Iberian coast, from where the species has been frequently identified in sympatry with *A. simplex* (s.s.) in several fish and cetacean species (Abollo et al., 2001; Cipriani et al., 2015; Mattiucci et al., 1997, 2004, 2007, 2014a, b, 2015a; Marques et al., 2006; Pontes et al., 2005).

A. pegreffii, on the other hand, seems to occur only occasionally in northern waters of NE Atlantic Ocean. Rare specimens of *A. pegreffii* were recorded in migratory Atlantic mackerels, *Scomber scombrus*, sampled in the southern Norwegian Sea (Levsen et al., 2017a,b). The occurrence of *A. pegreffii* far away from its geographical range has been hypothesized as related to the observed changing migratory route of the southern stock of *S. scombrus* in the NE Atlantic waters (Levsen et al., 2017a,b). *A. pegreffii* distribution is widespread also in the Austral region, between 30°S and 60°S where the species was identified, at both adult and larval stages (Mattiucci et al., 2014a, 2017a). In this geographical area, to date, it has been recorded at high abundance at adult stage in dolphins, mainly belonging to the family Delphinoidea (Section 4), and in a species of Neobalaenidae (*Caperea marginata*) from the South East Atlantic coast (South African coast) (Mattiucci et al., 2017b) (see also Sections 3.1 and 5.1).

A. pegreffii is recorded as a parasite, at the adult stage, in 6 species of dolphins and whales (Table 2); while, at its third larval stage, it occurs in 60 pelagic and benthopelagic fish species, and in 4 squid species throughout its geographical range (Table 2). Among the hosts involved in its life cycle, six definitive hosts were found to be shared by *A. pegreffii* with *A. simplex* (s.s.) in the contact areas between the two species (Iberian Atlantic coast, eastern Mediterranean Sea, and Japan Sea waters) (Fig. 3) (Abe et al., 2006; Quiazon et al., 2011; Umehara et al., 2006, 2008, 2010). Identification of *A. pegreffii* larvae in fish from East China Sea waters, was also reported (Zhu et al., 2007); more recently, from this geographical area, larvae of *A. pegreffii* were identified in *Pseudorhombus cinnamoneus* from Yellow Sea (Li et al., 2017). High prevalence by *A. pegreffii* larvae was seen in fish species from East China Sea, Taiwanese Sea, and Korean Sea. From these areas, the species has been often found in coinfection with *A. typica* (Bak et al., 2014; Chen and Shih, 2015; Chou et al., 2011; Du et al., 2010; Kong et al., 2015; Setyobudi et al., 2011).

Three definitive and some intermediate/paratenic hosts (Table 2) are shared by *A. pegreffii* with *A. berlandi* in the Austral waters off New Zealand, the South African coast, Falkand Island, and the southern Chilean coast (Fig. 3). Interestingly, the southern occurrence of the *A. pegreffii* throughout the geographical range in the Austral region is represented by its genetic detection at larval stage in migratory myctophids around the South Shetland Islands (Klimpel et al., 2010). Shamsi et al. (2017) have identified the occurrence of some larval stage of *A. pegreffii* in some unusual hosts (Table 2). *A. pegreffii* frequently coinfects with *A. simplex* (s.s.) in the same individual definitive or intermediate/paratenic hosts in some sympatric areas, including the Spanish Atlantic coast and the Alboran Sea in the Mediterranean basin (Fig. 3). In these sympatric areas, F1 hybrids (around 5.0%—Cipriani et al., 2015; Mattiucci et al., 2016) between the two *Anisakis* species have been recorded in several commercially important fish hosts by using a multilocus genetic; however, no backcross individuals have so far been detected (Mattiucci et al., 2016). Coinfection by *A. pegreffii* and *A. berlandi* has been found in both cetacean and fish hosts from the New Zealand coast (Mattiucci et al., 2014a). Therefore, despite the wide distribution of this species in waters of the world's oceans, according to the biogeographical data so far acquired, the species *A. pegreffii* cannot be considered as a '... cosmopolitan and generalist species ...' as recently stated (Mladineo et al., 2017b).

3.1.3 **A. berlandi** *(See Mattiucci et al., 2014a)* **(= A. simplex** *C of Mattiucci et al., 1997)*

A. berlandi currently exhibits a discontinuous distribution. This includes the Austral region: the Chilean Pacific, the South Shetland Islands, New Zealand waters, and the South African Atlantic coast (Klimpel et al., 2010; Mattiucci et al., 2014a, unpublished data; Mattiucci and Nascetti, 2008). This species has been identified, at the adult stage, in sympatry and syntopy with *A. pegreffii* in *Globicephala melas* and *Grampus griseus* from the New Zealand, and in *G. melas* from south west (South African coast) and south east (Chilean coast) Pacific waters (Table 2). Very few L4 stage specimens belonging to *A. berlandi* were identified in the pigmy sperm whale *Kogia sima* in south Pacific waters. In addition, adults of this *Anisakis* species have been rarely identified also in the pinniped *Mirounga leonina* from the subantarctic area (South Shetland Islands). Actually, its Type I larvae were identified in nine fish species from Austral waters off New Zealand (Mattiucci et al., 2014a), the South African coast (Mattiucci et al., 2014a), Southern Shetland Islands (Klimpel et al., 2010), the Southern Chilean coast, and in some unusual hosts from the New Caledonian waters (Shamsi et al., 2017) (Table 2). Klimpel et al. (2010) stated that the occurrence of few larval specimens of *A. berlandi* and *A. pegreffii* in myctophids from the southern waters of the Southern Ocean (i.e. South Shetland Islands, Antarctic area), could be related to the introduction of those two parasites species from outside the Antarctic, through their migrating fish intermediate hosts. Indeed, also the very low prevalence found in *M. leonina* from South Shetland Islands (Mattiucci and Nascetti, 2008) could be explained as an accidental infection when that pinniped host preyed upon on infected migratory fish species (Mattiucci et al., 2017a).

3.2 *Anisakis* spp. Included in Clade 2

The concatenated inference, based on the combined nuclear and mitochondrial sequence datasets, resulted in identical topologies in supporting, with high posterior probability and bootstrap values (Fig. 2) (Mattiucci et al., 2014a) a second clade formed by the two species, *A. ziphidarum* and *A. nascettii*. According to this phylogenetic analysis, the two species are sister taxa. Here, their known geographical range and hosts are reported.

3.2.1 A. ziphidarum *Paggi, Nascetti, Webb, Mattiucci, Cianchi, and Bullini, 1998*

It was first described, both genetically and morphologically, as an adult in the beaked whales *Mesoplodon layardii* and *Z. cavirostris* from the South Atlantic Ocean (off the South African coast). After its first morphological description and genetic characterization (Paggi et al., 1998) (see Section 4), this species has been recently identified genetically as an adult in other species of beaked whales, such as *Mesoplodon mirus*, *Mesoplodon densirostris*, and *Mesoplodon grayi* from the South Atlantic waters (off South Africa coast), from *Mesoplodon europaeus* of Caribbean sea (Cavallero et al., 2011; Colón-Llavina et al., 2009), from *Mesoplodon bowdoini* of New Zealand waters (Mattiucci et al., 2009), as well as from other geographical areas, such as its occurrence in *Mesoplodon* sp. and *Z. cavirostris* from Chilean waters (Mattiucci and Nascetti, 2008, unpublished data). *A. ziphidarum* was also found in a specimen of *K. sima* from Philippine waters (Quiazon et al., 2013). Thus, its geographical range appears to be wide (Fig. 4) and mainly related to the distribution of its definitive hosts (see Section 5). Up to date, still scanty data are available concerning its infection at larval stage in fish (Di Azevedo et al., 2016; Mattiucci and Nascetti, 2008). It was reported in nine fish species from both Boreal and Austral regions (Table 2). Quiazon et al. (2011) found rare infection in fish species from Japanese waters. Indeed, it seems that this species may involve other intermediate hosts in its life cycle, such as deep squid species: this would be in

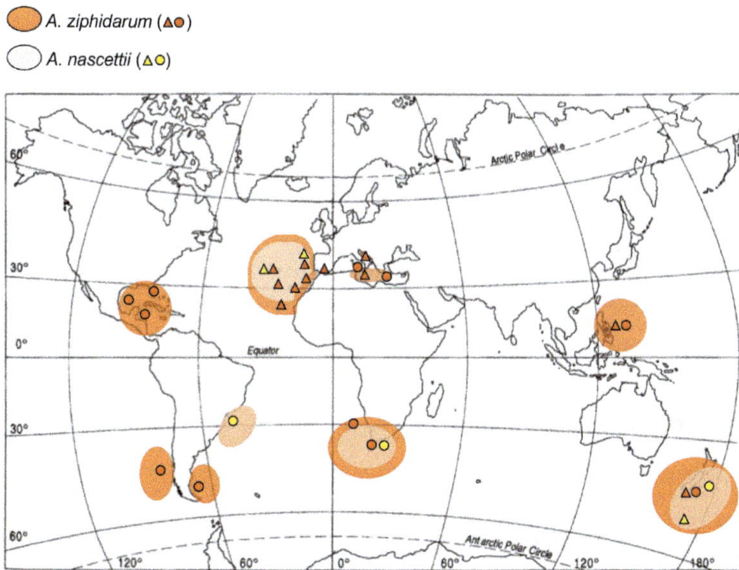

Fig. 4 World map showing range of distribution of *Anisakis ziphidarum* and *A. nascettii*. The *symbols* are related to the sampling localities of their definitive (*circle*) and intermediate hosts (*triangle*), so far identified by molecular markers.

agreement with the feeding habits of beaked whales. These cetacean species, in fact, mostly prey on squid belonging to the Families Onychoteuthidae and Histiotheuthidae, rather than fish, as those cephalopods represent the main food source of beaked whales all over the world (Leatherwood and Reeves, 1983; Ross, 1984; Santos et al., 2001a, b) (see Section 5).

3.2.2 A. nascettii *Mattiucci, Paoletti, and Webb, 2009*

This species has been described both genetically and morphologically (see also Section 4) from some specimens detected in the beaked whales *M. mirus* and *M. grayi* obtained from South African and New Zealand waters (Fig. 4 and Table 3). The gene pool of *A. nascettii* was found to be reproductively isolated from the sympatric species *A. ziphidarum*, occurring in the same hosts and geographical region. It is genetically very distinct from the other species of *Anisakis*, but is most closely related to *A. ziphidarum*. So far, it is a parasite of five species of ziphiids (see also Section 5). The third-stage larva of *A. nascettii* is apparently of Type I, showing a long ventricle and a mucron at the caudal end; however, it is a robust larva with respect to the Type I larva of the species included in Clade 1 (Mattiucci, personal observation). It has been genetically identified, at the larval stage, heavily infecting the squid *Moroteuthis ingens* in Tasman Sea waters (Mattiucci et al., 2009). This appears to support the hypothesis that this species, such as the sister taxa *A. ziphidarum*, involves squid—main food items of the beaked whales (Leatherwood and Reeves, 1982) all over the oceanographical areas—rather than fish, in the parasite species' life cycle. The sympatric occurrence of the *A. nascetti* and *A. ziphidarum*, at adult stage, in the same definitive hosts (i.e. the Andrews' beaked whale *M. bowdoini* and the strap-toothed whale *M. layardii*), and the absence of gene flow between them, reinforces the concept that they are two distinct biological species (Mattiucci et al., 2009).

3.3 *Anisakis* Species Included in Clade 3

According to the topology of the concatenated inference, based on combined nuclear (ITS rDNA) and mitochondrial (mtDNA *cox2* and *rrnS*) sequence datasets, a highly supported third clade exists, formed by the species *A. physeteris*, *A. brevispiculata*, and *A. paggiae* (Fig. 2) (data from Mattiucci et al., 2014a). Below, the geographic distribution and hosts of those species are reported.

3.3.1 A. physeteris *Baylis, 1920*

A. physeteris has been reported, at the adult stage, from the sperm whale, *Physeter macrocephalus* from the Tyrrhenian, Central Adriatic, Ionian, and Aegean Sea waters of the Mediterranean Sea (Mattiucci et al., 1986, 2001, unpublished data), from the Adriactic Croatian coast Blažeković et al., 2015), along the Iberian Atlantic waters (Mattiucci, unpublished data), from a sperm whale stranded along the Scottish coasts of the North Sea (Mattiucci, unpublished data) and from the Caribbean Sea (Cavallero et al., 2011; Colón-Llavina et al., 2009). Rare infection has been also recorded in the pygmy sperm whale, *Kogia* spp., from the Caribbean Sea and Gulf of Mexico (Cavallero et al., 2011). However, despite the large geographical range of its main definitive hosts of the Family Physeteridae, no data about genetic identification of adults in sperm whales from Pacific Ocean waters are so far reported.

Infection by Type II larvae belonging to this species (see also Section 4) has been rarely reported in some pelagic and demersal fish species from both the Atlantic and Pacific Oceans (Table 4). In contrast, it was found commonly in the swordfish, *Xiphias gladius*, in the Mediterranean Sea fish stock, and in the Atlantic stock of the same fish species (Mattiucci et al., 2014b, 2015a).

Infection in the commercial cephalopod *Illex coindetii* from the Mediterranean coast of Spain was recently reported (Picó-Durán et al., 2016).

3.3.2 A. brevispiculata *Dollfus, 1966*

The species, synonymized with *A. physeteris* by Davey (1971), was recognized as biological species on the basis of nuclear (Mattiucci et al., 2001) and molecular markers (Valentini et al., 2006), based on material from a pygmy sperm whale, *K. breviceps*, stranded on the South African coast. Indeed, its reproductive isolation from the morphologically closely related *A. physeteris* was demonstrated. Its recognition as a distinct phylogenetic lineage has been established as inferred from both nuclear and mitochondrial sequences datasets (Mattiucci et al., 2014a), also confirming that *A. brevispiculata* clusters well with those *Anisakis* species forming the third clade, and thus indicating that *A. physeteris* and *A. brevispiculata* represent sister taxa. Larval stages of this species were genetically identified from the swordfish *X. gladius*, from the Central Atlantic Ocean and Tropical Equatorial Atlantic waters (Mattiucci et al., 2011, 2014b) (Table 4 and Fig. 5). In addition, larval stages corresponding to this

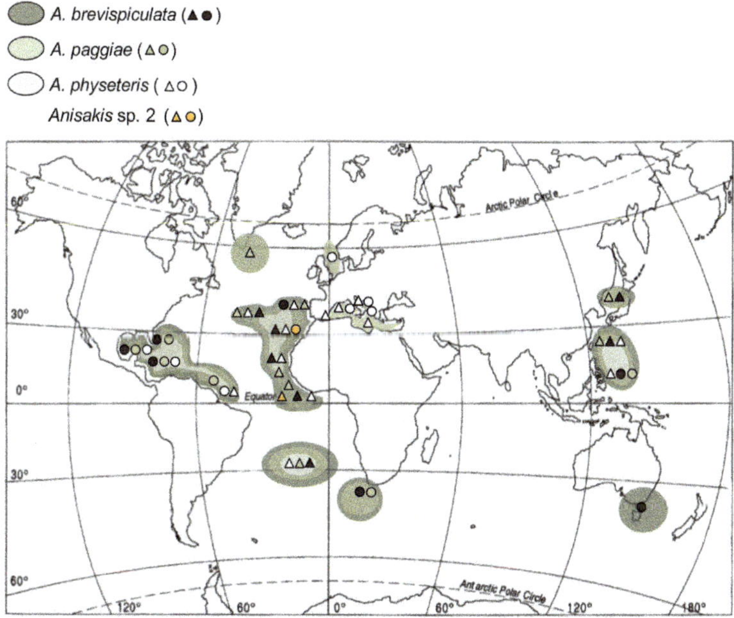

Fig. 5 World map showing the range of distribution of *Anisakis physeteris*, *A. brevispiculata*, and *A. paggiae*. The *symbols* are related to the sampling localities of their definitive (*circle*) and intermediate/paratenic (*triangle*) hosts, so far identified by molecular markers.

species were found in the fish *Beryx splendens* from Japanese waters (Murata et al., 2011). The same authors proved, by the phylogenetic analysis of the mtDNA *cox2* gene and morphological/merphometric analysis, the correspondence between the larval Type III (sensu Shiraki, 1974) corresponded to the species *A. brevispiculata* (Murata et al., 2011) (see also Section 4). Few larvae corresponding to *A. brevispiculata* were found in fish from Taiwan waters (Chen and Shih, 2015) and Japanese waters (Quiazon et al., 2011).

3.3.3 A. paggiae *Mattiucci, Nascetti, Dailey, Webb, Barros, Cianchi, and Bullini, 2005*

A. paggiae clusters with *A. physeteris* and *A. brevispiculata* in the third clade (Fig. 2). This species was first discovered by allozymes (Mattiucci et al., 2005) and mtDNA *cox2* sequence analysis (Valentini et al., 2006). It was also first described morphologically as an adult parasite of the pygmy sperm whale, *K. breviceps*, and the dwarf sperm whale, *K. sima*, from the South African Atlantic coast (Mattiucci et al., 2005). Scanty data are available regarding the identification of the intermediate hosts in the life cycle of *A. paggiae* from Austral waters. Several larvae of *A. paggiae* been identified in fish from Atlantic waters, i.e., *M. merluccius* and *X. gladius* (Mattiucci et al., 2014b). However, the infection in *X. gladius* remained confined to the stomach lumen of the fish host, likely acquiring the infection by squid, preferred prey items of the swordfish (Mattiucci et al., 2014b, personal observation), thus suggesting that other hosts, not yet detected, are involved in the life cycle of this *Anisakis* species.

Finally, a parasitological study carried out by Klimpel et al. (2011) on the meso- and bathypelagic fish *Anoplogaster cornuta*, from the North-East Atlantic Ocean, revealed the presence of *A. paggiae* at a high prevalence ($P = 57.1\%$) and mean intensity ($I_m = 2.2$). Because Kogiid whales (i.e. *K. breviceps* and *K. sima*) are the final hoṣts of this parasitic nematode (Mattiucci et al., 2005; Mattiucci and Nascetti, 2008), Klimpel et al. (2011) suggested that the finding of *A. paggiae* larvae in that fish host would have been introduced through migratory kogiid final hosts, and thus this finding represents an accidental host. Also, the finding in such a deep-sea fish seems to support an oceanic deep-water life cycle for *A. paggiae* in the North-East Atlantic Ocean (Klimpel et al., 2011). Interestingly, Murata et al. (2011) identified genetically some larvae corresponding to *A. paggiae* from the fish *B. splendens* from Japanese waters. The same authors also provide the evidence of correspondence between the larval Type IV (sensu Shiraki, 1974) with the species *A. paggiae* (Murata et al., 2011) (see also Section 4).

3.3.4 **Anisakis** *sp. 2 (See Mattiucci et al., 2014b)*

Anisakis Type II larvae of a taxon, indicated as *Anisakis* sp. 2, were first detected, as based on the allozyme data and by sequence analysis of the mtDNA *cox2* gene (Mattiucci et al., 2014b, unpublished data). It seems to represent a further taxon, genetically closely related to the species *A. physeteris*. Adult stages of this taxon have been so far found in a stranded sperm whale from off the Canary Islands (central Atlantic waters) (Garcia et al., 2011). This genotype, was also recognized as larvae of Type II from the swordfish *X. gladius* in the equatorial area (Mattiucci et al., 2007) by means of allozyme markers and mtDNA genes sequence analysis. *Anisakis* sp. 2 could represent a new taxon which appears to be phylogenetically closely related to *A. physeteris*; it clusters with the clade formed by *A. physeteris*, *A. brevispiculata*, and *A. paggiae*, thus suggesting that this new gene pool might represent an undescribed sibling species belonging to this same complex. However, more genetic/molecular analyses are needed to clarify the taxonomic status of this genotype.

3.4 *A. typica* (Diesing, 1860)

Finally, concatenated phylogenetic trees obtained from the combined nuclear and mitochondrial sequences identified *A. typica* (see Mattiucci et al., 2014a) as a separate lineage with respect to the other species. *A. typica* does not cluster in any previous clade, thus suggesting that it belongs to a separate clade (here indicated as Clade 4) of the *A. typica* species complex (Fig. 2). However, the precise positioning of the *A. typica* clade is not yet clearly resolved from a phylogenetic point of view. This is because although it represents a sister taxon to the other clades in the phylogenetic analyses based upon concatenated MP and BI, recently inferred from combined nuclear and mitochondrial datasets (Cavallero et al., 2012; Mattiucci et al., 2014a) *A. typica* is highly supported, as a sister group to the clades 1 and 2, as inferred from combined phylogenetic analysis of mitochondrial genes dataset (Fig. 2A). Analysis of further gene loci would be needed to clarify its phylogenetic relationship with respect to the other species of the genus *Anisakis*.

According to the biogeographical data so far reported for *A. typica* populations detected genetically, its range extends from 35–40°N to 36° S in warmer temperate and tropical waters (Table 5 and Fig. 6). In these areas, it was found as an adult in oceanic dolphins (Colón-Llavina et al., 2009; Iñiguez et al., 2009, 2010; Mattiucci and Nascetti, 2008) and as larva

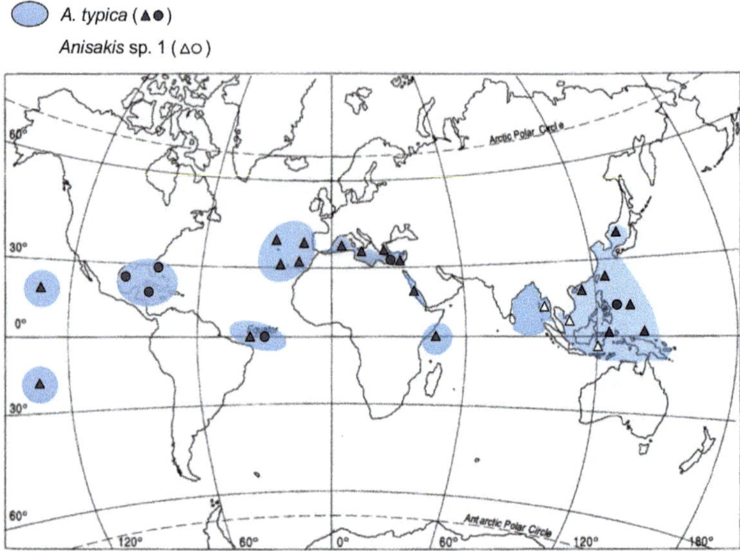

Fig. 6 World map showing range of distribution of *Anisakis typica* and *Anisakis* sp. 1. The *symbols* are related to the sampling localities for their definitive (*circle*) and intermediate/paratenic (*triangle*) hosts, so far identified by molecular markers.

in several fish species (Table 5 and Fig. 6). A case of infection by *A. typica* was also found in the pygmy sperm whale, *K. breviceps* from the Brazilian Atlantic coast (Borges et al., 2012; Iñiguez et al., 2011). *A. typica* has also been identified in the striped dolphin *S. coeruleoalba* from the eastern Mediterranean Sea (off Cyprus and the Aegean Sea), and in the European hake, *M. merluccius* (Mattiucci et al., 2002, 2014b). Because all the fish from the Indian Ocean (Somali coast) were found infected by *A. typica* larvae (Mattiucci et al., 2002), its presence in the eastern part of Mediterranean Sea waters could be the result of the 'Lesseptian migration' (through the Suez Canal) (Mattiucci et al., 2004) of its intermediate/paratenic hosts from the Indian Ocean. It was also rarely found in fish captured along the North African Coast of the Mediterranean Sea (Tunisian and Libyan coast) (Farjallah et al., 2008a), and in the flatfish *Platichthys flesus* captured in central Portuguese waters of the NE Atlantic Ocean (Hermida et al., 2012; Marques et al., 2006). Its larval stages were identified cooccurring in the same individual fish host, *Scomber japonicus* captured from the Atlantic Ocean off Madeira (Mattiucci et al., 2002) and from *Pagellus bogaraveo* (Hermida et al., 2012), as well as from the Morocco-Mauritania Atlantic coast (Farjallah et al., 2008a). The species has been also recorded from the Central Pacific (Hawaii and Moorea) by Kuhn et al. (2013) and Palm and Bray (2014), from Tursiops aduncus off Hurghada coast in the northern Red Sea, Egypt (Kleinertz et al., 2014) and, most recently, in some fish host of the Persian Gulf (Shamsi et al., 2017) (Table 5).

Finally, identification of *Anisakis* spp. larvae obtained from several fish species, detected *A. typica*, at lower proportion with respect to *A. pegrefffi* in fish from the East China Sea and Taiwan Sea (Chen and Shih, 2015; Chou et al., 2011; Kong et al., 2015; Quiazon et al., 2011; Umehara et al., 2010; Zhu et al., 2007) and Yellow Sea (Du et al., 2010) (Table 5 and Fig. 6). Anshary et al. (2014) identified molecularly, *A. typica* from several fish hosts from the Southern Makassari Strait, Indonesia with higher proportion in pelagic migratory fish species such as *Auxis thazard* and *Katsuwonus pelamis*, with high intensity of infection in the latter species. Palm et al. (2008) reported the high prevalence of infection by *A. typica* in Indonesia waters; the same author reported that *A. typica* larvae were found not only on the fish visceral body cavity, but it might also penetrate the muscle. Koinari et al. (2013) reported the occurrence of *A. typica* larvae in several ($N=7$) fish from Papua New Guinea, suggesting that the species involves epipelagic fish hosts in its life cycle. However, the occurrence of *A. typica* larvae was not reported in flesh of the infected hosts; the larvae have been always found infecting the fish viscera.

3.5 *Anisakis* sp. 1

A new gene pool, indicated as *Anisakis* sp. 1, was genetically detected by both allozyme and mtDNA analysis, at the larval stage, as a parasite of the fish *Nemipterus japonicus* caught off the Malaysian coast (Mattiucci and Nascetti, 2008). This new taxon is genetically distinct from all the known species of *Anisakis* but most closely related to *A. typica* from central Atlantic waters. Although known only at the larval stage, the third-stage larva of this undescribed taxon is a Type I larva (sensu Berland, 1961). The preliminary genetic results appear to indicate that this taxon may be a sibling species of *A. typica* occurring in central Pacific waters (Mattiucci, unpublished data). In the Austral region, while summarizing the knowledge about Australian *Anisakis* spp., Shamsi (2014) reported *A. typica* out of the occurrence of *A. brevispiculata*, *A. berlandi*, and *A. pegreffii* also in Pacific waters adjacent to Indonesia, mentioning that *A. typica* showed some genetic differences to the original genotypes from other areas. Likely, the same taxon was found and reported by molecular genotyping by Palm et al. (2008), who identified further different genotypes of *A. typica* in Balinese and Javanese waters (Indonesia); however in that paper only ITS was analysed. The same authors also suggested that the latter genotype, infecting two not-migratory fish species, could represent a sibling species of *A. typica*, only occurring in Indonesian waters (Palm et al., 2008). The taxon was indicated as *A. typica* var. *indonesiensis* (Palm et al., 2017). Despite the occurrence of this genotype in those basin waters, however, its occurrence at the adult stage has not yet been detected. As a consequence, a formal description of this provisional taxon, based on both genetic and morphological features has not yet been performed.

The taxon here indicated as *Anisakis* sp. 1, collected from fish of *Nemipterus* sp. of Malaysia, whose mtDNA cox2 sequences were here included in the phylogenetic tree (Fig. 2), was found to match 99%, 100%, the sequences deposited in GenBank as *A. typica*, under the accession number KC928269 (Anshary et al., 2014) from *A. thazard* of Indonesia, and JX648323 (Koinari et al., 2013), from *Pinjalo lewisi* of the Papua New Guinea, thus suggesting that they are belonging to the same genotype. It could represent a sibling species of *A. typica* (s.l.), to be also fully investigated and described.

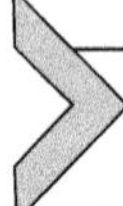

4. RECONCILING MOLECULAR AND MORPHOLOGICAL RESULTS

Can we find a link between systematics based on molecular and morphological results?

The high environmental heterogeneity represented by the different definitive hosts in the species of genus *Anisakis* could have also driven, through natural selection, to the evolution of phenotypes and developmental strategies which have led to maximize the match of some particular phenotypic traits to specific habitats in their definitive and intermediate hosts. Even if genetic speciation is not always accompanied by morphological changes in anisakid nematodes, stable habitat (site of infection) factors represented in different families of cetaceans, their definitive hosts (see Section 5) could have influenced the maintenance of quite fixed phenotypic (morphological) traits in those *Anisakis* species endoparasitic in those cetacean families. In other words, the same morphological traits could be found in those *Anisakis* spp. belonging to that specific clade, which are, in turn, parasitic on cetaceans species belonging to a specific family (see also Section 5). Those phenotypic traits maintained, at clade level, seem to be likely those that maximize parasite species fitness, and their developmental strategies in those definitive hosts. In turn, those morphological traits could be of help, in distinguishing adults of *Anisakis* spp., at least, at clade level, as below reported and discussed.

4.1 Diagnostic Morphological Features at Clade Level

The existence of distinct clades in the phylogeny of the species of the genus *Anisakis* seems to be also supported by some differential morphological features detected in the species clustering in those clades. Indeed, in parallel, the four major clusters defined by the phylogenetic analysis correspond and can also be morphologically delineated (characterized) with distinctive morphological traits in adult worms. Indeed, some morphological and morphometric features are shared between species belonging to the same clade, while they differ for those features with the other clades. Thus, the species belonging to Clade 1, which comprises the species *A. simplex* (s.s.), *A. pegreffii*, and *A. berlandi*, clustering together, share some common morphological characteristics: (i) the ventriculus is longer than broad and often sigmoid in shape and (ii) male spicules are long, thin, and unequal in length (Mattiucci et al., 2014a). This monophyletic group is clearly morphologically distinct from the *Anisakis* species clustering in other clades (Fig. 7 and Table 6). The monophyly of the group of the species included in Clade 2 (i.e. *A. ziphidarum* and *A. nascettii*), as proved by the molecular traits, is also supported by the specific morphological traits, in the adult worms, of those species, having: (i) a long ventriculus, not sigmoid in shape and (ii) male spicules long, thin, and equal in length (Mattiucci et al., 2009; Paggi et al., 1998) (Fig. 7 and Table 7). The existence of Clade 3, in which the three species *A. physeteris*, *A. brevispiculata*, and *A. paggiae* are included, is also supported by its own morphological traits: (i) the ventriculus, in the adult stage, is short, never sigmoid, and broader than long and (ii) male spicules are short, stout, and of similar length (Fig. 7 and Table 8) (Mattiucci et al., 2005). Finally, in *A. typica* adult worms (i) the ventriculus is long and (ii) male spicules are long, thin, and very marked unequal in

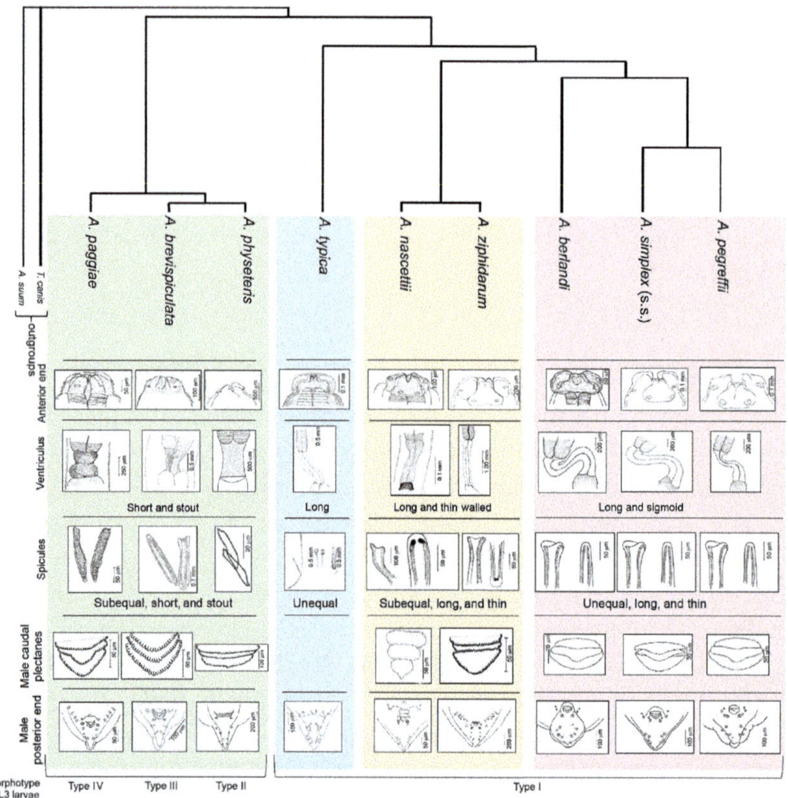

Fig. 7 Some morphological/morphometric diagnostic features among the genetically characterized species of *Anisakis*, mapped into their phylogenetic tree. Distinctive characters selected are: male cephalic end, ventriculus length and shape, spicules length and shape, ratio between the right and left spicule lengths (R/L), male caudal end and arrangement of caudal papillae, male caudal plates, and larval (L3 stage) morphotype.

length (mean ratio 1:3). In addition, the lips of *A. typica* are clearly distinct from those of the other species, having an anterior bilobed dorsal lip. Finally, the arrangement of male caudal papillae is quite peculiar, as it represents the only case in which the paracloacal papillae are not double as for other species of *Anisakis* (Fig. 7) (Davey, 1971; Mattiucci, personal observation).

In addition, the larval ontogeny seems to follow the *Anisakis* phylogeny, being the third-stage larva (L3) indicated as Types II, III, and IV morphology (sensu Berland, 1961; Shiraki, 1974)—characteristic of *A. physeteris*, *A. brevispiculata*, and *A. paggiae*, respectively (Murata et al., 2011)—which

Table 6 Morphometric Data of Adults of *Anisakis simplex* (s.s.), *A. pegreffii*, and *A. berlandi*

Parasite Species	*A. simplex* (s.s.)	*A. pegreffii*	*A. berlandi*
Specimen (*n*)	14 (12 males, 2 females)	23 (15 males, 8 females)	15 (11 males, 4 females)
TBL (mm)	70.0 (49.0–11.2)	41.6 (33.0–55.0)	50.0 (26.0–62.0)
Oesophagus length	—	4.30 (3.36–5.10)	4.69 (4.10–5.80)
VL	1.29 (1.08–1.44)	0.76 (0.55–0.90)	1.11 (0.80–1.35)
VL/TBL	0.018 (0.011–0.028)	0.018 (0.013–0.023)	0.023 (0.016–0.030)
Ventriculus shape	Long and sigmoid	Long and sigmoid	Long and sigmoid
LS	2.17 (1.70–2.80)	2.10 (1.50–2.65)	2.50 (2.10–2.90)
RS	1.44 (1.16–1.82)	1.30 (1.00–1.60)	1.35 (1.20–1.60)
R/L	0.68 (0.46–0.83)	0.64 (0.54–0.77)	0.58 (0.49–0.70)
LS/TBL	0.033 (0.029–0.036)	0.047 (0.037–0.056)	0.047 (0.039–0.058)
RS/TBL	0.023 (0.016–0.027)	0.029 (0.026–0.039)	0.026 (0.024–0.027)
TL	0.28 (0.22–0.37)	0.25 (0.19–0.32)	0.27 (0.22–0.29)
TL/TBL	0.004 (0.003–0.005)	0.006 (0.004–0.008)	0.005 (0.005–0.006)
WPL1	0.077 (0.066–0.093)	0.082 (0.077–0.087)	0.086 (0.068–0.11)
WPL2	0.076 (0.065–0.092)	0.081 (0.076–0.086)	0.083 (0.075–0.102)
WPL3	0.074 (0.061–0.088)	0.081 (0.074–0.089)	0.089 (0.076–0.114)
WPL1/WPL3	0.97 (0.94–1.00)	0.94 (0.87–1.03)	1.04 (1.01–1.11)
Largest WPL/TL	0.33 (0.26–0.50)	0.32 (0.27–0.41)	0.33 (0.26–0.39)

Ratio values are reported as a percentage. All measurements are given in millimetres. *n*: number of specimens analysed; LS: left spicule; RS: right spicule; TBL: total body length; TL: tail length; VL: ventriculus length; WPL: width of caudal plectanes.

Data from Mattiucci, S., Cipriani, P., Webb, S.C., Paoletti, M., Marcer, F., Bellisario, B., Gibson, D.I., Nascetti, G. 2014a. Genetic and morphological approaches distinguishing the three sibling species of the *Anisakis simplex* species complex, with a species designation as *Anisakis berlandi* n. sp. for *A. simplex* sp. C (Nematoda: Anisakidae). J. Parasitol. 15, 12–15.

could resemble an ancestral character of this group of species. Whereas, a Type I morphology (sensu Berland, 1961) can be found in all the most derived species, including the clade of the *A. simplex* (s.l.) complex (Clade I), *A. ziphidarum* and *A. nascettii* (Clade 2), and *A. typica*.

Table 7 Morphometric Data of Adults of *Anisakis nascettii* and *A. ziphidarum*

Parasite Species	*A. nascettii*	*A. ziphidarum*
Specimen (*n*)	5 (2 males, 3 females)	7 (4 males, 3 females)
TBL (mm)	88.0 (67.0–120.0)	195.0 (160.0–220.0)
Oesophagus length	6.1 (5.0–8.5)	9.9 (8.80–11.05)
VL	1.43 (0.92–2.3)	1.7 (1.23–2.09)
VL/TBL	0.016	0.008 (0.007–0.009)
Ventriculus shape	Long and thin walled	Long and thin walled
LS	1.9 (1.7–2.1)	1.49 (1.43–1.55)
RS	1.8 (1.6–2.0)	1.46 (1.40–1.53)
R/L	0.94 (0.94–0.95)	0.97 (0.97–0.98)
LS/TBL	0.021 (0.017–0.025)	0.0076 (0.007–0.89)
RS/TBL	0.020 (0.016–0.023)	0.0074 (0.006–0.008)
TL	0.29 (0.28–0.30)	0.52 (0.40–0.63)
TL/TBL	0.003 (0.002–0.004)	0.0026 (0.0025–0.0028)
WPL1	0.045	0.081 (0.077–0.085)
WPL2	0.035	0.071 (0.060–0.082)
WPL3	0.030	0.061 (0.059–0.063)
WPL1/WPL3	1.50	1.32 (1.30–1.34)
Largest WPL/TL	0.155	0.155 (0.13–0.19)

Ratio values are reported as a percentage. All measurements are given in millimetres. *n*: number of specimens studied; LS: left spicule; RS: right spicule; TBL: total body length; TL: tail length; VL: ventriculus length; WPL: width of caudal plectanes.

Data from Paggi, L., Nascetti, G., Webb, S.C., Mattiucci, S., Cianchi, R., Bullini, L. 1998. A new species of *Anisakis* Dujardin, 1845 (Nematoda: Anisakidae) from beaked whale (Ziphiidae): allozyme and morphological evidence. Syst. Parasitol. 40, 161–174; Mattiucci, S., Paoletti, M., Webb, S.C. 2009. *Anisakis nascettii* n. sp. (Nematoda: Anisakidae) from beaked whales of the southern hemisphere: morphological description, genetic relationships between congeners and ecological data. Syst. Parasitol. 74, 199–217.

4.2 Diagnostic Morphological Features Between 'Cryptic' Species of the Genus *Anisakis*

Nowadays, the importance of performing morphology-based alpha taxonomy is reemphasized, once cryptic species have been discovered using molecular data. Whenever possible, morphological diagnostic traits have to support the nomenclature designation of those biological species

Table 8 Morphometric Data of Adults of *A. physeteris*, *A. brevispiculata*, and *A. paggiae*

Parasite Species	*A. physeteris*	*A. brevispiculata*	*A. paggiae*
Specimen (*n*)	13 (6 males, 7 females)	4 (3 males, 1 females)	11 (6 males, 5 females)
TBL (mm)	87.5 (70.0–110.0)	47.7 (46.0–52.0)	35.5 (23.0–50.0)
Oesophagus length	6.75 (5.50–8.5)	3 (2.85–3.35)	2.4 (2.0–2.8)
VL	0.97 (0.89–1.10)	0.57 (0.56–0.60)	0.4 (0.35–0.45)
VL/TBL	0.011 (0.01–0.012)	0.011 (0.011–0.012)	0.011 (0.009–0.015)
Ventriculus shape	Short	Short	Short and violin shaped
LS	0.37 (0.35–0.40)	0.29 (0.26–0.32)	0.2 (0.18–0.22)
RS	0.33 (0.32–0.34)	0.27 (0.26–0.28)	0.19 (0.17–0.21)
R/L	0.89 (0.85–0.91)	0.93 (0.8–1.0)	0.95 (0.94–0.95)
LS/TBL	0.004 (0.003–0.005)	0.006 (0.005–0.006)	0.0056 (0.004–0.007)
RS/TBL	0.003 (0.003–0.004)	0.005 (0.0053–0.0056)	0.0053 (0.004–0.007)
TL	0.77 (0.71–0.83)	0.41 (0.22–0.59)	0.16 (0.15–0.17)
TL/TBL	0.0088 (0.007–0.01)	0.008 (0.004–0.011)	0.0045 (0.003–0.006)
WPL1	0.080–0.100	0.050–0.70	0.049–0.051
WPL2	0.080–0.090	0.070–0.080	0.030–0.040
WPL3	0.080–0.100	0.080–0.090	0.040–0.045
Shape of plectanes	Coarsely denticulated; large and flattened	Deeply denticulated; large and triangular	Denticulated, narrow, the middle is curved; the distal is semicircular

Ratio values are reported as a percentage. All measurements are given in millimetres. *n*: number of specimens studied; LS: left spicule; RS: right spicule; TBL: total body length; TL: tail length; VL: ventriculus length; WPL: width of caudal plectanes.

Data from Mattiucci, S., Nascetti, G., Dailey, M., Webb, S.C., Barros, N., Cianchi, R., Bullini, L. 2005. Evidence for a new species of Anisakis Dujardin, 1845: morphological description and genetic relationships between congeners (Nematoda: Anisakidae). Syst. Parasitol. 61, 157–171; unpublished data.

genetically identified, allowing a better definition of parasite biodiversity (Mattiucci et al., 2014a; Nadler and Pèrez-Ponce de León, 2011). Indeed, it has been suggested that 'cryptic' and 'sister taxa' detected by genetic data are always provisionally cryptic, until additional morphological study reveals diagnostic structural differences that permit a morphological diagnosis and a formal species description. Thus, performing a deeper morphological analysis of those individuals genetically characterized, as belonging to different sister taxa, could reveal some features and/or morphometric characters of diagnostic value (i.e. fixed alternative states) in different sister and/or, more generally, among cryptic species (Mattiucci et al., 2005, 2009, 2014a).

Actually, the 'cryptic' and 'sibling' species (i.e. the latter term means 'sister taxa') in the genus *Anisakis* included in the monophyletic clades, are considered to be also morphologically distinct, in some specific traits. Some morphological characters are known for distinguishing the sibling species of the *A. simplex* complex (Mattiucci et al., 2014a; Quiazon et al., 2008; Shamsi et al., 2012). Among those morphological features available in adults of *Anisakis* spp., the following key diagnostic traits between, for instance, *A. berlandi* and *A. simplex* (s.s.) were detected: ventriculus length, tail shape, tail length/total body length ratio, and left spicule length/total body length ratio; between *A. berlandi* and *A. pegreffii*: ventriculus length and plectane 1 width/plectane 3 width ratio; and, finally, between *A. simplex* (s.s.) and *A. pegreffii*: ventriculus length, left and right spicule length/total body length ratios, and tail length/total body length ratio (Mattiucci et al., 2014a) (Table 6 and Fig. 7). Allometric characters, such as the TL with respect to the TBL (TL/TBL), were also found to be significantly different among all three species.

In addition, the male caudal PLs (PL1, PL2, and PL3) differed in shape and width among the three cryptic species (Table 6 and Fig. 7). Indeed, for instance, in *A. simplex* (s.s.) the lateral extremities of the plectanes structures are rounded; conversely, they are more flattened and pointed in *A. pegreffii*. In addition, the ratio between the width of PL1 and PL3 (WPL1/WPL3) was significantly greater in *A. berlandi* than in *A. pegreffii* (Fig. 7 and Table 6). On the other hand, on the basis of the same morphological characters and allometric characters, it was possible to distinguish the two sister taxa, or sibling species belonging to Clade 2: i.e., *A. nascettii* and *A. ziphidarum*. Indeed, clear morphological characters were determined after the genetic identification of the male specimens belonging to *A. nascettii* and *A. ziphidarum*. These are: the longer spicule length in *A. nascettii* compared with *A. ziphidarum* and the spicule length/body length ratio. Furthermore, *A. nascettii* has the d1 papilla much closer to the paracloacal papilla (pc), the distal papilla d4 in *A. nascettii* is posterior to the d2 but lateral to it in *A. ziphidarum*, and the distance between the d1 and d4 is greater in *A. nascettii* (Mattiucci et al., 2009). Finally, the caudal plates (PL) (plectanes) are different in shape in *A. nascettii*, being narrower than those in *A. ziphidarum*, their margins are fringed to form spines and the PL3 is an inverted tear shape (Mattiucci et al., 2009) (Fig. 7 and Table 7).

Analogous studies allowed identification of diagnostic morphological features in male and female adult specimens used to help in distinguishing *A. paggiae* from *A. physeteris* and *A. brevispiculata* (see Mattiucci et al., 2005), as well as from the species *Anisakis oceanica* (Johnston and Mawson, 1951). The latter taxon, originally described as *Stomachus oceanicus* from *Globicephalus ventricosus* (now *G. melas*) from off the New South Wales coast (southern Pacific Ocean), was later synonymized by Davey (1971) with *A. physeteris*. Mattiucci et al. (2005) and Shamsi et al. (2012) raised the possibility that *A. oceanica* might be a valid species; however, its validity was not so far determined, based on genetic/molecular markers. Morphological keys for the recognized adult *Anisakis* spp. so far included in Clade 3 was provided by Mattiucci et al. (2005). *A. paggiae* can be distinguished from *A. physeteris* by the length of spicules, being shorter than the latter's 0.35–0.40 mm (according to the description given by Baylis, in Mozgovoi, 1951), and from *A. brevispiculata* by the length of the ventriculus, being shorter than the latter's 0.56–0.60 mm. In addition, *A. paggiae*, when morphologically compared with the type material of *A. oceanica*, exhibited a similar length of the spicules, a similar length and shape of the ventriculus, and a similar oesophagus/ventriculus ratio. However, according to the original description and figure given by Johnston and Mawson, *A. oceanica* has '... two pairs of large double papillae immediately postanally ...', whereas *A. paggiae* has a single pair of double paracloacal papilla. Moreover, male specimens among the type material of *A. oceanica* had three coarsely denticulate caudal plates, which are much larger and more flattened than those observed in the species *A. paggiae*. In *A. oceanica*, the absolute width of caudal plates 1, 2, and 3 are: wpl1 = 90–100, wpl2 = 80–90, and wpl3 = 80–90 μm. In *A. oceanica*, the tail is also longer (0.20–0.25 mm) than in *A. paggiae* (0.15–7 mm) and it has a pointed distal extremity. Furthermore, distances between distal papillae (d1d2: 70–90; d2d3: 50–60; and d2d4: 30–40 μm) are different from those observed in *A. paggiae*. Finally, a characteristic violin-shaped ventriculus is distinguishing morphological feature of *A. paggiae* (Mattiucci et al., 2005) (Fig. 7).

The application of morphological and genetic datasets combined in a multivariate discriminate analysis, such as the generalized Procrustes rotation (PR), was recently used for combining molecular/genetic and morphological datasets in the systematics of anisakid nematodes (Mattiucci et al., 2014a). PR is a multivariate technique developed to simultaneously compare several datasets. It is based on a traditional singular value decomposition to decompose a matrix into principal components. The main idea of PR has been to compare two or more spaces, where the same variables (here considering individuals from different cryptic species) are measured, by calculating a new set of factors (i.e. dimensions) that resemble all score subspaces. On the other hand, the versatility of the application of discriminant analysis in population genetics, and the molecular systematics of animal populations, including parasites, has been recently emphasized (De Meeûs et al., 2007; Jombart et al., 2010). The first two axes of the PCA ordinations were used in individuals belonging to the three cryptic species of the *A. simplex* (s.l.) complex, to compare the morphometric and genetic traits of the sampled specimens. A Procrustes test (also known as analysis of congruence) was used to estimate the significances of the Procrustes statistics, assessing similarities between different ordinations (Mattiucci et al., 2014a). The discriminant analysis of principal components inferred from both genetic and morphological datasets, based on distance methods, demonstrated that the three cryptic species *A. simplex* (s.s.), *A. pegreffii*, and *A. berlandi*, clustered in three distinct groups, respectively. The PR showed convergence between different ordinations based on the distance values obtained from both genetic and morphometric characteristics among specimens belonging to the three species. Those results suggest that both the morphometric and genetic traits convey similar information, thus providing data useful for the assessment of morphological and genetic variation between specimens of different cryptic species.

4.3 Larval Morphological Features in the Species of Genus *Anisakis*

The *Anisakis* Type I larval morphology (sensu Berland, 1961) is characterized by a long ventriculus, and the presence of a mucron at the tail end, features which may represent the apomorphic (derived) state within species, in the *Anisakis* phylogeny, forming the first (i.e. *A. simplex* (s.s), *A. pegreffii*, and *A. berlandi*), the second (i.e. *A. nascettii* and *A. ziphidarum*), and the *A. typica* major clades (Fig. 2). Actually, no morphological diagnostic features able to distinguish at the species level those Type I larvae have been found. Quiazon et al. (2008) proposed some morphological features at the larval stage, being represented by the ventriculus length between *A. simplex* (s.s.) and *A. pegreffii*.

A combination of molecular and morphological analysis has been also proposed by Murata et al. (2011), for larvae of those species included in

Clade 3. The authors showed by molecular identification of hundreds of *Anisakis* larvae morphologically indicated as Types I, II (sensu Berland, 1961), III, and IV (sensu Shiraki, 1974), which *Anisakis* Types II, III, and IV larvae could be identified genetically as belonging, respectively, to the species *A. physeteris*, *A. brevispiculata*, and *A. paggiae* (Table 9). According to those authors, *Anisakis* Types II, III, and IV were morphologically close to those reported by Shiraki (1974); they have a short ventriculus, tail without mucron but morphologically different (except two Type II larvae that showed a tiny spine-like mucron), being conical and tapering in Type II, rounded (Type II) short and pointed (Type IV) (Table 10). Thus, to conclude on the morphological traits, those three *Anisakis* species so far included in Clade 3 of *Anisakis* phylogeny (i.e. *A. physeteris*, *A. brevispiculata*, and *A. paggiae*) could be differentiated not only at genetic/molecular level, but also by morphological analysis at both larval and adult stages.

Table 9 Morphometric Data of Adults of *Anisakis typica*

Parasite Species	***A. typica***
Specimen (*n*)	7 (4 males, 3 females)
TBL (mm)	70.75 (55.0–80.0)
Oesophagus length	4.8 (3.70–5.60)
VL	1.36 (1.20–1.53)
VL/TBL	0.019
Ventriculus shape	Sinuous
LS	3.11 (2.83–3.40)
RS	0.845 (0.84–0.85)
R/L	0.27
LS/TBL	0.04
RS/TBL	0.011
TL	0.40 (0.22–0.60)
TL/TBL	0.005

Ratio values are reported as a percentage. All measurements are given in millimetres. *n*: number of specimens studied; LS: left spicule; RS: right spicule; TBL: total body length; TL: tail length; VL: ventriculus length; WPL: width of caudal plectanes (Mattiucci et al., 2002, unpublished data).

Table 10 Measurement of Third-Stage Larvae of *Anisakis* Types I, II, III, and IV

***Anisakis* Species**	***Anisakis* spp. (Clade 1[a], 2)**	***A. physeteris***	***A. brevispiculata***	***A. paggiae***
Type of Larvae	**Type I**	**Type II**	**Type III**	**Type IV**
Body length	28.4 (23.00–31.7)	29.14 (22.0–34.50)	32.17 (27.00–35.00)	18.22 (14.00–23.00)
		25.70 (21.60–31.90)	28.90 (23.80–38.40)	20.10 (14.80–26.40)
Body width	0.49 (0.45–0.52)	0.65 (0.51–0.75)	0.86 (0.75–0.95)	0.50 (0.40–0.60)
		0.61 (0.53–0.71)	0.84 (0.65–0.97)	0.54 (0.41–0.67)
Oesophagus length	2.06 (1.76–2.33)	2.11 (1.70–2.40)	1.96 (1.70–2.35)	1.46 (1.25–1.85)
		2.15 (1.80–2.51)	1.83 (1.46–2.33)	1.59 (1.24–2.18)
Ventriculus length	1.30 (1.08–1.46)	0.59 (0.50–0.72)	0.51 (0.45–0.56)	0.31 (0.22–0.40)
		0.58 (0.47–0.71)	0.57 (0.47–0.73)	0.38 (0.30–0.59)
Tail length	0.12 (0.10–0.13)	0.26 (0.16–0.38)	0.12 (0.11–0.15)	0.11 (0.07–0.17)
		0.28 (0.21–0.38)	0.16 (0.10–0.20)	0.15 (0.11–0.19)
Body length/body width	57.95 (51.11–60.96)	45.2 (37.7–50.8)	38.0 (28.4–46.7)	36.6 (30.0–46.0)
Body length/oesophagus length	13.7 (13.06–13.78)	13.9 (11.1–17.4)	16.8 (11.5–20.6)	12.6 (10.3–15.0)
Body length/ventriculus length	21.8 (21.29–22.6)	49.7 (39.0–62.0)	63.7 (49.1–77.8)	59.7 (46.8–81.8)
Body length/tail length	236.6 (230–243.8)	115.0 (84.0–171.0)	261.2 (206.7–291.7)	170.9 (124.1–242.9)

[a]Quiazon et al. (2008) detected some morphological characters in Type I larvae of *A. simplex* (s.s.) and *A. pegreffii*.
A. simplex: Body length (12.75–29.94), body width (0.45–0.75), distance of nerve ring to anterior end (0.21–0.35), oesophagus length (1.18–2.58), ventriculus length (0.90–1.50), ventriculus width (0.13–0.31), ratio between oesophagus and ventriculus length (1:0.9–1:2.3), tail length (0.04–0.14), and mucron length (0.02–0.03).
A. pegreffii: Body length (11.10–26.78), body width (0.38–0.60), distance of nerve ring to anterior end (0.20–0.31), oesophagus length (1.04–2.11), ventriculus length (0.50–0.78), ventriculus width (0.12–0.27), ratio between oesophagus and ventriculus length (1:1.5–1:3.1), tail length (0.05–0.12), and mucron length (0.02–0.03).
All measurements are given in millimetres.
Data from Shiraki, T. 1974. Larval nematodes of family Anisakidae (Nematoda) in the Northern Sea of Japan—as a causative agent of eosinophilic phlegmone or granuloma in the human gastro-intestinal tract. Acta Med. Biol. 22, 57–98; Murata, R., Suzuki, J., Sadamasu, K., Kai, A. 2011. Morphological and molecular characterization of *Anisakis* larvae (Nematoda: Anisakidae) in *Beryx splendens* from Japanese waters. Parasitol. Int. 60, 193–198; Mattiucci, personal observation.

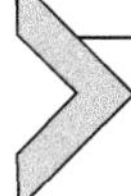

5. HOW DOES *ANISAKIS* SPP. DIVERSITY VARY ACROSS HOST SPECIES?

Marked differences in definitive host preferences have been detected among the species belonging to *Anisakis*. They show distinct host preferences with respect to their cetacean definitive hosts, often exhibiting differential distribution in different host species. The differential distribution patterns are likely the product of ecological processes and, therefore, interactions between parasite populations could promote and even enhance the complexity of multispecies assemblages in a given host (definitive and intermediate). Such differences in life history and host preferences between related species of anisakid nematodes appear to be due to differential host–parasite coadaptation and coevolution, as well as interspecific competition. This again may reduce the range of potential hosts, or promoting differential utilization of resources in a single individual host under sympatric conditions.

5.1 Host Preference vs Definitive Hosts

As described above, multigene sequence analysis has allowed to classify *Anisakis* spp. into distinct phylogenetic clades (Mattiucci et al., 2014a). The phylogenetic relationships between *Anisakis* spp. seem to be supported by ecological data and specific host–parasite relationships. The existence of host specificity among *A. simplex* (s.s.), *A. pegreffii*, and *A. berlandi* for 'oceanic dolphins' and whales, is supported by recent findings (Cavallero et al., 2011; Mattiucci et al., 2014a; Mattiucci and Nascetti, 2008). The three cryptic species of the *A. simplex* (s.l.) species complex are parasites of the cetacean families Delphinidae, Monodontidae, and Phoceanidae, from all the oceanic waters, and can be considered the only species of *Anisakis* parasitizing these cetacean taxa (Fig. 8). The *Anisakis* species share the same species of Delphinoidea as definitive host in different geographical areas, depending on the range of distribution of the actual host and parasite species. The parasites often occur in syntopy in the same definitive host species in areas where their geographical ranges overlap. This was apparently the case in *Delphinus delphis*, *G. melas*, *S. coeruleoalba*, *Phocoena phocoena*, and *T. truncatus*, stranded along the Iberian Atlantic coast, were *A. simplex* (s.s.) and *A. pegreffii* shared syntopically the same cetacean host. Basically the same was observed in *S. coerulealba* stranded along the Japanese coast. Both the Atlantic Iberian coast and the Pacific Japanese coast are considered to be sympatric areas of *A. simplex* (s.s.) and *A. pegreffii*.

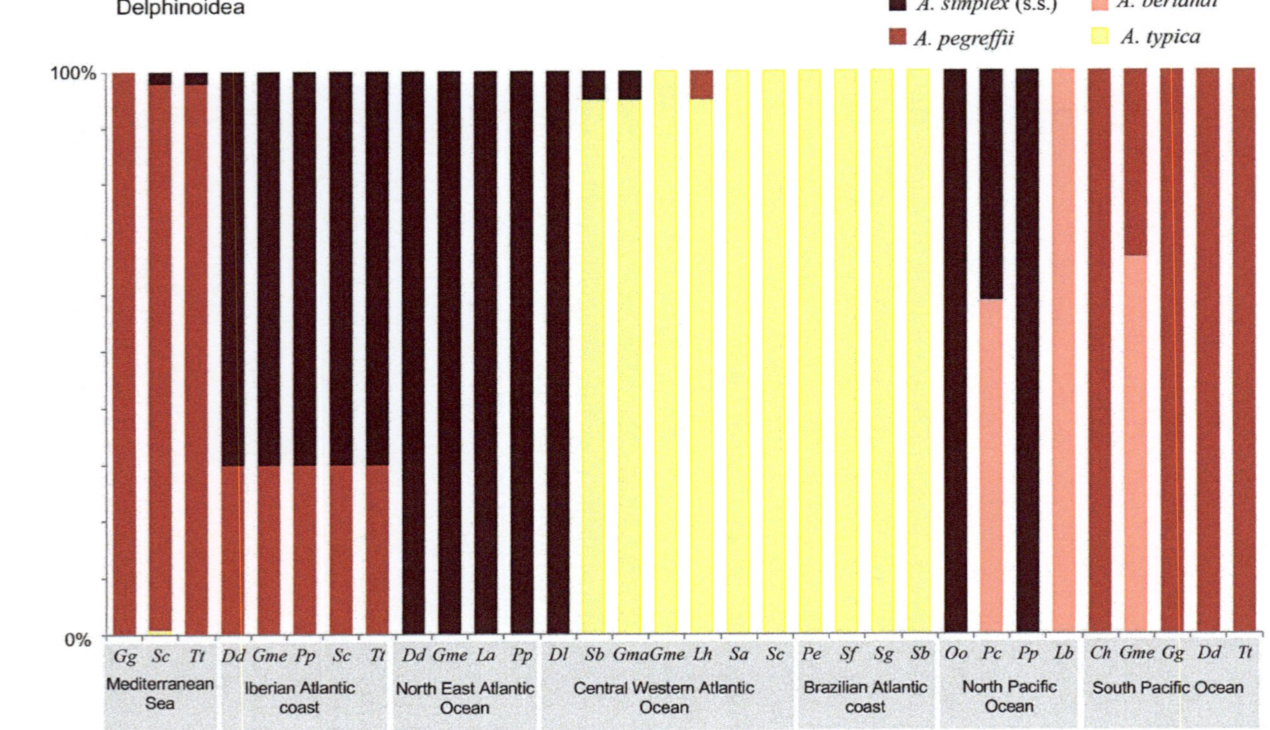

Fig. 8 Distribution of *Anisakis* spp. in various definitive hosts: relative proportions of *Anisakis* spp. in cetacean belonging to species of Delphinoidea. Host codes are reported by alphabetical order: *Ch*: *Cephalorhynchus hectori*, *Dd*: *Delphinus delphis*, *Dl*: *Delphinapterus leucas*, *Gg*: *Grampus griseus*, *Gma*: *Globicephala macrorhynchus*, *Gme*: *Globicephala melas*, *La*: *Lagenorhynchus albirostris*, *Lb*: *Lissodelphis borealis*, *Lh*: *Lagenodelphis hosei*, *Oo*: *Orcinus orca*, *Pc*: *Pseudorca crassidens*, *Pe*: *Peponocephala electra*, *Pp*: *Phocoena phocoena*, *Sa*: *Stenella attenuata*, *Sb*: *Steno bredanensis*, *Sc*: *Stenella coeruleoalba*, *Sf*: *Sotalia fluvialis*, *Sg*: *Sotalia guianensis*, *Tt*: *Tursiops truncatus*.

With regard to the other species of the complex, a coinfection of *A. simplex* (s.s.) and *A. berlandi* was recorded in *Pseudorca crassidens* from the North-East Pacific Ocean, which seems to be a sympatric area of the two *Anisakis* species in the Pacific Ocean. Moreover, *G. melas* was infected with *A. pegreffii* and *A. berlandi* in the South Pacific Ocean, which thus appears to be an overlapping area of these two species.

Some dolphin species, such as *G. melas*, have been found to host all three species of the *A. simplex* (s.l.) complex. Notably, *G. melas*, in South Pacific waters, *A. pegreffii* and *A. berlandi* were recorded in syntopy in the same individual host. On the other hand, the same definitive pilot whale species was found to be parasitized in sympatry by *A. pegreffii* and *A. simplex* (s.s.) in Spanish Atlantic waters (Mattiucci et al., 2014a). Interestingly, the existence of two subspecies of pilot whale was suggested (Rice, 1998), with subspecies *G. melas melas* primarily occurring in the Boreal region whereas subspecies *G. melas edwardii* (Smith) seems to live in the Austral region. The occurrence of *A. simplex* (s.s.) in Boreal individuals of pilot whales, along with the detection of *A. berlandi* in Austral specimens of pilot whales (Mattiucci et al., 2014a), seems to support this hypothesis. Thus, the use of *Anisakis* spp. to gain more insight in migration routes and population structure of their definitive hosts represents a promising perspective.

Some species belonging to the Delphinoidea superfamily such as *Sotalia* spp., *Stenella attenuata*, *Stenella clymene*, *Peponocephala electra*, and *Steno bredanensis*, typically preferring tropical and warmer waters of lower latitudes, are definitive hosts of *A. typica* (Fig. 8). This *Anisakis* species likely requires warmer waters to complete its life cycle. Indeed, the actual cetacean definitive hosts mostly inhabit the Atlantic, Indian, and Pacific Oceans where they are restricted to tropical, subtropical, and warmer temperate regions (Leatherwood and Reeves, 1982). However, it was reported that dolphin species such as *T. truncatus* and *S. coeruleoalba*, infected with both *A. simplex* (s.s.) and *A. pegreffii*, can be parasitized by *A. typica* as well whenever the actual dolphin species occur at warmer temperate latitudes, such as the southern eastern Mediterranean Sea (Mattiucci et al., 2002), or the Central Atlantic Ocean (Gulf of Mexico) (Cavallero et al., 2011).

Whales belonging to families Physeteridae, i.e., *Physeter macrocephalus*, and Kogiidae (i.e. *Kogia breviceps* and *K. sima*), are the main definitive hosts for the species included in clade 3, i.e., *A. physeteris*, *A. brevispiculata*, and *A. paggiae*. Indeed, irrespective of the geographical distribution of the parasite species, *A. physeteris* has been found to parasitize the sperm whale, *P. macrocephalus*, while *A. brevispiculata* and *A. paggiae* are parasites of the pigmy sperm whales *K. breviceps* and *K. sima* (Mattiucci and Nascetti, 2008) (Fig. 9). These whales rely primarily on mesopelagic squid as food source in open Atlantic and Pacific waters Oceans, as well as in the Mediterranean Sea. The rare occurrence of larval stages of *A. physeteris* in fish species seems to indicate that *A. physeteris*—the dominant species of *Anisakis* in sperm whales—has a life cycle involving squid rather than fish. In support of this hypothesis, stranded *P. macrocephalus* from South-eastern Italy, which was parasitized by several hundreds of adults of *A. physeteris* (Mattiucci et al., 2010), had stomachs filled with beaks of histioutheutid cephalopods belonging to *Ancistroteuthis lichtensteinii*, *Dosidicus gigas*, and *Histioteuthis bonnellii* (Mattiucci et al., personal observation). Similarly, the stomach of a stranded

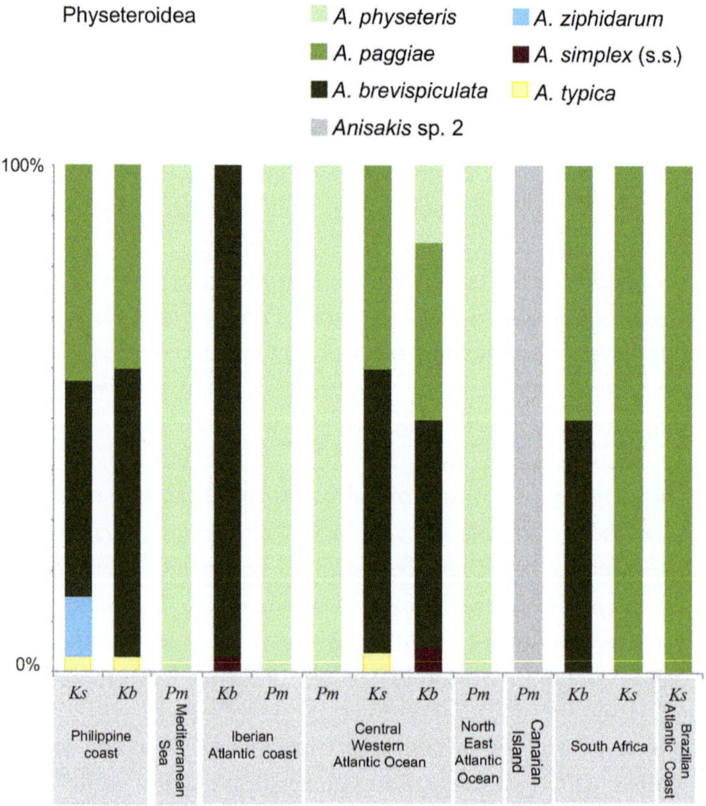

Fig. 9 Distribution of *Anisakis* spp. in various definitive hosts: relative proportions of *Anisakis* spp. in cetaceans belonging to species of the Family Physeteridae. Host codes are reported by alphabetical order: *Ks*: *Kogia sima*, *Kb*: *Kogia breviceps*, *Pm*: *Physeter macrocephalus*.

sperm whale in Greece, contained up to 30,000 beaks of 12 squid species, all known to occur in the Mediterranean Sea (Rendell and Frantzis, 2016). Additionally, larval *A. physeteris* occur in *Illex coindetii* from the Spanish Mediterranean Sea coast (Picó-Durán et al., 2016). Thus, the high parasitic burden of *P. macrocephalus* stranded at the Italian coast in recent years (2009–14) (Blažeković et al., 2015; Cipriani et al., 2017c; Mazzariol et al., 2011), indicates that *A. physeteris* can complete its life cycle in the Mediterranean Sea, involving a population of *P. macrocephalus* as definitive host, which seems to be genetically distinct from the adjacent Atlantic sperm whale populations (Podestà et al., 2016). Interestingly, the species indicated as *Anisakis* sp. 2 (see Section 3), genetically closely related to *A. physeteris*, found at the adult stage in *P. macrocephalus* stranded along the Canary Islands and at larval stages in the swordfish *X. gladius* from equatorial–tropical waters, seems to suggest that a sister taxon of *A. physeteris*, possibly related to the existence of Atlantic populations of *P. macrocephalus* (Podestà et al., 2016), could exist. However, the taxonomic status of that *Anisakis* genotype needs to be investigated at both genetic and morphological levels.

Similarly, the known geographical distribution of *A. paggiae* and *A. brevispiculata* mirrors the known distribution of their main definitive hosts represented by the pigmy sperm whales *K. breviceps* and *K. sima* (Fig. 9). *Kogia* spp. inhabits deep waters in tropical and warm/temperate waters all over the world. The whales feed primarily on deep-sea cephalopods, and less often on deep-sea fish. The report of *A. physeteris*, *A. brevispiculata*, and *A. paggiae* in the stomach of some fish species, such as the swordfish *X. gladius* (Mattiucci et al., 2014b), which primarily preys on squid, seems to support this hypothesis. In addition, its occurrence in deep-sea fish supports the existence of deep-sea life cycles of these *Anisakis* species (Fig. 9).

Finally, beaked whales belonging to the Family Ziphiidae, i.e., *Z. cavirostris* and several species belonging to the genus *Mesoplodon*, are hosts of *A. ziphidarum* and *A. nascettii*, which are partitioned in the distinct Clade 2 in the *Anisakis* parasite phylogenetic tree (Fig. 2). Interestingly, *A. nascettii* has been so far recorded in various species of ziphids in the genus *Mesoplodon*, while *A. ziphidarum* is so far the only species found in Cuvier's beaked whale, *Z. cavirostris* (Fig. 10). Thus, the distribution of these *Anisakis* spp. follows the zoogeography of their definitive hosts. For instance, *A. ziphidarum* is the only ziphid species present in the Mediterranean Sea waters. However, the Cuvier's beaked whale has a cosmopolitan distribution in all the oceans, with the exception of very high-latitude polar regions of both hemispheres (Podestà et al., 2016). Habitat modelling studies of

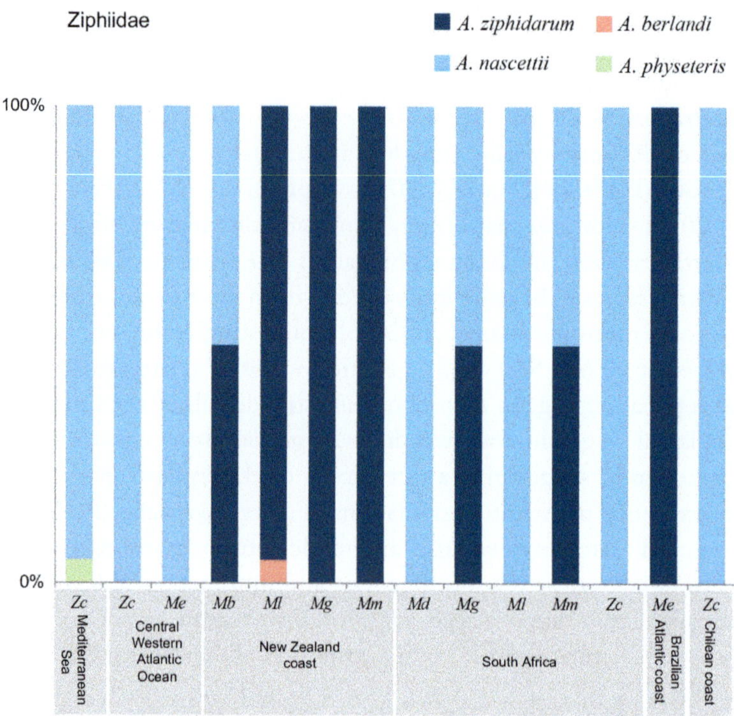

Fig. 10 Distribution of *Anisakis* spp. in various definitive hosts: relative proportions of *Anisakis* spp. in cetaceans belonging to species of the Family Ziphiidae. Host codes are reported by alphabetical order: *Mb*: *Mesoplodon bowdoini, Md*: *Mesoplodon densirostris, Me*: *Mesoplodon europaeus, Mg*: *Mesoplodon grayi, Ml*: *Mesoplodon layardii, Mm*: *Mesoplodon mirus, Zc*: *Ziphius cavirostris.*

Z. cavirostris in the Mediterranean Sea (Podestà et al., 2016) confirmed the species preference of deep waters (>200 m); the species has been often found over the continental slope, and with particular association with submarine canyons, as it has been also reported from other basin waters all over the world. In those areas, the main feeding items are constituted by the deep-sea squid species. Very few larvae belonging to the species *A. ziphidarum* were found in fish species sampled from those areas where the hosts are present (such as Alboran Sea, southern Adriatic/Ionian Sea). Interestingly, larvae of *A. ziphidarum* were genetically identified as only rarely occurring in some demersal fish species such as *M. merluccius* or *Lepidopus caudatus* from Alboran Sea (Cipriani et al., 2017b; Levsen et al., 2017a) and recently reported in *M. merluccius* from the Adriatic Sea (Mladineo et al., 2017b). Thus, taking into account the frequent presence of *Z. cavirostris* in the Mediterranean Sea, the finding of its larvae in fish, even rarely, supports the hypothesis that the life cycle of *A. ziphidarum*, in the Mediterranean Sea, could be maintained.

Indeed, the high burden of *A. nascettii* larvae found in the allotheuthid *Morotheutis ingens* from South Pacific areas supports the hypothesis that *Anisakis* spp. belonging to Clade 2 (i.e. *A. ziphidarum* and *A. nascettii*) also have a deep-sea life cycle (Figs 10–12).

The host–parasite associations between *Anisakis* spp. and different cetacean taxa show a clear pattern of correspondence. The distinct clades formed by *Anisakis* species have been suggested to 'mirror' their host preference, the clades observed in the phylogenetic relationships so far proposed for their definitive hosts. Indeed, according to the phylogenetic hypothesis proposed for the Cetacea by various authors (Arnason et al., 2004; Cassens et al., 2000; McGowen et al., 2014; Milinkovitch, 1995; Nikaido et al., 2001), the Mistycetes represent a monophyletic group, and the odontoceti diverged into the four extant lineages, Physeteroidea (sperm whales), the Ziphiidae (beaked whales), Platanistidae (Indian river dolphins), and Delphinoidea (encompassing the families Iniidae, Monodontidae, Phocoenidae, and Delphinidae). Phylogenetic trees provided by Nikaido et al. (2001), Arnason et al. (2004) and, more recently, by McGowen et al. (2014), were congruent in depicting the branching order of the extant cetacean lineages, where the families Physeteridae and Kogiidae (superfamily Physeteroidea)

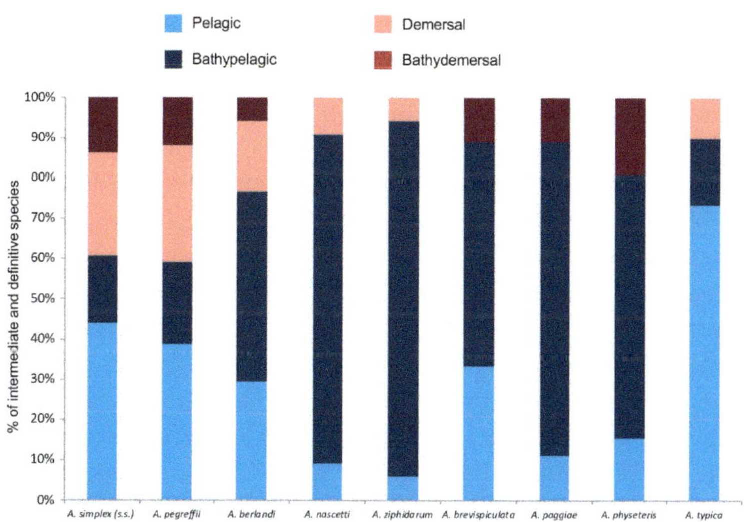

Fig. 11 Distribution (relative proportions) of *Anisakis* spp. according to the ecology of their hosts (fish and cetaceans hosts).

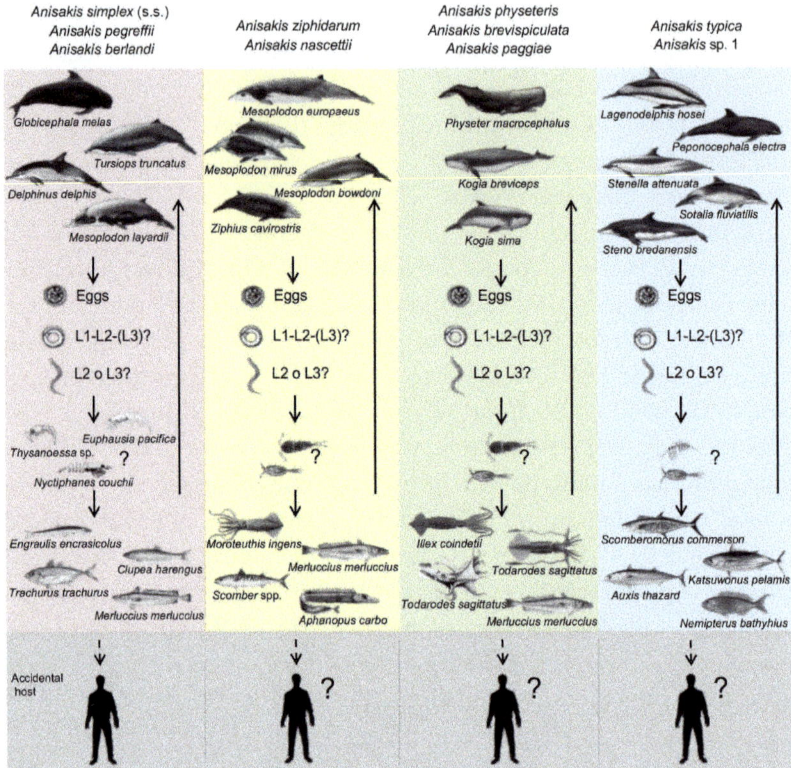

Fig. 12 Schematic hypothetic life cycle of species of the genus *Anisakis*, including their potential zoonotic role in humans. Some of the main definitive (marine mammals) and intermediate/paratenic hosts (fish) of the actual *Anisakis* spp. are shown.

represent basal taxa, followed by the Ziphiidae and the freshwater and the marine dolphins belonging to the superfamily Delphinoidea, as the most derived. In accordance with these analyses, the branching order proposed for the *Anisakis* taxa showed that nematodes from the sperm whale and pygmy sperm whales (i.e. *A. physeteris*, *A. brevispiculata*, and *A. paggiae*) always occupy a basal and well-supported lineage (Clade 3), followed by those parasitizing the beaked whales (*A. ziphidarum* and *A. nascettii*) (Clade 2). The species of the *A. simplex* complex (Clade I), parasites of delphinoids, are the most derived (Mattiucci and Nascetti, 2008). The clade represented by *A. typica*, species occurring mostly in tropical/equatorial dolphin species, still remains to be fully investigated. Also, the possible occurrence and validity of other *Anisakis* spp., e.g., *A. insignis* from the river dolphins of tropical/equatorial areas belonging to the Family Platanistidae, needs further elucidation.

Thus, the empirical results obtained in this present *Anisakis*-cetacean association, suggest that host–parasite codivergence events have been speculated by those *Anisakis* spp., at least at the cetaceans' family level (Mattiucci and Nascetti, 2008).

5.2 Host Preference vs Intermediate/Paratenic Hosts

Currently, according to the those records concerning the intermediate/paratenic hosts identified for the nine species and two genotypes belonging to the genus *Anisakis*, we could assess that a total of teleost fish species and cephalopod species of the marine realm are involved in the life cycles of these parasites.

However, fish species differentially occur as intermediate/paratenic hosts among the so far known species belonging to the genus *Anisakis*. This is in accordance with the ecology of the fish hosts, the geographical range of *Anisakis* spp. and their life cycles. Concerning the fish ecology, according to different ecosystems to which they are associated with in the marine realm, they could be pelagic or demersal fish. The first group includes species which are feeding and live in the open sea and are associated mainly with the surface. Demersal fish can be divided into two main types: strictly benthic fish which are living and feeding on the sea floor, and benthopelagic fish, which are effecting migration in the water column just above the sea floor (the demersal zone which is above the benthic zone); they are feeding both on the bottom, and in midwater.

The fish species mainly involved in the life cycle of the species included in Clade 1 (i.e. *A. simplex* (s.s.), *A. pegreffii*, and *A. berlandi*) are mainly of the pelagic, benthopelagic, and benthodemersal domains. Indeed, the three *Anisakis* species are widely occurring and at high prevalence in pelagic, benthopelagic, and demersal species of the Families Gadidae, Merlucciidae, Scombridae, Carangidae, and Salmonidae. More rarely, they have been found in bathydemersal species, such as the Pleuronectidae. Demersal fish from coastal waters can be found on, or nearby the continental shelf. In deep waters they are found on, or nearby the continental slope. Among those fish species, considered as demersal, there are some, such as the cod *Gadus morhua*, or the European hake *M. merluccius*, and other gadids (found infected by *Anisakis* spp.) which feed on both benthopelagic and pelagic organisms. Those organisms could also include crustacean invertebrates, which, in turn, are the first intermediate hosts in these species.

The oceanic and coastal dolphins in the Delphinoidea, the main definitive hosts of those *Anisakis* spp., are mainly associated with a pelagic and benthopelagic realm. We could assume that the life cycle of the those three *Anisakis* spp. includes pelagic and benthopelagic food webs in those boreal arctic, temperate, and austral temperate waters (Klimpel et al., 2011; Kuhn et al., 2011, 2016; Mattiucci et al., 2014a; Mattiucci and Nascetti, 2008), included in their differential ranges of distribution (Figs 8, 11, and 12).

Similarly, a pathway of transmission throughout a mainly pelagic trophic web can be suggested in the life cycle of the species *A. typica*, which has been found at the larval stage, in pelagic and reef-associated teleost species (Kuhn et al., 2013; Mattiucci et al., 2002), as well as adults in pelagic and coastal Delphinoidea of warmer temperate/tropical basin waters.

A low number of teleost species have so far been identified as intermediate/paratenic hosts of the two species so far included in Clade 2 (*A. ziphidarum* and *A. nascettii*); they mostly belong to demersal teleost species of the Families Merlucciidae, Carangidae, Scombridae, Oreosomatidae, Trachichytiidae, and Tichiuridae. However, in those fish hosts the two species very rarely occur; indeed, the records relative to those fish species consist of very low numbers of larvae identified ($N=1$–5) among the hundreds of other *Anisakis* spp. On the contrary, high abundance (hundreds of larvae) was found in deep-sea squid such as *M. ingens*, whose larvae corresponded genetically to *A. nascettii*. The findings seem to reflect the feeding behaviour of their main definitive hosts (Section 5.1) whose preferred prey are deep-sea squids. As a consequence, it seems to support the hypothesis that, even if only scanty data so far exist on the larval occurrence of the two *Anisakis* species in the oceanographic basin waters, a deep-sea ecology may have evolved in the life-history strategy suggested by the distribution of *A. ziphidarum* and *A. nascettii* (Figs 11 and 12).

Finally, the three species of Clade 3 represented by the genetically (Mattiucci and Nascetti, 2008) and morphologically highly differentiated taxa compared to other *Anisakis* species (*A. physeteris*, *A. brevispiculata*, and *A. paggiae*), were mostly found in bathypelagic and bathydemersal teleosts. However, still only scanty data exist with respect to the occurrence of the latter parasite species in squid, which likely represent the most suitable paratenic hosts in their life cycle (Figs 9, 11, and 12).

The finding of high prevalence and parasitic burdens with hundreds of larvae belonging to *A. physeteris*, *A. brevispiculata*, and *A. paggiae*, in the stomach of the swordfish *X. gladius* from the Central Atlantic Ocean (Garcia et al., 2011; Mattiucci et al., 2014b), and the co-occurrence of squid beaks in that fish species, are in accordance with the finding of several hundreds of *A. physeteris* in sperm whales which harboured also hundreds of cephalopod beaks in their stomach (Section 5.1). These findings seem to suggest that the parasites have evolved strategies to extend their life cycles throughout a bathypelagic food web which likely involves squid rather than fish, in all those oceanographic basin waters included in the geographical ranges of their definitive hosts (Figs 11 and 12).

6. MOLECULAR EPIDEMIOLOGY OF *ANISAKIS* SPP. IN FISHERIES

According to the World Review of Fisheries, wild marine fish catches have increased from 90.0 million metric tonnes, estimated in 2006 to 90.4 million tonnes in 2011. According to the Food and Agriculture Administration (FAO) (2012), also aquaculture production showed a great increase, from 47.3 million tonnes in 2006 to 63.6 million tonnes in 2011.

The latter includes both inland and marine fisheries (Fayer, 2015). Indeed, major changes in fish consumption have been taking place in that period: the per capita fish consumption nearly doubled in the past 4 decades, and there is also a higher demand for seafood products to be consumed raw (Fayer, 2015). These changes raise concern about seafood safety and quality. Among the parasites affecting seafood products, *Anisakis* spp. is of both health and economic importance to fishery products. The zoonotic implications associated with this parasite are a major concern, and the presence of this nematode in seafood products, even when worms are dead, may significantly lower its aesthetic appeal.

6.1 What Drivers Shape the Distribution of *Anisakis* spp. in Wild Fisheries?

As reported in Sections 3 and 5, *Anisakis* spp. larvae affect different species of teleost fishes. Most of these fish species are characterized by high commercial value in several fisheries of the world, including the NE Atlantic Ocean and the Mediterranean Sea, the western Atlantic Ocean, the southern Atlantic, and the northern and southern Pacific.

In the last decade, the new diagnostic methods to detect these nematodes in fish and squid, and the application of molecular/genetic markers in identifying *Anisakis* spp. larvae, have greatly increased our knowledge of their geographical distribution. These new tools permitted to highlight ecological characteristics responsible for shaping the distribution and infections with these parasites. Indeed, various biotic and abiotic factors (hereby called drivers) can explain the variation of the infection levels, in terms of prevalence and abundance data so far registered for these parasites from different fish species and oceanic waters.

Whenever considering the 'drivers' that may help to elucidate the epidemiological data recorded for *Anisakis*, it is important to bear in mind that the transmission pathways of *Anisakis* spp. are accessible throughout well-defined interactions among organisms at different levels of the trophic webs in marine ecosystems. Successful transmission requires a wide availability of hosts with a substantial population size, and the stability of marine trophic webs. These factors, in turn, maintain large effective parasite populations. Any successful completion of the *Anisakis* spp. life cycle requires therefore stable trophic webs and suitable environmental conditions (Mattiucci et al., 2015a, b; Mattiucci and Nascetti, 2008).

Consequently, the infection levels of *Anisakis* species observed in wild fisheries from a given geographic area may be also strongly affected by the whole asset of ecological conditions shaping the population size of any of the host organisms that are involved in the parasite life cycle. Not only biotic factors (in both definitive and intermediate hosts), but also abiotic environmental parameters such as water temperature and salinity, are influencing the biogeography and infection dynamics of *Anisakis* species.

In a modelling analysis of biotic and abiotic factors in the Baltic Sea, Podolska and Horbowy (2003) reported that both were responsible for the *Anisakis* infection levels recorded in Baltic herring, explaining 71% of the deviance for the prevalence, and 25% for the intensity. Recently, an integrated method based on the maximum entropy approach (maxent permutation analysis) was used to study the habitat suitability of endoparasites, such as *Anisakis* spp., in the marine realm (Kuhn et al., 2016). Based on this approach, the following variables were included as a measure of contribution to habitat suitability: depth, mean surface temperature and its range, salinity and primary production. These variables, together with the modelled suitability of definitive hosts (considered as biotic factor), were used to estimate, in a final model, the habitat suitability of the so far known *Anisakis* spp. The elaboration obtained from this analysis, mostly concerning anisakids species of genus *Anisakis* (Kuhn et al., 2016), has suggested that abiotic and biotic factors diversely affect the early larval stage and adult forms of endohelminth parasites of the marine realm. This first approach to integrate variables to explain the actual distribution of *Anisakis* spp. worldwide requires inclusion of other factors in the model. For instance, the distribution and demography of intermediate/paratenic hosts, whose role is relevant to maintain the distribution and infection levels by different species of *Anisakis*, would represent another relevant factor that should be taken into account.

Thus, we will here treat and discuss in detail some of these 'drivers'. However, no single variable alone can explain the differences in infection characteristics between different *Anisakis* spp. in wild fisheries throughout the world. Only a multivariate analysis of the major factors that shape the distribution of *Anisakis* spp. could offer an overall view on how they affect each other in a synergistic manner.

It is worth to note that most of the epidemiological estimates of the *Anisakis* spp. infections from the literature mainly concern fish species of major commercial importance in wild fisheries. Thus, most of the epidemiological drivers that are herein discussed deal with *Anisakis* spp. infecting fish species of economic importance in commercial fisheries, and mostly related to geographical areas from which cases of human anisakiasis have been reported (see also Section 8).

6.1.1 Temperature, Salinity, and Oceanographic Conditions

Temperature, salinity, and other abiotic factors could affect the hatching of eggs and the survival and dispersion of the first larval stages of anisakids (Marcogliese, 2001). In this respect, it was demonstrated experimentally by Højgaard (1997) that the hatching time and the survival of hatched larvae of *A. simplex* (s.l.) [which, according to the host used in the experiments and its geographical origin, likely corresponded to *A. simplex* (s.s.)] was inversely related to temperature (5–21°C), but did not vary significantly with salinity (0–28 psu). Larval survival increased slightly with salinity, but decreased with increasing temperature (Højgaard, 1997). The proportion of hatched eggs was the highest at 12°C and the lowest at 21°C (Højgaard, 1997). However, the experiments were not carried out at higher temperature (>21°C) or salinity (>28 psu). Unfortunately, these results could not explain the effect of higher temperatures and salinity, which, however, appear to be key abiotic characteristics of latitudes in which *A. simplex* (s.s.) does seemingly not find suitable ecological conditions to complete its life cycle. In other words, water temperature and salinity seem to be major limiting factors responsible for the presence or absence of different *Anisakis* species from different climatically or hydrologically defined water basins. Such condition could occur, for instance, in the Mediterranean Sea, where mean temperatures vary from 12.8–13.5°C in the western basin to 13.5–15.5°C in the east, and high mean salinities of 37.5–39.5 psu are generally recorded (Coll et al., 2010). Thus, in the

Mediterranean Sea *A. simplex* (s.s.) has only been found in fish and cetacean hosts from the Alboran Sea, which, from an oceanographic point of view, is more similar to Atlantic waters than to central or eastern parts of the Mediterranean Sea. The latter parasite species is only very rarely reported in other areas of the Mediterranean Sea, mainly related to migratory fish and cetacean species (see Section 7), thus suggesting that *A. simplex* (s.s.) cannot complete its life cycle in this particular Mediterranean basin. The data further indicate that both temperature and salinity could be relevant abiotic factors in shaping the geographical distribution of *A. simplex* (s.s.), and completing its life cycle in the northern hemisphere. Additionally, Højgaard (1997) found that light exposure tended to reduce hatching time, thus suggesting that the life cycle of *A. simplex* (s.l.) is adapted to off-shore pelagic marine environments (Højgaard, 1997). This hypothesis has been supported over the years by the genetic identification of adults belonging to the species *A. simplex* (s.s.) from several cetacean host species considered as 'oceanic dolphins', and in several fish and squid species of pelagic and benthopelagic domain, thus not associated with light-exposed habitats.

Finally, the recruitment of *A. simplex* (s.l.) larvae may be affected by oceanographic conditions, e.g., the presence of upwelling or downwelling conditions. Pulleiro-Potel et al. (2015) suggested that seawater depth has a relative importance on the distribution of these parasites. In a parasitological survey carried out on nematode infections in different mesozooplankton communities from the Galician coast of the NE Atlantic Ocean, Gregori et al. (2015) hypothesized that larval recruitment of *A. simplex* (s.l.) to the zooplankton communities could be directly conditioned by oceanography. They suggested the existence of a pattern in the Ría de Vigo upwelling–downwelling system, hypothesizing that the recruitment of *Anisakis* spp. in summer communities was higher than in autumn, when the upwelling conditions and strong water instability prevailed (Gregori et al., 2015). A similar trend was suggested by Pascual et al. (2007) who hypothesized that stability in water masses enhances parasite recruitment, especially for heteroxenous parasites having multiple-host life cycles, as for the *A. simplex* (s.l.) species complex. On a global geographical scale, the same authors (Pascual et al., 2007) concluded that in upwelling systems the parasite faunas are generally impoverished, whereas downwelling events provide optimal conditions for successful recruitment of the larval stage to pass to the successive intermediate/paratenic hosts. On the other hand, in a modelling study, Davis et al. (1991) found that the growth rate of predators is strongly dependent on the existence of patches in which prey density is well above the large-scale average. Physical turbulence leads to more frequent encounters between predators and prey and has the effect of restoring predator growth rates to low turbulence values (Mann and Lazier, 2006).

Characteristic upwelling phenomena that occur all along the western Portuguese coast, promoting high primary production, and directly feeding a wealth of organisms, including the first intermediate hosts of *Anisakis* spp. larvae, could offer an explanation for the high infection rates of *A. pegreffii* and *A. simplex* (s.s.) in *Pagellus bogaraveo* (Hermida et al., 2012).

In the Alboran Sea (Mediterranean Sea), the inflowing Atlantic water describes a quasi permanent anticyclonic gyre in the West, and a more variable one in the East. The particular water circulation in the Alboran Sea generates an ocenographic front from Spain (Almeria) to Oran (Morocco), known as 'Almeria-Oran front' (AOF) (Patarnello et al., 2007). This oceanographic front is characterized by a steep gradient of temperature (1.4°C) and salinity (2 ppt), and a water current speed of 40 cm/s (Tintore et al., 1988). The AOF has been considered the real boundary between Mediterranean Sea and Atlantic Ocean, instead of the Gibraltar Strait, and this discontinuity seems to involve both abiotic and biotic factors, such as stocks/populations of various marine fish species. As a consequence, the oceanic anticyclonic gyre at AOF has been considered responsible for the mixed infection by *A. pegreffii* and *A. simplex* (s.s.) so far evidenced in fish caught in these waters (Mattiucci et al., 2014b). Several intermediate/paratenic teleost host species, such as *M. merluccius*, *Trachurus trachurus*, *S. scombrus*, *Lophius piscatorius*, and *Micromesistius poutassou*, were found to harbour both the main predominant species of the Mediterranean Sea, i.e., *A. pegreffii*, and, at lower prevalence, *A. simplex* (s.s.) (Gómez-Mateos et al., 2016; Levsen et al., 2017a; Mattiucci, personal observation; Mattiucci et al., 2014b). Indeed, both first larval stages and/or first (crustacean) intermediate hosts, infected with *A. simplex* (s.s.), could be drawn into this area of the Mediterranean Sea from the Atlantic, along with inflowing water driven by the upwelling–downwelling system which again is generated by the anticyclonic gyre at AOF (Cipriani et al., 2017a, b; Mattiucci et al., 2014b).

Cipriani et al. (2017b) suggested that some peculiar oceanographic characteristics of the central area of the Adriatic Sea, a rather narrow basin (Russo and Artegiani, 1996), would represent one of the factors explaining the high level of infection recorded in two commercially important fish species, i.e., *M. merluccius* and *Engraulis encrasicolus* (Cipriani et al., 2017a, b). Circulation patterns in the Adriatic Sea are generally cyclonic; a northward current flows along the eastern coast, while a southward current flows along the western coast (Russo and Artegiani, 1996). Unlike the other subbasins, the central Adriatic Sea seems to have a constant nutrient supply, partly due to the proximity to river inputs, and partly due to cyclonic gyres, which provokes a constant upwelling of deep nutrient-rich waters to the surface. These conditions have a direct impact on the presence, quantity, and distribution of a conspicuous biomass of phyto- and zooplanktonic organisms—the latter acting as first intermediate hosts of *A. pegreffii* in that water basin. As a consequence, the large biomass of planktonic organisms could enhance the presence of small pelagic planktivorous fish, such as anchovies and sardines, which act as second intermediate or paratenic host for this nematode. On the other hand, both water circulation and intensity of primary production have been cited as regulating the intensity of anchovies spawning in this particular area of the Adriatic Sea (Regner, 1996). The abiotic conditions of the central Adriatic Sea, which directly influence the biotic conditions by favouring the presence of a stable food web, may result in enhancing the overlapped and abundant distribution of hosts belonging to the different trophic levels of the life cycle of *A. pegreffii*. In turn, this could explain the distribution and high infection levels of this parasite in this semienclosed and narrow basin. In other words, both abiotic (water circulation) and biotic factors could represent a direct advantage for the persistence of the parasite life cycle in this area of the Adriatic Sea.

In South American Atlantic waters, the water circulation is characterized by the southern extension of the Patagonian Shelf—which is one of the largest and most productive ecosystems of the southern hemisphere—and the northern extension of the Malvinas current. The two currents run close to the coast and meet at the continental slope, near 38°S, creating a strong frontal zone, which establishes the division between subtropical and subantarctic waters (Piola and Rivas, 1997). Interestingly, the Malvinas current produces a latitudinal gradient of temperature, that decreases southward, while in the northern part of the Argentine coast the warm Brazilian current becomes influential (Cantatore and Timi, 2015). This oceanographic front would be responsible, among the other abiotic variables, in shaping the current distribution and prevalence of *Anisakis* spp. in these waters. Indeed, while fish captured in warmer waters of coastal Brazil seem primarily to be infected with *A. typica* (Iñiguez et al., 2009, 2011), a species typically spot in warmer waters, while fish from northern Argentine waters have been found infected with *A. simplex* (s.l.) (likely corresponding to *A. pegreffii*) (Timi et al., 2014; Timi and Mattiucci, personal communication). According to these authors (Cantatore and Timi, 2015), the infection level of *A. simplex* (s.l.) in this area follows a latititudinal scheme, with a pronounced increase in prevalence, reaching up to 100% at southern latitudes (between the 44° and 46°S) in the gadoid species *Merluccius hubbsi*. However, below these latitudes, it seems that the life cycle of species of the *A. simplex* (s.l.) occurring in the southern hemisphere (i.e. *A. pegreffii* and *A. berlandi*) cannot take place. In support of this hypothesis, Klimpel et al. (2010) stated that the occurrence of a few larval specimens of *A. berlandi* (=*A. simplex* C) and *A. pegreffii* in myctophids from the southern waters of the southern Atlantic Ocean (i.e. South Shetland Islands) could be related to the introduction of the parasite species from outside Antarctica through their migrating teleost paratenic hosts. Indeed, the same authors found *A. pegreffii* and *A. berlandi* larvae in migratory myctophid fish, while the endemic Antarctic fish were *Anisakis* free (Klimpel et al., 2010), despite the fact that some cetacean species (such as *Balaenoptera physalus* and *P. macrocephalus*) commonly migrate into the subantarctic and high-antarctic oceanographic basin waters (Leatherwood and Reeves, 1983).

Thus, it seems that water temperature, which has been recognized as a good predictor of latitudinal gradients in diversity of parasite species (Rohde, 2005), seems to play an evolutionary adaptive role for the species of the genus *Anisakis*, contributing to shape their spatial distribution and diversity.

6.1.2 The Fishing Ground

The distribution of *Anisakis* spp. in a given fish host is directly linked to the abiotic and biotic factors that characterize a given fishing ground, ecologically, and hydrologically (Section 6.1.1). In some cases, both fish host and parasite population distributions are apparently shaped by the same environmental factors (Mattiucci et al., 2015a).

Several studies included multifactorial statistical analyses to identify which factors could better explain the *Anisakis* larvae burden in fish in relation to host size geographical catching area and sampling year and season. Fishing area turned out to be the dominant variable associated with parasite prevalence and abundance (Cipriani et al., 2017a, b; Levsen et al., 2017a; Levsen and Karl, 2014). These findings could be helpful to evaluate the influence of fishing ground as a major driver of the molecular epidemiology of *Anisakis* spp. in fish species.

Some authors have found significant differences in relative proportions, or in the abundance of infections with different *Anisakis* spp. from the same fish species but captured at different geographical fishing areas. Interestingly,

Cho et al. (2015) have found that in the East and South Korean Seas the relative proportion of *A. pegreffii* recorded in several fish species was constantly around the 86%, while the remaining anisakids were represented by *A. typica* larvae (8.6%) and *A. simplex* (s.s.) (5.4%). Chub mackerel fished from seven geographically different sampling areas of the Korean Sea showed the same distribution and relative percentage of *Anisakis* spp. as observed in the Tsushima Current stock of *S. japonicus* from Japan (Bak et al., 2014). The latter fish stock was dominated by *A. pegreffii*, while a very small proportion (2.3%) was represented by *A. simplex* (s.s.). The latter authors suggested that the findings are consistent with the existence of a Korean Sea stock population of chub mackerel, different from the Japanese chub mackerel population (Bak et al., 2014).

In Taiwanese waters, Chou et al. (2011) reported a difference in relative proportion of *A. pegreffii* and *A. typica*, with *A. pegreffii* prevailing in *Scomber australicus* from coastal waters off southwestern Taiwan. A similar trend was found in waters off northeastern Taiwan by Chen and Shih (2015), who analysed the distribution of five species of *Anisakis* (i.e. *A. pegreffii*, *A. physeteris*, *A. brevispiculata*, *A. paggiae*, and *A. typica*), where *A. pegreffii* had highest relative proportion in *S. australicus*. In both cases, the authors suggested the use of these parasite species as 'biomarkers' for further elucidation of the actual fish stock compositions.

Chen and Shih (2015) hypothesized that the ENSO (El Niño Southern Oscillation) phenomenon could have resulted in a more than 2.7°C increase in seawater temperature over the past 100 years, and to changing ocean currents, which in turn could have provoked a differential distribution of fish and cetaceans host species, thus explaining the differential distribution of *Anisakis* species.

Umehara et al. (2010) found a differential distribution of three *Anisakis* species (i.e. *A. pegreffii*, *A. simplex* (s.s.), and *A. typica*) occurring in hairtail fish (*Trichiurus* spp.) caught from off the coast of Taiwan and in Japan Sea waters. Interestingly, the hairtail captured from off the southeastern coast of Taiwan was primarily infected with *A. typica* (84%) and *A. pegreffii* (16%). However, the same fish species captured in coastal waters of the Sea of Japan was markedly different, with a higher relative proportion of *A. simplex* (s.s.) in hairtails from fishing grounds along the Pacific coast of Japan, while *A. pegreffii* dominated the fishing ground close to the Nagasaki Prefecture along the southwestern coast of Japan.

A large parasitological survey carried out by Quiazon et al. (2011) on several fish species from Japanese waters demonstrated that the distribution of genetically identified *Anisakis* spp. (i.e. *A. pegreffii*, *A. simplex* (s.s.), *A. physeteris*, *A. brevispiculata*, *A. paggiae*, and *A. typica*) was related to fishing ground rather than being host specific, particularly with regards to *A. pegreffii* and *A. simplex* (s.s.). The authors concluded that *A. simplex* (s.s.) was mainly present in fish from northern Japan including the Hokkaido area, while *A. pegreffii* predominated in fishes from the Sea of Japan to the East China Sea (Fig. 13). Concordant results were obtained by Kong et al. (2015) identifying *A. pegreffii* in fish species of commercial importance from the fishing grounds of the East China Sea. Similarly, fish sampled from the Yellow Sea (China) by Du et al. (2010) were dominated by *A. pegreffii*, while a few larvae from some pelagic fish species corresponded to *A. typica*; the latter finding was related to fish specimens migrating from the southern area of the Pacific Ocean.

The findings suggested that *A. simplex* (s.s.) infected *Oncorhynchus keta* likely belonged to a Japanese stock which members spend most of their life in Arctic Boreal region of the Pacific Ocean, up to the Bering Sea, where the dominant anisakid species is *A. simplex* (s.s.) (Quiazon et al., 2011). In accordance with the authors' hypothesis, *A. simplex* (s.s.), as the only

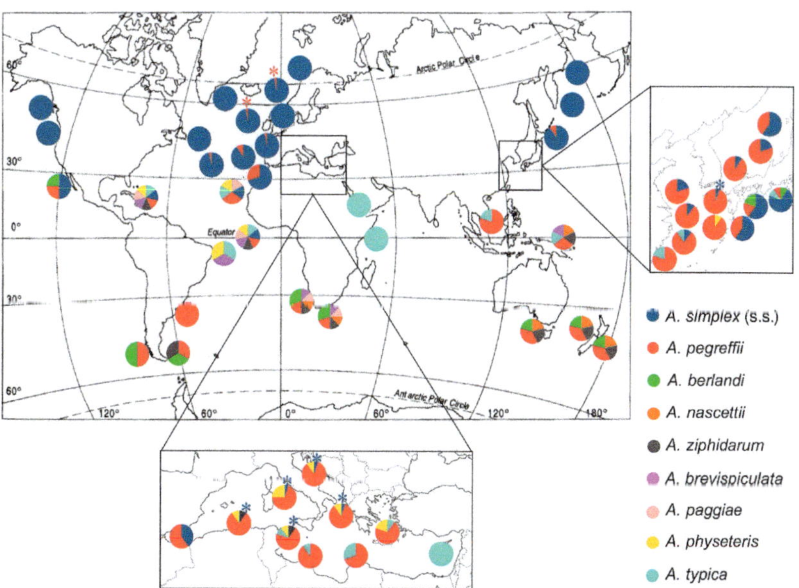

Fig. 13 Relative proportions of *Anisakis* spp., according to different fishing grounds (geographical area). Data acquired from the molecular epidemiology of *Anisakis* occurring in fish of commercial importance. The asterisk (*) indicates that finding is related to its host's migration.

species present in that fishing area, was previously reported to infect *O. keta* and several salmonid species from the Bering Sea and Hokkaido Island (Mattiucci et al., 1998; Quiazon et al., 2009).

Interestingly, a similar trend was seen in *S. japonicus* from the Sea of Japan and East China Sea of the Pacific Ocean (Suzuki et al., 2010). The highest proportion of larvae identified in fish captured off the Nagasaki Prefecture in the East China Sea to off the Ishikura Prefecture in the Sea of Japan was represented by *A. pegreffii* larvae, while *S. japonicus* fished along the Pacific coast of Japan was dominated by *A. simplex* (s.s.) (Suzuki et al., 2010). These findings were related to the existence of different stocks of *S. japonicus* from these geographical areas, and different stock-specific migration routes. Additionally, the role of one of the main definitive hosts of *A. simplex* (s.s.) in the Arctic Boreal region, *Balaenoptera acutorostrata* (Mattiucci and Nascetti, 2008; Umehara et al., 2006), which occurs in two genetically distinct populations in the North Pacific Ocean and in the North Atlantic Ocean waters (op. cit.), could play an important role in maintaining the life cycle of *A. simples* (s.s.) in Pacific Ocean waters off Japan.

It has been widely reported that fish species collected from the fishing grounds of the southern basin waters of the NE Atlantic Ocean, along the Atlantic Iberian coast, harbour mixed infections with the two zoonotic species *A. simplex* (s.s.) and *A. pegreffii*. The area likely represents the southern limit of distribution of *A. simplex* (s.s.), and the northern one of *A. pegreffii*. However, the relative proportions of the two species, and their distribution in demersal and pelagic fish species from those fishing grounds, could vary northwards and southwards off the coast of the Iberian Peninsula (Fig. 13). Mattiucci et al. (2004, 2008) found similar relative proportions of the two *Anisakis* species in two fish species, i.e., *M. merluccius* and *T. trachurus*. Indeed, in the North-East Atlantic hake samples from the northern Strait of Gibraltar, *A. simplex* (s.s.) was the most prevalent species, occurring in sympatry with *A. pegreffii* from the Spanish Atlantic coast. Hakes that exhibited mixed infections of *A. simplex* (s.s.) and *A. pegreffii* represented >20% of the fish examined from the Atlantic coast of Galicia, and 14% from the Bay of Biscay (Mattiucci et al., 2004). Moving south from the Strait of Gibraltar, in the Atlantic Ocean, mixed infections of different species of *Anisakis* in hake caught along the Atlantic coast of Morocco were observed. More than 22% of the fish examined were found to be parasitized by five species of *Anisakis*: the most common was *A. pegreffii* followed by *A. physeteris*, *A. ziphidarum*, and, at a lower percentage, *A. brevispiculata* and *A. paggiae*. Thus, according to the distribution and infection patterns of larval *Anisakis* spp. in hake, different populations of *M. merluccius* seem to exist in European waters. The larval distribution and abundance patterns of the different species of *Anisakis* in hake from different fishing grounds indicate that: (1) separate stocks of *M. merluccius* exist in the Mediterranean and in the Atlantic; (2) at least two distinct hake subpopulations exist in the North Atlantic; one north of the Gibraltar Strait and another one off the Atlantic coast of Morocco; and (3) some substructuring of the hake populations in western and eastern parts of the Mediterranean Sea has been observed, with the population from the Levantine Sea being distinct from other Mediterranean populations (Mattiucci et al., 2015a). Interestingly, the prominent role of selection in moulding the genetic stock structure of the fish hosts, between and within sea basins (Cimmaruta et al., 2005; Milano et al., 2014), seems to indicate that water temperature and salinity are the main drivers that shape the genetic structure of a number of outlier loci in the European hake. Those findings also suggested that the European hake represents a case study, showing how marine environmental variables, such as temperature and salinity, are correlated both to the genetic structure of the fish host species (Cimmaruta et al., 2005), and to the pattern of distribution of the larval parasite *Anisakis* spp. (Mattiucci et al., 2015a, b). Similarly, *A. pegreffii* and *A. simplex* (s.s.), showed significant differences in their relative proportions in the horse mackerel, *T. trachurus*, caught in different fishing grounds of the NE Atlantic and the Mediterranean Sea. Indeed, *A. pegreffii* was identified as the predominant species parasitizing horse mackerel in the Mediterranean Sea; it was the only species found in Adriatic and Ionian Sea samples while in the horse mackerel from the Aegean Sea, *A. pegreffii* occurred in mixed infections with *A. typica*, outnumbering the latter species. Similarly, *A. pegreffii* was the dominant species in the western Mediterranean Sea although *A. physeteris* occurred in mixed infections with the former in horse mackerel fished along the Tyrrhenian Sea coast. Mattiucci et al. (2008) found that the samples of horse mackerel fished in the Alboran Sea showed mixed infections of both sibling species, *A. pegreffii* and *A. simplex* (s.s.), almost in equal proportions. In Alboran Sea fishing grounds, the parasite species are even present in the squid *I. coindetii*, which harboured *A. pegreffii* and *A. physeteris* (Picó-Durán et al., 2016). The latter authors also discussed the lower parasite recruitment in the Almeria-Oran front (AOF), associated with the instability of trophic interactions which may be due to the instability in water masses caused by physical perturbation in the zone (Picó-Durán et al., 2016).

Moving to the NE Atlantic, the prevalence of *A. pegreffii* progressively declines from almost 87% off the Algarve coast to 30% in horse mackerel from off Spanish Galicia (Levsen et al., 2017a,b; Mattiucci et al., 2017b). *A. pegreffii* was also rarely identified from horse mackerel caught off the south coast of Ireland and from the Bay of Biscay, likely due to the migration route of the southern stock of horse mackerel in that basin water. The opposite trend was found in the occurrence of *A. simplex* (s.s.), which showed increasing proportions, from south to north, in horse mackerel fished off the Portuguese coast to off the coast of Spanish Galicia. The latter became the only species present in the fish species caught in the Norwegian Sea and along the western English channel (Levsen et al., 2017a,b; Mattiucci et al., 2017b). In the Alboran Sea, the two species were found in syntopy in almost 60% of the infected fish, while most (35%) of the remaining fish examined from this basin were found infected only by the species *A. pegreffii* (Mattiucci et al., 2008, 2015a) (Fig. 13).

Differential distribution of *Anisakis* spp. larvae in *P. bogaraveo* from different fishing grounds off the Portuguese coast of the NE Atlantic Ocean, was also detected (Hermida et al., 2012). Interestingly, the samples from the mainland were mainly found to be infected with *A. simplex* (s.s.) and *A. pegreffii* while those from the Azores and Madeira were found to be infected with four species, i.e., *A. pegreffii*, *A. physeteris*, *A. ziphidarum*, and *A. typica*, and the genotype indicated as *Anisakis* sp. B (likely corresponding to the genotype indicated as *Anisakis* sp. 2, in Garcia et al., 2011; Mattiucci et al., 2014b; in this review). A similar *Anisakis* spp. distribution was found in European hake from Morocco-Mauritania Atlantic coast (Mattiucci et al., 2004). Indeed, this fishing ground including Madeira and the Canary Islands is located at the confluence of different geographic areas, i.e., the North Atlantic, the Central Atlantic, and the Mediterranean Sea. This oceanic area, surrounding the Canary Islands, is included in the migration route of several cetacean species, which are definitive hosts of several *Anisakis* spp., in a latitude of temperate and warmer-tropical ecosystems, where occurs the overlapping of distribution of the those species of *Anisakis* whose life cycle is likely more adapted.

This hypothesis seems to be supported by the findings of Farjallah et al. (2008b), according to which the distribution and relative proportions of *A. pegreffii*, *A. typica*, *A. ziphidarum* in some demersal and pelagic fish species from off the Atlantic Morocco-Mauritania coast was quite similar to those reported by Mattiucci et al. (2004) and Hermida et al. (2012).

Similar findings have been reported from other fish species of the Atlantic Spanish coast, as well as from the cooccurrence of *A. pegreffii* and *A. simplex* (s.s.) in the Alboran Sea waters (Cipriani et al., 2017b; Gómez-Mateos et al., 2016; Levsen et al., 2017b). Recently, Bao et al. (2015) found that the fish length and the geographical location explained, respectively, 71.5% and 35.8% of the deviance for the infection of the two species *A. pegreffii* and *A. simplex* (s.s.) in *Alosa alosa* and *Alosa fallax*, from the Western Iberian coast of the Atlantic Ocean. The same authors found that levels of the infection by *A. pegreffii* were higher than those by *A. simplex* (s.s.) in all the Western Iberian fish. Being the species *A. pegreffii* the prevailing one of the southern coast of Iberian Peninsula, i.e., along the Portuguese coast, the authors suggested that fish species acquired a high infection rate with *A. pegreffii* during its feeding migration in those southern areas (Bao et al., 2015).

Similarly, while *A. simplex* (s.s.) and *A. pegreffii* seem to co-occur in mackerel, *S. scombrus,* caught off the NE Atlantic coast of Spain and Portugal, several *A. pegreffii* were also recorded in the viscera of some mackerel from the North Sea and southern Norwegian Sea (Levsen et al., 2017a). The findings imply that the actual mackerel started their feeding migration in areas south to the British Isles. Thus, *A. pegreffii* could act as biological marker for the assessment of stock origin of individual mackerel fished along its range of distribution (Levsen et al., 2017a).

The fishing ground of origin of a certain fish species can often explain the differences not only in the distribution of *Anisakis* spp., but also of their infection levels in the same fish host when sampled from different fishing areas. This is, for instance, the case of the mackerel *S. scombrus*, which shows significant differences in the infection values of the two species (i.e. *A. simplex* (s.s.) and *A. pegreffii*), in terms of both prevalence and abundance in the fish captured from the NE Atlantic Ocean vs those fished in the Mediterranean Sea. Indeed, mackerels caught in the NE Atlantic fishing grounds were markedly more infected with *A. simplex* (s.s.) than those mackerel belonging to Mediterranean populations (Levsen et al., 2017a). A similar trend was observed between hake (*M. merluccius*) collected from NE Atlantic Ocean, with respect to hake belonging to a Mediterranean stocks: the Atlantic fish showed considerably higher level of infection with *A. simplex* (s.s.) compared with the infection by *A. pegreffii* in the Mediterranean areas (Cipriani et al., 2015, 2017b).

Moreover, the fishing grounds play a role in defining different infection levels in terms of prevalence and intensity by the same species of *Anisakis*. In other words, the infection levels of the same parasite species could vary across different geographical areas, independently of the fish host species. For instance, Cipriani et al. (2015) found that the infection level by *A. pegreffii* in European hake, *M. merluccius*, fished in the Tyrrhenian Sea (Mediterranean Sea) was lower than that found in the same fish species, of the same length/size, fished in the NE Atlantic coast of Spain. However, the same parasite species reached very high levels of infection in hake samples fished in the central Adriatic Sea (Mediterranean Sea). This geographical area represents a 'hotspot' of infection with *A. pegreffii* in the Mediterranean Sea, for some pelagic fish species such as anchovies *E. encrasicolus* (Cipriani et al., 2017a), and in demersal fish species such as the European hake, *M. merluccius* (Cipriani et al., 2017b). Taking into account the whole sample of anchovies parasitologically examined, the results obtained by GLM_3 modelling showed that the main factors associated with the *A. pegreffii* parasite burden were fishing ground, and, secondly, fish length. In fact, in some fishing areas (Central Adriatic Sea), smaller-sized *E. encrasicolus* (<130 mm) showed higher levels of infection, while other samples (Ionian and Balearic Seas), characterized by bigger length (134 mm), were completely uninfected with *A. pegreffii* (Cipriani et al., 2017a). Similarly, remarkably high infection levels with *A. pegreffii* were recorded in hake from the Adriatic Sea/Ionian Seas (Mediterranean Sea) when compared with fish of similar length obtained from the western Mediterranean fishing grounds, with the samples from the Adriatic Sea showing the highest levels of the infection (Cipriani et al., 2017b).

Thus, according to the data so far available, it seems that a specific fishing ground, with all its ecological conditions, trophic web stability, presence of abundant host populations, could somehow predict the occurrence and the relative proportions of *Anisakis* spp. found in a fish host species. This has important implications in terms of surveillance programs of the molecular epidemiology of zoonotic species of the genus *Anisakis* from different geographical areas.

6.1.3 Fish Host Body Size

Many authors included either the length or the weight of fish as a possible explanatory driver of the infection with *Anisakis* spp. larvae. Indeed, while the fishing ground plays an important role in depicting the distribution of different species of *Anisakis*, the fish length represents, generally speaking, a major driver that explain different values of the parasitic burden observed in the same fish species. Indeed, the majority of the parasitological surveys carried out on different species of fish have found a generally strong positive correlation between fish size/length and *Anisakis* spp. prevalence and/or abundance values (Bao et al., 2015; Chou et al., 2011; Cipriani et al., 2015, 2017a, b, c; Hermida et al., 2012; Levsen and Lunestad, 2010; Meloni et al., 2011; Mladineo et al., 2012; Münster et al., 2015; Pierce et al., 2017; Quiazon et al., 2011; Setyobudi et al., 2011). Similar parasitic infection levels were also observed by different authors, in the same length/size of a specific fish species, thus suggesting that fish length could explain, in some cases, the majority of variance in a multifactorial analysis of the *Anisakis* sp. epidemiology. For instance, Cipriani et al. (2017a), in an epidemiological study on *A. pegreffii* in anchovies from different locations in the Mediterranean Sea reported that fish of a certain length were infected at similar rates as those previously reported by other authors from the same fishing ground in another timeframe (Fig. 14).

The positive correlation between parasitic burden and fish length is a direct consequence of the accumulation (bioaccumulation) of parasites throughout the fish's lifespan according to its feeding habits. Generally, larger fish accumulate anisakid larvae during their longer lifespan, showing higher infection rates than smaller fish. Obviously, this positively correlated infection/length rate is directly dependent on the age of the fish. The trend

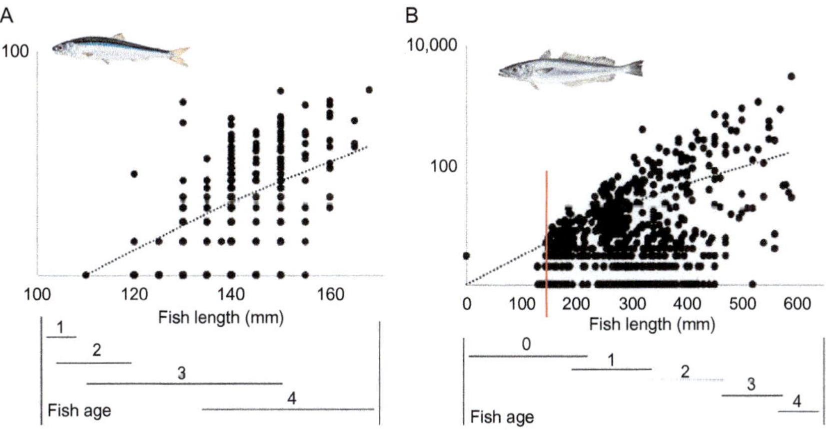

Fig. 14 Spearman correlation analysis ($r^2=0.08$, $P<0.01$) between fish length (*L*) and number of *A. pegreffii* (*n*), in two fish species, respectively, *Engraulis encrasicolus* (A) and *Merluccius merluccius* (B). Below, estimate of the fish age based on the fish length. The *vertical line* in the *M. merluccius* graphic, indicates the diet switch of that fish species, from small epiplanktonic crustaceans (Euphausiacea) to fish prey item (Clupeiformes) such as the anchovy *E. encrasicolus*.

of *Anisakis* spp. larvae to be superabundant in piscivourous fish species compared to planktivorous or mixed feeders, is also known to be responsible for this correlation. However, the ontogenetic shift in feeding behaviour of the fish species, with the changing of prey items from small crustaceans to small fish as prey for bigger sized fish, is also responsible of this positive correlation. Münster et al. (2015) found a clear relationship between total fish length/weight and parasitic burden in cod from East and West Greenland. In addition, they found some correlation between fish size and distinct prey items, thus suggesting that an ontogenetic shift could explain the positive correlation. On the other hand, a similar trend was also suggested by Mattiucci et al. (2004) in the European hake, where the fish longer than 30–35 cm showed the higher abundance values of the infection with *A. pegreffii* in the Mediterranean Sea, and with *A. simplex* (s.s.) from the Atlantic Spanish coast (Mattiucci et al., 2004). Similar results of positive correlation between host length and infection were found in horse mackerel, *T. trachurus* (Mattiucci et al., 2008) and in anchovies (Cavallero et al., 2012; Cipriani et al., 2017a; Mladineo et al., 2012).

Setyobudi et al. (2011) found a positive correlation between fish fork length and mean intensity of *A. simplex* (s.s.) in chum salmon from Korea, with the highest value occurring in chum salmon of 55.1–60 cm of body length, while the lowest level occurred in chum salmon having body length below 50 cm. The authors also reported that mean intensity values did not differ between males and females (Setyobudi et al., 2011). Chou et al. (2011) have demonstrated that the infection rate by *A. simplex* (s.l.) larvae from *S. australicus* fished off the eastern Taiwanese waters were positively correlated to the host size. Indeed, while the prevalence maintained the same values in the different fish size categories, and there was no sex effect, the abundance was higher in individuals weighing more than 300–450 g, corresponding to an age of 2–4 years (Chou et al., 2011).

The size of herring *Clupea harengus* from the Northern Norwegian Sea was highly significantly correlated with overall *A. simplex* (s.s.) larval abundance, while sampling year had only a secondary influence on the probability of the infection (Levsen and Lunestad, 2010). The same authors also highlighted that just in herring from the NE Atlantic Ocean, a temporal and spatial variation in *Anisakis* sp. larval occurrence emerged, between and within the North Sea and Norwegian Sea herring stocks (Levsen and Lunestad, 2010).

In other case study, such as in the haddock *Melanogrammus aeglefinus*, the presence and abundance of *A. simplex* (s.s.) in haddock appeared to be related to both size and fish and these effects were difficult to separate due to the low overlap in the size ranges of fish sampled around Scotland and in the Barents Sea (Pierce et al., 2017).

On the contrary, other authors did not find any relationship between fish size and infection. For instance, there was no significant statistical difference between the number of *Anisakis* larvae in the edible parts and total fish length of two Pacific salmon species, i.e., *Oncorhynchus nerka* and *O. keta* (Karl et al., 2011).

Conversely, a weak, but significantly positive correlation between fish host size and both overall larval *Anisakis* abundance and larval abundance in the flesh ($P < 0.0001$ in both cases) in mackerel, *S. scombrus* was found by Levsen et al. (2017a). Interestingly, no significant correlation between host body size and larval abundance in the flesh was observed in mackerel fished from the NE Atlantic Ocean, having the greatest length. These findings seem to be in accordance with Levsen and Berland (2012), who reported significantly higher prevalence and abundance of the infection with *A. simplex* (s.s.) in smaller mackerel in comparison with larger ones. The authors suggested that smaller (being younger) fish are less able to immunologically resist the parasite. This would mean that the infection becomes prevalent at a younger age in that fish species (i.e. *S. scombrus*); while, the opposite trend has been postulated for the bigger (older) fish specimens (Levsen and Berland, 2012). In support of this hypothesis, the authors frequently found dead or partially disintegrated larvae in the visceral organs and fillets of *S. scombrus*.

A similar trend in the *A. simplex* (s.s.) infection was found in saithe from the North-East Atlantic (Priebe et al., 1991). Thus, smaller and younger fish (3–4 years old) showed a significantly higher level of *A. simplex* larval abundance in the fish flesh, with respect to individuals of 5 years of age. Also in this case of infection, the most of larvae present in the fillet were dead, while only those infecting the flesh of younger fish were alive (Priebe et al., 1991). Interestingly, the analysis of ELISA response to the parasitic infection of saithe at these different age groups revealed that there was a moderate correlation between antibody titre and the age of prespawning and postspawning feeding periods. The authors also hypothesized that the migrating ability of larval *Anisakis* and their lifetime in the muscle of saithe could be influenced by a specific immune response, which increases with fish host age/size.

More recently, Gómez-Mateos et al. (2016) stated that the weight of *M. poutassou* collected from off the Gulf of Cadice (NE Atlantic waters) was not significantly associated with the general parasitic burden of the two species *A. simplex* (s.s.) and *A. pegreffii* ($P=0.61$), or with fish length ($P=0.47$) (Gómez-Mateos et al., 2016). According to their data, the prevalence of infection did not vary significantly with fish size, however, blue whiting ranging 21–22 cm in length showed the highest abundance of *A. simplex* (s.l.) larvae, while longer fish were less infected. This phenomenon could be related to an ontogenetic shift in preferred food items by *M. poutassou*, i.e., its diet includes gradually larger proportions of uninfected prey as they grow, or to an immune mechanism allowing elimination of the parasites, instead of accumulating them during its host life span.

A significantly negative correlation between fish size and *Anisakis* spp. infection level was reported by other authors, as well. Thus, small fish being more heavily infected than larger ones was observed for some fish species such as *G. morhua* (Horbowy et al., 2016; Strømnes and Andersen, 1998). These authors considered their findings as the result of increased mortality of large and more heavily infected specimens (due to a decrease of fish condition and/or more intensive predation on these fish). However, this hypothesis has yet not been confirmed or validated in immunological studies of the possible pathological effects of *Anisakis* spp. infection in their fish hosts.

However, an immune mechanism allowing the fish to at least partially eliminate the parasites was suggested by Levsen and Berland (2012). This last observation supports the hypothesis of Levsen and Berland (2012) according to which *A. simplex* (s.l.) infection, at least in some pelagic and demersal fish species, is not only related to the feeding habit and the size/age of the fish host, but may also be influenced by specific fish host immune responses. Consequently, it would suggest that some fish species are less adapted to cope with these parasites. Additionally, the cost of the metabolic rate to resist the parasite by smaller fish is relatively higher than the energy spent to grow to reach reproductive size. Such mechanisms may in turn have measurable effect on the body condition index (factor).

Concerning the latter aspect, some studies reported a correlation between Fulton's condition factor *K*, as a measure of an individual fish's health, and the parasitic burden by *Anisakis* spp. The aim of this correlation was to ascertain if the presence of *Anisakis* spp. in fish could eventually alter the health of the host, thus showing some minor pathological effect, especially in cases of massive parasite burden. However, the relationship between fish condition and parasite burden is more complex, in fact some authors reported a positive correlation between Fulton's *K* and parasite burden in some fish species, indicating that fish with a good health status, with proper feeding condition, harboured more larvae than fish with lower feeding activity.

Some change in the Fulton's condition factor *K* in sculpin, along with the increase of the *Anisakis* spp. intensity in the liver of the fish, was found (Margolis, 1970). More recently, Podolska and Horbowy (2003) noted that a significant positive correlation between Fulton's condition factor and larval prevalence emerged in Baltic herrings; the authors found that a better condition of the fish host was correlated with a higher level of infection with *Anisakis* spp. larvae. This could be explained as a good length/weight ratio of the fish host, resulting from more optimal feeding conditions, which in turn would increase the infection probability of *Anisakis* spp.

However, an important parasite burden may induce a weaker condition factor (Lagrue and Poulin, 2015). Again, different fish species may show differential immunological responses to *Anisakis* spp. infestation, spending some metabolic energy to reduce the overall parasite burden (Buchmann, 2012; Levsen and Berland, 2012; Levsen et al., 2017a).

Differences in the prevalence and abundance of infection by *A. simplex* (s.s.) in three closely related salmonid fish species, *Oncorhynchus mykiss*, *Salmo trutta*, and *Salmo salar*, showed different susceptibility to the parasite in experimental infection trials, indicating differential immunological response patterns among these fish species (Bahlool et al., 2012).

6.1.4 Host Population Structure, Demography, and Migration Routes

Basically, host population distribution, their demography, and their migration routes could affect parasite population dynamics. The transmission pathways of *Anisakis* spp. are displayed throughout well-defined interactions among organisms at different levels of trophic web in marine ecosystems. The transmission from a host to another requires a wide availability, thus a conspicuous population size of their hosts, and the stability of marine trophic webs represents the requirement to maintain such complex dynamics.

The population structure of many organisms representing definitive, intermediate, and paratenic hosts of *Anisakis* spp., results in a complex heterogeneous mosaics at both local and wider levels, influencing the endoparasite population structure and demography. These interspecific dynamics could also acquire a role in shaping the genetic diversity, population genetic structure, and gene flow among anisakid parasites populations (Mattiucci and Nascetti, 2008).

Representing a biotic factor, the distribution of those cetacean species acting as definitive hosts of *Anisakis* spp. has been suggested as good factor in modelling and in the shaping the habitat, of these parasites (Kuhn et al., 2016; Mattiucci and Nascetti, 2008). Higher abundance of the infection has been generally observed, for instance, in *A. pegreffii* from Southern Atlantic Ocean, where thousands of individual worms were typically collected from a single infected cetacean species as the main definitive hosts, compared with few 100 collected for the same species in cetaceans from the Mediterranean Sea (Mattiucci, unpublished data). These data are consistent with biotic factors, such as host density of those suitable definitive and intermediate hosts for *Anisakis* spp. and other anisakid nematodes in those Austral waters, also maintaining the high genetic diversity in the anisakid gene pools (Mattiucci et al., 2017a). A recent parasitological survey of anisakid nematodes obtained from antarctic fish species, catched in a 'pristine' ecosystem, not subjected to the population reduction from overfishing and other anthropogenic stressors, has evidenced an extremely high parasitic burden with the two antarctic anisakid nematodes, i.e., *Contracaecum osculatum* sp. D and *C. osculatum* sp. E. Accordingly, the two parasite species have shown also high levels of genetic variability at both nuclear and mitochondrial levels (Mattiucci et al., 2015b). These two parameters, infection levels in fish and genetic variability values, remained unaltered over a temporal scale of 20 years (Mattiucci et al., 2015b), probably because the antarctic food webs involved in the life cycles of those anisakid species remained unaltered over the years, as did the demography of their definitive and intermediate hosts.

Indeed, it was hypothesized that various causes of habitat disturbance (pollution, capture, overfishing, and global climate change) could probably provoke various degrees of stress on the trophic webs, involving different species of marine mammals, fish, and invertebrates (suitable as definitive and intermediate hosts of *Anisakis* spp. and other anisakid nematodes). This, in turn, could adversely affect also the population size of their endoparasitic anisakid nematodes and, consequently, their genetic diversity (Mattiucci and Nascetti, 2008). In fact, populations of anisakid nematodes characterized by small population demography tend to have a lower genetic diversity (variability) values, as smaller populations lose diversity more quickly than large populations due to genetic drift (Mattiucci et al., 2015b).

It was also observed that the demography of certain local populations of definitive hosts of *Anisakis* could play an important role in maintaining 'hot spots' of infection in local fisheries. For example, the central Adriatic Sea is inhabited by several species of cetaceans, mainly represented by the bottlenose dolphins *T. truncatus* and more rarely, by the striped dolphins *S. coeruleoalba*. Populations of bottlenose dolphins may represent a suitable host for *A. pegreffii* in this part of the Adriatic Sea since they prey heavily on anchovy, hake, and other fish (Holcer et al., 2014). Blažeković et al. (2015) reported high levels of *A. pegreffii* abundance in this mammal stranded along the Croatian coast. This dolphin species revealed a fine-scale genetic structure throughout the Adriatic Sea inferred by nuclear DNA markers, showing a differentiation between the northern population with respect to those from central south subbasin waters, as well as between the west and east coast (Gaspari et al., 2015). This suggests that local populations of *T. truncatus* tend to persist in different areas of the basin, feeding on populations of prey items of 'local' food webs. In turn, the population structure of this suitable definite host in the Adriatic Sea and its biology has been considered, coupled with an 'anthropogenic role' (see Section 9) and some oceanographic conditions (Section 6.1.1), to maintain also a high demography of *A. pegreffii* in local fisheries represented by two important commercially important fish species in Italy (i.e. *M. merluccius* and *E. encrasicolus*) (Cipriani et al., 2017a, b).

The high dispersal capacity of some definitive and intermediate/paratenic hosts of *Anisakis* spp. are responsible for maintaining the distribution of populations of a parasite species, even if they are geographically located thousands of kilometres apart, as is the case for Austral populations (for instance from New Zealand waters) of *A. pegreffii*, compared to its Mediterranean populations. These features are also likely responsible for providing a continuous gene flow between the populations, thus maintaining a high genetic heterogeneity between different populations. As we have illustrated above, these worms mature in several species of 'oceanic dolphins', whose large geographical range and their vagility, could maintain the high level of gene flow observed in these anisakid nematodes. Similarly, also the high mobility and dispersal capacity of their intermediate/paratenic hosts (fish) could contribute to this dispersal of the parasites and enhance the high degree of gene exchange. Indeed, levels of interpopulational gene flow, indirectly estimated from allele frequencies at nuclear level (allozymes) and microsatellites have been reported in these marine ascaridoid nematodes to be at high levels (Mattiucci and Nascetti, 2008; Mattiucci et al., unpublished data). More recently, *Anisakis* and mitochondrial haplotype distributions from both parsimony network and analysis of molecular variance, revealed a panmictic distribution of three *Anisakis* species (*A. simplex* (s.s.), *A. pegreffii*, and *A. berlandi*) from sardines *S. sagax*, distributed throughout the California current systems, thus indicating the absence of a population genetic substructuring at intraspecific level among those populations (Baldwin et al., 2011). According to the authors, the panmictic distribution of the larval population of *Anisakis* spp. may have been the result of the migratory pathways of the intermediate fish host (Pacific sardine) along the California coast (Baldwin et al., 2011). The authors also suggested that limited oceanographic barriers, and the complexity of the California current, are not preventing the mixing of *Anisakis* spp. or their populations. Moreover, the major biogeographic break in the California Current at Point Conception does apparently not limit the distribution of highly migratory fish or cetacean species (Checkley and Barth, 2009). No geographic separation has been identified in *A. simplex* (s.s.) using the mtDNA *cox2* marker to investigate the haplotype structure from three fishing grounds of the fish species *Sebastes mantella.* Main pairwise divergence and *Fst* values were low; thus, accordingly, the authors found a lack of genetic structuring in those parasite populations collected from that fish species (Klapper et al., 2016). Conversely, a low genetic differentiation has been observed in the *A. simplex* (s.s.) populations collected from *C. harengus* caught from four different fishing grounds of the NE Atlantic Ocean. Results indicate a certain genetic substructuring of *A. simplex* (s.s.) obtained from herring fished in different areas, with the population from the Norwegian Sea being the most differentiated in comparison with the others. The population genetic structure of *A. simplex* (s.s.) appeared to be in accordance with the existence of a Norwegian stock of herring in the NE Atlantic area, showing a significant differentiation from the other samples of the southern areas ($P < 0.05$). Results suggest that mitochondrial genes, such as mtDNA *cox2*, could represent a valuable genetic marker for population structure analysis of *Anisakis* spp. (Blažeković et al., 2015; Cross et al., 2007; Kijewska et al., 2009; Mattiucci et al., 2017c). Indeed, for instance, a Bayesian inference obtained from mtDNA *cox1* sequences analysis supported for the existence of a population of *A. simplex* (s.s.) from Atlantic Ocean, distinct from the populations investigated in the Pacific Ocean (Kijewska et al., 2009).

On the other hand, as we suggested above, oceanographic barriers could represent limits for the movement of populations of cetaceans and fish species infected with *Anisakis* spp. For example, the existence of different populations (stocks) of European hake, horse mackerel, and swordfish in the Mediterranean Sea and Atlantic Ocean, could mirror the presence of different species and populations of *Anisakis* spp. in populations (stocks) of the same fish species in other areas (Mattiucci et al., 2014b).

While a major effort has been made to investigate the presence of *Anisakis* spp. in different fish species, scanty data are available as to their occurrence and genetic identification at the larval stage in the first crustacean intermediate hosts. Consequently, the role, the importance, and the influence of these categories of marine organisms involved in the life cycle of species of *Anisakis* still cannot be used whenever modelling *Anisakis* spp. distribution. Typical first intermediate hosts in the life cycles of *Anisakis* spp. should be represented by invertebrates (Copepoda: Euphasiacea). They are a food item of both intermediate/paratenic hosts and definitive hosts (such as Mysticetes) of several *Anisakis* spp. Thus, they are responsible for transferring the *Anisakis* infection throughout the trophic webs. Several species of Euphausiids and Mysids have been reported as first intermediate hosts of *A. simplex* (s.l.) in the Northern Hemisphere. Smith and Wootten (1978) reported the occurrence of *A. simplex* (s.l.) larvae in *Thysanoessa* sp. and *Meganyctiphanes norvegica* from the Scottish waters. Højgaard (1995) conducted infection experiments with *Anisakis* sp. larvae in the latter crustacean species. More recently, Smith and Snyder (2005) reported the infection by *A. simplex* (s.l.) in *Euphausia pacifica* (Prevalence, P (%) = 0.013) and *Thysanoessa raschii* (Prevalence, P (%) = 0.019) from Prince William Sound, Alaska. In Japanese waters, the euphasiids *Euphausia similis* and *E. pacifica* were experimentally infected by *Anisakis* spp. larvae (Oshima et al., 1968).

Larvae of *A. simplex* (s.l.) at third stage (L3) were found in intermediate hosts from the temperate NE Atlantic waters of the Spanish Galician waters (Gregori et al., 2015). In addition, they detected *A. simplex* (s.s.) and *A. pegreffii* as free-living larvae in the water column in this oceanographic area. However, the prevalence found in krill was very low for *Anisakis* species in this intermediate host. In the *Nyctiphanes couchii* population, the prevalence of the infection by *A. simplex* (s.l.) was 0.0019%, while the abundances of *A. simplex* (s.l.) L3 larvae found free in the water column ranged from 0.005 to 0.009 L3/m^3. The abundance of *A. simplex* (s.l.) among mysids was 0.001 L3/mysids (Gregori et al., 2015). The existence of those parasites in a variety of mesozooplankton organisms suggests that the transmission routes of *A. simplex* (s.s.) and *A. pegreffii* from the Northeast Atlantic Galician Waters, which is a sympatric area of the two species, are wider than expected. Those results suggest also that the two parasite species seem to be not specific to their intermediate hosts (Gregori et al., 2015). However, all those macroplanktonic crustaceans occur from surface to approximately 1000 m depth, depending by the species and diurnal cycle (Mauchline, 1980). *N. couchii* is also an important prey item of different cephalopods and fish species, which act as intermediate/paratenic hosts of *A. simplex* (s.l.) in that area (Gregori et al., 2015). Mysids are the most abundant crustaceans in the benthos, which are able to undertake vertical migration, thus becoming prey item of both demersal and benthopelagic fish and squid species, filling a gap in the trophic link between benthic and pelagic environments. This movement increases the probability of transmission of anisakid nematodes to the next host (Marcogliese, 2002). A crucial factor could be represented by the time that occurs before the encounter between *Anisakis* spp. larvae and euphasiacean hosts. Taking into account the feeding behaviour of euphasiaceans and sinking rate of eggs and larvae, it was calculated that eggs and larvae require about 1 month to pass throughout the water column to encounter the relevant host, and the survival of the larvae could be around 3 months (Højgaard, 1995). It was also suggested that there is a peak season with favourable conditions for infections with *Anisakis* spp. larvae of the first intermediate host in NE Atlantic waters. It was postulated that, for instance, around the Faroe Islands, the peak presence of the long-finned pilot whale is around July–August and, consequently, the infection of the intermediate hosts could occur between October–November when eggs and larvae have been released by these cetaceans and when the temperature is around 2–9°C (Højgaard, 1997). At these temperatures, the egg hatching should occur within 2–3 weeks (Højgaard, 1997). He hypothesized that in some periods, the infected euphasiaceans migrate extensively in NE Atlantic waters (Højgaard, 1997). It was further hypothesized that if organic matter is trapped in the thermocline in oceans with turbulence close to zero, the density of eggs and larvae could increase also increasing the probability of transmission to the euphasiaceans hosts (Højgaard, 1997). This phenomenon might be rare in the North Atlantic, but may occur more frequently in other basins with closed waters such as the Adriatic Sea (Cipriani et al., 2017a, b). On the other hand, it was suggested that the primary and secondary production can be enhanced where permanent upwelling of nutrients occurs, thus allowing the accumulation of prey for predators in the halo/thermocline fronts, which would then favour the probability of transmission of *Anisakis* spp. larvae to the next host (Cipriani et al., 2017a; Kuhn et al., 2016).

6.2 *Anisakis* spp. in Farmed Fish

In cultured fish, the larvae of *Anisakis* spp. have no significance as disease causing parasites. However, any detection of anisakids in products derived from farmed fish could seriously affect consumer confidence, not only due to the possible health risk posed by anisakid larvae, but also related to the larvae's most unappealing appearance, especially when still alive. These aspects have gained importance during recent years due to the increased popularity of Asian inspired dishes based on raw or only lightly processed fish products such as sushi and sashimi. Sea-caged salmonid fish species including Atlantic salmon (*S. salar*), coho salmon (*Oncorhynchus kisutch*) and rainbow trout (*O. mykiss*) have generally not been regarded at risk to acquire *Anisakis* larvae, although Marty (2008) recorded a single anisakid larva (not specifically identified) in the intestinal caecum of one farmed Atlantic salmon from British Columbia, Canada. The assumption that maricultured salmonids are free of parasitic nematodes rests on the results of a number of surveys from different countries or areas. Some more recent studies included 166 sea-caged rainbow trout from western Zealand, Denmark, and 720 sea-reared Atlantic salmon from various Scottish farming sites, but no infected fish were found (Skov et al., 2009; Wootten et al., 2010). Cultured European sea bass (*Dicentrarchus labrax*) and gilt-head sea bream (*Sparus aurata*) in Southeast Spain apparently did not carry any anisakid larvae (Peñalver et al., 2010).

The apparent absence of *Anisakis* spp. larvae in sea-reared fish was mainly explained by the widespread application of pelleted and heat-treated compound feed. It was argued that the use of artificial diets seems to reduce the risk of opportunistic feeding of farmed fish on free-living wild prey that may occasionally enter the pens (EFSA, 2010). Based on these considerations, the current EU legislation (Regulation (EU) 1276/2011) states that farmed fish may be excepted from the so-called freezing requirement provided that they are cultured from embryos and fed exclusively on a diet that cannot contain viable parasites. However, recent investigations revealed the presence of *A. simplex* (s.s.) in runts (loser fish) of cultured Atlantic salmon in southern and western Norway (Levsen and Maage, 2016; Mo et al., 2014) and in runts of cultured rainbow trout from a northern Norwegian locality (Roiha et al., 2017).

In the two most recent studies, 4184 and 1038 farmed Atlantic salmon and rainbow trout, originating from all salmon and trout producing countries, were examined for anisakid nematodes between January 2014 and September 2016 by applying the UV-Press method (Karl and Leinemann, 1993; Levsen and Karl, 2014). Most of the samples consisted of harvest quality fish (85%–90%) processed for human consumption while a smaller portion consisted of discarded fish including runts (10%–15%). Although a few *A. simplex* (s.s.) larvae were found in the viscera or musculature of single salmon and trout runts from southern and northern Norway, respectively, none of the harvest quality fish was infected. However, the findings showed that *A. simplex* (s.s.) larvae are evidently able to enter the sea cages and may subsequently infect cultured fish. Interestingly, the two farming localities that lodged the infected runts were located at the outer coast (in contrast to more sheltered fjord localities), in close vicinity to open waters where marine mammals including migrating cetaceans such as minke whales, frequently occur. This again may have facilitated the dissemination of *Anisakis* spp. larvae within the trophic web of the actual areas, including planktonic and semiplanktonic crustaceans and various fish species. Thus, the parasites may have accessed the cages through copepods, euphausiids, or smaller wild fish, which occasionally may cross the side nettings. Hence, while healthy salmon or trout seem to feed exclusively on pelleted dry feed, runts have to rely on feeding opportunistically on whatever prey is available, in order to survive.

While modern sea-rearing techniques seem largely to prevent caged fish from acquiring parasitic nematodes, certain practices still exist which may increase the risk of infection of farmed fish with larval anisakid nematodes. For example, the presence of *A. simplex* third-stage larvae in the stomach lumen or abdominal cavity of sea-caged cobia (*Rachycentron canadum*) in Taiwan was linked to the occasional feeding of the cobias with chopped unfrozen raw fish or residuals thereof (Shih et al., 2010).

Another farming method that puts farmed fish at risk to acquire anisakid larvae is the capture of juvenile wild fish for subsequent on-growing and fattening in captivity. The practice is or has been used in culturing facilities of both, cod *G. morhua* and Bluefin tuna (*T. thynnus*) in Europe. Although currently not practised on an industrial scale, cod for on-growing has traditionally been captured at sizes of 3–5 g juveniles and 1–2 kg adults. When comparing the parasite fauna of wild and farmed cod from four localities in western and northern Norway, Heuch et al. (2011) recorded 100% prevalence of *A. simplex* (s.l.) larvae (genetic or molecular species identification was not performed) in the viscera of 35 wild caught and subsequently sea-caged cod in northern Norway (no other infection descriptors were reported in the study). At the time of capture, the body weight of the actual cod was around 400 g, indicating that the fish acquired the infection while still free living. Similarly, Smrzlić et al. (2012) reported the presence of 226 *A. typica* larvae (identified by sequence analysis of the ITS rDNA and *cox2* mDNA genes) in a single tuna kept for on-growing and fattening in a Croatian fish farm located in the mid-Adriatic Sea. In this area, the tuna on-growing and fattening cycle typically lasts for 1.5–2 years during which they are fed a diet of local fresh baitfish or imported frozen herring and sardines. However, since none of the fresh baitfish seemed to be infected with *A. typica*, the authors (Smrzlić et al., 2012) suggested that the actual tuna was already parasitized by the time it was caught and subsequently released into the net-pen for on-growing purposes.

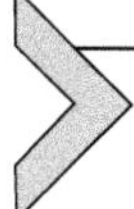

7. DO *ANISAKIS* SPP. ALWAYS OCCUPY THE SAME SITE IN FISH?

Actually, the answer is: NO

7.1 Detection of *Anisakis* spp. Larvae in Fishery Products

The methods used to inspect fishery products for the presence of *Anisakis* larvae have been improved in the last years, reaching a greater efficiency in terms of timely and labour resources they require while yielding higher detection score.

So far, the nematode inspection methods that are largely applied by the fish processing industry, in public routine spot-test laboratory analyses or in large-scale scientific surveys, are all based on visual inspection techniques, targeting whole larvae. These methods are all destructive in nature, which implies that the actual fillets or products are destroyed during the inspection process. Attempts to design nondestructive nematode inspection technologies, which would allow real-time screening of fish products/fillets during processing, have to our knowledge not yet been successful. Thus, nondestructive technologies which have been tested for their applicability under real-time industrial conditions, include ultrasound (Hafsteinsson et al., 1989; Nilsen et al., 2008), X-rays (Nilsen et al., 2008), conductivity (Nilsen et al., 2008), electromagnetism (Choudhury and Bublitz, 1994; Haagensen, 1993), magnetometry (Jenks et al., 1996), real-time FRET (fluorescence resonance energy transfer) (Intapan et al., 2008; Monis et al., 2005), imaging spectroscopy (Heia et al., 2007), and magnetic resonance (MR) (Bao et al., 2017a).

The methods based on visual inspection include light microscopy (Rijpstra et al., 1988), candling on a light table (Butt et al., 2004; EFSA, 2010; Wold et al., 2001), artificial peptic digestion (Llarena-Reino et al., 2013; Lunestad, 2003; Lysne et al., 1995; Thien et al., 2007; Thu et al., 2007), UV-Press (Karl and Leinemann, 1993). The amount of laboratory work associated with these methods varies greatly and often restricts the applicability of these methods in routine examinations of pelagic fishes (Levsen et al., 2005).

The most widely used *Anisakis* larval detection methods based on visual inspection of fish tissues are the following: (1) plain visual inspection (candling) of fish fillets on light tables; (2) artificial peptic digestion (Gómez-Mateos et al., 2016; Llarena-Reino et al., 2013; Lunestad, 2003; Lysne et al., 1995; Thien et al., 2007; Thu et al., 2007), and (3) inspection under UV light of pressed and frozen fish samples (i.e. UV-Press method—Fig. 15) (Cipriani et al., 2015, 2016, 2017a, b; Karl and Levsen, 2011; Levsen and Karl, 2014; Levsen and Lunestad, 2010; Levsen et al., 2017a; Mladineo et al., 2017b).

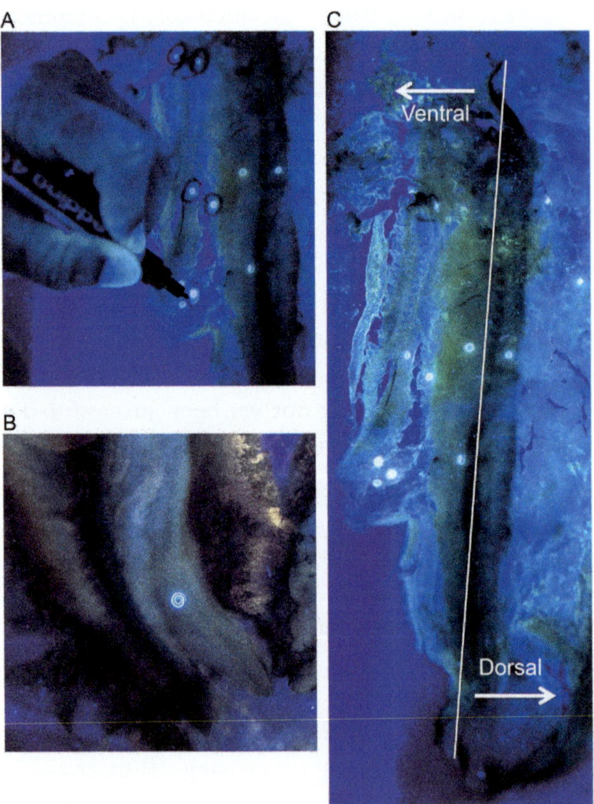

Fig. 15 Photographs of *Anisakis* spp. larvae detected by the UV-Press method in different fish species (A) marking fluorescent larvae of *A. simplex* (s.s.) spotted in the flesh of grey gurnard *Eutrigla gurnardus* under UV light; (B) fluorescent coiled larva of *A. pegreffii* in the flesh of anchovy, *Engraulis encrasicolus*, viewed under UV light; and (C) Distribution of several *A. simplex* (s.s.) larvae in the ventral and dorsal parts of the left flesh side of an anchovy.

Current EU legislation (Commission Regulation EC No. 2074/2005) points out that visual inspection of the whole fish abdominal cavity (including liver, gonad, and egg mass) should be done by fish operators to control the risk of visible parasites, thus ensuring from the catch to the plate that no contaminated fish reach the consumer (Llarena-Reino et al., 2012). Commission Regulation EC No. 2074/2005 states that: 'The visual inspection of fish fillets or fish slices must be carried out by qualified persons during trimming and after filleting or slicing. Where an individual examination is not possible because of the size of the fillets or the filleting operations, a sampling plan must be drawn up and kept available for the competent authority in accordance with Chapter II (4) of Section VIII of Annex III to Regulation (EC) No 853/2004. Where candling of fillets is necessary from a technical viewpoint, it must be included in the sampling plan'.

Candling, i.e., plain visual inspection of whole fillets or products on a light table in a darkened room, is nondestructive, and is the recommended method for parasite inspection in processing plants, even though the accuracy of a visual inspection method in the fish industry largely depends on the training and skills of the inspector (Levsen et al., 2005). The method has been mostly used in the industry in routine controls to detect and remove visible nematodes in fillets, mainly of larger and more valuable fish such as salmonids or certain gadoid species, before further processing (Karl and Leinemann, 1993).

However, the efficiency of candling is significantly lower compared to the other two visual inspection methodologies (peptic digestion and UV-Press method) (Llarena-Reino et al., 2012). Levsen et al. (2005) reported that only 7%–10% of the nematode larvae present in the fillets of the three fish species they examined were detected by candling in comparison with those found by UV-Press. Analogously, in a comparison between two different methodologies (i.e. candling and UV-Press) performed during a study of the occurrence of *A. pegreffii* in *E. encrasicolus*, the UV-Press method was more accurate in the detection of larvae in the flesh of anchovies, allowing to disclose 50.9% of the larvae which would have been undetected by candling and plain visual inspection (Cipriani and Mattiucci, personal observation) (Fig. 15). Thus, nondestructive visual candling methods are less efficient to detect anisakid larvae which are embedded deeply within the fish tissue, than destructive methods such as artificial peptic digestion and UV-Press (EFSA, 2010; Karl and Leinemann, 1993; Levsen et al., 2005).

Acidified pepsin solution used to digest fish tissue, that leave larvae alive and easier to spot, has been largely applied as a confirmatory invasive protocol to detect absence or presence of larval nematodes in fish products (Lunestad, 2003). This method represents a valid tool to quantify parasitic infections and to estimate the number of parasites in the fish musculature (Lysne et al., 1995; Thien et al., 2007; Thu et al., 2007). Additional variations of the peptic digestion method from CODEX STAN 244-2004 protocol have been developed by some authors, with the aim to improve the method (Dixon, 2006). Llarena-Reino et al. (2013) developed a novel procedure based on a liquid pepsin form, which was more effective; it also offered easier handling procedures than other pepsin formulations. Liquid pepsin, or LP protocol, was more sensitive, efficient, and accurate, compared with the standard CODEX STAN 244-2004 protocol (Llarena-Reino et al., 2013). The digestion procedure is based on a complete destruction of the fish tissue by means of digestive enzymes. The remaining nematodes can easily be separated and counted, after the digested tissue has been passed through a sieve. Most of the *Anisakis* larvae seem to survive the procedure. The technique can yield quantitative results and can be applied to all fishery products. When applied to fresh fillets, this technique does keep the parasites alive and available for identification using genetic/molecular methodologies; it also allows the recovery of dead worms from frozen material (Cipriani et al., 2016).

However, a limitation in the use of digestion procedure, even if largely improved by liquid pepsin form, is given by the relatively small number of samples which can be digested, and the comparatively long time period required to digest the fish tissue (Karl and Leinemann, 1993). The only limits of pepsin digestion method are that it is time consuming and difficult to adopt for the commercial mass sampling of huge amounts of fish (European Food Safety Authority—EFSA, 2010).

Conversely, the UV-Press method is rapid, thus allowing the processing of a larger sample number per time unit, compared with the former procedure (Levsen and Lunestad, 2010). The method utilizes the fluorescence of frozen nematodes (Pippy, 1970) and is based on visual inspection under UV light of flattened/pressed and deep-frozen fish fillets and viscera (Karl and Leinemann, 1993). Combining the pressing technique and candling under UV light leads to a simple, fast and quantitative detection method for nematodes in all kinds of frozen and then thawed fish fillets and fishery products (Karl and Leinemann, 1993). The fillets and viscera are pressed in a commercially available pressing device inside transparent bags, before flattened to 1–2 mm thickness. These bags are then deep frozen (−18°C) for at least 12 h, prior to visual inspection under a 366 nm UV light source

(Levsen and Lunestad, 2010). The *Anisakis* spp. larvae appear bright and fluorescent (Fig. 15) and can be easily detected, counted, picked and stored for further identification. Another advantage of this method is that it allows the determination of the exact infection site of larvae in the fish host, both in the visceral organs and in the flesh of the fish. Thus, it does not permit collection of live larvae as does peptic digestion, because the freezing of the bags prior to UV light exposure, kills the *Anisakis* larvae.

This method has been recently validated throughout a ring trial carried out in the framework of a European Project (PARASITE), in comparison with the peptic digestion method. The obtained results indicated that the two methods share same accuracy in the *Anisakis* spp. larval detection (Gómez-Morales et al., 2017).

Recently, Bao et al. (2017a) proposed an innovative method based on magnetic resonance imaging to investigate the presence of anisakid nematodes. According to the results of this study, an accurate localization of *A. simplex* (s.l.) larvae in the body cavity of herrings obtained from the North Sea was provided. The authors did not report any larva naturally infecting the fish flesh, but experimentally infected some muscle portion to positively validate that the magnetic resonance imaging was able to detect larvae in the fish flesh.

Worth to mention is the rapid development in the last decade of DNA-based methodologies to detect anisakid DNA in fish fillets. This method is thus not targeting whole larvae in fish tissues, but aims to detect parasite DNA in homogenized samples of fish products. A primer-probe system on a real-time PCR (RT-PCR) assay based on the mtDNA *cox2* gene sequence was first attempted on species of the genus *Anisakis* (Lopez and Pardo, 2010), however the procedure was not able to discriminate between the two zoonotic species *A. simplex* (s.s.) and *A. pegreffii*. Another primer-probe system was based on the ITS region of the rDNA, allowing the distinction between *A. simplex* (s.s.) and *A. pegreffii* (Mossali et al., 2010). Similarly, the ITS region of the rDNA was used to detect anisakids in the seafood products (Espiñeira et al., 2010), but the method was not able to identify the larvae at their species level. Furthermore, a RT-PCR method for the detection in situ was proposed, capable to amplify a region of the ITS-2 rDNA, and detect only the species *A. pegreffii* (Fang et al., 2011). A further probe based on the mtDNA *cox1* was also proposed (Herrero et al., 2011); however, it was not able to distinguishing between different species of anisakid nematodes. More recently, Paoletti et al. (2017) have proposed species-specific RT-PCR primers/probe systems to identify fish parasites of the

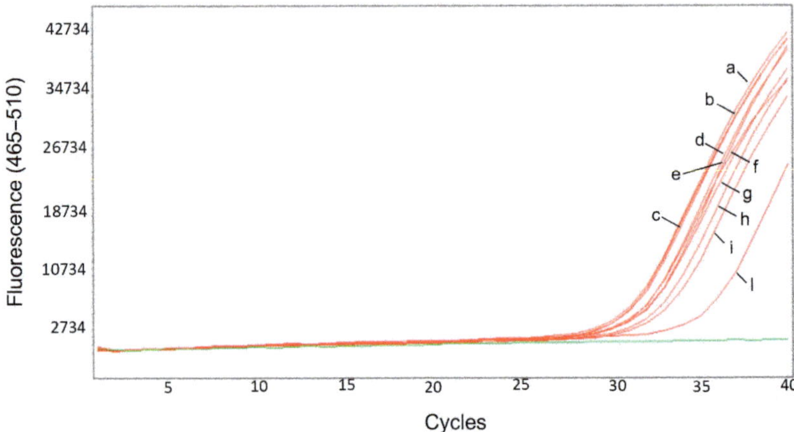

Fig. 16 Detection of DNA of *A. pegreffii* larvae in fish fillet using a species-specific primers probe system in RT-PCR. Fluorescent signals (6-FAM dye; 510 nm) [reported in the figure with alphabetical letters from (a) to (l)] can be observed in the RT-PCR assay obtained by using 10 distinct homogenates of constant fish tissue (10 g), with a decrescent number of *A. pegreffii* larvae. In particular, five whole larvae (i); four whole larvae and one half (l); four whole larvae (a); three whole larvae and half (b); three whole larvae (f); two whole larvae and half (c); two whole larvae (h); one larva and half (d); one larva (g); and, finally, half larva of *A. pegreffii*. *Green line*: negative control.

genera *Anisakis* and *Pseudoterranova*. The RT-PCR system allows the identification of those widespread species (i.e. *A. simplex* (s.s.), *A. pegreffii*, *P. decipiens* (s.s.), *P. krabbei*, and *P. bulbosa*) (Paoletti et al., 2017), infecting the muscle of fish species from European fishing grounds (Fig. 16).

Nevertheless, although all these methods have been used and are being applied by fishery operators or laboratories as integrated strategies in official and self-control tests, still none of them has been accepted as the international standard meeting industrial requirements (Llarena-Reino et al., 2013), even if they have been recently validated through ring trial (Gomez et al., 2017).

7.2 Site of Infection by *Anisakis* spp. Larvae in Fish

The different *Anisakis* species seem to have preferential infection sites in their fish hosts, which appear dependent not only on the *Anisakis* species, but also on the fish host species.

Anisakis larvae, especially when present at high infection intensities, tend to locate in defined parts of the fish host, mostly in the body cavity, encysted on visceral organs. These are often encapsulated and coiled up in tight spirals on the organs' surface. In some fish species the larvae tend to embedded in the liver, over the gonads, in the mesentery, or attached to the intestinal wall. Some papers report in detail the localization within the body cavity of the fish. Bao et al. (2017b) investigated the presence of anisakid nematodes with an innovative method based on magnetic resonance imaging. They reported an accurate site detection of *A. simplex* (s.l.) larvae (likely, *A. simplex* (s.s.), according to the geographical origin of 209 herrings obtained from the North Sea. The majority of larvae (54%) were embedded in the hind stomach, 21% on the pyloric caeca, 15% over the hind intestine, and smaller number of larvae (<3%) over the fore intestine, fore stomach, liver, and gonad. The authors did not report any larvae naturally infecting the fish muscle.

For European hake, a detailed distribution of *A. pegreffii* larvae in four different infection sites was provided by Cipriani et al. (2017b): 28.3% was detected as embedded on the liver, 62.9% in the rest of the host's body cavity, 6.6% in the ventral part of the fish flesh, while only 2.1% were located in the dorsal flesh of hake (Cipriani et al., 2017b). A small sample of 50 horse mackerel *T. trachurus*, sampled in the central Adriatic Sea, showed a slightly different distribution of *A. pegreffii* larvae: 0.5% was detected on the liver, 95.2% in the rest of the host's body cavity, 3.7% in the ventral part of the fish flesh, while only 0.6% occurred in the dorsal flesh of horse mackerel (Mattiucci and Cipriani, personal observation) (Fig. 17).

Cipriani et al. (2017a) reported the occurrence of *A. pegreffii* in the flesh of anchovies *E. encrasicolus* sampled in several areas of the Mediterranean Sea.

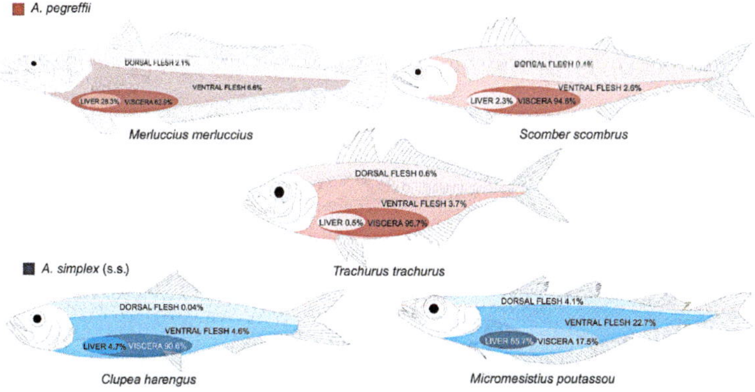

Fig. 17 Site of infection of *Anisakis simplex* (s.s.) and *A. pegreffii* larvae in different fish species. The relative frequencies of the larval *Anisakis* spp. are reported by different site of infection: liver, rest of viscera/body cavity, ventral flesh, and dorsal flesh.

The fish, frozen immediately after capture, showed a mean percentage of *A. pegreffii* larvae (4.0%) occurring in the flesh (Fig. 17). A similar proportion of larvae, compared with the whole nematode burden, was recorded also in the flesh of the sardine *Sardina pilchardus* (6.9%) sampled in the Adriatic Sea (Fig. 18) (Mattiucci and Cipriani, personal observation). Considering fish sampled in the Mediterranean Sea, *S. scombrus* and *S. japonicus* sampled in different locations, showed constant proportions (relatively, 1.0% and 2.8%) of larvae of *A. pegreffii* detected in the flesh of the fish host (Fig. 18) (Levsen et al., 2017b; Mattiucci and Cipriani, personal observation).

Pierce et al. (2017) observed moderate to high infection levels with *A. simplex* (s.s.) in haddock (*M. aeglefinus*) and withing (*Merlangius merlangus*) in Northeast Atlantic waters in the visceral organs and body cavity, and rather low infection in the musculature; the last mostly found in anterior ventral area of the fish species (Pierce et al., 2017). Some more recent studies consider also the distribution of larvae in different parts of the fish musculature, dividing it in two (ventral or hypaxial, dorsal or epaxial) (Cipriani et al., 2015, 2017b; Levsen and Lunestad, 2010), or in four parts (anterior ventral, anterior dorsal, posterior ventral, and posterior dorsal) (Karl et al., 2011; Levsen and Karl, 2014). Levsen and Lunestad (2010) investigated the occurrence and spatial distribution of *A. simplex* (s.s.) larvae in Norwegian spring spawning herring caught in the north-eastern Norwegian Sea. They reported that most of the larvae (96%) were located in the body cavity of the fish, while 3.5% of the larvae were found in the musculature of the belly flaps, i.e., the ventral portion of the body musculature covering the visceral cavity on both sides, while 0.5% occurred in the dorsal part of the fillets. This is in general agreement with Karl (2008) who found that in North Sea herring averagely 1.8% of all larvae were located in the belly flaps, while 0.6% of the larvae resided in the dorsal part of the flesh.

Levsen and Karl (2014) reported in detail the relative distribution of *A. simplex* (s.l.) larvae in the different body muscle sections of 188 specimens of Grey gurnards *Eutrigla gurnardus* (Fig. 19). The majority were located in the belly flap area (93.8%), 4.4% and 0.6% were found in the anterior and posterior epaxial (upper) muscle sections, respectively, while 1.2% lodged in the posterior hypaxial (lower) muscle section (Fig. 19). As reported for herring by Levsen and Lunestad (2010) again these findings show that by far most of the larvae are situated in the belly flaps, regardless of fish host body size (Levsen and Karl, 2014). Also Cipriani et al. (2016), in a study of differential migration rates of *A. pegreffii* and *A. simplex* (s.s.) in flesh of the European hake, reported that the most larvae of both species

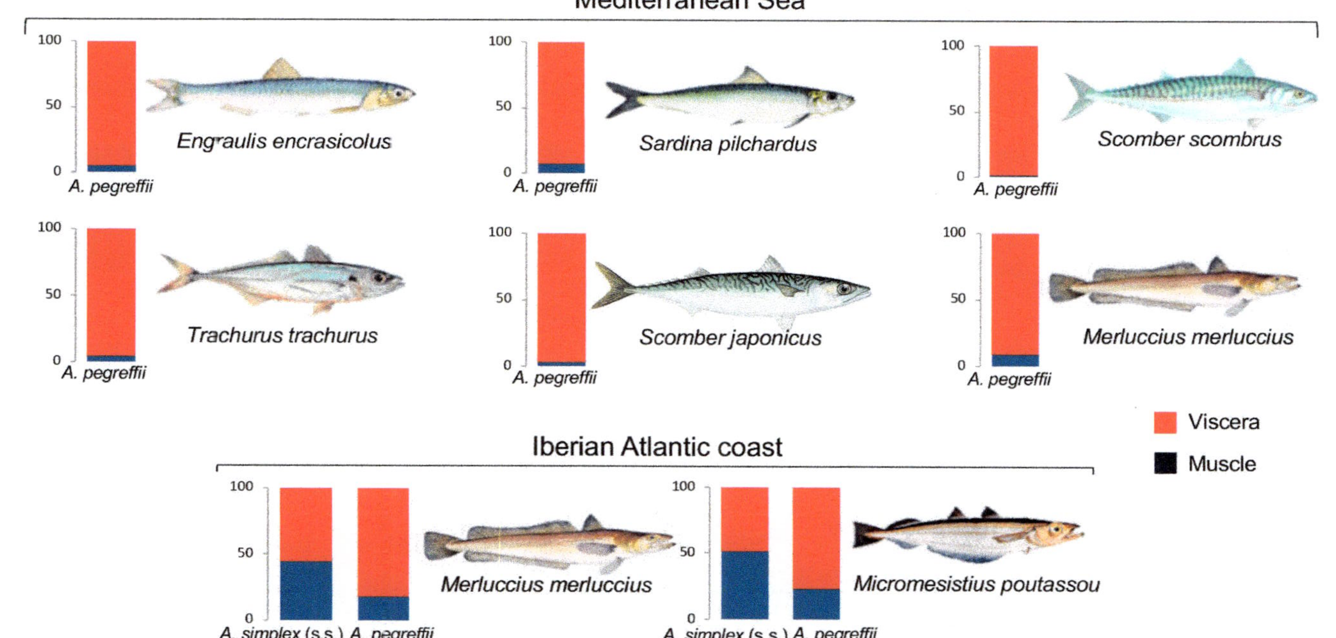

Fig. 18 Distribution of larvae of the two zoonotic species, i.e., *Anisakis simplex* (s.s.) and *A. pegreffii*, in viscera and muscle of some fish species from the Mediterranean Sea and the Iberian Atlantic coast.

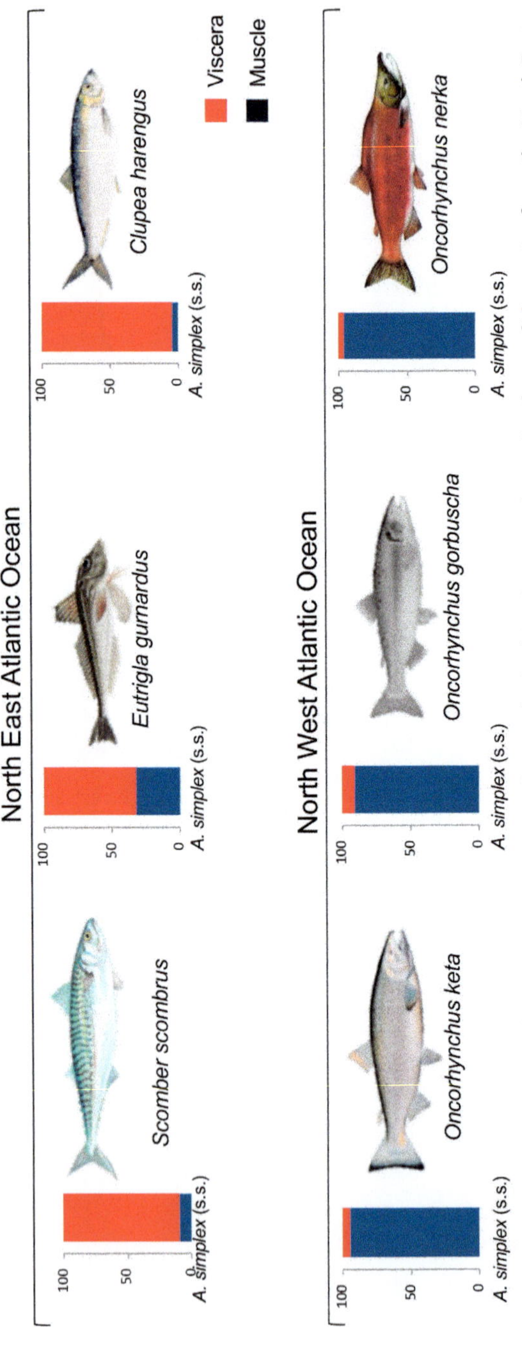

Fig. 19 Distribution of larvae of the zoonotic species *Anisakis simplex* (s.s.) in viscera and muscle of some fish species from the North East- and North West Atlantic Ocean.

penetrating the fish flesh were located in the ventral fillet (also called 'belly flap'), which represents the musculature surrounding the visceral organs (Figs 17 and 18).

A particular trend of localization seems to occur in several species of salmonids, where *A. simplex* (s.s.) infects with high frequency the flesh of the fish host. Setyobudi et al., 2011 reported that the 98% of *A. simplex* (s.s.) larvae isolated from the chum salmon *O. keta* were located in the flesh of the fish, while the 2% were in the body cavity, pyloric caeca, and on the liver. Similar results were observed in four species of *Oncorhynchus*, by Karl et al. (2011), who subdivided trimmed freshly caught Alaska salmon fillets in six portions to study the distribution of *Anisakis* larvae. All the three fish species showed a comparable distribution pattern of the larvae, that were devided into upper epaxial muscle, lower epaxial, tail muscle, upper belly flap, lower belly flap, and belly flap tail part. Approximately 80% of the *Anisakis* larvae were found in the belly flap area where the larvae were more or less equally distributed. A slightly different picture was observed in the epaxial muscle. The upper parts of the *O. nerka* and *O. keta* fillets were less infected than the lower epaxial muscle and the tail parts. These data are in full agreement with the distribution pattern in fillets of adult *O. keta* returning to the Chitose River, as reported by Sugawara et al. (2004) (Fig. 19).

An overall look on the infection site of some *Anisakis* species, with the relative distribution between visceral cavity and the flesh of various fish host, is reported in Figs 17 and 18. Only studies conducted by peptic digestion and UV-Press are reported here, as we consider them currently the only procedures capable of providing a standardized estimate of larval *Anisakis* spp. localization in the fish host. In fact, as fully reported in Section 6, the efficiency of candling and visual inspection in the detection of *Anisakis* spp. larvae tends to be low (Llarena-Reino et al., 2012). Levsen et al. (2005) reported that only 7%–10% of the all nematode larvae detected after peptic digestion that were present in the fillets of the three fish species they examined, could be detected by candling.

In some geographical areas, coinfection by *Anisakis* species with a differential localization in the same fish host can occur. Larval *Anisakis* belonging to different species show different relative frequencies of localization, even if overall parasite burden results similar. For instance, Suzuki et al. (2010) observed significant differences in infection patterns between two *Anisakis* species recovered in *S. japonicus*, sampled along Japanese coast. Despite the parasite nematodes species occurred with similar infection levels in all fish examined, the number of *A. simplex* (s.s.) larvae present in the musculature of the fish *S. japonicus* was 12 times higher than *A. pegreffii*, which represented only the 1.7% of the larvae in the muscle.

Similar results were obtained by Cipriani et al. (2015) in the European hake, *M. merluccius*: *A. simplex* (s.s.) outnumbers *A. pegreffii* in the musculature (flesh) of the same fish host (Fig. 18). Despite this marked difference in the relative proportions of the two species in the fish muscle, the two parasites showed the same proportion of the infection in the viscera of the same fish species, thus highlighting that at similar infection rate, *A. simplex* (s.s.) displays a better capacity of migrating into the muscular tissue of the fish host (Cipriani et al., 2015). This phenomenon was observed also in other fish species, such as blue whiting *M. poutassou* (Gómez-Mateos et al., 2016), where *A. simplex* (s.s.) doubling the infection rate of *A. pegreffii* in the musculature.

Several hypotheses could be made to explain the differential site preferences of *A. simplex* (s.s.) and *A. pegreffii* in their fish hosts. When the two species are in syntopic conditions, these mechanisms could be the result of a differential use of host resources, as a result of interspecific competition between them for the same trophic resource, theoretically represented by the visceral cavity and/or organs (especially the liver). In addition, concerning to the higher migration capacity exhibited by *A. simplex* (s.s.), this finding could be the result of a differential response to the host immune system, the immune response of the fish host being lower in its musculature (Cipriani et al., 2016). Some experimental studies support the hypothesis that *A. pegreffii* has a lower capacity to invade both fish tissues and laboratory animals (Arizono et al., 2012; Quiazon et al., 2008; Romero et al., 2013). It has been also suggested that the propensity of *Anisakis* spp. larvae to migrate to the flesh could be related to differences in the nature [such as fatty acid content (Smith, 1983)] of the flesh across fish species. The coiling habit shown by larvae in small fish hosts like anchovies, appeared different from the coiled/embedded larvae seen in larger fish species, such as the European hake, *M. merluccius*, or the blue whiting, *M. poutassou*. In these fish species, *Anisakis* spp. larvae were more encapsulated and embedded in the surfaces of the visceral organs, showing a thicker capsule and maintaining the coiled shape even if exposed to higher temperatures (Cipriani et al., 2016).

To date, the only *Anisakis* species capable of migrating into the fish flesh are *A. simplex* (s.s.) and *A. pegreffii*. When found in syntopy with *A. pegreffii* and *A. simplex* (s.s.) in European hakes, *A. physeteris* was never found in the flesh of the fish, while showing a preference for the liver of its host, with the 41% of larvae detected (59% was found in the fish visceral cavity) (Cipriani et al., 2017b).

Palm et al. (2008) reported that the musculature of *Auxis rochei rochei* harboured a single specimen of *A. typica* in 1 of the 40 sampled hosts; so far, this is the only report of this parasite species in the muscular tissue of the fish host.

7.3 *Intra Vitam* and *Post Mortem* Larval Migration

Several *Anisakis* species larvae showed the ability to reach the muscular part of their fish hosts. The larvae seem to be able to reach this site both during their host life span (*intra vitam* migration, Cipriani et al., 2016; Karl et al., 2011; Roepstorff et al., 1993), and after the death of the fish host (postmortem migration, Cipriani et al., 2016; Šimat et al., 2015). Some experimental studies conducted on fish samples stored at different conditions described these two behaviours of *Anisakis* larvae.

Smith and Wootten (1975) reported that: '… A large scale migration of *Anisakis* larvae from the viscera to the flesh of the herring occurred in both our experiments, so that almost 20 per cent of the total worm burden was present in the flesh after 37 h'. Roepstorff et al. (1993) observed that no migration of *Anisakis* larvae in the flesh of herring occurred when the fish was kept over a range of temperatures (0–10°C). Later, Karl et al. (2002) described a significant increase in the nematode burden of fish flesh stored 'ungutted' in ice.

Suzuki et al. (2010), in an experimental comparison of larval migration on purchased fish stored for 20 h at 4°C, observed that the penetration rate was 0% for *A. pegreffii*, but 9.3% for *A. simplex* (s.s.). At different storing conditions, after 20 h at 20°C, the penetration rate of *A. simplex* (s.s.) increased to 19.2%, while only 1.8% of *A. pegreffii* larvae did move (Suzuki et al., 2010). Thus, these authors demonstrated that *post mortem* migration could occur in that fish species (*S. japonicus*), and that it was temperature dependent. The authors observed a similar trend also in experimental agar penetration, with the 66% of *A. simplex* (s.s.) larvae migrating into the agar after 72 h (the temperature was not reported by the authors), while at the same time interval only the 33% of *A. pegreffii* larvae migrated in the agar support.

Concerning larval migration capacity, Cipriani et al. (2016, 2017a) reported the occurrence of *A. pegreffii* at low percentage (4.3%) in the fillets of anchovies, *E. encrasicolus*, frozen immediately after catch, thus indicating *intra vitam* migration of larvae through the fish flesh during the host lifespan. In the same study, after experimental storage of different batches of fish, Cipriani et al. (2016) recorded also a *post mortem* migration of *A. pegreffii* larvae in anchovies. The observed larval movements were strongly dependent on both temperature and time of storage of the fish. When stored at 2°C, *A. pegreffii* larvae exhibited a slight variation in the mean abundance values recorded both in fish visceral cavity and flesh during the complete storage interval (from 24 h until 72 h), thus at this temperature the nematodes probably stayed in a sort of undetermined latency period. Conversely, with an increased storage temperature, at both 5 and 7°C, the larvae became active and started moving. As consequence, the mean abundance of the infection of *A. pegreffii* sequently recorded in the viscera tended to decrease with time, while the same parameter increased in the musculature, proportionally to the time of storage. After 48 h at 7°C, the relative proportion of *A. pegreffii* larvae detected in the flesh of the anchovies increased to 12.7%, a value three times higher that recorded as *intra vitam* migration (4.3%). Indeed, this temperature-dependent motility of the *A. pegreffii* larvae was even more evident when the fish were stored at higher temperature (14°C). In less than 4 h, most of larvae changed their location, crawling out from the fish viscera, moving into the flesh, or out of the fish host body (Fig. 20).

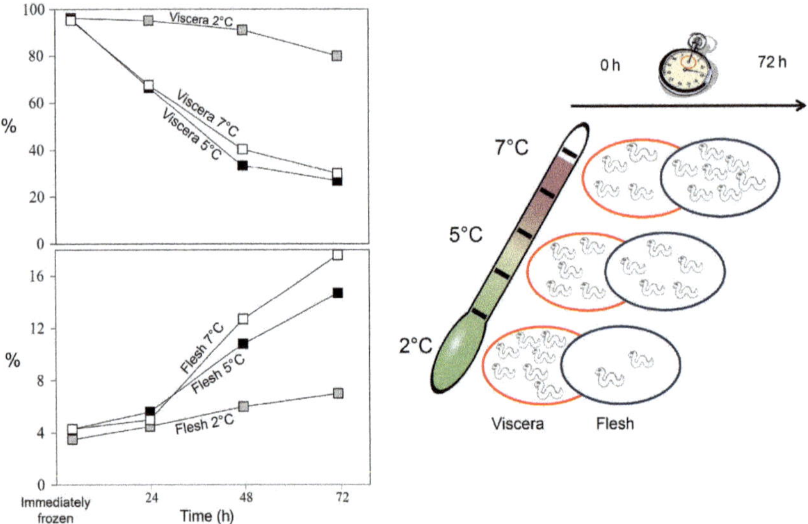

Fig. 20 Relative frequencies of distribution of *A. pegreffii* larvae between viscera and flesh of *Engraulis encrasicolus*, at different fish storage temperatures (2, 5, and 7°C) and at different time intervals (immediately frozen 'control' batch, 24, 48, and 72 h). *Data from Cipriani, P., Acerra, V., Bellisario, B., Sbaraglia, G. L., Cheleschi, R., Nascetti, G. Mattiucci, S. 2016. Larval migration of the zoonotic parasite* Anisakis pegreffii *(Nematoda: Anisakidae) in European anchovy,* Engraulis encrasicolus*: implications to seafood safety. Food Control 59, 148–157.*

Therefore, after fish capture, storage temperature and time play an important role in larval motility. The *post mortem* migration of *A. pegreffii* larvae in the flesh of anchovies can thus result in a different localization and thus a different zoonotic potential of this parasite in this fish host (Cipriani et al., 2016).

Šimat et al. (2015) have also observed a *post mortem* migration in anchovies fillets stored at 0 and 4 °C after 3 and 5 days, respectively, but not at 22 and −18°C. In case of storage at 22°C for 2 days, at the onset of putrefaction of the visceral organs, larvae migrated out of the visceral cavity towards the fish surface (Šimat et al., 2015). Measures of pH and biogenic amine profile observed during storage indicate that certain biochemical conditions may trigger larval migration into fillets (Šimat et al., 2015).

Anyway, temperature appears to be the most important variable determining the activation of *Anisakis* larvae, at least as observed from experimental and in vitro studies (Cipriani et al., 2016; Šimat et al., 2015). This is not surprising, since according to Smith and Wootten (1975), '… a rise in temperature might be expected to cause increased activity since the third stage larvae of *Anisakis* complete their life cycle in warm-blooded mammals'. Recently, it was also suggested that the level of mRNA expression of the heat shock protein (*Hsp90*) of *A. pegreffii* did not change significantly under cold shock (−4°C) (Chen et al., 2014). In cold stress experiments, carried out by those authors, the larvae remained alive, even if they were less active. In contrast, after the larvae were incubated at 37°C for 24 h, they showed increased motility (Chen et al., 2014). However, parallel significant increase of the mRNA expression levels of some proteins involved in the immune response in human anisakiasis have also been found to occur in larvae maintained in vitro culture after 24 h at 37°C (Colantoni et al., 2016).

The factors that cause excapsulation and migration of *Anisakis* larvae after death of the host fish are still unknown, but may be related to physicochemical changes in the viscera of the fish (Smith and Wootten, 1975). Cipriani et al. (2016) hypothesized that some physical or chemical cues associated with the increase of the temperature may be responsible in stimulating the motility of the coiled *A. pegreffii* larvae, both in the viscera and in the fillets. On the other hand, it was suggested that some signal molecules are able to regulate the dispersal behaviour and migration capacity in nematodes (Choe et al., 2012; Kaplan et al., 2012). Specific blends of ascararosides have been found to be associated with the strong dispersing activity in free-living nematodes, i.e., *Caenorhabditis elegans*. The dispersal blend is recognized by other nematodes; indeed, the authors hypothesized that many nematodes species are able to respond to signals released by other nematodes, as an avoidance signal, or as an aggregation pheromone blend (Kaplan et al., 2012). Thus, it seems that this conserved family of molecules could have evolved differentially in nematodes, partially with phylogeny as well as with lifestyle and/or ecological niche. Additionally, the authors found that ascaroside production is life stage dependent (Choe et al., 2012). Thus, the investigation of species-specific ascaroside profiles, also in the nematodes of the genus *Anisakis*, may permit the detection of any species-specific signal that would possibly be responsible also for their adaptive ecology in different site of infection in their hosts.

8. HUMAN ANISAKIASIS: WHICH ARE THE *ANISAKIS* SPP. INFECTIVE TO HUMANS?

Anisakis spp. larvae do not mature in humans, but are able to provoke the fish-borne zoonosis, commonly known as 'anisakiasis'. On the basis of the gastrointestinal site reached by living larvae, if accidentally ingested, the disease is often subdivided into: gastric anisakiasis (GA), intestinal anisakiasis (IA), ectopic anisakiasis (EA), and gastroallergic anisakiasis (GAA). In addition, anisakiasis is clinically speaking, classified as 'acute' or 'moderate', and 'invasive' or 'not invasive', depending on the state and vitality of actual larvae. The 'not invasive' form occurs when the larva is unable to penetrate the gastric or intestinal mucosa and it remains luminal; in this case, the patient could be asymptomatic and the larva could be eliminated with stools and/or vomit. The 'invasive' form is characterized by different steps of the parasite infection: (1) attachment to the gastric or intestinal mucosae and (2) active penetration of the larva into the tissues of the gastrointestinal tract. In the case of 'invasive anisakiasis', severe gastrointestinal symptoms during the acute phase of infection, such as epigastralgia, nausea, vomiting, and abdominal pain are described. The chronic infection of anisakiasis, is characterized by the *Anisakis* larvae deeply invading the gastric and intestinal walls, causing direct tissue damage, ulcers and, eventually, eosinophilic granuloma. Thus, the symptoms are correlated with the site of the infection of the larva in humans, and by the level of the histopathological lesions that the different zoonotic *Anisakis* species are able to produce.

In this section, the main results associated with the epidemiology and disease in humans are reported. The aspects have been revised and discussed taking into account that the main scope or connecting thread of this review is 'the biological parasite species'. Therefore, the different *Anisakis* species are considered in relation to their potential role as zoonotic agents in humans, pathogenicity, and diagnosis.

8.1 The Zoonotic Role of *Anisakis* spp.

Are all the Anisakis *spp. responsible for human anisakiasis*? According to the present knowledge not all among them. The actual evidence is based on the following findings:

While *A. simplex* (s.s.) and *A. pegreffii* have already demonstrated their ability to cause anisakiasis in humans, it is not yet clear if any other *Anisakis* species can do the same.

As we have presented in Section 7, *A. simplex* (s.s.) and *A. pegreffii*, are the species most frequently found in the edible parts of commercially important fish and squid; as a consequence, they are also the species with the highest risk to infect humans. Indeed, live larvae of the two species infecting the fillets of fish and the edible parts of squid to be consumed raw, undercooked, or in various traditional dishes, may cause anisakiasis. A number of fish dishes are considered at high risk for consumers to contract anisakiasis. They include, among others, Spanish boquerones and anchovies, Italian marinated anchovies, the Scandinavian gravlax, South American ceviche, Dutch salted and marinated herring, and Japanese sushi and sashimi. In this respect, the identification of larval anisakids becomes relevant also with respect to human medicine and pathology.

Are all the species of Anisakis *able to provoke 'invasive anisakiasis'?* Actually, based on the molecular diagnosis of the etiological agent in several clinical case reports, clear evidence for clinical, histopathological features of 'invasive anisakiasis' in humans has been described only for two species belonging to Clade 1 in the phylogeny of *Anisakis*: i.e., *A. simplex* (s.s.) and *A. pegreffii*.

What about the pathogenic role played by the remaining species belonging to the genus Anisakis? Concerning this aspect from a phylogenetic point of view, among the species of Clade 1, *A. berlandi* has never been associated with human disease. Although its occurrence in fish species has been documented, however, it is not yet clear if it parasitizes the musculature of the fish host species. As for the species of Clade 2, i.e., *A. ziphidarum* and *A. nascettii* (see also Sections 3 and 6), these parasites are only rarely found in commercial important fish species such as *M. merluccius*. However, in all cases, the larvae were found in or on the viscera of the fish rather than the flesh (Mattiucci et al., 2004). In addition, a higher level of infection was found in deep-sea squid species (e.g. *M. ingens*), which, however, are usually not regarded as 'sea-food'. Therefore, it seems that their role in transmission of the disease to humans is neglectable. Finally, the larvae of Type I of *A. typica* are only very rarely reported to occur in the fish muscle. Palm et al. (2008) reported the occurrence of one larvae of *A. typica* penetrating the muscle of a single fish. Anshary et al. (2014) did not record *A. typica* in the flesh of *K. pelamis* and *A. thazard* from Indonesia. However, according to Anshary et al. (2014) and Uga et al. (1996), there are reports on positive sero-prevalence for anisakiasis in humans in Indonesia; however, the etiological agents possibly responsible for that positive serodiagnosis were not identified. Generally, only scanty data exist concerning the occurrence of human anisakiasis from those geographic areas, where *A. typica* has been generally found infecting natural hosts (see also map of the geographical range of the species, Fig. 6). Similarly, no human cases related to the larvae of Types II, III, and IV (see Section 4) of those species included in Clade 3 (i.e. *A. physeteris*, *A. brevispiculata*, and *A. paggiae*). However, these species have never been found, so far, to infect the muscle of fish and squids, at least those of commercial importance. Mattiucci et al. (2004, and personal observation) recorded these species, as coiled larvae in the stomach and intestine wall of the European hake *M. merluccius*. Thus, it seems that the risk of infection to humans due to those species could be low. However, Romero et al. (2013) have carried out experimental infection of Winstar rats with *Anisakis* Type II larvae, which, according to the authors, belonged to *A. physeteris* and *A. paggiae*. After necroscopy, they found only 4 larvae out of the 56 used for the infection, attached to the gastrointestinal wall or penetrating it, thus suggesting that these two species are able to provoke 'invasive anisakiasis'. It should be underlined that, despite the fact that molecular methods to identify larval *Anisakis* spp. from fish are largely applied, the same cannot be asserted for all human case reports in the recent literature (see Table 11). Moreover, the notification of anisakiasis in humans is not mandatory so that, consequently, the disease is likely still underreported and misidentified.

8.2 Molecular Epidemiology of Human Anisakiasis

Anisakiasis has become increasingly more relevant as human health risk, in particular in countries or regions where the consumption of raw or only lightly processed fish and squid is frequent and/or has become increasingly popular. Consequently, cases of the disease are increasingly reported from Asia (mainly Japan, Korea, China, and Taiwan), representing more than 50% of all cases, while the remaining are from European countries, especially Italy, Spain,

Table 11 Published Reports of Human Anisakiasis Worldwide, With Focus on Those Described With the Larval Detection

	Origin of the Patient	Number of Patients	Clinical Features	Method Used for Larval Detection	Etiological Agent	References
European countries	Italy	1	Gastric	—	*Anisakis* sp.	Stallone et al. (1996)
	Italy	1	Extragastrointestinal	Laparotomy	*Anisakis* sp.	Cancrini et al. (1997)
	Italy	1	Gastric	Endoscopy	*Anisakis* sp.	Cancrini et al. (1998)
	Italy	1	Intestinal	Endoscopy	*Anisakis* sp.	Ioli et al. (1998)
	Italy	1	Gastric	Endoscopy	*A. pegreffii*[a]	D'Amelio et al. (1999)
	Italy	3	Gastric	—	*A. simplex*	Maggi et al. (2000)
	Italy	11	Gastric (2), intestinal (8), spleen (1)	Endoscopy, laparotomy	*Anisakis* sp.	Pampiglione et al. (2002)
	Italy	1	Intestinal	Histologic examination	*Anisakis* sp.	Caramello et al. (2003)
	Italy	1	Gastric, splenic	Laparotomy	*Anisakis* sp.	Testini et al. (2003)
	Italy	1	Intestinal	Laparotomy	*Anisakis* sp.	Moschella et al. (2005)
	Italy	1	Intestinal	Histologic examination	*Anisakis* sp.	De Nicola et al. (2005)
	Italy	1	Intestinal	Laparotomy	*A. simplex* (s.l.)	Montalto et al. (2005)
	Italy	1	Intestinal	Laparotomy	*A. simplex* (s.l.)	Pellegrini et al. (2005)
	Italy	3	Gastric	Endoscopy	*Anisakis* sp.	Fazii et al. (2006)
	Italy	1	Extragastrointestinal	Endoscopy	*Anisakis* sp.	Avellino et al. (2007)
	Italy	3	Gastric	Gastroscopy	*Anisakis* sp.	Ugenti et al. (2007)
	Italy	2	Gastric	Endoscopy	*A. pegreffii*[a]	Fumarola et al. (2009)
	Italy	1	Intestinal	Histologic examination	*A. simplex* (s.l.)	Marzocca et al. (2009)

Continued

Table 11 Published Reports of Human Anisakiasis Worldwide, With Focus on Those Described With the Larval Detection—cont'd

Origin of the Patient	Number of Patients	Clinical Features	Method Used for Larval Detection	Etiological Agent	References
Italy	1	Intestinal	Endoscopy, histologic examination	*A. simplex* (s.l.)	Aloia et al. (2011)
Italy	1	Intestinal	Histologic examination	*A. pegreffii*[a]	Mattiucci et al. (2011)
Italy	2	Gastric	Endoscopy	*A. simplex* (s.l.)	Pontone et al. (2012)
Italy	8	Gastric (6), (GAA) (2)	Endoscopy	*A. pegreffii*[a]	Mattiucci et al. (2013)
Italy	1	Intestinal	Colonoscopy	*A. simplex* (s.l.)	Mumoli and Merlo (2013)
Italy	1	Intestinal	Endoscopy, histologic examination	*A. simplex* (s.l.)	Andrisani et al. (2014)
Italy	1	Intestinal	Laparotomy	*Anisakis* sp.	Baron et al. (2014)
Italy	1	Intestinal	Laparotomy	*Anisakis* sp.	Piscaglia et al. (2014)
Italy	1	Gastric	Gastroscopy	*Anisakis* sp.	Mariano et al. (2015)
Italy	3	Gastric (1), intestinal (2)	Endoscopy, histologic examination	*A. pegreffii*[a]	Mattiucci et al. (2017d)
Italy	1	Gastric	Endoscopy	*Anisakis* sp.	Palma et al. (2018)
Croatia	1	Intestinal	Histologic examination	*A. pegreffii*[a]	Mladineo et al. (2016)
France	-	Intestinal granuloma	—	*Anisakis* sp.	Doby et al. (1975)
France	5	Intestinal	Microscopic analyses, gastroscopy	*Anisakis* sp.	Mudry et al. (1986)
France	4	Gastric	Gastroscopy	*Anisakis* sp.	Bouree et al. (1995)
France	6	Gastric, intestinal	Endoscopy	*Anisakis* sp.	Dupouy-Camet et al., 2016
Spain	2	—	—	*Anisakis* sp.	Acebes Rey et al. (1996)
Spain	1	Gastric	Endoscopy	*A. simplex* (s.l.)	Romeo Ramirez et al. (1997)
Spain	13	Gastric	Gastroscopy	*A. simplex* (s.l.)	Daschner et al. (1998)

	Spain	2	Intestinal	Histologic examination	*A. simplex* (s.l.)	del Olmo Escribano et al. (1998)
	Spain	5	GAA	Gastroscopy	*A. simplex* (s.l.)	Alonso et al. (1999)
	Spain	1	Gastric	—	*Anisakis* sp.	Barriga et al. (1999)
	Spain	20	GAA	Gastroscopy	*A. simplex* (s.l.)	Daschner et al. (2000)
	Spain	22	GAA	Gastroscopy	*A. simplex* (s.l.)	López-Serrano et al. (2000)
	Spain	2	Gastric	Endoscopy	*Anisakis* sp.	del Olmo Martínez et al. (2000)
	Spain	1	Intestinal	Histologic examination	*A. simplex* (s.l.)	López Peñas et al. (2000)
	Spain	9	Gastric (5), intestinal (3)	Endoscopy or intestinal resection	*A. simplex* (s.l.)	Repiso Ortega et al. (2003)
	Spain	24	Gastric, intestinal	Gastroscopy	*A. simplex* (s.l.)	Alonso-Gómez et al. (2004)
	Spain	1	Gastric	Endoscopy	*A. simplex* (s.l.)	Carrascosa et al. (2015)
	Spain	1	Intestinal	Colonoscopy	*A. simplex* (s.l.)	Riu Pons et al. (2015)
	The Netherlands	1	Gastric	—	*Anisakis* sp.	Van Thiel et al. (1960)
	Belgium	—	Intestinal	—	*Anisakis* sp.	Fain et al. (1969)
	Norway	—	Gastric	—	*Anisakis* sp.	Jacobsen and Berland (1969)
	Denmark	—	—	—	*Anisakis* sp.	Andreassen (1970)
	Belgium	6	Intestinal	Histologic examination	*Anisakis* sp.	Verhamme and Ramboer (1988)
	Belgium	1	Gastric	Endoscopy	*Anisakis* sp.	Vercammen et al. (1997)
	Germany	1	Gastric	—	*Anisakis* sp.	Plath et al. (2001)

Continued

Table 11 Published Reports of Human Anisakiasis Worldwide, With Focus on Those Described With the Larval Detection—cont'd

	Origin of the Patient	Number of Patients	Clinical Features	Method Used for Larval Detection	Etiological Agent	References
	Germany	1	Intestinal	Laparotomy	*Anisakis* sp.	Lock et al. (2008)
	Austria	2	Intestinal	Laparotomy	*A. simplex* (s.s.)[a]	Auer et al. (2007)
	Russia	—	Intestinal	—	*Anisakis* sp.	Karmanova et al. (2002)
USA	Boston	1	Intestinal	Histologic examination	*Anisakis* sp.	Roselyn et al. (1973)
	Hawaii	1	Gastric	—	*A. simplex* (s.l.)	Deardorff et al. (1986)
	Hawaii	1	—	X-ray	*Anisakis* sp.	Hiramoto and Tokeshi (1991)
	Quebec	1	Intestinal	Laparotomy	*Anisakis* sp.	Couture et al. (2003)
	USA	—	Intestinal	Histologic examination	*Anisakis* sp.	Pinkus et al. (1975)
	USA	—	Intestinal	Laparotomy	*Anisakis* sp.	Appleby et al. (1982)
Asian countries	Japan	2	Gastric granuloma	Histologic examination	*Anisakis* sp.	Asami et al. (1965)
	Japan	11	Intestinal	—	*Anisakis* sp.	Otsuru et al. (1965)
	Japan	—	Intestinal	—	*Anisakis* sp.	Ishikura et al. (1967)
	Japan	—	Pharynx mucosa	—	*Anisakis* sp.	Tanaka et al. (1968)
	Japan	1	Gastric	Endoscopic, histologic examination	*Anisakis* sp.	Hsiu et al. (1986)
	Japan	2	Intestinal	Colonoscopy	*Anisakis* sp.	Minamoto et al. (1991)
	Japan	1	Gastric	Gastroscopy	*A. simplex* (s.l.)	Kagei and Isogaki (1992)
	Japan	4	Intestinal	Colonoscopy	*Anisakis* sp.	Matsumoto et al. (1992)
	Japan	1	Esophageal	Endoscopy	*Anisakis* sp.	Urita et al. (1997)

Japan	1	Intestinal	Laparotomy	*Anisakis* sp.	Sasaki et al. (2003)
Japan	1	Intestinal	Laparotomy	*Anisakis* sp.	Ishida et al. (2007)
Japan	1	Gastric	Laparotomy	*Anisakis* sp.	Ito et al. (2007)
Japan	1	Intestinal	Laparotomy	*A. simplex* (s.l.)	Takei and Powell (2007)
Japan	85	Gastric	Endoscopy	*A. pegreffii* (1)[a] *A. simplex* (s.s.) (84)[a]	Umehara et al. (2007)
Japan	1	Intestinal	Laparotomy	*Anisakis* sp.	Sugita et al. (2008)
Japan	1	Gastric, intestinal	Gastroscopy, colonoscopy	*Anisakis* sp.	Kim et al. (2013)
Japan	1	Intestinal	Colonoscopy	*Anisakis* sp.	Yorimitsu et al. (2013)
Japan	44	Gastric (42), intestinal (2)	Endoscopy	*A. simplex* (s.s.)[a]	Arai et al. (2014)
Japan	1	Gastric	Endoscopy	*Anisakis* sp.	Sonoda et al. (2015)
Japan	1	Extragastrointestinal	Laparotomy	*A. simplex* (s.l.)	Takamizawa and Kobayashi (2015)
Japan	1	Gastric	Endoscopy	*Anisakis* sp.	Hamada et al. (2016)
Japan	1	Intestinal	Endoscopy	*Anisakis* sp.	Hashimoto et al. (2017)
Japan	1	Gastric	Esophagogastroduodenoscopy	*Anisakis* sp.	Hashimoto and Chonan (2016)
Japan	1	Intestinal	—	*Anisakis* sp.	Ikuta et al. (2016)
Japan	1	Gastric	Endoscopy	*Anisakis* sp.	Shimamura et al. (2016)
Japan	1	Intestinal	Endoscopy	*Anisakis* sp.	Tsukui et al. (2016)

Continued

Table 11 Published Reports of Human Anisakiasis Worldwide, With Focus on Those Described With the Larval Detection—cont'd

Origin of the Patient	Number of Patients	Clinical Features	Method Used for Larval Detection	Etiological Agent	References
Thailand	1	Intestinal	Histologic examination	*Anisakis* sp.	Hemsrichart (1993)
Korea	—	Palatin tonsil	Endoscopy	*Anisakis* sp.	Kim et al. (1971)
Korea	1	Intestinal	Laparotomy	*Anisakis* sp.	Kim et al. (1991)
Korea	1	Gastric	Gastroscopy	*A. simplex* (s.l.)	Noh et al. (2003)
Korea	1	Intestinal	Laparotomy	*Anisakis* sp.	Kang et al. (2010)
Korea	1	Gastric	Laparotomy	*Anisakis* sp.	Kang et al. (2014)
Korea	16	Gastric (4), intestinal (12)	Gastroduodenoscopy	*A. pegreffii* (15)[a] *A. simplex* (s.s.) (1)[a]	Lim et al. (2015)
Korea	15	Gastric	Endoscopy	*Anisakis* sp.	Sohn et al. (2015)
Taiwan	1	Gastric	Endoscopy	*A. simplex* (s.l.)	Li et al. (2015)

[a]Reports in which the etiological agent was identified, to the species level, by molecular methodologies.

France and Croatia, whereas there are much fewer cases in Germany, the Netherlands, United Kingdom and Norway, despite comparatively high per capita fish consumption rates in these countries, as well (Fayer, 2015).

Starting with the reviews by Audicana and Kennedy (2008) and Chai et al. (2005), there has been a marked increased in prevalence of human anisakiasis throughout the world in the past 10 years. In Italy, too, the home of most authors of this review, the first case of anisakiasis was described in 1996 (Stallone et al., 1996), since followed by numerous other reports of this zoonosis.

These findings are attributable to a number of factors. One is represented by the increase in knowledge about this disease among medical personnel, which has allowed increased collaboration between clinicians and biologists leading to better reports and descriptions of the disease. Another explanation is found among the improved methods for the detection of the disease in humans (i.e. gastric and intestinal endoscopy), which have facilitated early diagnosis and removal of *Anisakis* spp. larvae (Fig. 21). Molecular methods have greatly improved the identification of larvae and their fragments (Mattiucci et al., 2011). Finally, increased seafood consumption has led to higher per capita fish demand, in parallel with increased popularity of raw fish, not only as part of local traditions, but also among younger people (Fayer, 2015). A variety of seafood dishes are considered to be of high risk for the contraction of human anisakiasis in some countries, including *boquerones* and *marinated anchovies* in Spain and in Italian marinated anchovies, etc. (Bao et al., 2017b; Mattiucci et al., 2013).

This review has revised the literature concerning only those human cases reporting the finding of the *Anisakis* larvae, as well as the symptoms provoked by them (Table 11). According to the collected datasets so far, a total of 384 records of patients who were infected by *Anisakis* spp. larvae, are here listened. It was only after 1999 that molecular identification of the larvae in human anisakiasis cases has become possible. Thus, all cases of anisakiasis before 1999, where the larva was removed surgically or by endoscopy, could not be identified to species level (Table 11 and Fig. 22). In other cases, the histopathological findings of larvae embedded in eosinophilic granulomas, in which some morphological features of an *Anisakis* Type I larva were recognizable, was possible (Fig. 23). In Table 11, only those cases of anisakiasis are described in which the larva was recognized, starting with the first case reported from the Netherlands by Van Thiel et al. (1960) and Smith and Wootten (1978, and references therein). Most of the reports (72.0%) describe the occurrence of *Anisakis* spp. larvae associated with GA, while

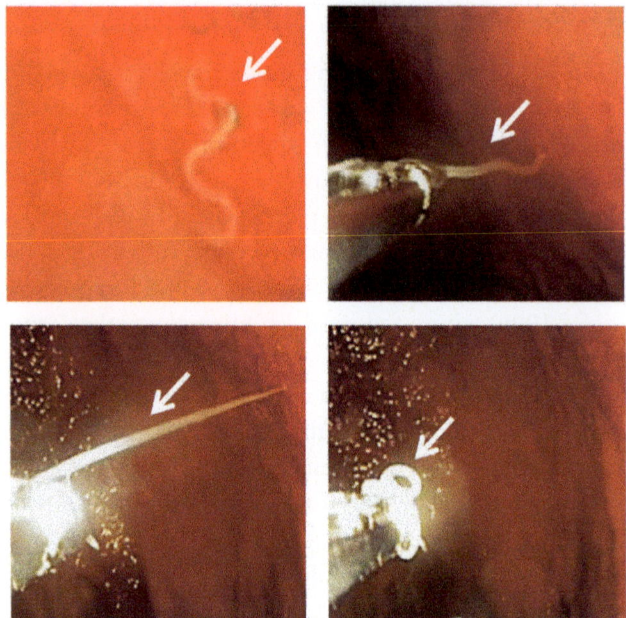

Fig. 21 Steps of the removal of *Anisakis* sp. larva from human gastric mucosa by biopsy forceps. The *arrow* indicates the larva, during the removal.

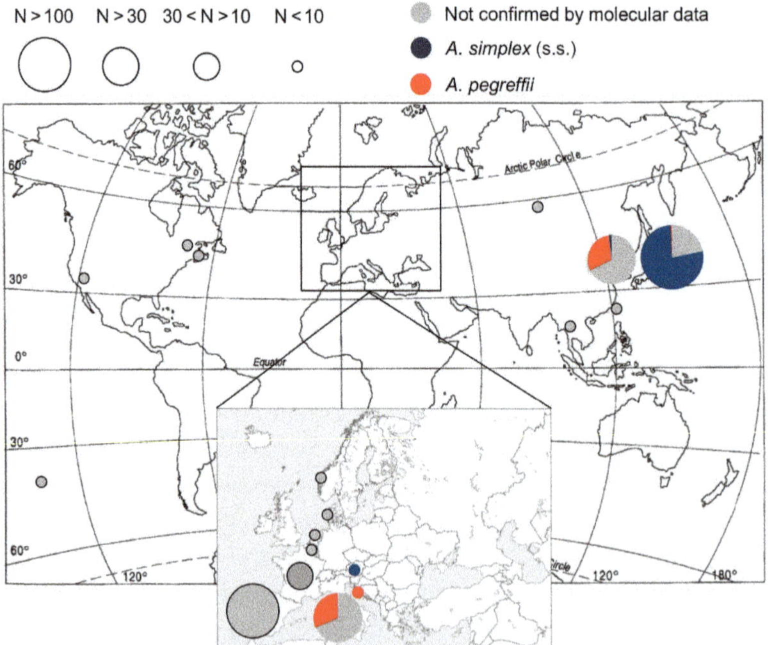

Fig. 22 World map of human anisakiasis, based on the 1960–2017 published literature: distribution of human cases which included larval identification. Cases of IgE-*Anisakis* hypersensitization are not reported. *N*= number of cases. The *pie graphs* show relative proportion of cases due to the two zoonotic species (i.e. *A. pegreffii* and *A. simplex* (s.s.)), since 1999 up to now.

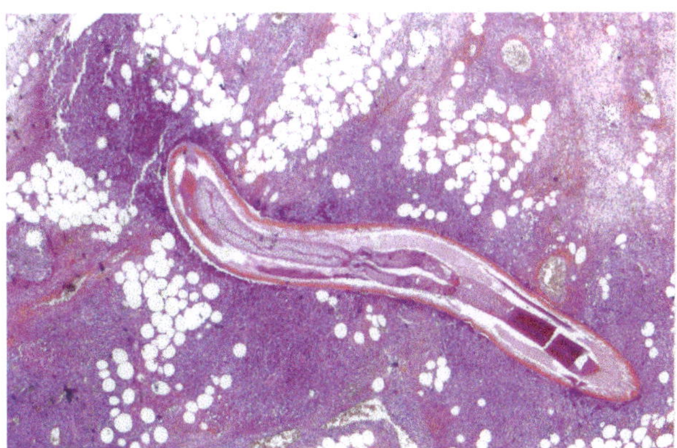

Fig. 23 Histological section of a surgically removed eosinophilic granuloma showing an *Anisakis* sp. larva in sagittal view.

26% were invasive and not invasive intestinal cases of anisakiasis (IA), with only 2% being extragastrointestinal (Table 11). However, the epidemiology of human anisakiasis, despite several reported cases over recent years, remains an underestimated fish-borne zoonosis. Endoscopic examination has been often used to diagnose cases of GA (Fig. 21).

Literature dealing with records of human cases of 'allergic anisakiasis' due to *Anisakis*-IgE hypersensitivity, without proper identification of the larva involved, was excluded from this review. The only exceptions are reports of *Anisakis* allergy associated with *Anisakis* spp. larvae, i.e., in GAA cases. In other words, here we will not consider all the literature on the occurrence of IgE-*Anisakis* hypersensitization associated with chronic urticaria (CU+), nor *Anisakis* antigens/allergens as possible 'seafood-allergens' (EFSA, 2010).

Actually, the species *A. pegreffii* was reported to cause GA, IA, and GAA in 32 patients (Table 11). Indeed, larval stages have been identified in cases of 'invasive' anisakiasis, after observing viable larvae during endoscopy (Arai et al., 2014; D'Amelio et al., 1999; Lim et al., 2015; Mattiucci et al., 2013, 2017d; Umehara et al., 2007), colonscopy (Mattiucci et al., 2017d), or in surgically removed eosinophilic granulomas (Mattiucci et al., 2011, 2017d; Mladineo et al., 2016). Those cases, molecularly identified as caused by *A. pegreffii*, were mostly from Italy (D'Amelio et al., 1999; Mattiucci et al., 2011, 2013, 2017d), Croatia (Mladineo et al., 2016), from Korea (Lim et al., 2015) and, more rarely, from Japan (Umehara et al., 2007). These findings are in accordance with the geographic distribution of the parasite species and its infection in those fish collected within the parasite's range of distribution.

A. simplex (s.s.) has been identified (by molecular methods) to cause 'invasive' GA and IA in Japan (Arai et al., 2014; Auer et al., 2007; Umehara et al., 2007). But in some European countries, such as Spain, despite widely reported cases of 'allergic anisakiasis', the etiological agent associated to those cases is not being identified to species level by molecular identification (Audicana and Kennedy, 2008 and references therein) (Daschner et al., 2012). Indeed, in the literature concerning 'allergic anisakiasis' from Spain, most of them report *A. simplex* (s.l.), without any molecular identification. In other words, the diversity of species included in that complex has not been named. Taking into account that the fish carrying larvae causing the human infections could be from a sympatric area of the two zoonotic species (see Sections 3 and 6), the etiological agent could be either *A. simplex* (s.s.), or *A. pegreffii*, or both.

8.3 Clinical Manifestation of Human Anisakiasis Caused by *A. simplex* (s.s.) and *A. pegreffii*

8.3.1 GA and IA

Clinical symptoms of GA provoked by *A. simplex* (s.s.) and *A. pegreffii* are similar: epigastric pain, symptoms are nausea, vomiting, abdominal fullness or distension, anorexia, and chest pain. As for the localization in the gastric mucosa, the majority of larvae were found in the greater curvature, followed by the posterior wall. Shibata et al. (1989) described endoscopic findings such as oedematous hypertrophic gastric folds, increase in gastric secretion and mucosal lesions, including oedema, redness, coagulation, haemorrhage, and ulceration. Cases of GA due to *A. pegreffii* described recently from Italy (Mattiucci et al., 2013) were characterized, clinically, by epigastric pain just after 2 h following the ingestion of raw seafood, vomiting, and other digestive symptoms. Endoscopic findings have showed that, in most of the cases, the larvae were mainly located in the lumen of the stomach penetrating the submucosal layer of the gastric wall (Fig. 21). Arai et al. (2014) have found that in GA due to *A. simplex* (s.s.), the parasite may prefer to invade and penetrate the normal intact gastric mucosa, instead of an atrophic mucosa. They hypothesized that this finding can be related to the fact that pH value is lower in the normal mucosa than in atrophic mucosa and that the parasite species seems to possess a higher penetration rate in agar at low pH (Arizono et al., 2012).

An experimental model of infection of Wistar rats using the two species, *A. pegreffii* and *A. simplex* (s.s.) (Romero et al., 2013), was performed to find possible differences in the pathogenicity elicited by the two species.

The authors found the 87% of the larvae penetrated or attached to the gastrointestinal wall of the infected rats. Among them, they observed that the penetration rate of *A. pegreffii* was 63% lower than that observed for *A. simplex* (s.s.), also confirming the 'in vitro' agar penetration and 'in vivo' (in natural hosts, i.e. fish) observations previously obtained (Arizono et al., 2012; Quiazon et al., 2011). They also concluded that the pathogenic role of *A. simplex* (s.s.) was higher than that of *A. pegreffii*.

Despite of the findings in experimental infection of laboratory animals, it cannot be inferred that *A. simplex* (s.s.) is a more pathogenic species than *A. pegreffii*, as suggested by Romero et al. (2013). This is actually based on the histopathological findings and clinical manifestation of GA cases associated to the two *Anisakis* species, molecularly identified in humans.

Cases of IA due to *A. pegreffii* and *A. simplex* (s.s.) include a 'mild form' characterized by eosinophilic granulomas forming 'tumour-like' formations in the intestinal wall, and a 'fulminant form' has the symptoms of acute ileus, acute appendicitis, acute abdomen, or regional ileitis. Cases of IA reported from Japan (Arai et al., 2014; Hashimoto et al., 2017; Ikuta et al., 2016; Ishida et al., 2007; Ishikura et al., 1967; Kim et al., 2013; Matsumoto et al., 1992; Minamoto et al., 1991; Otsuru et al., 1965; Sasaki et al., 2003; Sugita et al., 2008; Takei and Powell, 2007; Tsukui et al., 2016; Yorimitsu et al., 2013), Korea (Kang et al., 2010; Kim et al., 1991; Lim et al., 2015), and from Italy (Mattiucci et al., 2011, 2017d; Moschella et al., 2004) were characterized by a clinical picture of acute abdominal pain, acute appendicitis, or acute abdomen.

Pathological changes occurring within the gastrointestinal tract during an infection by larval stage of the two zoonotic parasite species (i.e. *A. simplex* (s.s.) and *A. pegreffii*) are quite similar. The combined result of the direct invasive capacity of the larva and its antigens and the interaction between the host's immune response during tissue invasion, causes the pathological findings associated with the infection. However, most of those antigenic proteins so far isolated and characterized as excretory/secretory (E/S) antigens from are from larvae indicated as *A. simplex* (s.l.), without any indication about the specific identification of those larvae (Caballero et al., 2008; Daschner et al., 2012; Kobayashi et al., 2011; Moneo et al., 2000; Rodriguez-Perez et al., 2008; Shimakura et al., 2004) (Table 12). Apparently, so far, the two species show a similar panel of antigens/allergens, which are recognized by IgE antibody response in human anisakiasis (Mattiucci et al., 2013). The first comparative molecular characterization of four antigens/allergens, among those of the two species, i.e., *A. simplex* (s.s.) and *A. pegreffii*,

Table 12 Characterized Antigens/Allergens of *Anisakis* sp. Larvae

Allergen	MW (kDa)	Localization in Larval Body	Function	Major Allergen	Panallergen	References
Ani s 1	24	ESP	Kunitz-type trypsin inhibitor	Yes	—	Moneo et al. (2000)
Ani s 2	97	Somatic	Paramyosin	Yes	Yes	Pérez-Pérez et al. (2000)
Ani s 3	41	Somatic	Tropomyosin	—	Yes	Asturias et al. (2000)
Ani s 4	9	ESP	Cystatin	—	—	Rodriguez-Mahillo et al. (2007)
Ani s 5	15	ESP	SXP/RAL protein	—	—	Kobayashi et al. (2007)
Ani s 6	7	ESP	Serpin	—	—	Kobayashi et al. (2007)
Ani s 7	139	ESP	Glycoprotein	Yes	—	Anadón et al. (2009)
Ani s 8	15	ESP	SXP/RAL protein	—	—	Kobayashi et al. (2007)
Ani s 9	14	ESP	SXP/RAL protein	—	—	Rodriguez-Perez et al. (2008)
Ani s 10	22	Somatic?	?	—	—	Caballero et al. (2011)
Ani s 11	55	Somatic?	?	—	—	Kobayashi et al. (2011)
Ani s 11-li	?	Somatic?	?	—	—	Kobayashi et al. (2011)
Ani s 12	31	?	?	Yes	—	Kobayashi et al. (2011)
Ani s 13	37	ESP	Haemoglobin	Yes	—	González-Fernández et al. (2015)
Ani s 14	27	?	?	Yes	—	Kobayashi et al. (2015)

ESP: excretory/secretory products; ? indicates unknown function, source, or molecular weight.

has been performed by Quiazon et al. (2013). Recently, 36 and 29 putative allergens were identified, respectively, in *A. simplex* (s.s.) and *A. pegreffii* by mean of transcriptomic analysis (Baird et al., 2016).

Both humoral and cellular responses are involved in infections with *A. pegreffii* and *A. simplex* (s.s.) larvae. Th2 cytokine production and the resulting mastocytosis, and the IgE response and eosinophilia characterize local inflammatory lesions produced by *Anisakis* spp. larvae. Eosinophilic infiltration in the tissues surrounding the parasite has been reported in both acute and chronic anisakiasis. The high concentration of eosinophilic cells in the invaded areas, seems to be related not only to the production of chemotactic factors released by T lymphocytes, mast cells, and basophils, but also to some chemotactic substances produced directly by *Anisakis* spp. larvae which attract these cells. Eosinophilic infiltration is the most effective process in the destruction of larvae at the gastrointestinal tract level, as immune hypersensitivity in response to the *Anisakis* infection (Nieuwenhuizen, 2016).

A. pegreffii and *A. simplex* (s.s.) show similar histopathological features. The first stage is the 'phlegmon formation', and the second is the 'abscess formation'. The last is rather frequent in GA and is characterized by abundant necrotic tissue around the larvae and by a rich population of eosinophils. The third stage is the 'abscess–granuloma formation', which corresponds to the flogistic evolution of the disease from one to several months after ingestion of the larva. At this stage, the larva is in the form of few remnants which are invaded by eosinophils, surrounded by giant cells and abundant inflammatory parvicellular infiltrate. Finally, the most advanced stage is 'granuloma formation', characterized by a further decrease in the presence of eosinophils, but with abundance of lymphocytes, giant cells, and significant collagenization (Kikuchi et al., 1990).

Although it is likely that the pathogenic effect of the two species is similar, the molecular mechanisms which underlie the human host–parasite species interactions, have not yet been fully elucidated. In a recent study (Messina et al., 2016), to test the mechanisms by which human nonimmune cells respond to *A. pegreffii* larvae, fibroblast cell line HS-68 cells were exposed to ES and crude extract (CE) products of the parasite, to evaluate molecular markers related to stress response, oxidative stress, inflammation, and apoptosis. It was observed that *A. pegreffii* products led to increased production of reactive oxygen species (ROS). In addition, the elevated induction of fibroblasts treated with those products, suggests a significantly negative effect on the human host DNA (Messina et al., 2016).

A study to characterize the effects of both ES and CE of *A. pegreffii* on dendritic cells (DCs) biological behaviour was also carried out (Napoletano et al., 2016, submitted). Preliminary results indicated that exposure to *A. pegreffii* CE impaired DC viability, differentiation and maturation, and modulating the ERK1, 2 signalling. These retard DCs at an immature stage, reducing their ability as antigen-presenting cells (APCs) to trigger an efficacious Th1 response. The interaction between *A. pegreffii* and DC precursors seems to be relevant for inducing a 'crippled' immune response, thus suggesting the DC impairment induced by *A. pegreffii* as a possible mechanism of evasion of the host immune system, evolved by this parasite (Mattiucci, personal communication; Napoletano et al., 2016, submitted).

8.3.2 GAA

The term GAA was advanced by Daschner et al. (2000) to indicate an acute allergic reaction in the context of an acute gastric presence of an *Anisakis* sp. larva, when the live parasite attempts to invade the submucosal layer of the gastric wall. GAA could be characterized by urticaria, angioedema, and generally consists of an acute IgE-mediated, generalized reaction. In this type of anisakiasis, the allergic reactions take place starting from 2 or 3 h to 2 or 3 days after the ingestion of infected fish (Daschner et al., 2011; Mattiucci et al., 2013). Recently, some GAA cases, due to *A. pegreffii*, characterized by urticaria and oedema of the oral mucosa, were recognized by molecular methods in Italian patients after they had consumed 'marinated anchovies' (Mattiucci et al., 2013). Endoscope observations showed *Anisakis* larvae invading the submucosa layer of the gastric wall. In addition, serum samples from the patients showed IgE reactivity in WB analysis against *Ani s 1-like*, *Ani s 7-like*, and *Ani s 13-like* antigens of *A. pegreffii* (Mattiucci et al., 2017e) (Table 12 and Section 8.4.3).

GAA has been described as a simultaneously primary and secondary immunological reaction, as the immunoglobulin isotypes are present from the first day of the infection. After 1 month, only IgE, IgA, IgG, and IgG4 generally show high levels of production. This polyclonal production is stimulated by active live *Anisakis* larvae, even after removal by endoscopy (Daschner et al., 2002). This immunological production could be of importance for distinguishing between 'sea food allergy' and the CU+ and *Anisakis* allergy (Daschner et al., 2002). Daschner and Pascual (2005) also suggested that the allergic reaction related to the invasion of an *Anisakis* sp. larva is a protective immunological reaction by the human host to avoid the progressive evolvement in the chronic disease, i.e., the granuloma formation (Gonzalez-Quijada et al., 2005).

Thus, according to the available literature, human infection with *A. pegreffii* are clearly associated with GAA, however, the same cannot be definitively asserted for *A. simplex* (s.s.) since any specific molecular identification of other GAA cases reported, has not yet been fully demonstrated.

8.4 Diagnosis of *Anisakis* spp. in Humans

Symptoms of human anisakiasis are often too vague and aspecific to allow quick and reliable diagnosis. The available diagnostic methodologies are reviewed below.

8.4.1 Histological Diagnosis

Eosinophic granulomas removed surgically from gastric and intestinal sites can reveal the presence of worms with the characteristic morphological features of *Anisakis* sp. larvae. Histological preparations (H&E) may reveal relevant features by using conventional bright light microscopy.

However, this can only be achieved if the nematode in the nodule is in a good state of preservation. In histological transverse sections, the following characters can be referred to as positively identifying *Anisakis* larva: a thin cuticle lacking lateral alae; polymyarian muscle cells, separated into four quadrants by chords with two wing-like distal lobes; circular intestine with a triangular lumen and 50–70 μm tall columnar epithelial cells; and banana-shaped excretory cells (renette cells), situated ventrally to the intestine. In histological sagittal section (Fig. 23), the following features are typical of *Anisakis* sp. larvae: the muscular part of the oesophagus followed by the glandular part (ventriculus) and absence of a ventricular appendix and/or intestinal caecum. These microscopic observations permitted the identification of *Anisakis* larva in several GA and IA reported in the literature (see Table 11). However, in histological sections of granulomas examined after intestinal surgery, it can be hard even to recognize the presence of a nematode. Additionally, the identification of the actual etiological agent to species level was not possible in several of these cases. To improve the specific identification of the parasite from surgically removed granulomas, a molecular method has been developed that is also efficient for the identification of the etiological agent, even if the larva in the granuloma is spoiled and only present at very low quantity (Mattiucci et al., 2011, 2017e; Paoletti et al., 2017). Such methods have been recently used to identify *A. pegreffii* in other IA cases (Mattiucci et al., 2017d; Mladineo et al., 2016) (see also Section 8.4.2).

8.4.2 Molecular Diagnosis

Endoscopy is often used to remove *Anisakis* spp. larvae from the stomach and intestine. However, the very limited specific diagnostic features of individual *Anisakis* spp. larvae available on the basis of morphological examination, allow identification only to genus level. If the removed larva is spoiled or fragmented rendering them useless for microscopy, no identification may be possible. It is fortunate that the application of molecular methodologies has advanced our knowledge of the causative agents of human anisakiasis. This has allowed the identification of *A. simplex* (s.s.) in 84.1% of cases and *A. pegreffii* in 15.9% (Table 11).

One of the most challenging issues in molecular diagnosis of anisakiasis in humans is the detection of *Anisakis* DNA in low quantities, which can occur in biopsy tissues. Thus access to a diagnostic tool based on a molecular technique that is specific, rapid, and of high sensitivity is crucial. In this context, the RT-PCR has overcome some limitations related to DNA quantity and the time, with respect to the conventional PCR-DNA direct sequencing of target genes. Thus, a sensitive method for the accurate and rapid identification of the etiological agent of human anisakiasis, based on RT-PCR hydrolysis probe, was recently developed (Paoletti et al., 2017). This method detects up to 0.0006 ng/μL of parasite DNA. The RT-PCR DNA probe assay was trialled, for the first time, in the identification of *A. pegreffii*, from an intestinal eosinophilic granuloma (Mattiucci et al., 2017d). The etiological agent would have been undetectable as inferred by direct DNA sequencing, due to the low quantity of DNA available. This could also occur when biopsy tissues are collected by endoscopy from ulcers caused by the parasite in the gastric and/or intestinal mucosa, without direct observation of the parasite itself. Indeed, since the symptoms of anisakiasias are not pathognomonic, the acute form could be misdiagnosed. Consequently, the infection could become chronic by inducing to the formation of granuloma. Therefore, early removal and identification of the worm seems to be the best preventive measure to avoid eosinophilic granulomatosis caused by the allergic reactions to the degenerated larva in chronic infections.

8.4.3 Serodiagnosis of Human Anisakiasis and IgE Sensitization

Currently, most serodiagnostic tests for *Anisakis* reactivity include the use of immunoCAP (iCAP) assay, immunoblotting (WB), and ELISA. Among these, it should especially be noted that specific IgE detection using iCAP assay can overestimate the number of human hypersensitive patients to *Anisakis* allergens. In other words, the sensitivity of these tests can be exaggerated and may yield false-positive results. In recent years, some authors have preferred to use IgE and IgG detection by immunoblotting (WB) assay, mostly using E/S antigens (see Table 12) to differentiate, between GAA and hypersensitive patients (Daschner et al., 2012). Among the antigens related to *Anisakis*, *Ani s 1*, *Ani s* 7, and *Ani s 13* could be considered major antigens in the diagnosis of allergic anisakiasis (Daschner et al., 2012; Mattiucci et al., 2017e). Nevertheless, the iCAP assay is still used in routine laboratory diagnosis for '*Anisakis* allergy'. In a recent study, IgE sensitization to *A. pegreffii* in Italian patients suffering from GAA or showing CU+ after fish consumption, IgE response was analysed by immunoblotting (WB), using both

E/S products (ESPs) and CE of *A. pegreffii* larvae. The results were compared with those obtained through conventional iCAP immunological method for *Anisakis* allergy. Among 110 subjects, 28 showed IgE positivity when applying both WB and iCAP methods; 13 proved IgE reactivity in WB assay, to EPS antigens of *A. pegreffii* (i.e. *Ani s 1-like*, *Ani s 7-like*, and *Ani s13-like*); whereas, 15 sera reacted only to *Ani s 7* and *Ani s 13*. The two methods, i.e., iCAP and WB exhibited a high concordance value at iCAP value <0.35 (negative result) and >50.0 (positive result) (Mattiucci et al., 2017e). The authors also found that some of the apparently positive sera in iCAP, were actually false positives, due to (i) crossreactive antibodies directed against nonparasitic epitopes; in this case, the iCAP usually shows a pronounced reactivity against panallergenic protein (e.g. tropomyosin) and (ii) an underlying atopic state of the patients who may have some of their polyclonal IgE directed against nonspecific antigens binding IgE (Mattiucci et al., 2017e).

However, the comparatively high *Anisakis* sensitization recently found in Mediterranean populations (Audicana and Kennedy, 2008; Mattiucci et al., 2017e; Mladineo et al., 2014; Moreno et al., 2006) may also be explained by a possible genetic susceptibility to antigens/allergens of both species (i.e. *A. simplex* (s.s.) and *A. pegreffii*). Indeed, a significant association between sensitization to *Anisakis* spp. and the alleles found in the human leucocyte antigen system (HLA, the major histocompatibility complex), at loci DRB1*1502-DQB1*0601, was shown in some Spanish populations (Sánchez-Velasco et al., 2000). Generally, human populations from Mediterranean countries have experienced a long relationship with geohelminths, such as *Ascaris lumbricoides*. As a consequence, genetic variability in the HLA alleles related to IgE-binding epitopes against antigens of *A. lumbricoides*, could have been promoted by natural selection (Mattiucci et al., 2017e; Reddy and Fried, 2008). However, among these, some IgE-binding proteins, such as tropomyosin, have also been found to induce crossreactive IgE antibody response against *A. simplex* (s.l.) (Lin et al., 2014; Lozano et al., 2004). It has been also postulated that those human populations which have not experienced geohelminth infections might have activation of allergic response against similar antigens (such as tropomyosin), with an exaggerated IgE-mediated immune response (Mattiucci et al., 2017e). On the other hand, a number of epidemiological studies have suggested that an increase in the prevalence of allergic disorders, which has been observed over the past decades, is attributable to a reduced microbial and parasitic exposure during childhood, as a consequence of the westernized life style (Bruschi et al., 2013; Reddy and Fried, 2008). This so-called hygiene hypothesis needs further investigation over a larger number of human populations, at least, from those geographical areas where high IgE-*Anisakis* hypersensitization has been recorded.

9. WHAT KEY QUESTIONS FOR FUTURE RESEARCH CHALLENGES?

Our journey is at its end: did we get lost? We have tried to maintain a direction among the prominent literature on *Anisakis* spp. and the disease these parasites provoke in humans. We have used the 'biological species' concept as a compass to show us the direction in which to go.

Indeed, *Anisakis* spp., their identification to species level, biology, evolutionary history, ecology, as well as differential adaptations to varying host-related conditions, and, finally, their role as zoonotic agent, have been the avenues to travel in order to point out patterns and trends in the *Anisakis* species distribution and epidemiology in both natural (cetaceans and fish) and accidental hosts (humans).

Have we found an answer to the scientific questions we initially raised in Section 1? We have attempted to find the answers, enlightening them with the existing knowledge. Some of the questions have been addressed, some other key questions, at least, only partially. Paradoxically, some more questions may have arisen during the journey, and starting from the results so far obtained, some others priorities have been detected. Thus, we do not want to close this chapter with any conclusive or summarizing statement. On the contrary, this review should rather be seen as guideline for other travellers along the infinite path, to gain more insights into the different aspects of *Anisakis* and anisakiasis.

9.1 What Molecular/Genetic Tools to Use in Future Studies of *Anisakis*?

The choice of the marker to be used in *Anisakis* spp. must be related to the research question: Is the objective an investigation into their phylogeny, gene flow, genetic variability, hybridization between divergent lineages, cophylogeography, host–parasite cophylogeny, or molecular ecology in a specific host–parasite relationship, including humans? As we have presented in Section 2, several markers can be used to investigate those processes, but for others, more and diverse genetic/molecular datasets are needed. As we have also shown (Section 2), a single locus DNA approach may lead to misdiagnosis of some species of the genus *Anisakis*. In addition, different markers can have accumulated homoplasy, as the result of response to selective forces, which make them sometimes not suitable to unravel phylogenetic relationships. Basically, throughout the identification of *Anisakis* spp. we have checked that analyses based on multiple different markers types result in the same congruent pattern (for instance, allozymes, mtDNA gene loci, and EF-1α nDNA) (Mattiucci et al., 2014a, 2016). Whereas, for some others markers, results are contrasting each other, e.g., inflicted from the ITS region of rDNA and those from other nuclear markers (i.e. EF-1α nDNA and allozymes) (Mattiucci et al., 2016). Incongruence may also be the outcome when running nuclear (ITS region of rDNA) and mitochondrial (mtDNA *cox2*) genes analysis (Mladineo et al., 2017b). Furthermore, high level of divergence can exist between mtDNA and nuclear markers, which is the case of the phylogenetic position of the clade formed by *A. typica*, compared to the other clades. This could lead also to ambiguous genealogical history.

So, how can we proceed? A multilocus approach could pave the road. Results could improve knowledge on the genetic structure of *Anisakis* spp., at both intraspecific and interspecific levels. Large genome analyses will provide, in future research, a further useful tool for moving away from single or low number markers in the study of these parasitic nematodes. The generation of multilocus genetic datasets, like nuclear sequences, SNP, and DNA microsatellite libraries will improve the investigation on molecular ecology of these parasites in the years to follow. With the advances in genomics we will be able to better elucidate molecular ecological questions, which again could dramatically widen the horizon of our journey. By applying these methodologies, it is possible to assess hundreds of loci in numerous individuals of target species of *Anisakis*. To date, the production of large-scale nuclear datasets based on DNA microsatellites has already been applied on populations of *A. pegreffii A. simplex* (s.s.) and *A. berlandi* (Mattiucci et al., unpublished data).

However, the next-generation DNA sequencing technology is not yet available for all *Anisakis* species. If this could be achieved, in parallel, for both host and *Anisakis* species, then our knowledge of host–parasite cophylogeographic and cophylogenetic aspects would be largely improved. Indeed, phylogenomic approaches that examine the evolutionary relationships between parasites and their hosts, would facilitate to investigate patterns of codivergence, thus allowing better insight into the driving forces behind the speciation processes in this group of marine parasites. In addition, at a microevolutionary level, the phylogeographic analysis of populations of *Anisakis* spp. could be tested over a large number of nuclear markers, in order to obtain a deeper knowledge on the genetic architecture of *Anisakis* species.

So far, very few studies (Baldwin et al., 2011; Blažeković et al., 2015; Klapper et al., 2015; Mattiucci et al., 2015b; 2017a, b, c; Mattiucci and Nascetti, 2007) have used molecular tools (allozymes and mtDNA *cox2*) to understand genetic variability in populations and species of *Anisakis*. Genetic and demographic datasets of these parasites have been correlated, and results indicate that some parasite populations expand while others contract, at both spatial and temporal scales (Mattiucci et al., 2015a) (see also Section 9.2). Furthermore, high values of genetic variability of anisakid nematodes at allozymes level, have been found to be positively correlated with the high environmental heterogeneity they experience during their life cycles (Bullini et al., 1986). It was indeed observed that anisakid nematodes of the genera *Anisakis*, *Pseudoterranova*, and *Contracaecum*, exhibit high level of genetic variability (Mattiucci and Nascetti, 2007). A major role of this phenomenon in natural selection was suggested (Bullini et al., 1986; Mattiucci and Nascetti, 2007). Molecular ecologists have been challenged to explain the large amount of genetic variability found in natural populations, and the degree to which this variation can be explained by adaptive evolution. On the other hand, evolutionary and ecological genomics emerged from the effort to understand the genomic mechanisms underlying the organisms' responses, including the host–parasite dynamics, to biotic and abiotic environmental changes (Andrew et al., 2013; Mostowy and Engelstädter, 2011; Zarlenga et al., 2014). Elucidating the genomic architecture of ecologically important traits in parasites would be achieved by large-scale genomic data. Emerging technologies could provide insight into the genome of species, allowing investigation of species-specific genomic architecture of adaptive evolution and ecological speciation, elucidating the genetic bases of microevolutionary changes in *Anisakis* spp. populations. Currently, there is a gap in our knowledge regarding the genetic bases that express the phenotypic traits involved in differential adaptations of *Anisakis* spp. to different intermediate/paratenic hosts, or those that could contribute to the fitness of the parasite species in the actual definitive host. The development of technologies of genome-wide analyses will in future years have a large applicability range across populations and species of *Anisakis* collected from both intermediate/paratenic and definitive hosts, over a wide geographical scale, enabling us to produce large genetic/molecular datasets.

Furthermore, recent advances in nucleic and sequencing technologies can now provide opportunities for analyses of transcriptomes in some parasite species (Cantacessi et al., 2012). Thus, deep explorations of transcriptomic and proteomic datasets of *Anisakis* spp. will have implications in improving the understanding of their interactions with different natural and accidental (humans) hosts. That approach will allow to disclose patterns and processes involved in differential host adaptations, tissues migration, pathogenicity, immunobiology, also with respect to the zoonotic diseases in humans. A first attempt of transcriptome and bioinformatic analyses explored in *A. pegreffii* and *A simplex* (s.s.), provided molecular resources to disclose putative proteins to be investigated as possibly implicated as allergens for humans (Baird et al., 2016; Cavallero et al., 2018).

9.2 What Approaches for Future Analysis of Distribution and Epidemiology?

9.2.1 At Fisheries Level

The bulk of data which so far exist on the epidemiology (see Section 7) strongly indicates that not a single variable may explain, alone, the differences so far observed in the infection by different *Anisakis* spp. in wild fisheries throughout the world. Thus, a multifactorial analysis of those major factors shaping the observed distribution of *Anisakis* spp., could really offer an overall view of how the parameters, in synergy, may be responsible for the distribution and infection values observed in a fish species. As we have presented for some fish species, the major 'driver' explaining the high prevalence and burden of the *Anisakis* spp. larvae infection is represented by the fishing ground, irrespective of the length/size of the hosts species. This influence of the ecological characteristics of the water basin, comprising the abiotic chemical and physical aspects, and the interactions and availability of organisms populating the environment, as biotic actors, seem to shape the basis of the complex life cycle of these parasites, determining their presence and their number.

9.2.2 At Human Level

How the monitoring of human anisakiasis could be improved?

As we have presented in Section 8, the main source of human infections is raw fish (i.e. not previously deep frozen or cooked). In some Mediterranean countries, several seafood dishes include raw or lightly marinated anchovies as the main ingredient. Recently, a quantitative risk assessment (QRA) model was used to determine the probability of contracting the disease through the consumption of homemade, untreated raw, or marinated anchovy meals in Spain (Bao et al., 2017b). According to the infection levels observed in Spain, along with information on the general consumption habits of Spanish people, the expected human cases of anisakiasis for the whole Spanish population was estimated (Bao et al., 2017b). Indeed, as we have underlined, the risk of larval infection may increase when considering that postmortem migration of *Anisakis* spp. from the fish viscera into muscle commonly occurs at storage temperatures >5°C (Cipriani et al., 2016). Monte Carlo simulations estimated that the postmortem migration of *Anisakis* from viscera to flesh increases the disease burden by >1000%, while an educational campaign to freeze anchovy before consumption may reduce cases by 80% (Bao et al., 2017b). The QRA tool can be used by policy makers to inform fishery industry, health professionals, and consumers about this underdiagnosed zoonosis. As a possible control of the parasitic infection at humans level would be a targeted public health education campaign.

Have we actually a real map of anisakiasis worldwide?

Studies on the zoonotic potential should be extended to other *Anisakis* species, and in other part of the world. Despite the fact that several factors (as presented in Section 8) are responsible for the probability of the infection to humans, however, cases of anisakiasis are most likely underreported or misdiagnosed, also due to aspecific symptoms related to acute and chronic infections. Indeed, although human infection is the highest in countries where eating raw fish is widespread, the molecular identification of human cases remains very limited, especially in those countries where allergic symptoms and IgE hypersensitivity associated with *Anisakis* are frequently reported. Yet, surprisingly, at least in Europe, the mandatory notification of the human anisakiasis does not exist. The accurate diagnosis of human anisakiasis by applying sensitive, rapid, and specific molecular methodologies, such as RT-PCR-based probe systems (Paoletti et al., 2017), Mattiucci et al. (2017d) would also greatly improve our knowledge on the epidemiology of this fish-borne disease. As for serodiagnosis of anisakiasis, we would suggest to consider those sera showing iCAP values lower than 0.35 kU_A/L as truly negative, in agreement with data so far acquired (Mattiucci et al., 2017e) and previously suggested (Anadón et al., 2010). On the other hand, the positivity of sera with higher iCAP values should need to be confirmed using IgE–WB, or ELISA, employing ESP antigens rather than CE, especially in case of patients suffering allergic signs and symptoms following the consumption of fish. Further research work in this respect by using more refined methods and serodiagnostic tools are needed, also considering crossreactivity with other antigens/allergens from other anisakid nematodes such as *Pseudoterranova* and/or *Contracaecum* infecting the edible parts (i.e. fillets, liver, and roe) of fish (Levsen et al., 2017a; Mattiucci et al., 2017b).

Finally, the role played by the genetic susceptibility in some human populations showing high IgE-*Anisakis* hypersensitization should be investigated.

The simultaneous use of both sensitive and specific methods for the molecular diagnosis of human anisakiasis, and serodiagnosis assays should be encompassed, in order to enlarge our knowledge about its occurrence in human populations that regularly consume raw or undercooked fish.

9.2.3 At the Ecosystem Level

Do climate changes affect the distribution of Anisakis spp.?

Currently, there is still an incomplete understanding of the ecological drivers of transmission, and thus of the magnitude of the effects provoked by climate and anthropogenic changes on *Anisakis* spp. distribution. Over the past decades, climates are changing worldwide at oceanographic level, with consequences on increasing number of species affected by those changes. In marine ecosystem, environmental changes include variation in oceanographic structure, ending primarily in oscillation from cold to warm conditions. Ocean temperatures are warming and records of loss of sea ice in the Arctic have been already reported (ACIA, 2005). Directional climate change interacts with long-term oceanographic regime shifts, such as the Pacific Decadal Oscillation; this acts as determinants of trophic webs structure and ecology of invertebrates, fish, and marine mammals, as observed for instance, in the North Pacific Ocean (Chavez et al., 2003). Associated ecological perturbations are modifying the structure of hosts involved in the life cycle of marine parasites in local climates areas (Marcogliese, 2008).

The magnitude of the effects of climate change on the marine temperature, and effects on the population size of the hosts involved in the parasites life cycles, cannot be monitored as an 'environmental driver' in the transmission of those zoonotic nematodes, due to a lack of knowledge about the biodiversity and distribution of these parasites in a particular region. Thus, enlarging the database on the general biodiversity at both species and gene level in these parasites could represent the first step for monitoring the effect of global changes (both climatic and anthropogenic ones) on their epidemiology, at both temporal and spatial levels. These changes could have effects on the pattern of distribution, seasonal timing of feeding, and spawning migration of intermediate (i.e. fish) and definitive (cetaceans) hosts in *Anisakis* spp. life cycle, in a given geographical area. The synchrony between the occurrence and availability of larval stages and the susceptible first intermediate hosts is a crucial aspect in the transmission of *Anisakis*. Climate change could result in a mismatch between development of the eggs, first stage larvae of *Anisakis*, and the availability of zooplankton in a local area, or provoke an asynchrony between the availability of the zooplankton blooming and a susceptible fish host. This could lead also to a shift to alternative preys of that fish stock in local basin water. As a consequence, mismatches due to climate change could provoke decline or increase, or modify the actual biodiversity distribution of *Anisakis* spp.

It has been suggested that climate changes can influence the occurrence and abundance of anisakid directly influencing their free-living larval stages and, indirectly, their invertebrate and vertebrate hosts in polar regions (Rokicki, 2009). Bak et al. (2014) have found that the prevalence of infection in chub mackerel fluctuated in the eastern area of Korea; the higher infection found was related to the changes of availability of invertebrates in the area, which, in turn, has been influenced by the changes of the sea water temperature in that basin oceanographic area (Bak et al., 2014). The ENSO (El Nino Southern Oscillation) phenomenon caused a drastic fluctuation in ocean currents and sea water temperature. Liu and Zhang (2013) reported that the surface temperature around Taiwan increased more than 2.7°C. In turn, this event has led to an increased number of fish species migrating into Taiwan Sea waters, resulting also in changing dynamics of *Anisakis* spp. infections (Bak et al., 2014).

Changes of migration routes of fish stocks, as consequence of ecosystem changes, can also influence the distribution and infection level of *Anisakis* spp., infecting that intermediate/paratenic host. In the NE Atlantic Ocean, southern and western NE Atlantic mackerel (*S. scombrus*) stock components undertake annually extensive northward feeding migrations. In accordance with the stock's behaviour, the occurrence of few larvae of *A. pegreffii* in coinfection with *A. simplex* (s.s.) in the mackerel fished in the North and Norwegian Sea was detected (Levsen et al., 2017b). This indicates that the fish coinfected with *A. pegreffii* and *A. simplex* (s.s.) likely belonged to the southern or western spawning components which started a probable new route of their feeding migration in waters south or southwest to the British Isles, which include parts of the sympatric area of the two *Anisakis* sibling species in the NE Atlantic Ocean (Levsen et al., 2017b). On the other hand, in these concerns, biodiversity and distribution of larval *Anisakis* spp. are widely considered as biomarkers of stocks of fish hosts (Mattiucci et al., 2015a, 2017b). Thus, further long-term investigations of the *Anisakis* species diversity across the entire distribution of their different fish species, may be proved as a useful supplementary marker to detect changing in the migration routes and geographical origin of fish stock components, also as consequence of global and/or anthropogenic changes.

Would anthropogenic impacts affect the epidemiology of Anisakis*?*
Paradoxically, just humans can play a 'key' role in the facilitation of the transmission pattern of the infection with *Anisakis*, increasing, as a 'cat chasing its tail', the risk of infection to himself. It has indeed been observed that the central Adriatic Sea represents a 'hot spot' for *A. pegreffii* in different fish species captured in these waters (Cipriani et al., 2017a, b). In the past, fish species collected from the same area (i.e. Adriatic Sea) did not show high infection rate by *A. pegreffii* (Mattiucci et al., 2004, 2008). The actual changes in the parasite demography from that fishing ground of the Mediterranean Sea could be related not only to ecological, but also to an 'anthropogenic shortcut' in the life cycle of the parasite. There is a quite widely used routine, followed by local fishermen, who eviscerate fish directly on board of the vessels and discard at sea those viscera, together with *A. pegreffii* larvae. This practice seems to be related to the need of removal of viscera in fish species heavily infected by *Anisakis* larvae (such as hakes and scabbard fish), whose presence would compromise the value of the fish itself at landing.

Actually, the Italian legislation has put in place an initial regulatory approach of the Ministry of Health (Circular No. 10 of 11/03/1992 and 12/05/1992 subsequent ordinance), which recommends to fishermen and operators involved in the fresh fish management, to eviscerate the fish larger than 18 cm belonging to several species. The same regulation also recommended that fishermen do not throw overboard the viscera removed, but discard them properly at land, in order to interrupt the parasite's life cycle.

Despite these guidelines, often viscera discards, originating from various commercial fish species of that area, are dropped at sea after fishing activity, together with the common by-catch discards. They likely include still alive *Anisakis* larvae that, once at sea, may represent a food source for both fish and cetaceans, which may eventually swallow both food and consequently acquire *Anisakis* larvae. This phenomenon that discards from fishing, attracting piscivorous birds, and marine mammals as final hosts may result as an increase in the parasites transmission, has been previously described for other parasite species (Arcos et al., 2001; Bozzano and Sardà, 2002; Morton and Yuen, 2000; Oro and Ruiz, 1997). Aware of this dolphin–fisheries interactions and considering the existence in that geographical area of the Mediterranean Sea of a population of *T. truncatus*, a species that remains the mostly involved in cases of interactions with coastal fisheries, we have collected several observations and reports by local fishermen of the constant presence of *T. truncatus* following trawlers after their activity at sea, eventually feeding on the discards dropped from the nets and the deck. Thus, dolphins directly feeding on these viscera and larvae discarded could acquire greater amount of *Anisakis* individuals, resulting in a sort of 'anthropogenic shortcut' in the parasite cycle (Fig. 24) (Cipriani et al., 2017a, b). This, in turn, maintains a high level of density of the parasite in that area of the Adriatic Sea and involves not just the final hosts of *A. pegreffii*, but can also enhance the presence of the parasite in other fish species important for fishermen (Fig. 24).

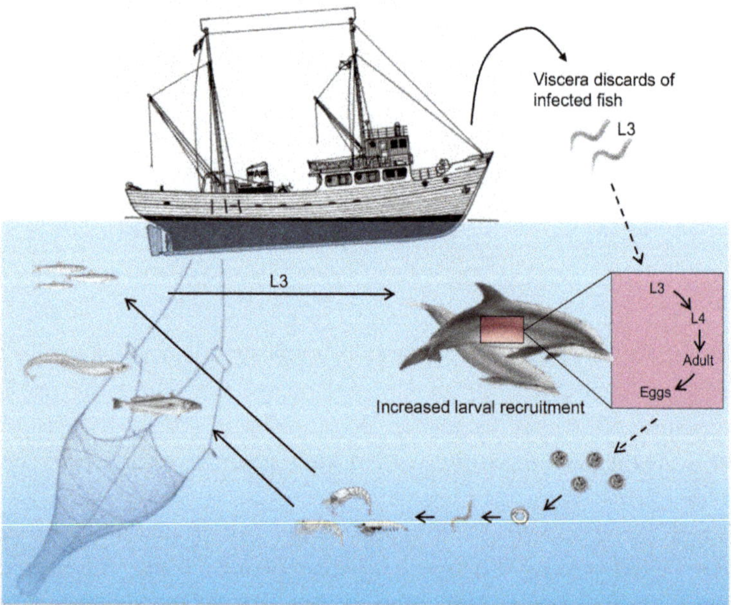

Fig. 24 Schematic illustration of the hypothetic 'anthropogenic shortcut' in the life cycle of the parasite, occurring in the area defined as 'hot spot' for *A. pegreffii* infection in the central Adriatic Sea. The fishermen may eviscerate the fish directly on board of the vessels, and discard the viscera at sea. Those viscera discards, including alive *Anisakis* larvae, once at sea, may represent a food source for both fish and cetaceans and, thus, may result in more efficient parasite recruitment in the actual water basin.

Thus, control measures should include sanitary education for fish fleeting operating in any given hot spot fishing ground, to avoid the parasite discarding at sea, as a practice that enhances parasite transmission in the ecosystem. An advance to reduce the zoonotic potential load in fish stocks, improving the environmental health of a fishing ground, was recently proposed by the use of a technological device (Gonzales et al., 2017). The Technological Device for Avoiding Parasite Discarding at Sea (TEDEPAD) is an industrial equipment designed to process the viscera of parasitized fish by anisakids, on board of commercial fishing vessels (Gonzales et al., 2017).

Comparison of historical data on the actual or future distribution and demography *Anisakis* spp., from a certain fishing ground and fish species, cannot be ignored. It represents the base for monitoring the influence of changes at ecosystem level in the pattern of transmission and efficiency of larval recruitment in the life cycle of *Anisakis* spp., from any fishing ground.

Can estimates of parasite demography and genetic variability of Anisakis *spp. populations be helpful when monitoring their epidemiology in given fishing grounds?*

The transmission routes of *Anisakis* spp. follow closely the trophic relationships among their successive hosts. Therefore, any successful completion of the *Anisakis* spp. life cycle requires stable trophic webs (Mattiucci and Nascetti, 2008). Consequently, the infection levels of anisakid species in a particular geographic area may be strongly affected by the population size of any of the host organisms that are involved in the parasite' life cycle. As a result, the life cycle of anisakid nematodes in marine ecosystems with various degrees of habitat disturbance could be affected by changes in host population size. Indeed, when the population size of the hosts participating in the life cycle of these parasites is reduced due to different causes (by-catch, viral diseases of marine mammals, overfishing of intermediate/paratenic hosts, etc.), the population size of their endoparasites (i.e. *Anisakis* spp.) could also be reduced. As a consequence of the demographic decline of a given population, low level of higher genetic variability (polymorphism) in its gene pool, could occur. For example, observations we made on *A. pegreffii* over a temporal scale level indicates the existence of the opposite trend. Despite the large effective population size and high gene flow estimates among its populations (Mattiucci et al., 1997), *A. pegreffii* from the Mediterranean Sea, showed a significant loss of genetic polymorphism and a decrease in the mean expected heterozygosity (*He*) value at some allozyme loci, over the past 20 years. In contrast, *A. pegreffii* from Austral regions, supported by larger populations of suitable definitive and intermediate/paratenic hosts, has apparently maintained stable values of genetic variability (Mattiucci and Nascetti, 2008). This finding was associated with the overall lower population density and geographic distribution of *A. pegreffii* in the Mediterranean Sea. Interestingly, a low level of nucleotide diversity was found at level of mtDNA *cox2* sequences analysis of the Mediterranean populations of *A. pegreffii* (Blažeković et al., 2015; Mattiucci et al., personal communication) vs its populations from Austral regions (Mattiucci et al., unpublished data).

This would suggest that lower disturbance of marine ecosystems, which permits the maintenance of more stable trophic webs, maintains also high infection levels of anisakid nematodes. This also suggests that data on anisakid abundance and genetic variability could be used for monitoring, at space and time level, the status of various marine trophic webs and, generally, of the 'environmental health' status (Mattiucci et al., 2015a, b, 2017a; Mattiucci and Nascetti, 2007; Zarlenga et al., 2014). Thus, we have found some other avenues to travel through the epidemiological aspects of anisakid nematodes from different fishing grounds.

Would an ecosystem with high level of parasitic load in its organisms represent a major risk to humans?

To address this question, the 'crosstalk' between 'environmental health' and 'human health' datasets will be fundamental. Paradoxically, in '*Anisakis*—epidemiology', the two datasets would be in contrast. Indeed, as we have underlined marine ecosystem at high level of 'environmental health', is just that one characterized by high population density of marine mammals, of fish species as suitable hosts, high zooplankton populations and, by a high stability of food webs. These factors also maintain high level of transmission and demography (i.e. population size) of *Anisakis* just in those hosts from a fishing ground (hot spot). Fish hosts, in turn, would potentially transmit the infection to humans (anisakiasis). Thus, the particular 'healthy environment' in a fishing ground, cannot be considered as the best one for the 'human health' perspective, because 'sea-food' resource from that ecosystem could be at high risk of the parasite accidental infection to humans. Once again, human behaviour, together with feeding habits, would be responsible for the human exposure to this zoonotic disease. Thus, sanitary education, using different strategies and measures, at both individual and public levels, should be the avenue to follow. These education plans should be enlarged to those human populations having major habits to consume raw fish, and targeted to those geographical areas that are at major risk of exposure to zoonotic species of *Anisakis*. This approach would help, in turn, to maintain a 'balance' between 'human health' and 'conservation of fisheries resources'. The latter, indeed, faces the greatest negative impact through the presence of these parasites. However, the economic impact of the parasites on fisheries, for instance in European countries, is difficult to assess. A direct consequence could be the rejection of entire shipments of fishery products due to the presence of nematodes, which again may result in loss of revenues for the actual exporters or retailers. An indirect, but perhaps more severe effect in a longer perspective, could be a significant decline in the per capita fish consumption rate in countries or regions where *Anisakis*-infected fish products are common (for instance, anchovies from Adriatic Sea, Cipriani et al., 2016, 2017a), or if *Anisakis* appears in the news, as happened in Germany in the late 1980s ('the German nematode crisis'; see Karl, 2008). Nevertheless, the benefits on public health of a generally high per capita fish consumption rate, by far, exceed any negative consequences the presence of *Anisakis* in fish may have. However, the fish processing industry should have the potentially severe quality-reducing effect of *Anisakis* spp. larvae in mind in order to maintain high consumer confidence, especially when producing and marketing fish products intended for consumption in a fresh, only lightly processed, or even raw state.

A last question arises: *shall a 'holistic approach', encompassing evolutionary, ecological, immunobiological and economic aspects related to these parasites, be used as a 'driver', for sustainable monitoring of the epidemiology of* Anisakis, *in both natural and accidental (human) hosts?* This would help to maintain a *'safe, but also a sustainable ocean for food, for tomorrow'*.

ACKNOWLEDGEMENTS

We thank Dr. S.C. Webb (Cawthron Institute, New Zealand) for providing comments and corrections, which helped to improve the manuscript; Prof. D. Canestrelli (Tuscia University) for his suggestions to the manuscript. We are very grateful to Dr. Marialetizia Palomba and Dr. Alessandra Colantoni, for their valuable help in preparing figures and tables. The authors are very grateful to anonymous referees whose suggestions and comments helped to improve the manuscript.

The authors acknowledge financial supports from: Grant no. C626H15SBJ4 of 'Sapienza—University of Rome' (year 2015–16); the European Union's Seventh Framework Programme for research, technological development and demonstration, under the grant agreement no. 312068 'PARASITE' (2013–16); Italian Ministry of University and Research (MIUR)—Research Programme in Antarctica (PRNA_00279).

REFERENCES

Abe, N., Tominaga, K., Kimata, I., 2006. Usefulness of PCR-restriction fragment length polymorphism analysis of the internal transcribed spacer region of rDNA for identification of *Anisakis simplex* complex. Jpn. J. Infect. Dis. 59, 60–62.

Abollo, E., Gestal, C., Pascal, S., 2001. *Anisakis* infestation in marine fish and cephalopods from Galician waters: and updated perspective. Parasitol. Res. 87, 492–499.

Abollo, E., Paggi, L., Pascual, S., D'Amelio, S., 2003. Occurrence of recombinant genotypes of *Anisakis simplex* s.s. and *Anisakis pegreffii* (Nematoda: Anisakidae) in an area of sympatry. Infect. Genet. Evol. 3, 175–181.

Acebes Rey, J.M., Fernández Orcajo, P., Díaz González, G., Velicia Llames, R., González Hernández, J.M., Citores González, R., 1996. 2 cases of anisakiasis in the del Río Hortega hospital (Valladolid). Rev. Esp. Enferm. Dig. 88, 59–60.

ACIA, 2005. Arctic Climate Impact Assessment. Cambridge University Press, Cambridge.

Agatsuma, T., Arakawa, Y., Iwagami, M., Honzako, Y., Cahyaningsih, U., Kang, S., Hong, S., 2000. Molecular evidence of natural hybridization between *Fasciola hepatica* and *F. gigantica*. Parasitol. Int. 49, 231–238.

Aloia, A., Carlomagno, P., Gambardella, M., Schiavo, M., Pasquale, V., 2011. Accidental endoscopic finding of *Anisakis simplex* in human colon. Microbiol. Med. 26, 209–211.

Alonso, A., Moreno-Ancillo, A., Daschner, A., López-Serrano, M.C., 1999. Dietary assessment in five cases of allergic reactions due to gastroallergic anisakiasis. Allergy 54, 517–520.

Alonso-Gómez, A., Moreno-Ancillo, A., López-Serrano, M.C., Suarez-de-Parga, J.M., Daschner, A., Caballero, M.T., Barranco, P., Cabañas, R., 2004. *Anisakis simplex* only provokes allergic symptoms when the worm parasitises the gastrointestinal tract. Parasitol. Res. 93, 378–384.

Anadón, A.M., Romarís, F., Escalante, M., Rodríguez, E., Gárate, T., Cuéllar, C., Ubeira, F.M., 2009. The *Anisakis simplex* Ani s 7 major allergen as an indicator of true *Anisakis* infections. Clin. Exp. Immunol. 156, 471–478.

Anadón, A.M., Rodríguez, E., Gárate, M.T., Cuéllar, C., Romarís, F., Chivato, T., Rodero, M., González-Díaz, H., Ubeira, F.M., 2010. Diagnosing human anisakiasis: recombinant Ani s 1 and Ani s 7 allergens *versus* the UniCAP 100 fluorescence enzyme immunoassay. Clin. Vaccine Immunol. 17, 496–502.

Anderson, T.J.C., 2001. The dangers of using single locus markers in parasite epidemiology: *Ascaris* as a case study. Trends Parasitol. 17, 183–188.

Andreassen, J., 1970. The first known case of anisakinosis in Denmark. Norw. J. Zool. 18, 105.

Andrew, R.L., Bernatchez, L., Bonin, A., Buerkle, C.A., Carstens, B.C., Emerson, B.C., Garant, D., Giraud, T., Kane, N.C., Rogers, S.M., Slate, J., Smith, H., Sork, V.L., Stone, G.N., Vines, T.H., Waits, L., Widmer, A., Rieseberg, L.H., 2013. A road map for molecular ecology. Mol. Ecol. 22, 2605–2626.

Andrews, R.H., Chilton, N.B., 1999. Multilocus enzyme electrophoresis: a valuable technique for providing answers to problems in parasite systematics. Int. J. Parasitol. 29, 213–253.

Andrisani, G., Spada, C., Petruzziello, L., Costamagna, G., 2014. An unusual colonic "tumour" Dig. Liver Dis. 46, 477–478.

Anshary, H., Sriwulan, Freeman, M.A., Ogawa, K., 2014. Occurrence and molecular identification of Anisakid Dujardin, 1984 from marine fish in southern Makassar Strait, Indonesia. Korean J. Parasitol. 1, 9–19.

Appleby, D., Kapoor, W., Karpf, M., Williams, S., 1982. Anisakiasis: nematode infestation producing small-bowel obstruction. Arch. Surg. 117, 836.

Arai, T., Akao, N., Seki, T., Kumagai, T., Ishikawa, H., Ohta, N., Hirata, N., Nakaji, S., Yamauchi, K., Hirai, M., Shiratori, T., Kobayashi, M., Fujii, H., Ishii, E., Naito, M., Saitoh, S., Yamaguchi, T., Shibata, N., Shimo, M., Tokiwa, T., 2014. Molecular genotyping of *Anisakis* larvae in middle eastern Japan and endoscopic evidence for preferential penetration of normal over atrophic mucosa. PLoS ONE 9, e89188.

Arcos, J., Oro, D., Sol, D., 2001. Competition between the yellow-legged gull *Larus cachinnans* and Audouin's gull *Larus audouinii* associated with commercial fishing vessels: the influence of season and fishing fleet. Mar. Biol. 139, 807–816.

Arizono, N., Yamada, M., Tegoshi, T., Yoshikawa, M., 2012. *Anisakis simplex* sensu stricto and *Anisakis pegreffii*: biological characteristics and pathogenetic potential in human anisakiasis. Foodborne Pathog. Dis. 9, 517–521.

Arnason, U., Gullberg, A., Janke, A., 2004. Mitogenomic analyses provide new insights into cetacean origin and evolution. Gene 333, 27–34.

Asami, K., Watanuki, T., Sakai, H., Imano, H., Okamoto, R., 1965. Two cases of stomach granuloma caused by *Anisakis*-like larval nematodes in japan. Am. J. Trop. Med. Hyg. 14, 119–123.

Asturias, J.A., Eraso, E., Martínez, A., 2000. Cloning and high level expression in *Escherichia coli* of an *Anisakis simplex* tropomyosin isoform. Mol. Biochem. Parasitol. 108, 263–267.

Audicana, M., Kennedy, M.W., 2008. *Anisakis simplex*: from obscure infectious worm to inducer of immune hypersensitivity. Clin. Microbiol. Rev. 21, 360–379.

Auer, H., Leskowschek, H., Engler, J., Leitner, G., Wentzel, C., Wolkerstorfer, W., Schneider, R., 2007. Epidemiology and nosology of anisakiosis, a rather rare helminthozoonosis in Central Europe—two case reports. Wien. Klin. Wochenschr. 119, 106–109.

Avellino, P., Farjallah, S., Di Giulio, E., Farina, C., Milione, M., Cipriani, P., Modiano, D., D'Amelio, S., 2007. Diagnosi di un caso di anisakidosi esofagea mediante PCR-RFLP. Microbiol. Med. 22, 226.

Avise, J.C., 1994. Molecular Markers, Natural History and Evolution. Chapman & Hall, New York.

Avise, J.C., Ball, R.M., 1990. Principles of genealogical concordance in species concepts and biological taxonomy. Oxford Surv. Ecol. Evol. 7, 47–67.

Avise, J.C., Saunders, N.C., 1984. Hybridization and introgression among species of sunfish (*Lepomis*): analysis by mitochondrial DNA and allozyme markers. Genetics 108, 237–255.

Bahlool, Q.Z., Skovgaard, A., Kania, P., Haarder, S., Buchmann, K., 2012. Microhabitat preference of *Anisakis simplex* in three salmonid species: immunological implications. Vet. Parasitol. 190, 489–495.

Baird, F.J., Su, X., Aibinu, I., Nolan, M.J., Sugiyama, H., Otranto, D., Lopata, A.L., Cantacessi, C., 2016. The *Anisakis* transcriptome provides a resource for fundamental and applied studies on allergy-causing parasites. PLoS Negl. Trop. Dis. 10, e0004845.

Bak, T.J., Jeon, C.H., Kim, J.H., 2014. Occurrence of anisakid nematode larvae in chub mackerel (*Scomber japonicus*) caught off Korea. Int. J. Food Microbiol. 191, 149–156.

Baldwin, R.E., Rew, M.B., Johansson, M.L., Banks, M.A., Jacobson, K.C., 2011. Population structure of three species of *Anisakis* nematodes recovered from Pacific sardines (*Sardinops sagax*) distributed throughout the California current system. J. Parasitol. 97, 545–554.

Bao, M., Roura, A., Mota, M., Pascual, S., 2015. Macroparasites of allis shad (*Alosa alosa*) and twaite shad (*Alosa fallax*) of the western Iberian Peninsula Rivers: ecological, phylogenetic and zoonotic insight. Parasitol. Res. 114, 3721–3739.

Bao, M., Strachan, N.J., Hastie, L.C., MacKenzie, K., Seton, H.C., Pierce, G.J., 2017a. Employing visual inspection and magnetic resonance imaging to investigate *Anisakis simplex* sl infection in herring viscera. Food Control 75, 40–47.

Bao, M., Pierce, G.J., Pascual, S., González-Muñoz, M., Mattiucci, S., Mladineo, I., Cipriani, P., Bušelić, I., Strachan, N.J.C., 2017b. Assessing the risk of an emerging zoonosis of worldwide concern: anisakiasis. Sci. Rep. 7, 43699.

Baron, L., Branca, G., Trombetta, C., Punzo, E., Quarto, F., Speciale, G., Barresi, V., 2014. Intestinal anisakidosis: histopathological findings and differential diagnosis. Pathol. Res. Pract. 210, 746–750.

Barriga, J., Salazar, F., Barriga, E., 1999. Anisakiasis report of a case and review of the literature. Rev. Gastroenterol. Peru 19, 317–323.

Berland, B., 1961. Nematodes from some Norwegian marine fishes. Sarsia 2, 1–50.

Blažeković, K., Pleić, I.L., Đuras, M., Gomerčić, T., Mladineo, I., 2015. Three *Anisakis* spp. isolated from toothed whales stranded along the eastern Adriatic Sea coast. Int. J. Parasitol. 45, 17–31.

Borges, J.N., Cunha, L.F.G., Santos, H.L.C., Monteiro-Neto, C., Santos, C.P., 2012. Morphological and molecular diagnosis of anisakid nematode larvae from cutlassfish (*Trichiurus lepturus*) off the coast of Rio de Janeiro, Brazil. PLoS One 7, e40447.

Bouree, P., Paugam, A., Petithory, J.C., 1995. Anisakidosis: report of 25 cases and review of the literature. Comp. Immunol. Microbiol. Infect. Dis. 18, 75–84.

Bozzano, A., Sardà, F., 2002. Fishery discard consumption rate and scavenging activity in the northwestern Mediterranean Sea. ICES J. Mar. Sci. 59, 15–28.

Bruschi, F., Araujo, M.I., Harnett, W., Pinelli, E., 2013. Allergy and parasites. J. Parasitol. Res. 2013, 502562.

Buchmann, K., 2012. Fish immune responses against endoparasitic nematodes—experimental models. J. Fish Dis. 35, 623–635.

Buchmann, K., Mehrdana, F., 2017. Effects of anisakid nematodes *Anisakis simplex* (s. l.), *Pseudoterranova decipiens* (s. l.) and *Contracaecum osculatum* (s. l.) on fish and consumer health. Food Waterb. Parasitol. 4, 13–22.

Bullini, L., Nascetti, G., Paggi, L., Orecchia, P., Mattiucci, S., Berland, B., 1986. Genetic variation of ascaridoid worms with different life cycles. Evolution 40, 437–440.

Butt, A.A., Aldridge, K.E., Sanders, C.V., 2004. Infections related to the ingestion of seafood part I: viral and bacterial infections. Lancet Infect. Dis. 4, 201–212.

Caballero, M.L., Moneo, I., Gómez-Aguado, F., Corcuera, M.T., Casado, I., Rodríguez-Pérez, R., 2008. Isolation of Ani s 5, an excretory-secretory and highly heat-resistant allergen useful for the diagnosis of *Anisakis* larvae sensitization. Parasitol. Res. 103, 1231–1233.

Caballero, M.L., Umpierrez, A., Moneo, I., Rodriguez-Perez, R., 2011. Ani s 10, a new *Anisakis simplex* allergen: cloning and heterologous expression. Parasitol. Int. 60, 209–212.

Cancrini, G., Magro, G., Giannone, G., 1997. First case of extra-gastrointestinal anisakiasis in a human diagnosed in Italy. Parassitologia 39, 13–17.

Cancrini, G., Macri, G., Ridolfi, G., 1998. Caso di anisakiosi gastrica in Sicilia, diagnosticato e risolto mediante endoscopia. Parassitologia 40, 26.

Canestrelli, D., Porretta, D., Lowe, W.H., Bisconti, R., Carere, C., Nascetti, G., 2016. The tangled evolutionary legacies of range expansion and hybridization. Trends Ecol. Evol. 31, 677–688.

Cantacessi, C., Campbell, B.E., Jex, A.R., Young, N.D., Hall, R.S., Ranganathan, S., Gasser, R.B., 2012. Bioinformatics meets parasitology. Parasite Immunol. 34, 265–275.

Cantatore, D.M.P., Timi, J.T., 2015. Marine parasites as biological tags in South American Atlantic waters, current status and perspectives. Parasitology 142, 5–24.

Caramello, P., Vitali, A., Canta, F., Caldana, A., Santi, F., Caputo, A., Lipani, F., Balbiano, R., 2003. Intestinal localization of anisakiasis manifested as acute abdomen. Clin. Microbial. Infect. 9, 734–737.

Carrascosa, M.F., Mones, J.C., Salcines-Caviedes, J.R., Román, J.G., 2015. A man with unsuspected marine eosinophilic gastritis. Lancet Infect. Dis. 15, 248.

Cassens, I., Vicario, S., Waddell, V.G., Balchowsky, H., Van Belle, D., Ding, W., Fan, C., Mohan, R.S.L., Simoes-Lopes, P.C., Bastida, R., Meyer, A., Stanhope, M.J., Milinkovitch, M.C., 2000. Independent adaptation to riverine habitats allowed survival of ancient cetacean lineages. PNAS 97, 11343–11347.

Cavallero, S., Nadler, S.A., Paggi, L., Barros, N.B., D'Amelio, S., 2011. Molecular characterization and phylogeny of anisakid nematodes from cetaceans from southeastern Atlantic coasts of USA, Gulf of Mexico and Caribbean Sea. Parasitol. Res. 108, 781–792.

Cavallero, S., Ligas, A., Bruschi, F., D'Amelio, S., 2012. Molecular identification of *Anisakis* spp. from fishes collected in the Tyrrhenian Sea (NW Mediterranean). Vet. Parasitol. 187, 563–566.

Cavallero, S., Costa, A., Caracappa, S., Gambetta, B., D'Amelio, S., 2014. Putative hybrids between two *Anisakis* cryptic species: molecular genotyping using high resolution melting. Exp. Parasitol. 146, 87–93.

Cavallero, S., Lombardo, F., Xiaopei, S., Salvemini, M., Cantacessi, C., D'Amelio, S., 2018. Tissue-specific transcriptomes of *Anisakis simplex* (*sensu stricto*) and *Anisakis pegreffii* reveal potential molecular mechanisms involved in pathogenicity. Parasit. Vectors 11, 31.

Ceballos-Mendiola, G., Valero, A., Polo-Vico, R., Tejada, M., Abattouy, N., Karl, H., De las Heras, C., Martin-Sánchez, J., 2010. Genetic variability of *Anisakis simplex* s.s. parasitizing European hake (*Merluccius merluccius*) in the Little Sole Bank area in the Northeast Atlantic. Parasitol. Res. 107, 1399–1404.

Chai, J.Y., Darwin Murrell, K., Lymbery, A.J., 2005. Fish-borne parasitic zoonoses: status and issues. Int. J. Parasitol. 35, 1233–1254.

Chaligiannis, I., Lalle, M., Pozio, E., Sotiraki, S., 2012. Anisakidae infection in fish of the Aegean Sea. Vet. Parasitol. 184, 362–366.

Chavez, F.P., Ryan, J., Lluch-Cota, S.E., Ñiquen, M., 2003. From anchovies to sardines and back: multidecadal change in the Pacific Ocean. Science 299, 217–221.

Checkley, D.M., Barth, J.A., 2009. Patterns and processes in the California current system. Prog. Oceanogr. 83, 49–64.

Chen, H.Y., Shih, H.H., 2015. Occurrence and prevalence of fish-borne *Anisakis* larvae in the spotted mackerel *Scomber australasicus* from Taiwanese waters. Acta Trop. 145, 61–67.

Chen, H.Y., Cheng, Y.S., Grabner, D.S., Chang, S.H., Shih, H.H., 2014. Effect of different temperatures on the expression of the newly characterized heat shock protein 90 (Hsp90) in L3 of *Anisakis* spp. isolated from *Scomber australasicus*. Vet. Parasitol. 205, 540–550.

Chilton, N.B., Huby-Chilton, F., Koehler, A.V., Gasser, R.B., Beveridge, I., 2016. Detection of cryptic species of *Rugopharynx* (Nematoda: Strongylida) from the stomachs of Australian macropodid marsupials. Int. J. Parasitol. 5, 124–133.

Cho, J., Lim, H., Jung, B.K., Shin, E.H., Cha, J.Y., 2015. *Anisakis pegreffii* larvae in sea eels (*Astroconger myriaster*) from the South Sea, Republic of Korea. Korean J. Parasitol. 53, 49–353.

Choe, A., von Reuss, S.H., Kogan, D., Gasser, R.B., Platzer, E.G., Schroeder, F.C., Sternberg, P., 2012. Ascaroside signaling is widely conserved among nematodes. Curr. Biol. 22, 772–780.

Chou, Y.Y., Wang, C.S., Chen, H.G., Chen, H.Y., Chen, S.N., Shih, H.H., 2011. Parasitism between *Anisakis simplex* (Nematoda: Anisakidae) third-stage larvae and the spotted mackerel *Scomber australasicus* with regard to the application of stock identification. Vet. Parasitol. 177, 324–331.

Choudhury, G.S., Bublitz, C.G., 1994. Electromagnetic method for detection of parasites in fish. J. Aquat. Food Prod. Technol. 3, 49–63.

Cimmaruta, R., Bondanelli, P., Nascetti, G., 2005. Genetic structure and environmental heterogeneity in the European hake (*Merluccius merluccius*). Mol. Ecol. 14, 2577–2591.

Cipriani, P., Smaldone, G., Acerra, V., D'Angelo, L., Anastasio, A., Bellisario, B., Palma, G., Nascetti, G., Mattiucci, S., 2014. Genetic identification and distribution of the parasitic larvae of *Anisakis pegreffii* and *A. simplex* (s. s.) in European hake *Merluccius merluccius* from the Tyrrhenian Sea and Spanish Atlantic coast: implications for food safety. Int. J. Food Microbiol. 198, 1–8.

Cipriani, P., Smaldone, G., Acerra, V., D'Angelo, L., Anastasio, A., Bellisario, B., Palma, G., Nascetti, G., Mattiucci, S., 2015. Genetic identification and distribution of the parasitic larvae of *Anisakis pegreffii* and *A. simplex* (s. s.) in European hake *Merluccius merluccius* from the Tyrrhenian Sea and Spanish Atlantic coast: implications for food safety. Int. J. Food Microbiol. 198, 1–8.

Cipriani, P., Acerra, V., Bellisario, B., Sbaraglia, G.L., Cheleschi, R., Nascetti, G., Mattiucci, S., 2016. Larval migration of the zoonotic parasite *Anisakis pegreffii* (Nematoda: Anisakidae) in European anchovy, *Engraulis encrasicolus*: implications to seafood safety. Food Control 59, 148–157.

Cipriani, P., Sbaraglia, G., Palomba, L., Giulietti, L., Bellisario, B., Buselic, I., Mladineo, I., Cheleschi, R., Nascetti, G., Mattiuci, S., 2017a. *Anisakis pegreffii* (Nematoda: Anisakidae) in European anchovy *Engraulis encrasicolus* from the Mediterranean Sea: considerations in relation to fishing ground as a driver for parasite distribution. Fish. Res. https://doi.org/10.1016/j.fishres.2017.03.020. (in press).

Cipriani, P., Sbaraglia, G., Paoletti, M., Giulietti, L., Bellisario, B., Palomba, L., Buselic, I., Mladineo, I., Nascetti, G., Mattiuci, S., 2017b. The Mediterranean European hake, *Merluccius merluccius*: detecting drivers influencing the *Anisakis* spp. larvae distribution. Fish. Res. https://doi.org/10.1016/j.fishres.2017.07.010. (in press).

Cipriani, P., Mattiucci, S., Paoletti, M., Giulietti, L., Marcer, F., Bello, E., Palomba, L., Frantzis, A., Brownlow, A., Davison, N., McGovern, B., Dougnac, C., Covelo, P., Santos, M.B., Webb, S.C., Nascetti, G., 2017c. Updating the biodiversity of nematodes of the genus *Anisakis* in ceataceans from different oceanographic basins. Abstract book, 31st Annual Conference of the European Cetacean Society.

Colantoni, A., Palomba, M.L., Crisafi, B., Cipriani, P., Nascetti, G., Mattiucci, S., 2016. The effect of temperature on the migration capacity and release of excretory/secretory products by *Anisakis pegreffii* larvae cultured in vitro (Nematoda: Anisakidae): a molecular approach. XXIX Congresso Nazionale della Società italiana di Parassitologia (Bari, 21–24 giugno). *Abstract book*, 10.

Coll, M., Piroddi, C., Steenbeek, J., Kaschner, K., Lasram, F.B.R., Aguzzi, J., Ballesteros, E., Bianchi, C.N., Corbera, J., Dailianis, T., Danovaro, R., Estrada, M., Froglia, C., Galil, B.S., Gasol, J.M., Gertwagen, R., Gil, J., Guilhaumon, F., Kesner-Reyes, K., Kitsos, M.S., Koukouras, A., Lampadariou, N., Laxamana, E., López-Fé de la Cuadra, C.M., Lotze, H.K., Martin, D., Mouillot, D., Oro, D., Raicevich, S., Rius-Barile, J., Saiz-Salinas, J.I., San Vicente, C., Somot, S., Templado, J., Turon, X., Vafidis, D., Villanueva, R., Voultsiadou, E., 2010. The biodiversity of the Mediterranean Sea: estimates, patterns, and threats. PLoS One 5, e11842.

Colón-Llavina, M.M., Mignucci-Giannoni, A.A., Mattiucci, S., Paoletti, M., Nascetti, G., Williams, E.H., 2009. Additional records of metazoan parasites from Caribbean marine mammals, including genetically identified anisakid nematodes. Parasitol. Res. 105, 1239.

Costa, A., Cammilleri, G., Graci, S., Buscemi, M.D., Vazzana, M., Principato, D., Giangrosso, G., Ferrantelli, V., 2016. Survey on the presence of *A. simplex* s.s. and *A. pegreffii* hybrid forms in Central-Western Mediterranean Sea. Parasitol. Int. 65, 696–701.

Couture, C., Measures, L., Gagnon, J., Desbiens, C., 2003. Human intestinal anisakiosis due to consumption of raw salmon. Am. J. Surg. Pathol. 27, 1167–1172.

Criscione, C.D., Anderson, J.D., Sudimack, D., Peng, W., Jha, B., Williams-Blangero, S., Anderson, T.J., 2007. Disentangling hybridization and host colonization in parasitic roundworms of humans and pigs. Proc. R. Soc. Lond. Ser. B Biol. Sci. 274, 2669–2677.

Cross, M.A., Collins, C., Campbell, N., Watts, P.C., Chubb, J.C., Cunningham, C.O., Hatfield, E.M.C., MacKenzie, K., 2007. Levels of intra-host and temporal sequence variation in a large CO1 sub-units from *Anisakis simplex sensu stricto* (Rudolphi 1809) (Nematoda: Anisakidae): implications for fisheries management. Mar. Biol. 151, 695–702.

D'Amelio, S., Mathiopoulos, K.D., Santos, C.P., Pugachev, O.N., Webb, S.C., Picanco, M., Paggi, L., 2000. Genetic markers in ribosomal DNA for the identification of members of the genus *Anisakis* (Nematoda: Ascaridoidea) defined by polymerase chain reaction-based restriction fragment length polymorphism. Int. J. Parasitol. 30, 223–226.

D'Amelio, S., Mathiopoulos, K.D., Brandonisio, O., Lucarelli, G., Doronzo, F., Paggi, L., 1999. Diagnosis of a case of gastric anisakidosis by PCR-based restriction fragment length polymorphism analysis. Parassitologia 41, 591–593.

Daschner, A., Pascual, C.Y., 2005. *Anisakis simplex*: sensitization and clinical allergy. Curr. Opin. Allergy Clin. Immunol. 5, 281–285.

Daschner, A., Alonso-Gómez, A., Caballero, T., Barranco, P., Suarez-De-Parga, J.M., López-Serrano, M.C., 1998. Gastric anisakiasis: an underestimated cause of acute urticaria and angio-oedema? Br. J. Dermatol. 139, 822–828.

Daschner, A., Alonso-Gómez, A., Cabañas, R., Suarez-de-Parga, J.M., López-Serrano, M.C., 2000. Gastroallergic anisakiasis: borderline between food allergy and parasitic disease-clinical and allergologic evaluation of 20 patients with confirmed acute parasitism by *Anisakis simplex*. J. Allergy Clin. Immunol. 105, 176–181.

Daschner, A., Cuellar, C., Sánchez-Pastor, S., Pascual, C.Y., Martín-Esteban, M., 2002. Gastro-allergic anisakiasis as a consequence of simultaneous primary and secondary immune response. Parasite Immunol. 24, 243–251.

Daschner, A., Rodero, M., De Frutos, C., Valls, A., Vega, F., Blanco, C., Cuéllar, C., 2011. Different serum cytokine levels in chronic vs. acute *Anisakis simplex* sensitization-associated urticaria. Parasite Immunol. 33, 357–362.

Daschner, A., Cuéllar, C., Rodero, M., 2012. The *Anisakis* allergy debate: does an evolutionary approach help? Trends Parasitol. 28, 9–15.

Davey, J.T., 1971. A revision of the genus *Anisakis* Dujardin, 1845 (Nematoda: Ascaridata). J. Helminthol. 45, 51–72.

Davis, C.S., Flierl, G.R., Wiebe, P.H., Franks, P.J.S., 1991. Micropatchiness, turbulence and recruitment in plankton. J. Mar. Res. 49, 109–151.

De Meeûs, T., McCoy, K.D., Prugnolle, F., Chevillon, C., Durand, P., Hurtrez-Bousses, S., Renaud, F., 2007. Population genetics and molecular epidemiology or how to "débusquer la bête" Infect. Genet. Evol. 7, 308–332.

De Nicola, P., Napolitano, L.M., Di Bartolomeo, N., Waku, M., Innocenti, P., 2005. Su di un caso di anisachiasi con perforazione del cieco. G. Chir. 26, 375–377.

Deardorff, T.L., Fukumura, T., Raybourne, R.B., 1986. Invasive anisakiasis. A case report from Hawaii. Gastroenterology 90, 1047–1050.

Del Olmo Escribano, M., Cózar Ibáñez, A., Martínez de Victoria, J.M., Ureña Tirao, C., 1998. Anisakiasis a nivel ileal. Rev. Esp. Enferm. Dig. 90, 120–123.

Del Olmo Martínez, L., González de Canales, P., Sanjosé González, G., 2000. Gastric anisakiasis diagnosed with endoscopy. Ann. Med. Interna 17, 429–431.

Detwiler, J.T., Criscione, C.D., 2010. An infectious topic in reticulate evolution: introgression and hybridization in animal parasites. Gen 1, 102–123.

Di Azevedo, M.I.N., Carvalho, V.L., Iñiguez, A.M., 2016. First record of the anisakid nematode *Anisakis nascettii* in the Gervais' beaked whale *Mesoplodon europaeus* from Brazil. J. Helminthol. 90, 48–53.

Dixon, B.R., 2006. Isolation and Identification of Anisakid Roundworm Larvae in Fish. Laboratory Procedure, OPFLP-2. HPFB, Government of Canada.
Doby, J.M., Le Masson, J.M., Babin, P., 1975. Granulome èosinophilique du colon par larve de nèmatode Anisakidae. Bull. Soc. Pathol. Exot. 67, 522–536.
Du, C., Zhang, L., Shi, M., Ming, Z., Hu, M., Gasser, R.B., 2010. Elucidating the identity of *Anisakis* larvae from a broad range of marine fishes from the Yellow Sea, China, using a combined electrophoretic-sequencing approach. Electrophoresis 31, 654–658.
Dunams-Morel, D.B., Reichard, M.V., Torretti, L., Zarlenga, D.S., Rosenthal, B.M., 2012. Discernible but limited introgression has occurred where *Trichinella nativa* and the T6 genotype occur in sympatry. Infect. Genet. Evol. 12, 530–538.
Dupouy-Camet, J., Touabet-Azouzi, N., Fréalle, E., Van Cauteren, D., Yera, H., Moneret-Vautrin, A., 2016. Incidence de l'anisakidose en France. Enquête rétrospective 2010–2014. Bull Epidémiolog Hebdomadaire, 5–6, 64–70. http://www.invs.sante.fr/beh/2016/5-6/2016_5-6_1.html.
Dybdahl, M.F., Jokela, J., Delph, L.F., Koskella, B., Lively, C.M., 2008. Hybrid fitness in a locally adapted parasite. Am. Nat. 172, 772–782.
EFSA, 2010. Scientific opinion on risk assessment of parasites in fishery products. EFSA J. 8, 1543.
Elder Jr., J.F., Turner, B.J., 1995. Concerted evolution of repetitive DNA sequences in eukaryotes. Q. Rev. Biol. 70, 297–320.
Espiñeira, M., Herrero, B., Vieites, J.M., Santaclara, F.J., 2010. Detection and identification of anisakids in seafood by fragment length polymorphism analysis and PCR–RFLP of ITS-1 region. Food Control 21, 1051–1060.
Fain, A., van Roy, M., Fontenelle, E., 1969. Un phlegmon du jéjunum produit par une larve d'*Anisakis*. Louvain Méd. 88, 579–587.
Fang, W., Liu, F., Zhang, S., Lin, J., Xu, S., Luo, D., 2011. *Anisakis pegreffii*: a quantitative fluorescence PCR assay for detection in situ. Exp. Parasitol. 127, 587–592.
FAO, 2012. The State of World Fisheries and Aquaculture. Food and Agriculture Organization of the United Nations, Rome. ISBN: 978-92-5-107225-7. http://www.fao.org/docrep/016/i2727e/i2727e.
Farjallah, S., Busi, M., Mahjoub, M.O., Slimane, B.B., Paggi, L., Said, K., D'Amelio, S., 2008a. Molecular characterization of larval anisakid nematodes from marine fishes off the Moroccan and Mauritanian coasts. Parasitol. Int. 57, 430–436.
Farjallah, S., Slimane, B.B., Busi, M., Paggi, L., Amor, N., Blel, H., Said, K., D'Amelio, S., 2008b. Occurrence and molecular identification of *Anisakis* spp. from the North African coasts of Mediterranean Sea. Parasitol. Res. 102, 371.
Farris, J.S., Kàllersió, M., Kluge, A.G., Bult, C., 1994. Testing significance of incongruence. Cladistics 10, 315–319.
Fayer, R., 2015. Introduction and public health importance of foodborne parasites. In: Xiao, L., Ryan, U., Feng, Y. (Eds.), Foodborne Parasites. CRC Press, NW, pp. 3–19.
Fazii, P., Neri, M., Bucci, E., Pistola, F., Laterza, F., Colagrande, E., Cosentino, L., Caldarella, M.P., Clerico, L., Stella, M., Pelatti, A., Riaro-Sforza, G., 2006. Diagnosi e terapia dell'anisakidosi mediante endoscopio. Descrizione di tre casi in Abruzzo. Microbiol. Med. 21, 213.
Fumarola, L., Monno, R., Ierardi, E., Rizzo, G., Giannelli, G., Lalle, M., Pozio, E., 2009. *Anisakis pegreffii* etiological agent of gastric infections in two Italian women. Foodborne Pathog. Dis. 6, 1157–1159.
Ganley, A.R.D., Kobayashi, T., 2007. Highly efficient concerted evolution in the ribosomal DNA repeats: total rDNA repeat variation revealed by whole-genome shotgun sequence data. Genome Res. 17, 184–191.

Garcia, A., Mattiucci, S., Damiano, S., Santos, M.N., Nascetti, G., 2011. Metazoan parasites of swordfish, *Xiphias gladius* (Pisces: Xiphiidae) from the Atlantic Ocean: implications for host stock identification. ICES J. Mar. Sci. 68, 175–182.

Gaspari, S., Holcer, D., Mackelworth, P., Fortuna, C., Frantzis, A., Genov, T., Vighi, M., Natali, C., Rako, N., Banchi, E., Chelazzi, G., 2015. Population genetic structure of common bottlenose dolphins (*Tursiops truncatus*) in the Adriatic Sea and contiguous regions: implications for international conservation. Aquat. Conserv. 25, 212–222.

Gilabert, A., Wasmuth, J.D., 2013. Unravelling parasitic nematode natural history using population genetics. Trends Parasitol. 29, 438–448.

Gómez-Mateos, M., Valero, A., Morales-Yuste, M., Martín-Sánchez, J., 2016. Molecular epidemiology and risk factors for *Anisakis simplex* s.l. infection in blue whiting (*Micromesistius poutassou*) in a confluence zone of the Atlantic and Mediterranean: differences between *A. simplex ss* and *A. pegreffii*. Int. J. Food Microbiol. 232, 111–116.

Gómez-Morales, M.A., Martinez Castro, C., Lalle, M., Fernandez, R., Pezzotti, P., Abollo, E., Pozio, E., 2017. UV-press method versus artificial digestion method to detect Anisakidae L3 in fish fillets: comparative study and suitability for the industry. Fish. Res. (in press).

Gonzales, A.F., Gracia, J., Minino, I., Romon, J., Larsson, C., Maroto, J., Regueira, M., Pascual, S., 2017. Approach to reduce the zoonotic parasite load in fish stocks: when science meets technology. Fish. Res. 10/1016/J.fish.res.2017.08.016. (in press).

González-Fernández, J., Daschner, A., Nieuwenhuizen, N.E., Lopata, A.L., De Frutos, C., Valls, A., Cuéllar, C., 2015. Haemoglobin, a new major allergen of *Anisakis simplex*. Int. J. Parasitol. 45, 399–407.

Gonzalez-Quijada, S., Gonzalez Escudero, R., Arias Garcia, L., Gil Martin, A.R., Vicente Serrano, J., Corral-Fernandez, E., 2005. Anisakiasis gastrointestinal manifestations: description of 42 cases. Rev. Clin. Esp. 205, 311–315.

Gregori, M., Roura, Á., Abollo, E., González, Á.F., Pascual, S., 2015. *Anisakis simplex* complex (Nematoda: Anisakidae) in zooplankton communities from temperate NE Atlantic waters. J. Nat. Hist. 49, 755–773.

Haagensen, P., 1993. U.S. Patent No. 5,241,365. U.S. Patent and Trademark Office, Washington, DC.

Hafsteinsson, H., Parker, K., Chivers, R., Rizvi, S.S., 1989. Application of ultrasonic waves to detect sealworms in fish tissue. J. Food Sci. 54, 244–247.

Hamada, K., Uedo, N., Tomita, Y., Iishi, H., 2016. A bleeding gastric ulcer caused by anisakiasis. Ann. Gastroenterol. 29, 378.

Hashimoto, R., Chonan, A., 2016. Gastric anisakiasis with a gastric ulcer. Intern. Med. 55, 3681.

Hashimoto, R., Matsuda, T., Nakahori, M., 2017. Small bowel anisakiasis detected by capsule endoscopy. Dig. Endosc. 29, 122–130. https://doi.org/10.1111/den.12738.

Heia, K., Sivertsen, A.H., Stormo, S.K., Elvevoll, E., Wold, J.P., Nilsen, H., 2007. Detection of nematodes in cod (*Gadus morhua*) fillets by imaging spectroscopy. J. Food Sci. 72, E011–E015.

Hemsrichart, V., 1993. Intestinal anisakiasis: first reported case in Thailand. J. Med. Assoc. Thai. 76, 117–121.

Hermida, M., Mota, R., Pacheco, C.C., Santos, C.L., Cruz, C., Saraia, A., Tamagnini, P., 2012. Infection levels and diversity of anisakid nematodes in blackspot seabream, *Pagellus bogaraveo*, from Portuguese waters. Parasitol. Res. 110, 1919–1928.

Herrero, B., Vieites, J.M., Espiñeira, M., 2011. Detection of anisakids in fish and seafood products by real-time PCR. Food Control 22, 933–939.

Heuch, P.A., Jansen, P.A., Hansen, H., Sterud, E., MacKenzie, K., Haugen, P., Hemmingsen, W., 2011. Parasite faunas of farmed cod and adjacent wild cod populations in Norway: a comparison. Aquac. Environ. Interact. 2, 1–13.

Hiramoto, J.T., Tokeshi, J., 1991. Anisakiasis in Hawaii: a radiological diagnosis. Hawaii Med. J. 50, 202–203.

Højgaard, D.P., 1995. Experimental infection of macroplankton from Faroese waters with newly hatched *Anisakis simplex* larvae. Fróðskaparrit 43, 115–121.

Højgaard, D.P., 1997. Seasonal changes in the infection of young saithe, *Pollachius virens*, with *Anisakis simplex* and other helminths. Fróðskaparrit 45, 57–68.

Holcer, D., Fortuna, C.M., Mackelworth, P.C., 2014. Status and conservation of Cetaceans in the Adriatic Sea. Draft internal report for the purposes of the Mediterranean Regional Workshop to Facilitate the Description of Ecologically or Biologically Significant Marine Areas, Malaga, Spain, 7–11 April.

Horbowy, J., Podolska, M., Nadolna-Ałtyn, K., 2016. Increasing occurrence of anisakid nematodes in the liver of cod (*Gadus morhua*) from the Baltic Sea: does infection affect the condition and mortality of fish? Fish. Res. 179, 98–103.

Hsiu, J.G., Gamsey, A.J., Ives, C.E., D'Amato, N.A., Hiller, A.N., 1986. Gastric anisakiasis: report of a case with clinical, endoscopic, and histological findings. Am. J. Gastroenterol. 81, 1185–1187.

Hu, M., D'Amelio, S., Zhu, X.Q., Paggi, L., Gasser, R., 2001. Mutation scanning for sequence variation in three mitochondrial DNA regions for members of the *Contracaecum osculatum* (Nematoda: Ascaridoidea) complex. Electrophoresis 22, 1069–1075.

Ikuta, R., Kitazawa, H., Matsuda, K., Tashiro, M., 2016. Anisakiasis of the Anorectum. Intern. Med. 55, 2513–2514.

Iñiguez, A.M., Santos, C.P., Vicente, A.C.P., 2009. Genetic characterization of *Anisakis typica* and *Anisakis physeteris* from marine mammals and fish from the Atlantic Ocean off Brazil. Vet. Parasitol. 165, 350–356.

Iñíguez, M., Masello, J.F., Arcucci, D., Krohling, F., Belgrano, J., 2010. On the occurrence of sei whales, *Balaenoptera borealis*, in the south-western Atlantic. Mar. Biodivers. Rec. 3, e68.

Iñiguez, A.M., Carvalho, V.L., Motta, M.R.A., Pinheiro, D.C.S.N., Vicente, A.C.P., 2011. Genetic analysis of *Anisakis typica* (Nematoda: Anisakidae) from cetaceans of the north-east coast of Brazil: new data on its definitive hosts. Vet. Parasitol. 178, 293–299.

Intapan, P.M., Thanchomnang, T., Lulitanond, V., Pongsaskulchoti, P., Maleewong, W., 2008. Detection of *Opisthorchis viverrini* in infected bithynid snails by real-time fluorescence resonance energy transfer PCR-based method and melting curve analysis. Parasitol. Res. 103, 649.

Ioli, A., Leonaldi, R., Gangemi, C., Lo Giudice, L., Bottari, M., Petithory, J.C., 1998. A propose of 1 case of anisakiasis contracted in Sicily. Bull. Soc. Pathol. Exot. 91, 232–234.

Ishida, M., Harada, M.A., Egawa, S., Watabe, S., Ebina, N., Unno, M., 2007. Three successive cases of enteric anisakiasis. Dig. Surg. 24, 228–231.

Ishikura, H., Hayasaka, H., Kikuchi, Y., 1967. Acute regional ileitis at Iwanai in Hokkaido with special reference to intestinal Anisakiasis. Sapporo Igaku Zasshi 32, 183–196.

Ito, Y., Ikematsu, Y., Yuzawa, H., Nishiwaki, Y., Kida, H., Waki, S., Uchimura, M., Ozawa, T., Iwaoka, T., Kanematsu, T., 2007. Chronic gastric anisakiasis presenting as pneumoperitoneum. Asian J. Surg. 30, 67–71.

Jacobsen, K.B., Berland, B., 1969. Fish nematodes as a cause of acute and chronic gastroenteritis with tissue eosinophilia. Nord. Med. 82, 1104–1111.

Jenkins, E.J., Castrodale, L.J., de Rosemond, S.J., Dixon, B.R., Elmore, S.A., Gesy, K.M., Hoberg, E.P., Polley, L., Schurer, J.M., Simard, M., Thompson, R.C., 2013. Tradition and transition: parasitic zoonoses of people and animals in Alaska, northern Canada, and Greenland. Adv. Parasitol. 82, 33–204.

Jenks, W.G., Bublitz, C.G., Choudhury, G.S., Ma, Y.P., Wikswo, J.P., 1996. Detection of parasites in fish by superconducting quantum interference device magnetometry. J. Food Sci. 61, 865–869.

Johnston, T.H., Mawson, P.M., 1951. Additional nematodes from Australian fish. Trans. R. Soc. S. Aust. 74, 18–34.

Jombart, T., Devillard, S., Balloux, F., 2010. Discriminant analysis of principal components: a new method for the analysis of genetically structured populations. BMC Genet. 11, 94.

Kagei, N., Isogaki, H., 1992. A case of abdominal syndrome caused by the presence of a large number of *Anisakis* larvae. Int. J. Parasitol. 22, 251–253.

Kang, D.B., Oh, J.T., Park, W.C., Lee, J.K., 2010. Small bowel obstruction caused by acute invasive enteric anisakiasis. Korean J. Gastroenterol. 56, 192–195.

Kang, D.B., Park, W.C., Lee, J.K., 2014. Chronic gastric anisakiasis provoking a bleeding gastric ulcer. Ann. Surg. Treat Res. 86, 270–273.

Kaplan, F., Alborn, H.T., Von Reuss, S.H., Ajjredini, R., Ali, J.G., Akyazi, F., Stelinski, L.L., Edison, A.S., Schroeder, F.C., Teal, P.E., 2012. Interspecific nematode signals regulate dispersal behavior. PLoS One 7, e38735.

Karl, H., 2008. Nematode larvae in fish on the German market—20 years of consumer related research. Arch. Lebensmittelhyg 59, 107–116.

Karl, H., Leinemann, M., 1993. A fast and quantitative detection method for nematodes in fish fillets and fishery products. Arch. Lebensmittelhyg. 44, 124–125.

Karl, H., Levsen, A., 2011. Occurrence and distribution of anisakid nematodes in grey gurnard (*Eutrigla gurnardus* L.) from the North Sea. Food Control 22, 1634–1638.

Karl, H., Meyer, C., Banneke, S., Jark, U., Feldhusen, F., 2002. The abundance of nematode larvae *Anisakis* sp. in the flesh of fishes and possible post mortem migration. Arch. Lebensmittelhyg. 53, 118–120.

Karl, H., Baumann, F., Ostermeyer, U., Kuhn, T., Klimpel, S., 2011. *Anisakis simplex* (s. s.) larvae in wild Alaska salmon: no indication of post-mortem migration from viscera into flesh. Dis. Aquat. Organ. 94, 201–209.

Karmanova, I.V., Plashkova, V.V., Nechaeva, O.I., Gubina, V.V., 2002. A case of human anisakiasis in Kamchatka. Med. Parazitol 2, 32–33.

Kijewska, A., Dzido, J., Rokicki, J., 2009. Mitochondrial DNA of *Anisakis simplex* s.s. as a potential tool for differentiating populations. J. Parasitol. 95, 1364–1370.

Kikuchi, Y., Ishikura, H., Kikuchi, K., Hyasaka, H., 1990. Pathology of gastric anisakiasis. In: Ishikura, H., Kikuchi, K. (Eds.), Gastric Anisakiasis in Japan. Springer-Verlag, Tokyo, pp. 129–143.

Kim, C.H., Chung, B.S., Moon, Y.I., Chun, S.H., 1971. A case report on human infection with *Anisakis* sp. in Korea. Kisaengchunghak Chapchi 9, 39–43.

Kim, L.S., Lee, Y.H., Kim, S., Park, H.R., Cho, S.Y., 1991. A case of anisakiasis causing intestinal obstruction. Kisaengchunghak Chapchi 29, 93–96.

Kim, S.H., Park, C.W., Kim, S.K., Won, S., Park, W.K., Kim, H.R., Nam, K.W., Lee, G.S., 2013. A case of anisakiasis invading the stomach and the colon at the same time after eating anchovies. Clin. Endosc. 46, 293–296. https://doi.org/10.5946/ce.2013.46.3.293.

Klapper, R., Kuhn, T., Münster, J., Levsen, A., Karl, H., Klimpel, S., 2015. Anisakid nematodes in beaked redfish (*Sebastes mentella*) from three fishing grounds in the North Atlantic, with special notes on distribution in the fish musculature. Vet. Parasitol. 207, 72–80.

Klapper, R., Kochmann, J., O'Hara, R.B., Karl, H., Kuhn, T., 2016. Parasites as biological tags for stock discrimination of beaked redfish (*Sebastes mentella*): parasite infra-communities vs. limited resolution of cytochrome markers. PLoS One 11, e0153964.

Kleinertz, S., Hermosilla, C., Ziltener, A., Kreicker, S., Hirzmann, J., Abdel-Ghaffar, F., Taubert, A., 2014. Gastrointestinal parasites of free-living indo-Pacific bottlenose dolphins (*Tursiops aduncus*) in the Northern Red Sea, Egypt. Parasitol. Res. 113, 1405–1415.

Klimpel, S., Busch, M.W., Kuhn, T., Rohde, A., Palm, H.W., 2010. The *Anisakis simplex* complex off the South Shetland Islands (Antarctica): endemic populations versus introduction through migratory hosts. Mar. Ecol. Prog. Ser. 403, 1–11.

Klimpel, S., Kuhn, T., Busch, M.W., Karl, H., Palm, H.W., 2011. Deep-water life cycle of *Anisakis paggiae* (Nematoda: Anisakidae) in the Irminger Sea indicates kogiids distribution in the north Atlantic waters. Polar Biol. 34, 899–906.

Kobayashi, Y., Ishizaki, S., Shimakura, K., Nagashima, Y., Shiomi, K., 2007. Molecular cloning and expression of two new allergens from *Anisakis simplex*. Parasitol. Res. 100, 1233–1241.

Kobayashi, Y., Ohsaki, K., Ikeda, K., Kakemoto, S., Ishizaki, S., Shimakura, K., Nagashima, Y., Shiomi, K., 2011. Identification of novel three allergens from *Anisakis simplex* by chemiluminescent immunoscreening of an expression cDNA library. Parasitol. Int. 60, 144–150.

Kobayashi, Y., Kakemoto, S., Shimakura, K., Shiomi, K., 2015. Molecular cloning and expression of a new major allergen, Ani s 14, from *Anisakis simplex*. Shokuhin Eiseigaku Zasshi 56, 194–199.

Koinari, M., Karl, S., Elliot, A., Ryan, U., Lymbery, A.J., 2013. Identification of *Anisakis* species (Nematoda: Anisakidae) in marine fish hosts from Papua New Guinea. Vet. Parasitol. 193, 126–133.

Kong, Q., Fan, L., Zhang, J., Akao, N., Dong, K., Lou, D., Ding, J., Tong, Q., Zheng, B., Chen, R., Ohta, N., Lu, S., 2015. Molecular identification of *Anisakis* and *Hysterothylacium* larvae in marine fishes from the East China Sea and the Pacific coast of central Japan. Int. J. Food Microbiol. 199, 1–7.

Kuhn, T., García-Màrquez, J., Klimpel, S., 2011. Adaptive radiation within marine anisakid nematodes: a zoogeographical modelling of cosmopolitan, zoonotic parasites. PLoS One 6, e28642.

Kuhn, T., Hailer, F., Palm, H.W., Klimpel, S., 2013. Global assessment of molecularly identified *Anisakis* Dujardin, 1845 (Nematoda: Anisakidae) in their teleost intermediate hosts. Folia Parasitol. 60, 123–134.

Kuhn, T., Cunze, S., Kochmann, J., Klimpel, S., 2016. Environmental variables and definitive host distribution: a habitat suitability modelling for endohelminth parasites in the marine realm. Sci. Rep. 6, 30246. https://doi.org/10.1038/srep30246.

Lagrue, C., Poulin, R., 2015. Bottom–up regulation of parasite population densities in freshwater ecosystems. Oikos 124, 1639–1647.

Lalev, A.I., Nazar, R.N., 1998. Conserved core structure in the internal transcribed spacer 1 of the Schizosaccharomyces pombe precursor ribosomal RNA. J. Mol. Biol. 284, 1341–1351.

Leatherwood, D., Reeves, R.R., 1982. Bottlenose dolphin *Tursiops truncatus* and other toothed cetaceans. In: Chapman, , Feldhamer, (Eds.), Wild Mammals of North America. In: Biol. Manag. Econom The Johns Hopkins University Press, Baltimore, MD, pp. 369–414.

Leatherwood, S., Reeves, R., 1983. The Sierra Club Handbook of Whales and Dolphins. Sierra Club Books, San Francisco.

Lee, M.H., Cheon, D., Choi, C., 2009. Molecular genotyping of *Anisakis* species from Korean sea fish by polymerase chain reaction–restriction fragment length polymorphism (PCR-RFLP). Food Control 20, 623–626.

Levsen, A., Berland, B., 2012. *Anisakis* species. In: Fish Parasites: Pathobiology and Protection. CABI Press, UK, p. 298.

Levsen, A., Karl, H., 2014. *Anisakis simplex* (sl) in Grey gurnard (*Eutrigla gurnardus*) from the North Sea: food safety considerations in relation to fishing ground and distribution in the flesh. Food Control 36, 15–19.

Levsen, A., Lunestad, B.T., 2010. *Anisakis simplex* third stage larvae in Norwegian spring spawning herring (*Clupea harengus* L.), with emphasis in larval distribution in the flesh. Vet. Parasitol. 171, 247–253.

Levsen, A., Maage, A., 2016. Absence of parasitic nematodes in farmed, harvest quality Atlantic salmon (*Salmo salar*) in Norway—results from a large scale survey. Food Control 68, 25–29.

Levsen, A., Lunestad, B.T., Berland, B., 2005. Low detection efficiency of candling as a commonly recommended inspection method for nematode larvae in the flesh of pelagic fish. J. Food Prot. 68, 828–832.

Levsen, A., González, A.F., Mattiucci, S., Cipriani, P., Paoletti, M., Gay, M., Højgaard, D.P., Joensen, M.M., Hastie, L.C., Bao, M., MacKenzie, K., Pierce, G.J., Karl, H., Ostermeyer, U., Buchmann, K., Bušelić, I., Mladineo, I., Pascual, S., 2017a. A survey of zoonotic nematodes of commercial key fish species from major European fishing grounds—introducing the FP7 PARASITE exposure assessment study. Fish. Res. doi.org/10.1016/j.fishres.2017.09.009. (in press).

Levsen, A., Cipriani, P., Mattiucci, S., Gay, M., Hastie, L.C., Pierce, G.J., Svanevik, C.S., Højgaard, D.P., Nascetti, G., González, A.F., Pascual, S., 2017b. *Anisakis* species composition and infection characteristics in Atlantic mackerel, *Scomber scombrus*, from major NE Atlantic and Mediterranean fishing grounds. Fish. Res. doi.org/10.1016/j.fishres.2017.07.030. (in press).

Li, S.W., Shiao, S.H., Weng, S.C., Liu, T.H., Su, K.E., Chen, C.C., 2015. A case of human infection with *Anisakis simplex* in Taiwan. Gastrointest. Endosc. 82, 757–758.

Li, L., Zhao, J.Y., Chen, H.X., Ju, H.D., An, M., Xu, Z., Zhang, L.P., 2017. Survey for the presence of ascaridoid larvae in the cinnamon flounder *Pseudorhombus cinnamoneus* (Temminck & Schlegel) (Pleuronectiformes: Paralichthyidae). Int. J. Food Microbiol. 241, 108–116.

Lim, H., Jung, B., Cho, J., Yooyen, T., Shin, E., Chai, J., 2015. Molecular diagnosis of cause of anisakiasis in humans, South Korea. Emerg. Infect. Dis. 21, 342–344.

Lin, A.H., Nepstad, I., Florvaag, E., Egaas, E., Van Do, T., 2014. An extended study of Seroprevalence of anti-*Anisakis simplex* IgE antibodies in Norwegian blood donors. Scand. J. Immunol. 79, 61–67.

Liu, Q., Zhang, Q., 2013. Analysis on long-term change of sea surface temperature in the China Seas. J. Ocean Univ. China 12, 295–300.

Llarena-Reino, M., González, Á.F., Vello, C., Outeiriño, L., Pascual, S., 2012. The accuracy of visual inspection for preventing risk of *Anisakis* spp. infection in unprocessed fish. Food Control 23, 54–58.

Llarena-Reino, M., Piñeiro, C., Antonio, J., Outeriño, L., Vello, C., González, A.F., Pascual, S., 2013. Optimization of the pepsin digestion method for anisakids inspection in the fishing industry. Vet. Parasitol. 191, 276–283.

Lock, G., Ehresmann, J., Jöntvedt, E., 2008. Severe segmental colitis due to anisakiasis. Unusual manifestation of a rare infection in Germany. Dtsch. Med. Wochenschr. 133, 1779–1782.

Lopez, I., Pardo, M.A., 2010. Evaluation of a real-time polymerase chain reaction (PCR) assay for detection of *Anisakis simplex* parasite as a food-borne allergen source in seafood products. J. Agric. Food Chem. 58, 1469–1477.

López Peñas, D., Ramírez Ortiz, L.M., del Rosal Palomeque, R., López Rubio, F., Fernández-Crehuet Navajas, R., Miño Fugarolas, G., 2000. Study of 13 cases of anisakiasis in the province of Cordoba. Med. Clin. 114, 177–180.

López-Serrano, M.C., Gomez, A.A., Daschner, A., Moreno-Ancillo, A., de Parga, J.M., Caballero, M.T., Barranco, P., Cabañas, R., 2000. Gastroallergic anisakiasis: findings in 22 patients. J. Gastroenterol. Hepatol. 15, 503–506.

Lozano, M.J., Martin, H.L., Diaz, S.V., Manas, A.I., Valero, L.A., Campos, B.M., 2004. Cross-reactivity between antigens of *Anisakis simplex* (s. l.) and other ascarid nematodes. Parasite 11, 219–223.

Lunestad, B.T., 2003. Absences of nematodes in farmed Atlantic salmon (Salmo salar L.) in Norway. J. Food Prot. 66, 122–124.

Lysne, D.A., Hemmingsen, W., Skorping, A., 1995. Pepsin digestion reveals both previous and present infections of metacercariae in the skin of fish. Fish. Res. 24, 173–177.

Maggi, P., Caputi-Iambrenghi, O., Scardigno, A., Scoppetta, L., Saracino, A., Valente, M., Pastore, G., Angarano, G., 2000. Gastrointestinal infection due to *Anisakis simplex* in southern Italy. Eur. J. Epidemiol. 16, 75–78.

Mann, K.H., Lazier, J.R.N., 2006. Vertical structure of the open ocean: biology of the mixed layer. Dyn. Mar. Ecosyst. 68, 117.

Marcogliese, D.J., 2001. Distribution and abundance of sealworm (*Pseudoterranova decipiens*) and other anisakid nematodes in fish and seals in the Gulf of St. Lawrence: potential importance of climatic conditions. NAMMCO Sci. Pub. 3, 113–128.

Marcogliese, D.J., 2002. Food webs and the transmission of parasites to marine fish. Parasitology 124, S83–S99.

Marcogliese, D.J., 2008. The impact of climate change on the parasites and infectious diseases of aquatic animals. Rev. Sci. Tech. 27, 467–484.

Margolis, L., 1970. Nematode Diseases of Marine Fishes. Am. Fish. Soc., vol. 5. pp. 190–208.

Mariano, E., Fioranelli, M., Roccia, M.G., Onorato, M., Di Nardo, V., Bianchi, M., 2015. Anectodal report of acute gastric anisakiasis and severe chest discomfort. J. Integr. Cardiol. 1, 210–212.

Marques, J.F., Cabral, H.N., Busi, M., D'Amelio, S., 2006. Molecular identification of *Anisakis* species from Pleuronectiformes off the Portuguese coast. J. Helminthol. 80, 47–51.

Martín-Sánchez, J., Artacho-Reinoso, M.E., Díaz-Gavilán, M., Valero-López, A., 2005. Structure of *Anisakis simplex* s.l. populations in a region sympatric for *A. pegreffii* and *A. simplex* s.s. Absence of reproductive isolation between both species. Mol. Biochem. Parasitol. 141, 155–162.

Marty, G.D., 2008. Anisakid larva in the viscera of a farmed Atlantic salmon (*Salmo salar*). Aquaculture 279, 209–210.

Marzocca, G., Rocchi, B., Lo Gatto, M., Polito, S., Varrone, F., Caputo, E., Sorbellini, F., 2009. Acute abdomen by anisakiasis and globalization. Ann. Ital. Chir. 80, 65–68.

Matsumoto, T., Iida, M., Kimura, Y., Tanaka, K., Kitada, T., Fujishima, M., 1992. Anisakiasis of the colon: radiologic and endoscopic features in six patients. Radiology 183, 97–99.

Mattiucci, S., Nascetti, G., 2006. Molecular systematics, phylogeny and ecology of anisakid nematodes of the genus *Anisakis* Dujardin, 1845: an update. Parasite 13, 99–113.

Mattiucci, S., Nascetti, G., 2007. Genetic diversity and infection levels of anisakid nematodes parasitic in fish and marine mammals from Boreal and Austral hemispheres. Vet. Parasitol. 148, 43–57.

Mattiucci, S., Nascetti, G., 2008. Advances and trends in the molecular systematics of *Anisakis* nematodes, with implications for their evolutionary ecology and host-parasite co-evolutionary processes. Adv. Parasitol. 66, 47–148.

Mattiucci, S., Nascetti, G., Bullini, L., Orecchia, P., Paggi, L., 1986. Genetic structure of *Anisakis physeteris* and its differentiation from the *Anisakis simplex* complex (Ascaridida: Anisakidae). Parasitology 93, 383–387.

Mattiucci, S., Nascetti, G., Cianchi, R., Paggi, L., Arduino, P., Margolis, L., Brattey, J., Webb, S.C., D'Amelio, S., Orecchia, P., Bullini, L., 1997. Genetic and ecological data on the *Anisakis simplex* complex with evidence for a new species (Nematoda, Ascaridoidea, Anisakidae). J. Parasitol. 83, 401–416.

Mattiucci, S., Paggi, L., Nascetti, G., Ishikura, H., Kikuchi, K., Sato, N., Cianchi, R., Bullini, L., 1998. Allozyme and morphological identification of shape *Anisakis*, *Contracaecum* and *Pseudoterranova* from Japanese waters (Nematoda, Ascaridoidea). Syst. Parasitol. 40, 81–92.

Mattiucci, S., Paggi, L., Nascetti, G., Abollo, E., Webb, S.C., Pascual, S., Cianchi, R., Bullini, L., 2001. Genetic divergence and riproductive isolation between *Anisakis brevispiculata* and *Anisakis physeteris* (Nematoda: Anisakidae). Int. J. Parasitol. 31, 9–14.

Mattiucci, S., Paggi, L., Nascetti, G., Portes Santos, C., Costa, G., Di Beneditto, A.P., Ramos, R., Argyrou, M., Cianchi, R., Bullini, L., 2002. Genetic markers in the study of *Anisakis typica* (Diesing, 1860): larval identification and genetic relationships with other species of *Anisakis* Dujardin, 1845 (Nematoda: Anisakidae). Syst. Parasitol. 51, 159–170.

Mattiucci, S., Abaunza, P., Ramadori, L., Nascetti, G., 2004. Genetic identification of *Anisakis* larvae in European hake from Atlantic and Mediterranean waters for stock recognition. J. Fish Biol. 65, 495–510.

Mattiucci, S., Nascetti, G., Dailey, M., Webb, S.C., Barros, N., Cianchi, R., Bullini, L., 2005. Evidence for a new species of *Anisakis* Dujardin, 1845: morphological description and genetic relationships between congeners (Nematoda: Anisakidae). Syst. Parasitol. 61, 157–171.

Mattiucci, S., Abaunza, P., Damiano, S., Garcia, A., Santos, M.N., Nascetti, G., 2007. Distribution of *Anisakis* larvae identified by genetic markers and their use for stock characterization of demersal and pelagic fish from European waters: an update. J. Helminthol. 81, 117–127.

Mattiucci, S., Farina, V., Campbell, N., Mackenzie, K., Ramos, P., Pinto, A.L., Abaunza, P., Nascetti, G., 2008. *Anisakis* spp. larvae (Nematoda: Anisakidae) from Atlantic horse mackerel: their genetic identification and use as biological tags for host stock identification. Fish. Res. 89, 146–171.

Mattiucci, S., Paoletti, M., Webb, S.C., 2009. *Anisakis nascettii* n. sp. (Nematoda: Anisakidae) from beaked whales of the southern hemisphere: morphological description, genetic relationships between congeners and ecological data. Syst. Parasitol. 74, 199–217.

Mattiucci, S., Paoletti, M., Marcer, F., Gazzonis, A., Manfredi, M.T., Fernandez, A.J., Gonzales, F., Nascetti, G., 2010. Genetic heterogeneity within *Anisakis physeteris* (sensu lato) (Nematoda: Anisakidae) from sperm whales, *Physeter macrocephalus*, from Mediterranean Sea (Apulian coast) and Atlantic Ocean (Canaries coast), Abstract, XXVI Congresso Nazionale SoIPa, Perugia, 22–25 Giugno. Parassitologia 52, 357.

Mattiucci, S., Paoletti, M., Borrini, F., Palumbo, M., Palmieri, R.M., Gomes, V., Casati, A., Nascetti, G., 2011. First molecular identification of the zoonotic parasite *Anisakis pegreffii* (Nematoda: Anisakidae) in a paraffin-embedded granuloma taken from a case of human intestinal anisakiasis in Italy. BMC Infect. Dis. 11, 82.

Mattiucci, S., Fazii, P., De Rosa, A., Paoletti, M., Megna, A.S., Glielmo, A., De Angelis, M., Costa, A., Meucci, C., Calvaruso, V., Sorrentini, I., Palma, G., Bruschi, F., Nascetti, G., 2013. Anisakiasis and gastroallergic reactions associated with *Anisakis pegreffii* infection, Italy. Emerg. Infect. Dis. 19, 496–499.

Mattiucci, S., Cipriani, P., Webb, S.C., Paoletti, M., Marcer, F., Bellisario, B., Gibson, D.I., Nascetti, G., 2014a. Genetic and morphological approaches distinguishing the three sibling species of the *Anisakis simplex* species complex, with a species designation as *Anisakis berlandi* n. sp. for *A. simplex* sp. C (Nematoda: Anisakidae). J. Parasitol. 15, 12–15.

Mattiucci, S., Garcia, A., Cipriani, P., Santos, M.N., Nascetti, G., Cimmaruta, R., 2014b. Metazoan parasite infection in the swordfish, *Xiphias gladius,* from the Mediterranean Sea and comparison with Atlantic populations: implications for its stock characterization. Parasite 21, 35.

Mattiucci, S., Cimmaruta, R., Cipriani, P., Abaunza, P., Bellisario, B., Nascetti, G., 2015a. Integrating *Anisakis* spp. parasites data and host genetic structure in the frame of a holistic approach for stock identification of selected Mediterranean Sea fish species. Parasitology 142, 90–108.

Mattiucci, S., Cipriani, P., Paoletti, M., Nardi, V., Santoro, M., Bellisario, B., Nascetti, G., 2015b. Temporal stability of parasite distribution and genetic variability values of *Contracaecum osculatum* sp. D and *C. osculatum* sp. E (Nematoda: Anisakidae) from fish of the Ross Sea (Antarctica). Int. J. Parasitol. Parasites Wildl. 4, 356–367.

Mattiucci, S., Acerra, V., Paoletti, M., Cipriani, P., Levsen, A., Webb, S.C., Canestrelli, D., Nascetti, G., 2016. No more time to stay 'single' in the detection of *Anisakis pegreffii*, *A. simplex* (s. s.) and hybridization events between them: a multi-marker nuclear genotyping approach. Parasitology 143, 998–1011.

Mattiucci, S., Paoletti, M., Cipriani, P., Webb, S.C., Timi, J.T., Nascetti, G., 2017a. Inventorying biodiversity of anisakid nematodes from the austral region: a hotspot of genetic diversity? In: Biodiversity and Evolution of Parasitic Life in the Southern Ocean. Springer International Publishing, pp. 109–140.

Mattiucci, S., Cipriani, P., Paoletti, M., Levsen, A., Nascetti, G., 2017b. Reviewing biodiversity and epidemiological aspects of anisakid nematodes from the North East Atlantic Ocean. J. Helminthol. 91, 422–439. https://doi.org/10.1017/S0022149X1700027X. (in press).

Mattiucci, S., Giulietti, L., Paoletti, M., Cipriani, P., Gay, M., Levsen, A., Klapper, R., Karl, H., Bao, M., Pierce, G.J., Nascetti, G., 2017c. Population structure of the parasite *Anisakis simplex* (s. s.) collected in *Clupea harengus* L. from North East Atlantic fishing grounds. Fish. Res. doi.org/10.1016/j.fishres.2017.08.002. (in press).

Mattiucci, S., Paoletti, M., Colantoni, A., Carbone, A., Gaeta, R., Proietti, A., Frattaroli, S., Fazii, P., Bruschi, F., Nascetti, G., 2017d. Invasive anisakiasis by the parasite *Anisakis pegreffii* (Nematoda: Anisakidae): diagnosis by real-time PCR hydrolysis probe system and immunoblotting assay. BMC Infect. Dis. 17, 530. https://doi.org/10.1186/s12879-017-2633-0.

Mattiucci, S., Colantoni, A., Crisafi, B., Mori-Ubaldini, F., Caponi, L., Fazii, P., Nascetti, G., Bruschi, F., 2017e. IgE sensitization to *Anisakis pegreffii* in Italy: comparison of two methods for the diagnosis of allergic anisakiasis. Parasite Immunol. https://doi.org/10.1111/pim.12440, in press.

Mauchline, J., 1980. The biology of mysids and euphausiids. Adv. Mar. Biol. 18, 1–369.

Mayr, E., 1963. Animal Species and Evolution. Belknap Press, Harvard University Press, Cambridge, MA.

Mazzariol, S., Di Guardo, G., Petrella, A., Marsili, L., Fossi, C.M., Leonzio, C., Zizzo, N., Vizzini, S., Gaspari, S., Pavan, G., Podestà, M., Garibaldi, F., Ferrante, M., Copat, C., Traversa, D., Marcer, F., Airoldi, S., Frantzis, A., De Bernaldo Quirós, Y., Cozzi, B., Fernández, A., 2011. Sometimes sperm whales (*Physeter macrocephalus*) cannot find their way back to the high seas: a multidisciplinary study on a mass stranding. PLoS One 6, e19417.

McGowen, M.R., Gatesy, J., Wildman, D.E., 2014. Molecular evolution tracks macroevolutionary transitions in Cetacea. Trends Ecol. Evol. 29, 336–346.

Meloni, M., Angelucci, G., Merella, P., Siddi, R., Deiana, C., Orrù, G., Salati, F., 2011. Molecular characterization of *Anisakis* larvae from fish caught off Sardinia. J. Parasitol. 97, 908–914.

Messina, C.M., Pizzo, F., Santulli, A., Bušelić, I., Boban, M., Orhanović, S., Mladineo, I., 2016. *Anisakis pegreffii* (Nematoda: Anisakidae) products modulate oxidative stress and apoptosis-related biomarkers in human cell lines. Parasit. Vectors 9, 607.

Milano, I., Babbucci, M., Cariani, A., Atanassova, M., Bekkevold, D., Carvalho, G.R., Espiñeira, M., Fiorentino, F., Garofalo, G., Geffen, A.J., Hansen, J.H., Helyar, S.J., Nielsen, E.E., Ogden, R., Patarnello, T., Stagioni, M., FishPopTrace Consortium, Tinti, F., Bargelloni, L., 2014. Outlier SNP markers reveal fine-scale genetic structuring across European hake populations (*Merluccius merluccius*). Mol. Ecol. 23, 118–135.

Milinkovitch, M.C., 1995. Molecular phylogeny of cetaceans prompts revision of morphological transformations. Trends Ecol. Evol. 10, 328–334.

Minamoto, T., Sawaguchi, K., Ogino, T., Mai, M., 1991. Anisakiasis of the colon: report of two cases with emphasis on the diagnostic and therapeutic value of colonoscopy. Endoscopy 23, 50–52.

Mladineo, I., Simat, V., Mileti, C.J., Beck, R., Poljak, V., 2012. Molecular identification and population dynamic of *Anisakis pegreffii* (Nematoda: Anisakidae Dujardin, 1845) isolated from the European anchovy (*Engraulis encrasicolus* L.) in the Adriatic Sea. Int. J. Food Microbiol. 157, 224–229.

Mladineo, I., Poljak, V., Martinez-Sernàndez, V., Ubeira, F.M., 2014. Anti-*Anisakis* IgE seroprevalence in the healthy Croatian coastal population and associated risk factors. PLoS Negl. Trop. Dis. 8, e2673.

Mladineo, I., Popović, M., Drmić-Hofman, I., Poljak, V., 2016. A case report of *Anisakis pegreffii* (Nematoda, Anisakidae) identified from archival paraffin sections of a Croatian patient. BMC Infect. Dis. 16, 42.

Mladineo, I., Trumbić, Ž., Radonić, I., Vrbatović, A., Hrabar, J., Bušelić, I., 2017a. *Anisakis simplex* complex: ecological significance of recombinant genotypes in an allopatric area of the Adriatic Sea inferred by genome-derived simple sequence repeats. Int. J. Parasitol. 47, 215–223. https://doi.org/10.1016/j.ijpara.2016.11.003.

Mladineo, I., Bušelić, I., Hrabar, J., Vrbatović, A., Radonić, I., 2017b. Population parameters and mito-nuclear mosaicism of *Anisakis* spp. in the Adriatic Sea. Mol. Biochem. Parasitol. 212, 46–54.

Mo, T.A., Gahr, A., Hansen, H., Hoel, E., Oaland, Ø., Poppe, T.T., 2014. Presence of *Anisakis simplex* (Rudolphi, 1809 det. Krabbe, 1878) and *Hysterothylacium aduncum* (Rudolphi, 1802) (Nematoda: Anisakidae) in runts of farmed Atlantic salmon, *Salmo salar* L. J. Fish Dis. 37, 135–140.

Molina-Fernández, D., Malagón, D., Gómez-Mateos, M., Benítez, R., Martín-Sánchez, J., Adroher, F.J., 2015. Fishing area and fish size as risk factors of *Anisakis* infection in sardines (*Sardina pilchardus*) from Iberian waters, southwestern Europe. Int. J. Food Microbiol. 203, 27–34.

Moneo, I., Caballero, M.L., Gómez, F., Ortega, E., Alonso, M.J., 2000. Isolation and characterization of a major allergen from the fish parasite *Anisakis simplex*. J. Allergy Clin. Immunol. 106, 177–182.

Monis, P.T., Giglio, S., Keegan, A.R., Thompson, R.A., 2005. Emerging technologies for the detection and genetic characterization of protozoan parasites. Trends Parasitol. 21, 340–346.

Montalto, M., Miele, L., Marcheggiano, A., Santoro, L., Curigliano, V., Vastola, M., Gasbarrini, G., 2005. *Anisakis* infestation: a case of acute abdomen mimicking Crohn's disease and eosinophilic gastroenteritis. Dig. Liver Dis. 37, 62–64.

Moreno, A.D.R., Valero, A., Mayorga, C., Gómez, B., Torres, M.J., Hernández, J., Ortiz, M., Maldonado, J.L., 2006. Sensitization to *Anisakis simplex* sl in a healthy population. Acta Trop. 97, 265–269.

Morton, B., Yuen, W.Y., 2000. The feeding behaviour and competition for carrion between two sympatric scavengers on a sandy shore in Hong Kong: the gastropod, *Nassarius festivus* (Powys) and the hermit crab, *Diogenes edwardsii* (De Haan). J. Exp. Mar. Biol. Ecol. 246, 1–29.

Moschella, C.M., Mattiucci, S., Mingazzini, P., De Angelis, G., Assenza, M., Lombardo, F., Monaco, S., Paggi, L., Modini, C., 2004. Intestinal anisakiasis in Italy: case report. J. Helminthol. 78, 271–273.

Moschella, C.M., Mattiucci, S., Mingazzini, P., Mongardini, M., Chein, A., Miccolis, D., Modini, C., 2005. Intestinal anisakiasis in Italy: a case treated by emergency surgery. G. Chir. 26, 201–205.

Mossali, C., Palermo, S., Capra, E., Piccolo, G., Botti, S., Bandi, C., D'Amelio, S., Giuffra, E., 2010. Sensitive detection and quantification of anisakid parasite residues in food products. Foodborne Pathog. Dis. 7, 391–397.

Mostowy, R., Engelstädter, J., 2011. The impact of environmental change on host-parasite coevolutionary dynamics. Proc. Biol. Sci. 278 (1716), 2283–2292. https://doi.org/10.1098/rspb.2010.2359.

Mudry, J., Lefebvre, P., Dei-Cas, E., Vernes, A., Poirriez, J., Débat, M., Marti, R., Binot, P., Cortot, A., 1986. Human anisakiasis: 5 cases in northern France. Gastroenterol. Clin. Biol. 10, 83–87.

Mumoli, N., Merlo, A., 2013. Colonic anisakiasis. Can. Med. Assoc. J. 185, e652. https://doi.org/10.1503/cmaj.120909.

Münster, J., Klimpel, S., Fock, H.O., MacKenzie, K., Kuhn, T., 2015. Parasites as biological tags to track an ontogenetic shift in the feeding behaviour of *Gadus morhua* off West and East Greenland. Parasitol. Res. 114, 2723–2733.

Murata, R., Suzuki, J., Sadamasu, K., Kai, A., 2011. Morphological and molecular characterization of *Anisakis* larvae (Nematoda: Anisakidae) in *Beryx splendens* from Japanese waters. Parasitol. Int. 60, 193–198.

Nadler, S.A., Hudspeth, D.S.S., 2000. Phylogeny of the Ascaridoidea (Nematoda: Ascaridida) based on three genes and morphology: hypotheses of structural and sequence evolution. J. Parasitol. 86, 380–393.

Nadler, S.A., Pèrez-Ponce de León, G., 2011. Integrating molecular and morphological approaches for characterizing parasite cryptic species: implications for parasitology. Parasitology 138, 1688–1709.

Nadler, S.A., D'Amelio, S., Dailey, M.D., Paggi, L., Siu, S., Sakanari, J.A., 2005. Molecular phylogenetics and diagnosis of *Anisakis, Pseudoterranova*, and *Contracaecum* from Northern Pacific marine mammals. J. Parasitol. 91, 1413–1429.

Napoletano, C., Colantoni, A., Nuti, M., Rughetti, A., Mattiucci, S., 2016. How *Anisakis pegreffii* (Nematoda: Anisakidae) modulates dendritic cells differentiation. XXIX Congresso Nazionale della Società Italiana di Parassitologia. Abstract book, 37.

Napoletano C., Mattiucci S., Colantoni A., Battisti F., Zizzari I.G., Rahimi H., Nuti M., Rughetti A., *Anisakis pegreffii* impacts differentiation and function of human dendritic cells. Parasite Immunol., submitted.

Nascetti, G., Paggi, L., Orecchia, P., Smith, J.W., Mattiucci, S., Bullini, L., 1986. Electrophoretic studies on *Anisakis simplex* complex (Ascaridida: Anisakidae) from the Mediterranean and North East Atlantic. Int. J. Parasitol. 16, 633–640.

Nascetti, G., Cianchi, R., Mattiucci, S., D'Amelio, S., Orecchia, P., Paggi, L., Brattey, J., Berland, B., Smith, J.W., Bullini, L., 1993. Three sibling species within *Contracaecum osculatum* (Nematoda, Ascaridida, Ascaridoidea) from the Atlantic Arctic-Boreal region: reproductive isolation and host preferences. Int. J. Parasitol. 23, 105–120.

Nieuwenhuizen, N.E., 2016. *Anisakis*–immunology of a foodborne parasitosis. Parasite Immunol. 38, 548–557.

Nikaido, M., Matsuno, F., Hamilton, H., Brownell, R.L., Cao, Y., Ding, W., Zuoyan, Z., Shedlock, A.M., Ewan Fordyce, R., Hasegawa, M., Okada, N., 2001. Retroposon analysis of major cetacean lineages: the monophyly of toothed whales and the paraphyly of river dolphins. PNAS 98, 7384–7389.

Nilsen, H., Heia, K., Sivertsen, A., 2008. Detection of parasites in fish: developing an industrial solution. Infofish Int. 3, 26–35.

Noh, J.H., Kim, B., Kim, S.M., Ock, M., Park, M.I., Goo, J.Y., 2003. A case of acute gastric anisakiasis provoking severe clinical problems by multiple infection. Korean J. Parasitol. 41, 97–100.

Oro, D., Ruiz, X., 1997. Exploitation of trawler discards by breeding seabirds in the northwestern Mediterranean: differences between the Ebro Delta and the Balearic Islands areas. ICES J. Mar. Sci. 54, 695–707.

Oshima, T., Kobayashi, A., Kumada, M., Koyama, T., Kagei, N., Nemoto, T., 1968. Experimentat infection with sccond stage larvaof *Anisakis* sp. on *Euphausia similis* and *Euphausia pacifica*. Japan. J. Parasitol. 17, 585.

Otsuru, M., Hatsukanu, T., Oyanagi, T., Kenmotsu, M., 1965. The visceral migrans of gastro-intestinal tract and its vicinity caused by some larval nematode. Jpn. J. Parasitol. 14, 542–555.

Paggi, L., Nascetti, G., Webb, S.C., Mattiucci, S., Cianchi, R., Bullini, L., 1998. A new species of *Anisakis* Dujardin, 1845 (Nematoda: Anisakidae) from beaked whale (Ziphiidae): allozyme and morphological evidence. Syst. Parasitol. 40, 161–174.

Palm, H.W., Bray, R.A., 2014. Marine Fish Parasitology in Hawaii. Westarp & Partner Digitaldruck, Hohenwarsleben, p. XII.

Palm, H.W., Damriyasa, I.M., Linda Oka, I.B.M., 2008. Molecular genotyping of *Anisakis* Dujardin, 1845 (Nematoda: Ascaridoidea: Anisakidae) larvae from marine fish of Balinese and Javanese waters, Indonesia. Helminthologia 45, 3–12.

Palm, H.W., Theisen, S., Damriyasa, I.M., Kusmintarsih, E.S., Oka, I.B., Setyowati, E.A., Suratma, N.A., Wibowo, S., Kleinertz, S., 2017. *Anisakis* (Nematoda: Ascaridoidea) from Indonesia. Dis. Aquat. Org. 123 (2), 141–157. https://doi.org/10.3354/dao03091.

Palma, R., Mattiucci, S., Panetta, C., Raniolo, M., Magliocca, F.M., Pontone, S., 2018. Paucisymptomatic gastric anisakiasis: endoscopical removal of *Anisakis* sp. Larva. Mini-invasive Surg. 2, 1.

Pampiglione, S., Rivasi, F., Criscuol, M., De Benedittis, A., Gentile, A., Russo, S., Testini, M., Villan, M., 2002. Human anisakiasis in Italy: a report of eleven new cases. Pathol. Res. Pract. 198, 429–434.

Paoletti, M., Mattiucci, S., Colantoni, A., Levsen, A., Gay, M., Nascetti, G., 2017. Development of species-specific real time-PCR primers/probe systems to identify fish parasites of the genera *Anisakis* and *Pseudoterranova* (Nematoda: Anisakidae). Fish. Res. https://doi.org/10.1016/j.fishres.2017.07.015, in press.

Pascual, S., Gonzáez, A., Guerra, A., 2007. Parasites and cephalopod fisheries uncertainty: towards a waterfall understanding. Rev. Fish Biol. Fish. 17, 139–144.

Patarnello, T., Volckaert, F.A., Castilho, R., 2007. Pillars of Hercules: is the Atlantic-Mediterranean transition a phylogeographical break? Mol. Ecol. 16, 4426–4444.

Pekmezci, G.Z., Onuk, E.E., Bolukbas, C.S., Yardimci, B., Gurler, A.T., Acici, M., Umur, S., 2014. Molecular identification of *Anisakis* species (Nematoda: Anisakidae) from marine fishes collected in Turkish waters. Vet. Parasitol. 201, 82–94.

Pellegrini, M., Occhini, R., Tordini, G., Vindigni, C., Russo, S., Marzocca, G., 2005. Acute abdomen due to small bowel anisakiasis. Dig. Liver Dis. 37, 65–67.

Peñalver, J., Dolores, E.M., Muñoz, P., 2010. Absence of anisakid larvae in farmed European sea bass (*Dicentrarchus labrax* L.) and gilthead sea bream (*Sparus aurata* L.) in Southeast Spain. J. Food Prot. 73, 1332–1334.

Pérez-Pérez, J., Fernández-Caldas, E., Marañón, F., Sastre, J.I.N., Bernal, M.L., Rodríguez, J., Bedate, C.A., 2000. Molecular cloning of paramyosin, a new allergen of *Anisakis simplex*. Int. Arch. Allergy Immunol. 123, 120–129.

Pérez-Ponce de León, G., Nadler, S.A., 2010. What we don't recognize can hurt us: a plea for awareness about cryptic species. J. Parasitol. 96, 453–464.

Picó-Durán, G., Pulleiro-Potel, L., Abollo, E., Pascual, S., Muñoz, P., 2016. Molecular identification of *Anisakis* and *Hysterothylacium* larvae in commercial cephalopods from the Spanish Mediterranean coast. Vet. Parasitol. 220, 47–53.

Pierce, G.J., Bao, M., Mackenzie, K., Dunser, A., Giulietti, L., Cipriani, P., Mattiucci, S., Hastie, L., 2017. Ascaridoid nematode infection in haddock (*Melanogrammus aeglefinus*) and whiting (*Merlangius merlangus*) in Northeast Atlantic waters. Fish. Res. https://doi.org/10.1016/j.fishres.2017.09.008. (in press).

Pinkus, G.S., Coolidge, C., Little, M.D., 1975. Intestinal anisakiasis. First case report from North America. Am. J. Med. 59, 114–120.

Piola, A.R., Rivas, A.L., 1997. Corrientes en la plataforma continental. In: Boschi, E. (Ed.), El Mar Argentino y sus Recursos Pesqueros. Inst. Nac. de Invest. y Desarrollo Pesquero, Mar del Plata, Argentina, pp. 119–132.

Pippy, J.H., 1970. Use of ultraviolet light to find parasitic nematodes in situ. J. Fish. Res. Board. Can. 27, 963–965.

Piscaglia, A.C., Ventura, M.T., Landolfo, G., Giordano, M., Russo, S., Landi, R., Zulian, V., Forte, F., Stefanelli, M.L., 2014. Chronic anisakidosis presenting with intestinal intussusception. Eur. Rev. Med. Pharmacol. Sci. 18, 3916–3920.

Plath, F., Holle, A., Zendeh, D., Moller, F.W., Barten, M., Reisinger, E.C., Liebe, S., 2001. Anisakiasis of the stomach—a case report from Germany. Z. Gastroenterol. 39, 177–180.

Podestà, M., Azzellino, A., Cañadas, A., Frantzis, A., Moulins, A., Rosso, M., Tepsich, P., Lanfredi, C., 2016. Cuvier's beaked whale, *Ziphius cavirostris*, distribution and occurrence in the Mediterranean Sea: high-use areas and conservation threats. Adv. Mar. Biol. 75, 103–140.

Podolska, M., Horbowy, J., 2003. Infection of Baltic herring (*Clupea harengus membras*) with *Anisakis simplex* larvae, 1992–1999: a statistical analysis using generalized linear models. ICES J. Mar. Sci. 60, 85–93.

Pontes, T., D'Amelio, S., Costa, G., Paggi, L., 2005. Molecular characterization of larval anisakid nematodes from marine fishes of Madeira by a PCR-based approach, with evidence for a new species. J. Parasitol. 91, 1430–1434.

Pontone, S., Leonetti, G., Guaitoli, E., Mocini, R., Manfredelli, S., Catania, A., Pontone, P., Sorrenti, S., 2012. Should the host reaction to anisakiasis influence the treatment? Different clinical presentations in two cases. Rev. Esp. Enferm. Dig. 104, 607–610.

Priebe, K., Huber, C., Märtlbauer, E., Terplan, G., 1991. Detection of antibodies against the larva of *Anisakis simplex* in the pollock *Pollachius virens* using ELISA. Zbl. Vet. Med. 38, 209–214.

Pulleiro-Potel, L., Barcala, E., Mayo-Hernández, E., Muñoz, P., 2015. Survey of anisakids in commercial teleosts from the western Mediterranean Sea: infection rates and possible effects of environmental and ecological factors. Food Control 55, 12–17.

Quiazon, K.M.A., Yoshinaga, T., Ogawa, K., Yukami, R., 2008. Morphological differences between larvae and in vitro-cultured adults of *Anisakis simplex* (sensu stricto) and *Anisakis pegreffii* (Nematoda: Anisakidae). Parasitol. Int. 57, 483–489.

Quiazon, K.M.A., Yoshinaga, T., Santos, M.D., Ogawa, K., 2009. Identification of larval *Anisakis* spp. (Nematoda: Anisakidae) in Alaska pollock (*Theragra chalcogramma*) in northern Japan using morphological and molecular markers. J. Parasitol. 95, 1227–1232.

Quiazon, K.M., Yoshinaga, T., Ogawa, K., 2011. Distribution of *Anisakis* species larvae from fishes of the Japanese waters. Parasitol. Int. 60, 223–226.

Quiazon, K.M., Santos, M.D., Yoshinaga, T., 2013. *Anisakis* species (Nematoda: Anisakidae) of Dwarf Sperm Whale *Kogia simus* (Owen, 1866) stranded off the Pacific coast of southern Philippine archipelago. Vet. Parasitol. 197, 221–230.

Reddy, A., Fried, B., 2008. Atopic disorders and parasitic infections. Adv. Parasitol. 66, 149–191.

Regner, S., 1996. Effects of environmental changes on early stages and reproduction of anchovy in the Adriatic Sea. Sci. Mar. 60, 167–177.

Rendell, L., Frantzis, A., 2016. Mediterranean sperm whales, *Physeter macrocephalus:* the precarious sate of a lost tribe. Adv. Mar. Biol. 75, 37–74.

Repiso Ortega, A., Alcántara Torres, M., González de Frutos, C., de Artaza Varasa, T., Rodríguez Merlo, R., Valle Muñoz, J., Martínez Potenciano, J.L., 2003. Gastrointestinal anisakiasis. Study of a series of 25 patients. Gastroenterol. Hepatol. 26, 341–346.

Rice, D.W., 1998. Marine Mammals of the World: Systematics and Distribution. Soc. Mar. Mammal., Lawrence, KS. Special Publication Number 4.

Rijpstra, A.C., Canning, E.U., Van Ketel, R.J., Eeftinck Schattenkerk, J.K.M., Laarman, J.J., 1988. Use of light microscopy to diagnose small-intestinal microsporidiosis in patients with AIDS. J. Infect. Dis. 157, 827–831.

Riu Pons, F., Gimeno Beltran, J., Albero Gonzalez, R., Álvarez Gonzalez, M.A., Dedeu Cusco, J.M., Barranco Priego, L., Seoane Urgorri, A., 2015. An unusual presentation of anisakiasis in the colon (with video). Gastrointest. Endosc. 81, 1050–1051.

Rodriguez-Mahillo, A.I., Gonzalez-Muñoz, M., Gomez-Aguado, F., Rodriguez-Perez, R., Corcuera, M.T., Caballero, M.L., Moneo, I., 2007. Cloning and characterisation of the *Anisakis simplex* allergen Ani s 4 as a cysteine-protease inhibitor. Int. J. Parasitol. 37, 907–917.

Rodriguez-Perez, R., Moneo, I., Rodriguez-Mahillo, A., Caballero, M.L., 2008. Cloning and expression of Ani s 9, a new *Anisakis simplex* allergen. Mol. Biochem. Parasitol. 159, 92–97.

Roepstorff, A., Karl, H., Bloemsa, B., Huss, H.H., 1993. Catch handling and the possible migration of *Anisakis* larvae in herring, *Clupea harengus*. J. Food Prot. 56, 783–787.

Rohde, K., 2005. Marine Parasitolology. CABI Publishing, Wallingford, UK.

Roiha, I.S., Maage, A., Levsen, A., 2017. Nasjonal Undersøking av Førekomst av *Anisakis Simplex* i Norsk Oppdrettsaure (*Onchorhynchus Mykiss*). Nasjonait institutt for ernaerings og sjømatforskning (NIFES).

Rokicki, J., 2009. Effects of climatic changes on anisakid nematodes in polar regions. Pol. Sci. 3, 197–201.

Romeo Ramírez, J.A., Martínez-Conde López, A.E., Olivares Galdeano, U., Sancha Pérez, A., López de Torre Ramírez de la Piscina, J., Barros Ingerto, J., Echavarri Iñigo, J., 1997. Gastric anisakiasis diagnosed by endoscopy. Gastroenterol. Hepatol. 20, 306–308.

Romero, M., Valero, A., Navarro-Moll, M.C., Martín-Sánchez, J., 2013. Experimental comparison of pathogenic potential of two sibling species *Anisakis simplex* (s. s.) and *Anisakis pegreffii* in Wistar rat. Trop. Med. Int. Health 18, 979–984.

Roselyn, H., Richman, M.D., Ann, M., Lewicki, M.D., 1973. Right ileocolitis secondary to anisakiasis. Am. J. Roentgenol. 119, 329–331.

Ross, G.J., 1984. Smaller cetaceans of the South East Coast of Southern Africa. Ann. Cape Prov. Mus. (Nat. Hist.), 15, pp. 173–410.

Russo, A., Artegiani, A., 1996. Adriatic Sea hydrography. Sci. Mar. 60, 33–43.

Sánchez-Velasco, P., Mendizábal, L., Antón, E.M., Ocejo-Vinyals, G., Jerez, J., Leyva-Cobián, F., 2000. Association of hypersensitivity to the nematode *Anisakis simplex* with HLA class II DRB1* 1502-DQB1* 0601 haplotype. Hum. Immunol. 61, 314–319.

Santos, M.B., Pierce, G.J., Herman, J., López, A., Guerra, A., Mente, E., Clarke, M.R., 2001a. Feeding ecology of Cuvier's beaked whale (*Ziphius cavirostris*): a review with new information on the diet of this species. J. Mar. Biol. Assoc. UK 81, 687–694.

Santos, M.B., Pierce, G.J., Reid, R.J., Patterson, I.A.P., Ross, H.M., Mente, E., 2001b. Stomach contents of bottlenose dolphins (*Tursiops truncatus*) in Scottish waters. J. Mar. Biol. Assoc. UK 81, 873–878.

Sasaki, T., Fukumori, D., Matsumoto, H., Ohmori, H., Yamamoto, F., 2003. Small bowel obstruction caused by anisakiasis of the small intestine: report of a case. Surg. Today 33, 123–125.

Setyobudi, E., Jeon, C.H., Lee, C.H., Seong, K.B., Kim, J.H., 2011. Occurrence and identification of *Anisakis* spp. (Nematoda: Anisakidae) isolated from chum salmon (*Oncorhynchus keta*) in Korea. Parasitol. Res. 108, 585–592.

Shamsi, S., 2014. Recent advances in our knowledge of Australian anisakid nematodes. Int. J. Parasitol. Parasites Wildl. 3, 178–187.

Shamsi, S., Gasser, R., Beveridge, I., 2012. Genetic characterisation and taxonomy of species of *Anisakis* (Nematoda: Anisakidae) parasitic in Australian marine mammals. Invertebr. Syst. 26, 204–212.

Shamsi, S., Briand, M.J., Justine, J.L., 2017. Occurrence of *Anisakis* (Nematoda: Anisakidae) larvae in unusual hosts in Southern hemisphere. Parasitol. Int. 66, 837–840.

Shibata, O., Uchida, Y., Furusawa, T., 1989. Acute gastric anisakiasis with special analysis of the location of the worms penetrating the gastric mucosa. In: Ishikura, H. et al. (Ed.), Gastric Anisakiasis in Japan. Springer, Tokyo, Japan, pp. 53–57.

Shih, H.H., Ku, C.C., Wang, C.S., 2010. *Anisakis simplex* (Nematoda: Anisakidae) third-stage larval infections of marine cage cultured cobia, *Rachycentron canadum* L., in Taiwan. Vet. Parasitol. 171, 277–285.

Shimakura, K., Miura, H., Ikeda, K., Ishizaki, S., Nagashima, Y., Shirai, T., Kasuya, S., Shiomi, K., 2004. Purification and molecular cloning of a major allergen from *Anisakis simplex*. Mol. Biochem. Parasitol. 135, 69–75.

Shimamura, Y., Ishii, N., Ego, M., Nakano, K., Ikeya, T., Nakamura, K., Takagi, K., Fukuda, K., Fujita, Y., 2016. Multiple acute infection by *Anisakis*: a case series. Intern. Med. 55, 907–910.

Shiraki, T., 1974. Larval nematodes of family Anisakidae (Nematoda) in the Northern Sea of Japan—as a causative agent of eosinophilic phlegmone or granuloma in the human gastro-intestinal tract. Acta Med. Biol. 22, 57–98.

Šimat, V., Miletić, J., Bogdanović, T., Poljak, V., Mladineo, I., 2015. Role of biogenic amines in the post-mortem migration of *Anisakis pegreffii* (Nematoda: Anisakidae Dujardin, 1845) larvae into fish fillets. Int. J. Food Microbiol. 214, 179–186.

Skov, J., Kania, P.W., Olsen, M.M., Lauridsen, J.H., Buchmann, K., 2009. Nematode infections of maricultured and wild fishes in Danish waters: a comparative study. Aquaculture 298, 24–28.

Smith, H.K., 1983. Fishery and biology of *Nototodarus gouldi* (McCoy, 1888) in western Bass Strait. Mem. Natl. Mus. Victoria 44, 285–290.

Smith, J.W., Snyder, J.M., 2005. New locality for third-stage larvae of *Anisaks simplex* (sensu lato) (Nematoda: Ascaridoidea) in euphausiids *Euphausia pacifica* and *Thysanoessa raschii* from Prince William Sound, Alaska. Parasitol. Res. 97, 539–542.

Smith, J.W., Wootten, R., 1975. Experimental studies on the migration of *Anisakis* sp. larvae (Nematoda: ascaridida) into the flesh of herring, *Clupea harengus* L. Int. J. Parasitol. 5, 133–136.

Smith, J.W., Wootten, R., 1978. *Anisakis* and anisakiasis. Adv. Parasitol. 16, 93–163.

Smrzlić, I.V., Valić, D., Kapetanović, D., Kurtović, B., Teskeredžić, E., 2012. Molecular characterisation of Anisakidae larvae from fish in Adriatic Sea. Parasitol. Res. 111, 2385–2391.

Sohn, W.M., Na, B.K., Kim, T.H., Park, T.J., 2015. Anisakiasis: report of 15 gastric cases caused by *Anisakis* type I larvae and a brief review of Korean anisakiasis cases. Korean J. Parasitol. 53, 465.

Sonoda, H., Yamamoto, K., Ozeki, K., Inoye, H., Toda, S., Maehara, Y., 2015. An *Anisakis* larva attached to early gastric cancer: report of a case. Surg. Today 45, 1321–1325.

Stallone, O., Paggi, L., Balestrazzi, A., Mattiucci, S., Montinari, M., 1996. Gastric Anisakiasis in Italy: case report. Med. J. Sur. Med. 4, 13–16.

Steinauer, L.M., Agola, L.E., Mwangi, I.N., Mkoji, G.M., Loker, E.S., 2008. Molecular epidemiology of *Schistosoma mansoni*: a robust, high-throughput method to assess multiple microsatellite markers from individual miracidia. Infect. Genet. Evol. 8, 68–73.

Strømnes, E., Andersen, K., 1998. Distribution of whaleworm (*Anisakis simplex*, Nematoda, Ascaridoidea) L3 larvae in three species of marine fish; saithe (*Pollachius virens* (L.)), cod (*Gadus morhua* L.) and redfish (*Sebastes marinus* (L.)) from Norwegian waters. Parasitol. Res. 84, 281–285.

Sugawara, Y., Urawa, S., Kaeriyama, M., 2004. Infection of *Anisakis simplex* (Nematoda: Anisakidae) Larvae in Chum Salmon (*Oncorhynchus keta*) in the North Pacific Ocean, Bering Sea, and a River of Hokkaido. North Pacific AnadFish Comm. Doc. 791. pp. 1–13.

Sugita, S., Sasaki, A., Shiraishi, N., Kitano, S., 2008. Laparoscopic treatment for a case of ileal anisakiasis. Surg. Laparosc. Endosc. Percutan. Tech. 18, 216–218.

Suzuki, J., Murata, R., Hosaka, M., Araki, J., 2010. Risk factors for human *Anisakis* infection and association between the geographic origins of *Scomber japonicus* and anisakid nematodes. Int. J. Food Microbiol. 137, 88–93.

Swofford, D.L., 2003. PAUP*. Phylogenetic Analysis Using Parsimony (*and Other Methods). Sinauer Associates, Sunderland, MA.

Takamizawa, Y., Kobayashi, Y., 2015. Adhesive intestinal obstruction caused by extra-gastrointestinal anisakiasis. Am. J. Trop. Med. Hyg. 92, 675–676.

Takei, H., Powell, S.Z., 2007. Intestinal anisakidosis (anisakiosis). Ann. Diagn. Pathol. 11, 350–352.

Tanaka, H., Takata, S., Nishimura, T., Watanabe, S., 1968. A case report of *Anisakis* larva penetration in the pharynx mucosa. Japan. J. Parasitol. 17, 641.

Testini, M., Gentile, A., Lissidini, G., Di Venere, B., Pampiglione, S., 2003. Splenic anisakiasis resulting from a gastric perforation: an unusual occurrence. Int. Surg. 88, 126–128.

Thien, P.C., Dalsgaard, A., Thanh, B.N., Olsen, A., Murrell, K.D., 2007. Prevalence of fishborne zoonotic parasites in important cultured fish species in the Mekong Delta, Vietnam. Parasitol. Res. 101, 1277–1284.

Thu, N.D., Dalsgaard, A., Loan, L.T.T., Murrell, K.D., 2007. Survey for zoonotic liver and intestinal trematode metacercariae in cultured and wild fish in an Gang Province, Vietnam. Korean J. Parasitol. 45, 45–54.

Timi, J.T., Paoletti, M., Cimmaruta, R., Lanfranchi, A.L., Alarcos, A.J., Garbin, L., Nascimento, M.G., Rodríguez, D.H., Giardinof, G.V., Mattiucci, S., 2014. Molecular identification, morphological characterization and new insights into the ecology of larval *Pseudoterranova cattani* in fishes from the Argentine coast with its differentiation from the Antarctic species, *P. decipiens* sp. E (Nematoda: Anisakidae). Vet. Parasitol. 199, 59–72.

Tintore, J., La Violette, P.E., Blade, I., Cruzado, A., 1988. A study of an intense density front in the eastern Alboran Sea: the Almeria-Oran front. J. Phys. Oceanogr. 18, 1384–1397.

Tsukui, M., Morimoto, N., Kurata, H., Sunada, F., 2016. Asymptomatic anisakiasis of the colon incidentally diagnosed and treated during colonoscopy by retroflexion in the ascending colon. J. Rural Med. 11, 73–75.

Uga, S., Ono, K., Kataoka, N., Hasan, H., 1996. Seroepidemiology of five major zoonotic parasite infections in inhabitants of Sidoarjo, East Java, Indonesia. Southeast Asian J. Trop. Med. Public Health 27, 556–561.

Ugenti, I., Lattarulo, S., Ferrarese, F., De Ceglie, A., Manta, R., Brandonisio, O., 2007. Acute gastric anisakiasis: an Italian experience. Minerva Chir. 62, 51–60.

Umehara, A., Kawakami, Y., Matsui, T., Araki, J., Uchida, A., 2006. Molecular identification of *Anisakis simplex* sensu stricto and *Anisakis pegreffii* (Nematoda: Anisakidae) from fish and cetacean in Japanese waters. Parasitol. Int. 55, 267–271.

Umehara, A., Kawakami, Y., Araki, J., Uchida, A., 2007. Molecular identification of the etiological agent of the human anisakiasis in Japan. Parasitol. Int. 56, 211–215.

Umehara, A., Kawakami, Y., Araki, J., Uchida, A., 2008. Multiplex PCR for the identification of *Anisakis simplex* sensu stricto, *Anisakis pegreffii* and the other anisakid nematodes. Parasitol. Int. 57, 49–53.

Umehara, A., Kawakami, Y., Ooi, H.K., Uchida, A., Ohmae, H., Sugiyama, H., 2010. Molecular identification of *Anisakis* type I larvae isolated from hairtail fish off the coasts of Taiwan and Japan. Int. J. Food Microbiol. 143, 161–165.

Urita, Y., Nishino, M., Koyama, H., Kondo, E., Naruki, Y., Otsuka, S., 1997. Esophageal anisakiasis accompanied by reflux esophagitis. Intern. Med. 36, 890–893.

Valentini, A., Mattiucci, S., Bondanelli, P., Webb, S.C., Mignucci-Giannone, A., Colom-Llavina, M.M., Nascetti, G., 2006. Genetic relationships among *Anisakis* species

(Nematoda: Anisakidae) inferred from mitochondrial *cox2* sequences, and comparison with allozyme data. J. Parasitol. 92, 156–166.

Van Herwerden, L., Blair, D., Agatsuma, T., 1999. Genetic diversity in parthenogenetic triploid Paragonimus westermani. Int. J. Parasitol. 29, 1477–1482.

Van Thiel, P., Kuipers, F.C., Roskam, R.T., 1960. A nematode parasitic to herring, causing acute abdominal syndromes in man. Trop. Geogr. Med. 12, 97–113.

Vercammen, F., Kumar, V., Bollen, J., Lievens, C., Van den Bergh, L., Vervoort, T., 1997. Gastric involvement with *Anisakis* sp. larva in a Belgian patient after consumption of cod. Acta Gastroenterol. Belg. 60, 302–303.

Verhamme, M.A., Ramboer, C.H., 1988. Anisakiasis caused by herring in vinegar: a little known medical problem. Gut 29, 843–847.

Wheeler, Q.D. (Ed.), 2008. The New Taxonomy. In: Syst. Assoc. Special Vol. Series, CRC Press, Boca Raton, FL.

Wold, J.P., Westad, F., Heia, K., 2001. Detection of parasites in cod fillets by using SIMCA classification in multispectral images in the visible and NIR region. Appl. Spectrosc. 55, 1025–1034.

Wolpoff, M.H., 1989. The place of Neanderthals in human evolution. In: Trinkaus, E. (Ed.), The Emergence of Modern Humans, Cambridge University Press, New York, pp. 97–141.

Wootten, R., Yoon, G.H., Bron, J.E., 2010. A Survey of Anisakid Nematodes in Scottish Wild Atlantic Salmon. FSAS project, S14008.

Yorimitsu, N., Hiraoka, A., Utsunomiya, H., Imai, Y., Tatsukawa, H., Tazuya, N., Yamago, H., Shimizu, Y., Hidaka, S., Tanihira, T., Hasebe, A., Miyamoto, Y., Ninomiya, T., Abe, M., Hiasa, Y., Matsuura, B., Onji, M., Michitaka, K., 2013. Colonic intussusception caused by anisakiasis: a case report and review of the literature. Intern. Med. 52, 223–236.

Zarlenga, D.S., Hoberg, E., Rosenthal, B., Mattiucci, S., Nascetti, G., 2014. Anthropogenics: human influence on global and genetic homogenization of parasite populations. J. Parasitol. 100, 756–772.

Zhu, X.Q., D'Amelio, S., Paggi, L., Gasser, R.B., 2000. Assessing sequence variation in the internal transcribed spacers of ribosomal DNA within and among members of the *Contracaecum osculatum* complex (Nematoda: Ascaridoidea: Anisakidae). Parasitol. Res. 86, 677–683.

Zhu, X.Q., Podolska, M., Liu, J.S., Yu, H.Q., Chen, H.H., Lin, Z.X., Luo, C.B., Song, H.Q., Lin, R.Q., 2007. Identification of anisakid nematodes with zoonotic potential from Europe and China by single-strnd conformation polymorphism analysis of nuclear ribosomal DNA. Parasitol. Res. 101, 1703–1707.

Zuo, S., Huwer, B., Bahlool, Q., Al-Jubury, A., Christensen, D.N., Korbut, R., Kania, P., Buchmann, K., 2016. Host size-dependent anisakid infection in Baltic cod *Gadus morhua* associated with differential food preferences. Dis. Aquat. Organ. 120, 69–75.

FURTHER READING

Adams, A.M., Miller, K.S., Wekell, M.M., Dong, F.M., 1999. Survival of *Anisakis simplex* in microwave-processed arrowtooth flounder (*Atheresthes stomias*). J. Food Prot. 62, 403–409.

Balbuena, J.A., Míguez-Lozano, R., Blasco-Costa, I., 2013. PACo: a novel procrustes application to cophylogenetic analysis. PLoS One 8, e61048.

Cruz, C., Saraiva, A., Santos, M.J., Eiras, J.C., Ventura, C., Soares, J.P., Hermida, M., 2009. Parasitic infection levels by *Anisakis* spp. larvae (Nematoda: Anisakidae) in the black scabbardfish *Aphanopus carbo* (Osteichthyes: Trichiuridae) from Portuguese waters. Sci. Mar. 73, 115–120.

Domínguez-Ortega, J., Martínez-Alonso, J.C., Alonso-Llamazares, A., Argüelles-Grande, C., Chamorro, M., Robledo, T., Palacio, R., Martínez-Cócera, C., 2003. Measurement of serum levels of eosinophil cationic protein in the diagnosis of acute gastrointestinal anisakiasis. Clin. Microbial. Infect. 9, 453–457.

Gay, M., Bao, M., MacKenzie, K., Pascual, S., Buchmann, K., Bourgau, O., Couvreur, C., Mattiucci, S., Paoletti, S., Hastie, L.C., Levsen, A., Pierce, G.J., 2017. Infection levels and species diversity of ascaridoid nematodes in Atlantic cod, *Gadus morhua*, are correlated with geographic area and fish size. Fish. Res. https://doi.org/10.1016/j.fishres.2017.06.006, in press.

Irigoitia, M.M., Incorvaia, I.S., Timi, J.T., 2017. Evaluating the usefulness of natural tags for host population structure in chondrichthyans: parasite assemblages of *Sympterygia bonapartii* (Rajiformes: Arhynchobatidae) in the southwestern Atlantic. Fish. Res. 195, 80–90.

Køie, M., Fagerholm, H.P., 1995. The life cycle of *Contracaecum osculatum* (Rudolphi, 1802) sensu stricto (Nematoda, Ascaridoidea, Anisakidae) in view of experimental infections. Parasitol. Res. 81, 481–489.

Levsen, A., Paoletti, M., Cipriani, P., Nascetti, G., Mattiucci, S., 2016. Species composition and infection dynamics of ascaridoid nematodes in Barents Sea capelin (*Mallotus villosus*) reflecting trophic position of fish host. Parasitol. Res. 115, 4281–4291.

Pascual, S., González, A.F., 2017. The fish nematode problem in major European fish stocks. Fish. Res. https://doi.org/10.1016/j.fishres.2017.04.010. (in press).

Petrushevsky, G.K., Kogteva, E.P., 1954. Effects of parasitic diseases on the condition of fish. Zool. Zhurnal 33, 395–405.

Sequeira, V., Gordo, L.S., Neves, A., Paiva, R.B., Cabral, H.N., Marques, J.F., 2010. Macroparasites as biological tags for stock identification of the bluemouth, *Helicolenus dactylopterus* (Delaroche, 1809) in Portuguese waters. Fish. Res. 106, 321–328.

Strømnes, E., Andersen, K., 2000. "Spring rise" of whaleworm (*Anisakis simplex*: Nematoda, Ascaridoidea) third-stage larvae in some fish species from Norwegian waters. Parasitol. Res. 86, 619–624.

CHAPTER FIVE

Evolution, Systematics, and Biogeography of the Triatominae, Vectors of Chagas Disease

Fernando Araujo Monteiro[*,1], Christiane Weirauch[†], Márcio Felix[‡], Cristiano Lazoski[§], Fernando Abad-Franch[¶]

[*]Laboratório de Epidemiologia e Sistemática Molecular, Instituto Oswaldo Cruz, FIOCRUZ, Rio de Janeiro, Brazil
[†]University of California, Riverside, Riverside, CA, United States
[‡]Laboratório de Biodiversidade Entomológica, Instituto Oswaldo Cruz, FIOCRUZ, Rio de Janeiro, Brazil
[§]Instituto de Biologia, Universidade Federal do Rio de Janeiro, Rio de Janeiro, Brazil
[¶]Grupo Triatomíneos, Instituto René Rachou, FIOCRUZ, Belo Horizonte, Brazil
[1]Corresponding author: e-mail address: fam@ioc.fiocruz.br

Contents

1. Introduction 266
2. Evolution of the Triatominae: From Predators to Blood Feeders 267
   2.1 Are the Triatominae Monophyletic, Paraphyletic, or Polyphyletic? 267
   2.2 Putative Synapomorphies of the Triatominae 272
   2.3 Evolution of Haematophagy in the Triatominae 272
3. Systematics of the Triatominae 274
   3.1 Classical Taxonomy 274
   3.2 Molecular Systematics 280
   3.3 Uncovering and Sorting Out Hidden Diversity: Species Complexes 283
   3.4 Taxonomy: Describing and Sorting Out Newly Uncovered Diversity 293
4. Biogeography of the Triatominae 296
   4.1 The Tribe Triatomini 298
   4.2 The Tribe Rhodniini 318
   4.3 Other Tribes 321
5. Closing Thoughts and Conclusions 323
Acknowledgements 325
References 325

Abstract

In this chapter, we review and update current knowledge about the evolution, systematics, and biogeography of the Triatominae (Hemiptera: Reduviidae)—true bugs that feed primarily on vertebrate blood. In the Americas, triatomines are the vectors of *Trypanosoma cruzi*, the etiological agent of Chagas disease. Despite declining incidence and prevalence, Chagas disease is still a major public health concern in Latin America.

Advances in Parasitology, Volume 99
ISSN 0065-308X
https://doi.org/10.1016/bs.apar.2017.12.002

Triatomines occur also in the Old World, where vector-borne *T. cruzi* transmission has not been recorded. Triatomines evolved from predatory reduviid bugs, most likely in the New World, and diversified extensively across the Americas (including the Caribbean) and in parts of Asia and Oceania. Here, we first discuss our current understanding of how, how many times, and when the blood-feeding habit might have evolved among the Reduviidae. Then we present a summary of recent advances in the systematics of this diverse group of insects, with an emphasis on the contribution of molecular tools to the clarification of taxonomic controversies. Finally, and in the light of both up-to-date phylogenetic hypotheses and a thorough review of distribution records, we propose a global synthesis of the biogeography of the Triatominae. Over 130 triatomine species contribute to maintaining *T. cruzi* transmission among mammals (sometimes including humans) in almost every terrestrial ecoregion of the Americas. This means that Chagas disease will never be eradicated and underscores the fact that effective disease prevention will perforce require stronger, long-term vector control-surveillance systems.

1. INTRODUCTION

The Triatominae (Hemiptera: Reduviidae) are a diverse assemblage of true bugs that feed primarily on vertebrate blood. In the Americas, triatomines are the vectors of *Trypanosoma cruzi*, the etiological agent of Chagas disease (Lent and Wygodzinsky, 1979). Despite declining incidence and prevalence over the last decades, Chagas disease is still a major public health concern across Latin America (Rassi et al., 2010). Triatomines occur also in Asia and Oceania, where vector-borne transmission of *T. cruzi* is not known to occur. These large blood-feeding bugs are already mentioned in 16th century chronicles and captured the attention of many pioneer naturalists including Darwin (Darwin, 1839; Lent and Wygodzinsky, 1979). Widespread interest on their biology, however, arose only after Carlos Chagas discovered that they vector a human pathogen (Chagas, 1909).

Triatomines evolved from predatory reduviids, most likely in the New World, and diversified extensively across the Americas including the Caribbean and in parts of the Oriental region (Lent and Wygodzinsky, 1979). Here we present an overview of the evolution, systematics, and biogeography of these disease vectors. First we discuss our current understanding of how, how many times, and when haematophagy (the key trait defining the subfamily) might have evolved. Then we present a summary of recent advances in the systematics of this diverse group of insects, with an emphasis on the contribution of molecular tools to the clarification of taxonomic controversies. Finally, and in the light of both up-to-date phylogenetic hypotheses and a thorough review of distribution records, we propose a global synthesis of the biogeography of the Triatominae.

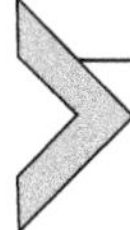

2. EVOLUTION OF THE TRIATOMINAE: FROM PREDATORS TO BLOOD FEEDERS

First recognized as a tribe by Jeannel (1919) and briefly elevated to family status by Pinto (1926), the Triatominae have been treated as a subfamily of the otherwise predatory Reduviidae since Usinger (1943). Modern catalogues, reference books, and subfamily level identification keys may disagree in the number and concept of reduviid subfamilies, but all classify the Triatominae as a distinct subfamily among assassin bugs (Maldonado Capriles, 1990; Putshkov and Putshkov, 1985; Schuh and Slater, 1995; Weirauch et al., 2014). Lent and Wygodzinsky (1979) were the first to explicitly refer to the Triatominae as a monophyletic group, proposing that haematophagy and the increased dorsal flexibility of the last labial segment compared to predatory Reduviidae likely represent synapomorphic characters. It was not until the late 1980s that the hypothesis of triatomine monophyly was challenged, followed by a series of phylogenetic analyses employing various optimality criteria including maximum parsimony (MP), maximum likelihood (ML), and Bayesian methods on datasets that have steadily increased with respect to character and taxon sampling (Fig. 1). Nonmonophyly hypotheses postulate either multiple origins of blood feeding or a reversal to predatory habits in some reduviids (de Paula et al., 2005; Hwang and Weirauch, 2012; Patterson, 2007; Schofield, 1988).

2.1 Are the Triatominae Monophyletic, Paraphyletic, or Polyphyletic?

Based on the observation that different triatomine species and genera show diverse ecological preferences, Schofield (1988) first posited that Triatominae may not represent a clade, but that instead different haematophagous lineages may have evolved independently from predatory ancestors that share ecological features such as a particular microhabitat. Schofield and Dujardin (1999) subsequently took morphological and molecular (e.g. cuticular hydrocarbons or salivary proteins) differences between species in the Rhodniini and Triatomini, the two largest and best studied triatomine tribes, as further evidence for the hypothesis of triatomine nonmonophyly. As pointed out by Schaefer (2003), this rather tentative working hypothesis on triatomine

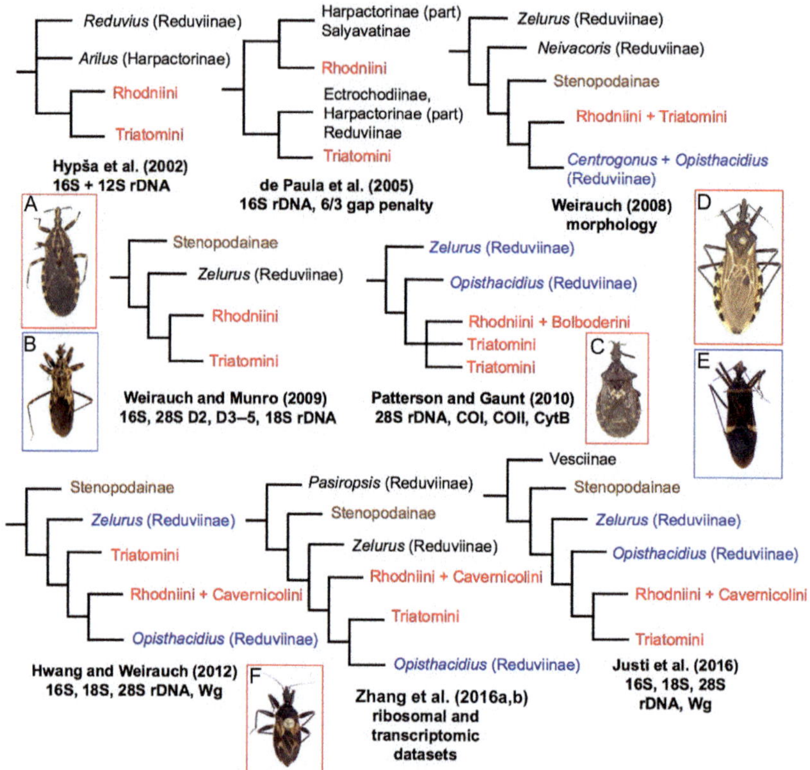

Fig. 1 Summary of published phylogenetic hypotheses of Reduviidae showing proposed relationships of Triatominae to specific predatory Reduviidae. Phylogenetic hypotheses show blood-feeding Triatominae, taxa belonging to the *Zelurus* clade of the polyphyletic Reduviinae, Stenopodainae, and other Reduviidae. Habitus *insets* illustrate Triatominae and taxa of the reduviine *Zelurus* clade. (A) *Rhodnius paraensis* (Triatominae: Rhodniini); (B) *Opisthacidius picturatus* (Reduviinae); (C) *Microtriatoma borbai* (Triatominae: Bolboderini); (D) *Triatoma melanocephala* (Triatominae: Triatomini); (E) *Zelurus* sp. (Reduviinae); and (F) *Cavernicola pilosa* (Triatominae: Cavernicolini).

polyphyly quickly morphed into the preferred hypothesis (Schofield, 2000), although tests in the form of phylogenetic analyses were unavailable at the time.

Since the early 2000s, morphology-based and molecular phylogenetic analyses with vastly different taxon and character sampling have aimed at shedding light on the relations among predatory and haematophagous Reduviidae—i.e., at testing the monophyly, paraphyly, or polyphyly of the Triatominae (Fig. 1). Hypša et al. (2002) provided the first phylogenetic test of Triatominae monophyly; they analysed 900 base pairs (bp) of mitochondrial ribosomal DNA (rDNA) for 62 taxa, including 57 Triatominae, 2 nontriatomine Reduviidae (the reduviine *Reduvius personatus* and the harpactorine *Arilus cristatus*), and 3 nonreduviid Hemiptera. Triatominae were represented by species of Rhodniini and Triatomini and were recovered as monophyletic although with low branch support. Given the small number of nontriatomine Reduviidae included in this analysis, combined with reliance on the relatively fast-evolving 16S and 12S rDNA genes, this analysis presented a first, but not particularly strong, test of Triatominae monophyly and relations (Fig. 1). de Paula et al. (2005) analysed a dataset including 72 species of Reduviidae (43 species of Triatomini, 14 species of Rhodniini, and 15 species of predatory Reduviidae in 5 subfamilies), but relied exclusively on 16S rDNA. In contrast to the analysis by Hypša et al. (2002), de Paula et al. (2005) did not find support for the monophyly of Triatominae. Instead, (i) the Salyavatinae, which specialize on termites (Gordon and Weirauch, 2016), and part of the Harpactorinae, which are mostly diurnal and vegetation dwelling (Zhang et al., 2016a), were recovered as sister to the Rhodniini and (ii) a clade comprising the Ectrichodiinae, thought to feed exclusively on millipedes (Forthman and Weirauch, 2012), plus part of the Reduviinae (*Reduvius*) and part of the Harpactorinae was sister to the Triatomini in one of two analyses (Fig. 1). The second analysis employed a different gap opening/gap extension penalty scheme and recovered the reduviine clade *Tapeinus* + *Tiarodes* as sister to Triatomini, with other relations similar to the first analysis (de Paula et al., 2005). Taken at the time as the strongest phylogenetic support for the polyphyly of Triatominae, critical taxa such as Stenopodainae and a more comprehensive sample of Reduviinae that had long been suspected to be a polyphyletic assemblage (Usinger, 1943) were omitted (de Paula et al., 2005). In addition, the use of only one gene limited the implications of these analyses.

Using 162 morphological characters and a dataset of 75 species of Reduviidae and outgroups, but including only four species of Triatominae (Rhodniini and Triatomini), Weirauch (2008) found support for the monophyly of Triatominae and placed the subfamily in a clade that also includes the predatory Stenopodainae and part of the Reduviinae, e.g., *Opisthacidius* (Fig. 1). Although sampling of predatory Reduviidae was fairly comprehensive (20 subfamilies), the number of triatomine terminals was limited and the analyses failed to include representatives of the smaller triatomine tribes that are poorly represented in natural history collections. Two subsequent multigene analyses that sampled more comprehensively either predatory Reduviidae (89 Reduviidae including 5 species of Triatominae and 5 outgroups; 4 rDNA regions; Weirauch and Munro, 2009) or Triatominae

(38 terminals, 3 triatomine tribes, 8 predatory reduviid subfamilies; 3 protein-coding mitochondrial genes and 1 nuclear ribosomal gene; Patterson and Gaunt, 2010) tested and recovered the monophyly of Triatominae, but yet again excluded all or some of the smaller tribes (Fig. 1). A more extensive analysis by Hwang and Weirauch (2012) found that the Triatominae were rendered paraphyletic by the predatory reduviine genus *Opisthacidius* in ML analyses (178 taxa including 13 Triatominae; 4 ribosomal gene regions and 1 protein-coding nuclear gene). Relations between *Cavernicola* + *Rhodnius*, *Opisthacidius*, and the remaining Triatominae (*Triatoma*, *Paratriatoma*, *Panstrongylus*, and *Eratyrus*) remained unresolved in MP analyses (Fig. 1).

Embracing the dawn of the genomic era, Zhang et al. (2016b) produced a dataset combining transcriptome (370 loci; 23 taxa) and ribosomal data (52 taxa, 29 of which with rDNA sequences only) to test relations among reduviid lineages. The aim was to boost support values, and thus confidence, in the relations along the previously weakly supported 'backbone' phylogeny of Higher Reduviidae. Of the five taxa representing Triatominae and putatively closely related predatory Reduviidae, transcriptome-level data were only available for *Triatoma protracta*. Similar to Hwang and Weirauch (2012), Triatominae (Cavernicolini, Rhodniini, and Triatomini) were rendered paraphyletic by *Opisthacidius* (Zhang et al., 2016b) (Fig. 1). This study yielded strong support for the backbone relations of Reduviidae, resolving a sister group relation of the reduviine *Pasiropsis* to the clade comprising Stenopodainae, Reduviinae (part), and Triatominae, which together were sister to a large group of predatory Reduviidae consisting of Harpactorinae, Physoderinae, Salyavatinae, and the *Acanthaspis* group of Reduviinae. The largest dataset with respect to taxon sampling published to date (229 Reduviidae including 70 triatomine species in the Cavernicolini, Rhodniini, and Triatomini), with gene sampling as in Hwang and Weirauch (2012), recovered Triatominae as monophyletic (99% posterior probability) and *Opisthacidius* as its sister taxon (Justi et al., 2016) (Fig. 1). Sampling of putatively closely related predatory Reduviidae in the genera *Opisthacidius* and *Zelurus* was identical to Hwang and Weirauch (2012), suggesting that differences in ingroup sampling density, analytical methods, or both may underlie the apparent support for these mutually exclusive hypotheses.

As a result of the above analyses, there are now testable hypotheses on the closest predatory relatives of the Triatominae (Fig. 1). In his 'phylogenetic key', Usinger (1944) grouped Triatominae with Reduviinae, Cetherinae, and Salyavatinae + Sphaeridopinae, but did not mention characters in support of this grouping. Davis (1961) treated the Triatominae as part of the Peiratine complex, an assemblage of 11 subfamilies that he diagnosed by the presence of a metacoxal comb. Lent and Wygodzinsky (1979) suspected a close relation between Triatominae and Physoderinae based on the straight labium found in both groups. Clayton (1990) was the first to propose a sister group relation between Triatominae and Stenopodainae, as recovered in some of her cladistic analyses. The more inclusive morphological analysis by Weirauch (2008) modified this hypothesis by including in this clade some taxa representing the polyphyletic Reduviinae, namely *Opisthacidius*, *Centrogonus*, *Zelurus*, and *Neivacoris*. Triatominae, Stenopodainae, and part of the Reduviinae, typically only represented by species of *Opisthacidius* and *Zelurus*, form a well-supported clade in almost all subsequent molecular analyses (Fig. 1). A recent ML analysis of complete mitochondrial genome data recovered *Triatoma* (*Triatoma dimidiata* + *Triatoma infestans*) as sister to Stenopodainae (represented by *Oncocephalus breviscutum*) in a well-supported clade; this analysis, however, lacks representatives of the Reduviinae (Pita et al., 2017).

In summary, the majority of morphological and molecular analyses refute the notion that Triatominae may represent a polyphyletic assemblage. Current evidence tentatively points towards the monophyly of the Triatominae, but the subfamily is rendered paraphyletic by a clade of predatory Reduviinae in two analyses (Fig. 1). A more definitive assessment will only be possible with dramatically improved taxon and character sampling. Future analyses should include representatives of the three small and somewhat enigmatic triatomine tribes Alberproseniini, Bolboderini, and Cavernicolini that are lacking in all (Alberproseniini) or several (Bolboderini and Cavernicolini) phylogenetic analyses published to date; additional predatory Reduviidae that may be closely related to the Triatominae (e.g. *Zeluroides* and *Gnistus*; Hwang and Weirauch, 2012) should also be included. Past molecular analyses have relied on a handful of frequently used genes, and morphological analyses on a relatively small number of characters; combined molecular and morphological analyses for predatory and haematophagous Reduviidae are unavailable. The generation and analysis of genome-level data (e.g. genomes, transcriptomes, or targeted exon capture) have become feasible and affordable for taxon-rich datasets and have the potential to resolve difficult nodes, such as those linking the Triatominae and their predatory close relatives.

2.2 Putative Synapomorphies of the Triatominae

The list of putative synapomorphies for the Triatominae is surprisingly short, given the decades-long interest in studying the morphology of the group. Lent and Wygodzinsky (1979) proposed the dorsal flexibility of the labium, haematophagy, and the loss of dorsal abdominal glands as synapomorphies. Clayton (1990) specified that the dorsal flexibility of the labium may be due to an enlarged membrane between the second and third visible labial segments. Weirauch (2008), based on microdissections of the labium and analyses of muscle origins and insertions, found that this flexibility is likely due to the more dorsal insertion of the head muscle 7 apodeme compared to predatory Reduviidae. Based on her cladistic analysis, Weirauch (2008) also found several additional putative synapomorphies, including a transversely subdivided labrum, the very small mandibular plate, lateral insertion of the antenna, straight labium, the unique structure and armature of the apex of the mandibles and maxillae, the fine structure and orientation of the tenant hairs that form the fossula spongiosa, and dermal glands with uniquely ornamented pores. Future examination of these characters in Alberproseniini, Bolboderini, and Cavernicolini, which were not included in that analysis, will provide a more rigorous assessment of this putative list of synapomorphies.

2.3 Evolution of Haematophagy in the Triatominae

Identifying ecological, morphological, and molecular modifications involved in the transition from a predatory to a haematophagous lifestyle in triatomine bugs is an intriguing endeavour. The lack of robust and densely sampled phylogenies currently impedes meaningful tests of any hypotheses regarding these issues. What follows is a selection of topics that will benefit from a better understanding of phylogenetic relations as well as more extensive ecological, morphological, and molecular data for the Triatominae and closely related predatory Reduviidae.

Opisthacidius was supported as the sister clade to Triatominae in the analyses by Weirauch (2008), Patterson and Gaunt (2010), and Justi et al. (2016) (Fig. 1). Should this hypothesis be correct, it would provide empirical evidence for at least part of a scenario first proposed by Schofield (1988). Schofield (1988) envisioned that the transition from obligate predatory lifestyle to haematophagy may have been a stepwise process that involved predatory assassin bugs seeking shelter in vertebrate nests an then feeding opportunistically on nest-dwelling arthropods; this was followed by facultative and then obligate (or almost so) haematophagy. Although little is known about the natural history of *Opisthacidius*, one species was collected from a parrot nest where it was suspected to feed on nest-dwelling arthropods (Lent and Wygodzinsky, 1956). Along the same lines, Hwang and Weirauch (2012), using ancestral-state reconstruction, identified vertebrate nests as the likely ancestral microhabitat for both Triatominae and *Opisthacidius*. Despite the fact that haematophagy has evolved multiple times independently in the Heteroptera (Schuh and Slater, 1995), and the observation that certain Triatominae species may survive on insect haemolymph (Lent and Wygodzinsky, 1979) while others engage in cleptohaematophagy (Sandoval et al., 2000), it remains questionable if transitions between predatory and haematophagous lifestyles are indeed as simple as was suggested by Schofield (2000).

Morphological transitions that seemingly coincided with the evolution of haematophagy in Triatominae may or may not represent adaptations to blood-feeding and associated behavioural changes. The dorsal shift of the insertion of the muscle 7 apodeme in Rhodniini and Triatomini (Weirauch, 2008) is likely the morphological basis for the greater flexibility (compared to predatory Reduviidae) of the last labial segment, which may facilitate the kissing bugs' stealthy feeding on vertebrates. The mandibular and maxillary stylets in Rhodniini and Triatomini are strongly modified compared to predatory Reduviidae (Weirauch, 2008). Modifications include one longitudinal row of hooks on the lateral surface of the mandible, the lack of ventral row of processes in the right maxillary stylet, and the right and left stylets together forming a structure that was described as a 'valve' by Cobben (1978). Together with modifications of salivary compounds injected before and during feeding (Ribeiro et al., 2012), the stylet structure may facilitate the largely painless bites of the Triatominae. In addition, the structure of the salivary glands is relatively uniform across predatory Reduviidae, but highly modified in Triatominae (Haridass and Ananthakrishnan, 1981; Lacombe, 1999; Louis and Kumar, 1973), suggesting that both structure and function of salivary glands have undergone a transition. This is in agreement with the discovery of several lineage-specific expansions in gene families related to blood feeding in the *Rhodnius prolixus* genome (Mesquita et al., 2015). Lastly, the fossula spongiosa, a hairy attachment structure of legs thought to facilitate prey capture in predatory Reduviidae (Weirauch, 2007), was likely present in the last common ancestor of the Triatominae, Stenopodainae, and part of the Reduviinae (Zhang et al., 2016b). This structure is retained in some but not all Triatominae, is particularly prevalent among males, and is often absent in nymphs, indicating a possible change in function to assist in locomotion on smooth surfaces and mating behaviours (Abad-Franch et al., 2015; Weirauch, 2007).

Even though the 'how' of the transitions between predatory and haematophagous lifestyles within Reduviidae still largely remains in the dark, recent analyses have made significant contributions towards clarifying *when* this transition may have happened. Vastly different divergence times have been estimated for the evolution of the Triatominae from their predatory ancestors, ranging from 4 to 107 million years ago (Mya) (Gaunt and Miles, 2000; Hwang and Weirauch, 2012; Patterson and Gaunt, 2010; Schofield and Galvão, 2009). However, recent analyses using fossil-calibrated relaxed clock models have started to converge on estimates around 35 Mya to just above 40 Mya (Hwang and Weirauch, 2012; Ibarra-Cerdeña et al., 2014; Justi et al., 2016). These estimates coincide with those for the radiations of Neotropical mammals and birds (e.g. Bininda-Emonds et al., 2007) and ecosystem diversification in South America (e.g. Graham, 2011), and could hence explain the diversification of the early haematophagous lineages within the Reduviidae.

3. SYSTEMATICS OF THE TRIATOMINAE

3.1 Classical Taxonomy

Taxonomy is one of the earliest and most fundamental disciplines in the biological sciences. It is traditionally dedicated to the study of morphological characters upon which biodiversity is described, named, and hierarchically classified (Godfray, 2002). Morphological studies of extant and extinct organisms have been carried out for centuries, contributing to the accumulation of the vast data record underlying classifications. The formal hierarchical system used to classify this diversity was established with the publication by Linnaeus of *Species Plantarum* (1753) and *Systema Naturae* (1758). There are, today, international nomenclature commissions and codes that attempt to ensure that each taxon has a single, exclusive, and valid accepted name so that it can be unmistakably linked to any kind of biological information (Kuntner and Agnarsson, 2006).

Early classifications typically relied on the observed degree of morphological similarity (and difference) among organisms, without reference to any phylogenetic framework. It is now clear, however, that morphologically more similar species are not necessarily more closely related. Explicit phylogenetic procedures are therefore needed to single out natural, monophyletic groups upon which to build realistic classification schemes (Kuntner and Agnarsson, 2006). Traditional morphology-based taxonomy has recently evolved to incorporate new types of data (e.g. molecular, ecological, biogeographic, or behavioural), giving rise to what has become known as 'integrative taxonomy'. With this approach, taxonomists search for congruence among different datasets to ultimately generate a formal description of a new taxon (Dayrat, 2005).

3.1.1 Overview: Triatomine Systematics From De Geer to the 21st Century

The subfamily Triatominae includes 150 extant and 2 extinct, formally named species classified in 16 genera and 5 tribes. Triatomines occur mainly in the Americas; the exceptions are eight Old World *Triatoma* species, including the cosmopolitan *Triatoma rubrofasciata*, and the Indian genus *Linshcosteus* (see Section 4). De Geer (1773) described the first triatomine species known to science as *Cimex rubrofasciatus*. In 1833, Laporte created the genus *Triatoma* with *T. rubrofasciata* as the type species (Galvão and de Paula, 2014; Laporte, 1832–1833). Latreille, Stål, Klug, Burmeister, Champion, Berg, Erichson, and Walker, among others, also published important early studies including the description of several genera and species; by 1907, the subfamily comprised 59 named species (Lent and Wygodzinsky, 1979). The discovery of Chagas disease, and in particular the identification of triatomines as the vectors of *T. cruzi* (Chagas, 1909), provided a strong incentive for taxonomic research, and Neiva, Pinto, Del Ponte, Usinger, Martínez, Barber, Larrousse, or Lent contributed many new taxa in the 20th century (Lent and Wygodzinsky, 1979). Herman Lent described 31 currently valid species and advanced Triatominae taxonomy more than any other author (Galvão et al., 2003); together with Pedro Wygodzinsky, he coauthored the seminal taxonomic revision of this group published in 1979 (see Section 3.1.2).

Despite all these advances, recognizing and describing triatomine species based on morphology alone was not always straightforward. Phenotypic variability, which is common in many triatomine species, may affect traits often used in classical taxonomy including colour patterns or the size and shape of bodies, heads, wings, and genital structures (Schofield and Galvão, 2009). Phenotypic plasticity in response to varying environmental conditions may partly underlie such morphological diversity (Dujardin et al., 1999a, 2009; see also Nattero et al., 2013; Nouvellet et al., 2011). Several triatomine species were described after phenotypic variation that turned out to be intraspecific and were later synonymized (e.g. Galvão et al., 2003; Garcia et al., 2005; Lent and Wygodzinsky, 1979; Patterson et al., 2009). In addition, the monophyly of taxa at the genus level and above remained untested. Some triatomine genera indeed differ in just a few morphological characters whose evolutionary significance is at best obscure—notably overall head shape in *Triatoma* vs *Panstrongylus* or *Rhodnius* vs *Psammolestes* (Lent and Wygodzinsky, 1979; Patterson, 2007).

The first molecular phylogenetic studies of Triatominae appeared in the late 1990s (e.g. Dujardin et al., 1999b; García and Powell, 1998; Stothard et al., 1998). Molecular systematics soon revealed that some phenotype-defined species might conceal genetically distinct taxa (e.g. domestic vs sylvatic Colombian *R.* '*prolixus*' or some *Triatoma* '*sordida*' populations), that morphologically similar species may be only distantly related (e.g. *T. sordida* and *Triatoma guasayana*), and that, at the same time, closely related species may sharply differ in phenotype (e.g. *T. guasayana* and *Triatoma rubrovaria*) (Dujardin et al., 1999b; García and Powell, 1998; Noireau et al., 1998; see also Section 3.3). At the genus level, Lyman et al. (1999) suggested that *Rhodnius* is paraphyletic with respect to *Psammolestes*, a hypothesis that subsequently received strong support (e.g. de Paula et al., 2007; Hypša et al., 2002; Monteiro et al., 2000). Similarly, DNA sequence data suggest that *Triatoma* is most likely paraphyletic with respect to several other genera including *Panstrongylus*, *Eratyrus*, *Dipetalogaster*, or *Linshcosteus* (e.g. Hwang and Weirauch, 2012; Hypša et al., 2002; Ibarra-Cerdeña et al., 2014; Justi et al., 2016; Marcilla et al., 2002).

The difficulties inherent to classical, phenotype-based triatomine taxonomy led Bargues et al. (2010) to recommend that species status should be granted to a putative new taxon only when genetic information agrees with morphological variation. This, however, does not solve the problem posed by genetically distinct species that nevertheless display very similar phenotypes—the so-called cryptic or sibling species. Integrative taxonomy aims at handling these difficulties by combining standard comparative morphology with information drawn from the diverse fields of molecular genetics, biogeography, phylogeography, behaviour, ecology, or development (Dayrat, 2005; Jörger and Schrödl, 2013).

3.1.2 Lent and Wygodzinsky: Setting the Standard

The revision of the Triatominae by Lent and Wygodzinsky (1979) is, to date, the most authoritative work published on the natural history, biogeography, morphology, and systematics of this group. It covers 111 species (including 6 new species in 5 genera) and provides identification keys for adults (down to species) and first- and fifth-instar nymphs (to genus). Many taxonomic issues that had accrued over 2 centuries were addressed, discussed, and resolved in this monograph—which was also the first to present phylogenetic cladograms hypothesizing relations among subgroups of Triatominae. Using the methods outlined by Hennig (1966), Lent and Wygodzinsky (1979) proposed putative monophyletic groups defined by their shared and derived traits. One cladogram focused on species of *Panstrongylus* based on 21 adult morphological characters; still, because outgroups were not specified, it is unclear how those characters were polarized. The second cladogram explored relations among the four genera in the tribe Bolboderini, based on 18 characters and with plesiomorphic character states inferred based on 'nonspecialized conditions found in other triatomines and nontriatomine reduviids' (Lent and Wygodzinsky, 1979, p. 437). These phylogenies are the first application of cladistic methods to triatomine systematics; although Usinger (1944) used what he called a 'phylogenetic key' to depict hypothetical relations among tribes and genera, these were only illustrative diagrams of unpolarized characters.

Usinger (1944) was also the first to suggest subgeneric groups for the North American and Mesoamerican *Triatoma*. Lent and Wygodzinsky (1979) extended this approach to the whole genus and (using characters of nymphs and adults) proposed a hierarchy of species groups, subgroups, and complexes (Table 1). They stressed, however, that this 'arrangement' was 'not intended to express cladistic relationships' and did not translate 'into a taxonomic scheme' (Lent and Wygodzinsky, 1979, p. 183). Thirty years later, Schofield and Galvão (2009) rearranged *Triatoma* into three groups and eight complexes—two of which, Phyllosoma and Infestans, were further divided into subcomplexes (Schofield and Galvão, 2009) (Table 2). Barrett (1991, p. 147) mentioned the '*R. prolixus* complex' within *Rhodnius*, implying that it encompasses *R. prolixus* itself plus the morphologically similar *Rhodnius dalessandroi*, *Rhodnius robustus*, *Rhodnius neglectus*, and *Rhodnius nasutus* (see also Schofield and Dujardin, 1999). Carcavallo et al. (2000) proposed 10 *Triatoma*, 1 *Panstrongylus*, and 4 *Rhodnius* complexes.

These arrangements all aimed at refining or extending Lent and Wygodzinsky's (1979) proposal, and each expressed, at least implicitly, a view of the patterns of common descent within clusters of closely related species. As such, these species groups and complexes have supplied useful hypotheses about triatomine systematics and evolution (see Sections 3.2

Table 1 Lent and Wygodzinsky (1979) Groups, Subgroups, and Complexes in the Genus *Triatoma (Comments Extracted From Lent and Wygodzinsky (1979))*

Group	Subgroup	Complex	Species
Triatoma protracta (species not necessarily closely related)		*protracta* (seems monophyletic)	*T. protracta*, *T. barberi*, *T. incrassata*, *T. peninsularis*, *T. sinaloensis*, *T. neotomae* (with doubts), *T. nitida* (with doubts)
		lecticularia (no evidence of monophyly)	*T. lecticularia*, *T. sanguisuga*, *T. indictiva* (close to *T. sanguisuga*)
Triatoma rubrofasciata (majority of *Triatoma*)	*infestans*	*circummaculata*	*T. circummaculata* (*circummaculata* appears also in the *infestans* complex; probably a typo), *T. limai*
		infestans	*T. infestans*, *T. arthurneivai*, *T. brasiliensis*, *T. costalimai* (likely), *T. delpontei*, *T. guasayana*, *T. lenti*, *T. maculata*, *T. matogrossensis*, *T. melanocephala* (likely), *T. patagonica*, *T. petrocchiae* (likely), *T. platensis*, *T. pseudomaculata*, *T. rubrovaria*, *T. sordida*, *T. tibiamaculata*, *T. vitticeps*, *T. williami*, *T. wygodzinskyi*
		dispar (assigned with 'less than complete confidence')	*T. dispar*, *T. carrioni*, *T. venosa*
	rubrofasciata	*rubrofasciata*	*T. rubrofasciata*, *T. amicitiae* (likely), *T. bouvieri* (likely), *T. leopoldi* (likely), *T. migrans* (likely), *T. pugasi* (likely), *T. sinica* (likely), *T. cavernicola*
		phyllosoma	*T. phyllosoma*, *T. longipennis*, *T. mazzottii*, *T. pallidipennis*, *T. picturata*, *T. mexicana* (likely), *T. dimidiata* (likely)
		spinolai	'*T*'. *spinolai* (now *Mepraia*), *T. eratyrusiformis*, *T. breyeri*
		flavida	*T. flavida*, *T. obscura*
		recurva	*T. recurva*, *T. gerstaeckeri*
		nigromaculata	*T. nigromaculata* ('very similar chromatically' to *T. dispar*)
Unassigned			*T. rubida* (not explicitly assigned; implicitly presented as close to *lecticularia*), *T. oliveirai* (may belong in the *infestans* complex), *T. guazu* (may belong in the *infestans* complex), *T. ryckmani*, *T. hegneri*

Table 2 Schofield and Galvão (2009) Groups, Complexes, and Subcomplexes in the Genus *Triatoma*

Group	Complex	Subcomplex	Species
Rubrofasciata (mainly North American and Old World)	Phyllosoma	Dimidiata	*Triatoma dimidiata*, *T. hegneri*, *T. brailovskyi*, *T. gomeznunezi*
		Phyllosoma (=*Meccus*)	*T. bassolsae*, *T. bolivari*, *T. longipennis*, *T. mazzottii*, *T. mexicana*, *T. pallidipennis*, *T. phyllosoma*, *T. picturata*, *T. ryckmani*
	Flavida (=*Nesotriatoma*)		*T. flavida*, *T. bruneri*, *T. obscura*
	Rubrofasciata		*T. amicitiae*, *T. bouvieri*, *T. cavernicola*, *T. leopoldi*, *T. migrans*, *T. pugasi*, *T. rubrofasciata*, *T. sinica*
	Protracta		*T. barberi*, *T. incrassata*, *T. neotomae*, *T. nitida*, *T. peninsularis*, *T. protracta*, *T. sinaloensis*
	Lecticularia		*T. gerstaeckeri*, *T. indictiva*, *T. lecticularia*, *T. recurva*, *T. rubida*, *T. sanguisuga*
Dispar (Andean)	Dispar		*T. boliviana*, *T. carrioni*, *T. dispar*, *T. nigromaculata*, *T. venosa*
Infestans (South American)	Infestans	Brasiliensis	*T. brasiliensis*, *T. juazeirensis*, *T. melanica*, *T. melanocephala*, *T. petrocchiae*, *T. lenti*, *T. sherlocki*, *T. tibiamaculata* (with doubts), *T. vitticeps* (with doubts)
		Infestans	*T. delpontei*, *T. infestans*, *T. platensis*
		Maculata	*T. arthurneivai*, *T. maculata*, *T. pseudomaculata*, *T. wygodzinskyi*
		Matogrossensis	*T. baratai*, *T. costalimai*, *T. deaneorum*, *T. guazu*, *T. jurbergi*, *T. matogrossensis*, *T. vandae*, *T. williami*
		Rubrovaria	*T. carcavalloi*, *T. circummaculata*, *T. klugi*, *T. limai*, *T. oliveirai*, *T. rubrovaria*
		Sordida	*T. garciabesi*, *T. guasayana*, *T. patagonica*, *T. sordida*
	Spinolai (=*Mepraia*)		*T. breyeri*, *T. eratyrusiformis*, *M. spinolai*, *M. gajardoi*

and 3.3). However, and as mentioned above, it has become progressively clear that the strong (or exclusive) reliance on morphological similarity and difference often led to invalid groupings. Lent and Wygodzinsky (1979) already emphasized the similarity of some *Rhodnius* (particularly *R. prolixus*, *R. robustus*, *R. neglectus*, and *R. nasutus*) and *Triatoma* species (e.g. *Triatoma peninsularis* and *Triatoma sinaloensis* or *Triatoma maculata* and *Triatoma pseudomaculata*), described colour variants in several taxa (e.g. *T. rubrofasciata*, *Triatoma rubida*, *T. dimidiata*, *Triatoma picturata*, *Triatoma brasiliensis*, *T. rubrovaria*, *T. sordida*, or *Panstrongylus geniculatus*), and paid special attention to the biosystematics of polytypic species (notably *T. protracta*; see also Ryckman, 1962; Usinger et al., 1966). Remarkably, Lent and Wygodzinsky (1979) also illustrate a striking instance of within-species phenotype divergence associated with adaptation to a unique microhabitat: cave-dwelling *T. dimidiata* differ in colour and head shape (including very small eyes and ocelli) from their already variable 'typical' conspecifics.

Building upon the foundations laid by pioneers such as Usinger, Ryckman, Mazzotti, Pellegrino, or Ábalos (see, e.g. Usinger et al., 1966), Lent and Wygodzinsky (1979) were therefore also anticipating the potential limitations of morphology as the sole (or main) tool for investigating the systematics and evolution of the Triatominae. They already argued that 'genetical evidence' was needed to resolve the relations within some problematic groups—e.g., the *phyllosoma* or *protracta* complexes. Such evidence was then represented by mating experiments measuring interfertility and, for more distantly related taxa, by cytogenetics (Lent and Wygodzinsky, 1979; Usinger et al., 1966). Although some early studies had yielded evidence of molecular differentiation among triatomine taxa (Adams and Ryckman, 1969; Benoit and van Sande, 1959; Brodie and Ryckman, 1967; van Sande and Karcher, 1960), it was not until the 1980s that molecular systematics and evolutionary genetics started to have an impact on triatomine research (Abad-Franch and Monteiro, 2005; Monteiro et al., 2001).

3.2 Molecular Systematics

3.2.1 The Pioneers From Allozymes to DNA

Important advances have been made since the publication of the first review article on triatomine molecular systematics 17 years ago (Monteiro et al., 2001). DNA sequence analysis was just beginning to be tested as a complementary methodology to address taxonomic issues in this group of vectors. The first papers to report DNA sequence variation in Triatominae were only published in 1998. In fact, DNA-based studies (using randomly amplified polymorphic DNA, single-strand conformational polymorphism, or sequencing) accounted for only 20% of all published work until 2001. This is because, at that time, allozymes were the method of choice in triatomine molecular systematics (Monteiro et al., 2001).

The earliest work on allozymes examined variation in *T. infestans* (Dujardin and Tibayrenc, 1985a, b; Dujardin et al., 1987; Tibayrenc, 1980) and *Rhodnius* spp. (e.g. Chávez et al., 1999; Dujardin et al., 1991; Harry, 1993; Harry et al., 1992; López and Moreno, 1995; Soares et al., 1999; Solano et al., 1996; see also Dujardin et al., 1999b; Monteiro et al., 2002). Other studies investigated, e.g., several Mexican *Triatoma* (Flores et al., 2001), *T. sordida* (García et al., 1995; Monteiro et al., 2009; Noireau et al., 1998; Panzera et al., 1997), or *T. brasiliensis* (Costa et al., 1997) and the closely related *Triatoma petrocchiae* (Monteiro et al., 1998). (See Monteiro et al. (2001) and Abad-Franch and Monteiro (2005) for a more detailed discussion of allozyme studies.)

The pioneers of mitochondrial DNA (mtDNA) sequence-based triatomine taxonomy and phylogeny were García and Powell (1998) and Stothard et al. (1998). García and Powell (1998) investigated the phylogenetic relations of eight *Triatoma* species based on three mtDNA markers (12S, 16S, and COI). Stothard et al. (1998) used a 400-bp fragment of the 16S gene to infer relations among nine species of *Triatoma*, *Panstrongylus*, and *Rhodnius*. Subsequently, Lyman et al. (1999) used two mtDNA loci (16S and cytochrome *b* [cyt *b*]) to study representatives of five triatomine genera. They found a faster evolutionary pace of the protein-coding cyt *b*; unexpectedly, *Psammolestes coreodes* was nested within *Rhodnius*, and both *Dipetalogaster maxima* and *Panstrongylus megistus* nested within *Triatoma*. The following year, Monteiro et al. (2000) published the first study that included both mtDNA and nuclear DNA sequence data, with tree topologies inferred through neighbour-joining (NJ) and MP. Focusing on the phylogeny of the Rhodniini, this report confirmed the paraphyly of the genus *Rhodnius* with respect to *Psammolestes* and validated *R. robustus* as a distinct (albeit very heterogeneous) species. Marcilla et al. (2001) were the first to use the second internal transcribed spacer of the rDNA (ITS-2) to assess relations among triatomines. The results of NJ, MP, and ML analyses supported an old divergence between South American and North-Central American Triatomini; *Dipetalogaster* and *Psammolestes* fell, respectively, within the *Triatoma* and *Rhodnius* clades, as previously observed with mtDNA markers. Hypša et al. (2002) published another important work that included the most extensive taxon sampling at the time. Phylogenetic analyses (NJ and MP) based on

12S and 16S mtDNA sequence data suggested a monophyletic origin of the Triatominae and paraphyly of *Triatoma* with respect to *Linshcosteus*, *Dipetalogaster*, *Eratyrus*, and *Panstrongylus*, as well as of *Rhodnius* with respect to *Psammolestes*.

On a finer systematic scale, Monteiro et al. (2003) tackled the taxonomic controversy involving *R. prolixus*, a major disease vector, and the near sibling but epidemiologically less relevant *R. robustus*. Based on NJ and MP analyses of mtDNA (cyt *b*) and nuclear rDNA (28S D2 region) sequences, this study revealed that both belong to a complex of several cryptic species (see Section 3.3.1). García et al. (2003) were the first to assess, using 12S and 16S mtDNA sequences, intraspecific variation and gene flow among natural *T. infestans* populations from Argentina. After pioneering research by Harry et al. (1998, 2008) on nuclear microsatellite loci in *Rhodnius* spp., Fitzpatrick et al. (2008) investigated the genetic structure of palm- and house-dwelling populations of *R. prolixus* in Venezuela. Both microsatellites and cyt *b* data suggested that sylvatic *R. prolixus* can colonize in houses, and hence that the long-term control of Chagas disease in Venezuela, and likely across the Orinoco basin, will depend on continuous surveillance to detect and eliminate new house infestations by wild *R. prolixus*. Further analyses of the mtDNA marker detected introgression between *R. robustus* and *R. prolixus*. Other early studies on microsatellite loci variation in triatomines targeted the major human disease vectors *T. dimidiata* (Anderson et al., 2002; Dumonteil et al., 2007) and *T. infestans* (García et al., 2004; Pérez de Rosas et al., 2007).

3.2.2 Genomics, Transcriptomics, and Other New Tools

Dotson and Beard (2001) published the first complete mitochondrial genome from a triatomine bug, *T. dimidiata*. Recently, Pita et al. (2017) sequenced the mitochondrial genome of *T. infestans*. The first (and so far only) complete nuclear genome of a triatomine, *R. prolixus*, was published by Mesquita et al. (2015). The *R. prolixus* genome resulted from the coordinated effort of scientists from 10 countries, mostly in the Americas. It revealed large and unique expansions of gene families involved in chemoreception, feeding, and digestion—which may have facilitated the bug's adaptation to its blood-feeding lifestyle. In addition, a peculiar immune signalling network and unchanged replication of *T. cruzi* in bugs in which key immune genes were experimentally silenced suggested that the parasite might have evolved evasion or tolerance mechanisms (Mesquita et al., 2015).

Transcriptomic approaches have recently begun to be applied to triatomines including *T. brasiliensis* (Marchant et al., 2015; Santos et al., 2007),

T. dimidiata (Kato et al., 2010), *R. prolixus* (Ribeiro et al., 2014), and *T. infestans* (Gonçalves et al., 2017; Traverso et al., 2016). Zhang et al. (2016b) were the first to use transcriptome data to investigate phylogenetic relationships of the Reduviidae including Triatominae. At a much finer physiological scale, the development of RNA interference (RNAi) technology is allowing for the dissection of individual gene function in many organisms including triatomines (Araujo et al., 2006, 2009; Paim et al., 2013); proof-of-concept research has shown, in addition, that RNAi might find practical application in novel strategies for the control of Chagas disease vectors (Taracena et al., 2015).

3.3 Uncovering and Sorting Out Hidden Diversity: Species Complexes

Here we use the term 'species' to describe any population that broadly satisfies the requirements of the biological species concept (BSC; Mayr, 1963) or the phylogenetic species concept (PSC; Cracraft, 1989; see also de Queiroz, 2007). The BSC states that species are groups of actually or potentially interbreeding individuals, with boundaries defined by intrinsic barriers to gene flow that have a genetic basis (Mayr, 1963). Under the PSC, a species is an irreducible cluster of organisms, diagnosably distinct from other such clusters, and within which there is a parental pattern of ancestry and descent (Cracraft, 1989). The term 'species complex' is commonly used in the triatomine systematics literature, but may have different meanings depending on the context. It has traditionally been used to designate subgeneric assemblages defined by morphological similarity (Lent and Wygodzinsky, 1979; Schofield and Galvão, 2009; Usinger, 1944). These groups were proposed at a time when there were no (or only very few) phylogenetic hypotheses postulated for lower taxonomic ranks (see Tables 1 and 2). We believe that those early similarity-based groups now need to be reevaluated and either corroborated or modified in the light of new information provided by molecular phylogenies. Here, we will use the term 'species complex' to identify any group of closely related, morphologically similar or even indistinguishable species usually disclosed as a result of molecular investigations (Table 3).

3.3.1 The R. prolixus + R. robustus *Complex*

Monteiro et al. (2003) presented a comprehensive phylogeographic study of the closely related species *R. prolixus* and *R. robustus*, which are almost sibling (Harry, 1993, 1994) and yield identical allozyme banding patterns

Table 3 Lineages, Clades, Species Groups, and Complexes in the Triatomini

Lineage	Clade (Broad Biogeographic Correspondence)	Species Group	Complex	Species
Triatoma dispar	*T. dispar* (Mesoamerican-Andes ranges)	*T. dispar*		*T. dispar*, *T. venosa*, ***T. nigromaculata***, ***T. carrioni***, ***T. boliviana***
'North American'	*Hermanlentia–Mepraia* (central-southern Andes)			*Hermanlentia matsunoi*
		Mepraia	*M. spinolai*	*M. spinolai*, *M. gajardoi*, *M. parapatrica*, *T. eratyrusiformis*, *T. breyeri* I, *T. breyeri* II
	North American (north of Tehuantepec)	*T. protracta*		*T. protracta*, *T. barberi*, *T. lecticularia*, *Paratriatoma hirsuta*, *Dipetalogaster maxima*
		T. nitida		*T. nitida*, *T. rubida*, *T. ryckmani*, *T. bolivari*
		(Uncertain)		***T. neotomae***, ***T. incrassata***, ***T. indictiva***, ***T. peninsularis***, ***T. sinaloensis***
	North-Mesoamerican + Old World	*T. phyllosoma*	*T. dimidiata*	*T. dimidiata* I, *T. dimidiata* II, *T. dimidiata* III, *T. dimidiata* IV, *T. hegneri*
			T. phyllosoma	***T. phyllosoma***, ***T. bassolsae***, ***T. brailovskyi***, ***T. gomeznunezi***, ***T. longipennis***, ***T. mazzottii***, ***T. pallidipennis***, ***T. picturata***
			(Uncertain)	***T. gerstaeckeri***, ***T. mexicana***, ***T. recurva***, ***T. sanguisuga***
		Linshcosteus		*L. carnifex*, *L. confumus*, *L. costalis*, *L. kali*, *L. karupus*, *L. chota*
		T. rubrofasciata		*T. rubrofasciata*, *T. amicitiae*, ***T. bouvieri***, *T. cavernicola*, *T. leopoldi/novaeguineae*, ***T. migrans***, *T. pugasi*, *T. sinica*

	Antillean *Triatoma* + *Panstrongylus*	'*Nesotriatoma*'		*T. flavida*, *T. bruneri*, *T. obscura*
		Panstrongylus geniculatus	*P. geniculatus*	***P. geniculatus***, ***P. martinezorum***, ***P. mitarakaensis***, ***P. diasi***, ***P. guentheri***, ***P. lenti***, ***P. lutzi/sherlocki***, ***P. tupynambai***
			(Uncertain)	*P. chinai*, *P. howardi*
		P. lignarius		*P. lignarius/herreri*, *P. humeralis*
		P. megistus		*P. megistus*, *T. tibiamaculata*
		(Uncertain)		***P. rufotuberculatus***
South American	Atlantic forest *Triatoma*	*T. vitticeps*		*T. vitticeps*, *T. melanocephala*
	Eratyrus (Andes-Amazon)	*Eratyrus*		*E. cuspidatus*, *E. mucronatus*
	Triatoma maculata (trans-Amazonian vicariant)	*T. maculata*		*T. maculata*
	Triatoma infestans (Chaco-Andes)	*T. infestans*		*T. infestans/melanosoma*, *T. delpontei*, *T. platensis*
	Triatoma brasiliensis (Caatinga)	*T. brasiliensis*	*T. brasiliensis*	*T. brasiliensis/macromelasoma*, *T. juazeirensis*, *T. melanica*, *T. sherlocki*, *T. lenti*, *T. bahiensis*, *T. petrocchiae*
	Triatoma sordida (open ecoregions: Cerrado-Caatinga-Pantanal-Chaco-Pampas)	*T. rubrovaria*	*T. rubrovaria*	*T. rubrovaria*, *T. circummaculata*, *T. carcavalloi*, *T. oliveirai*, *T. pintodiasi*, *T. klugi*, *T. patagonica*, *T. limai*, *T. guasayana*
		T. pseudomaculata	*T. pseudomaculata*	*T. pseudomaculata*, *T. arthurneivai*, *T. wygodzinskyi*, *T. deaneorum*, *T. williami*, *T. baratai*, *T. guazu*
			(Uncertain)	***T. costalimai***, ***T. jatai***
		T. sordida	*T. sordida*	*T. sordida* 1, *T. sordida* 2, *T. sordida/garciabesi*, *T. jurbergi*, *T. matogrossensis*, *T. vandae*

'Species complexes' are sets of closely related species whose relations are difficult to establish based on morphology and were usually resolved using molecular approaches. Species in **bold** are those whose relations we consider worth investigating.

(Harry et al., 1992; Monteiro et al., 2002). Monteiro et al. (2003) analysed mtDNA cyt *b* fragments from 26 populations sampled in Guatemala, Honduras, Colombia, Ecuador, Venezuela, French Guiana, and Brazil. Phylogenetic analyses clustered *R. prolixus* in a very homogeneous northern subclade including samples from the Orinoco basin (Venezuela) plus domestic bugs from Colombia and Central America. In contrast, *R. robustus s.l.* showed strong geographical structuring with two main subclades—one including wild bugs from the Orinoco basin in Venezuela (named *R. robustus* I) and another encompassing three reciprocally monophyletic, largely parapatric Amazonian clusters (*R. robustus* II–IV).

Surprisingly, *R. robustus* I was genetically closer to *R. prolixus* than to *R. robustus* from the Amazon region; divergence was low within subclades (Kimura 2-parameter distance [K2P] < 0.015) but large between *R. prolixus* and sympatric *R. robustus* I (K2P 0.030–0.033) and even larger when these were compared with the Amazonian subclades (K2P 0.056–0.085) (Monteiro et al., 2003). Fitzpatrick et al. (2008) recovered the same topology using cyt *b* sequences from 551 bugs sampled in palms, houses, and chicken coops of the Orinoco and Amazon basins. A further cyt *b* genotype, named *R. robustus* V, was discovered in the central–northern Amazon and shown to be closely related to the *R. prolixus* + *R. robustus* I subclade (Abad-Franch and Monteiro, 2007; see Fig. 2). The paraphyly of *R. robustus* with respect to *R. prolixus* was also supported by analysis of two nuclear markers (Monteiro et al., 2003; Pavan et al., 2013). These studies showed that *R. prolixus* and *R. robustus* are separate taxa, as previously suggested by Lyman et al. (1999) and Monteiro et al. (2000), but with the latter encompassing several divergent lineages (see also Sections 3.4.4 and 4.2.3). Pavan et al. (2016) reinforced this view of taxonomic separation by revealing that nymphs of *R. prolixus* and *R. robustus* II display different locomotor activity patterns.

3.3.2 The T. dimidiata *Complex*

T. dimidiata is a major Chagas disease vector from southern Mexico through Central America and in parts of Colombia, Venezuela, Ecuador, and Peru (Lent and Wygodzinsky, 1979). It shows substantial diversity in morphology, behaviour, and genetics throughout this extensive geographic range (Bargues et al., 2008; Bustamante et al., 2004; Dorn et al., 2007, 2009, 2016; Gómez-Palacio et al., 2013; Lent and Jurberg, 1985; Lent and Wygodzinsky, 1979; Monteiro et al., 2013; Panzera et al., 2006; Parra-Henao et al., 2016). The species was described in 1811 by Latreille as *Reduvius dimidiatus* and has since gone through a series of name changes

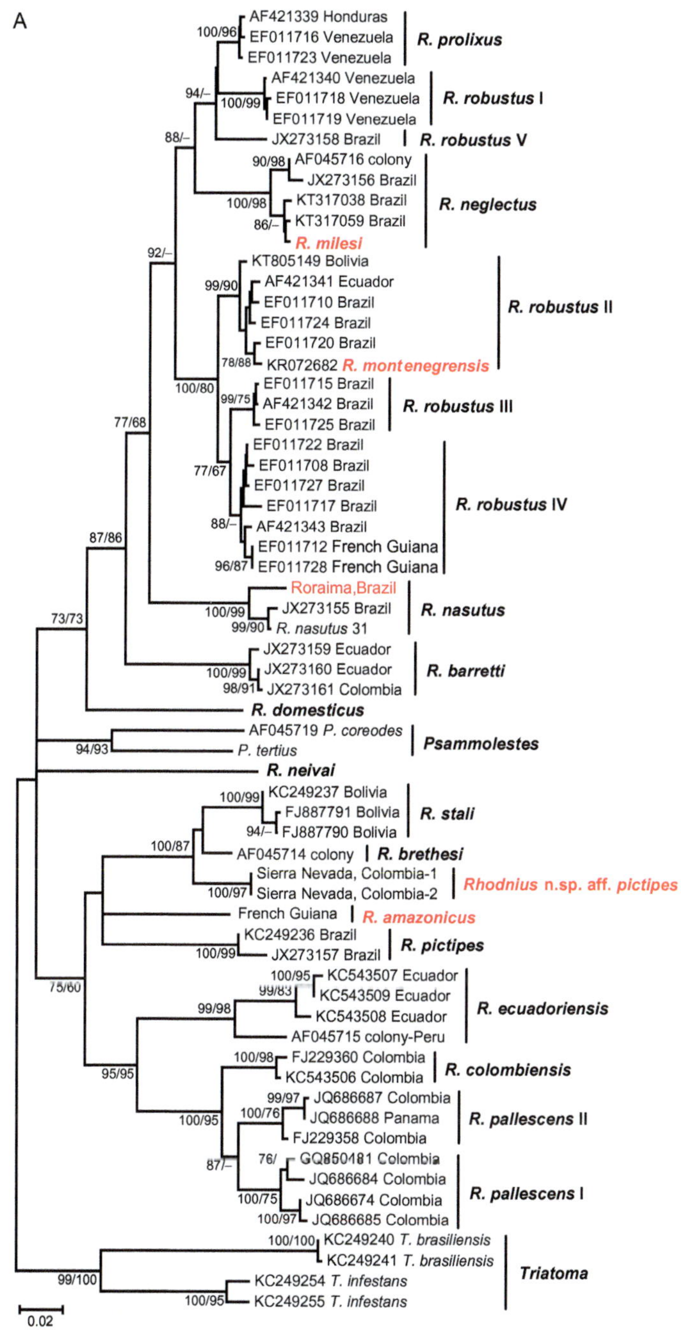
A
AF421339 Honduras
EF011716 Venezuela
EF011723 Venezuela
R. prolixus
AF421340 Venezuela
EF011718 Venezuela
EF011719 Venezuela
R. robustus I
JX273158 Brazil
R. robustus V
AF045716 colony
JX273156 Brazil
KT317038 Brazil
KT317059 Brazil
R. milesi
R. neglectus
KT805149 Bolivia
AF421341 Ecuador
EF011710 Brazil
EF011724 Brazil
EF011720 Brazil
KR072682 R. montenegrensis
R. robustus II
EF011715 Brazil
AF421342 Brazil
EF011725 Brazil
R. robustus III
EF011722 Brazil
EF011708 Brazil
EF011727 Brazil
EF011717 Brazil
AF421343 Brazil
EF011712 French Guiana
EF011728 French Guiana
R. robustus IV
Roraima,Brazil
JX273155 Brazil
R. nasutus 31
R. nasutus
JX273159 Ecuador
JX273160 Ecuador
JX273161 Colombia
R. barretti
R. domesticus
AF045719 P. coreodes
P. tertius
Psammolestes
R. neivai
KC249237 Bolivia
FJ887791 Bolivia
FJ887790 Bolivia
R. stali
AF045714 colony
R. brethesi
Sierra Nevada, Colombia-1
Sierra Nevada, Colombia-2
Rhodnius n.sp. aff. pictipes
French Guiana
R. amazonicus
KC249236 Brazil
JX273157 Brazil
R. pictipes
KC543507 Ecuador
KC543509 Ecuador
KC543508 Ecuador
AF045715 colony-Peru
R. ecuadoriensis
FJ229360 Colombia
KC543506 Colombia
R. colombiensis
JQ686687 Colombia
JQ686688 Panama
FJ229358 Colombia
R. pallescens II
GQ850481 Colombia
JQ686684 Colombia
JQ686674 Colombia
JQ686685 Colombia
R. pallescens I
KC249240 T. brasiliensis
KC249241 T. brasiliensis
KC249254 T. infestans
KC249255 T. infestans
Triatoma
100/96
94/–
100/99
88/–
90/98
100/98
86/–
92/–
99/90
78/88
100/80
99/75
77/68
77/67
88/–
87/86
96/87
100/99
99/90
73/73
100/99
98/91
94/93
100/99
94/–
100/87
100/97
100/99
75/60
100/95
99/83
99/98
100/98
95/95
99/97
100/76
100/95
87/–
76/
100/75
100/97
100/100
99/100
100/95
0.02
cyt b-NJ/ML

Fig. 2—Cont'd

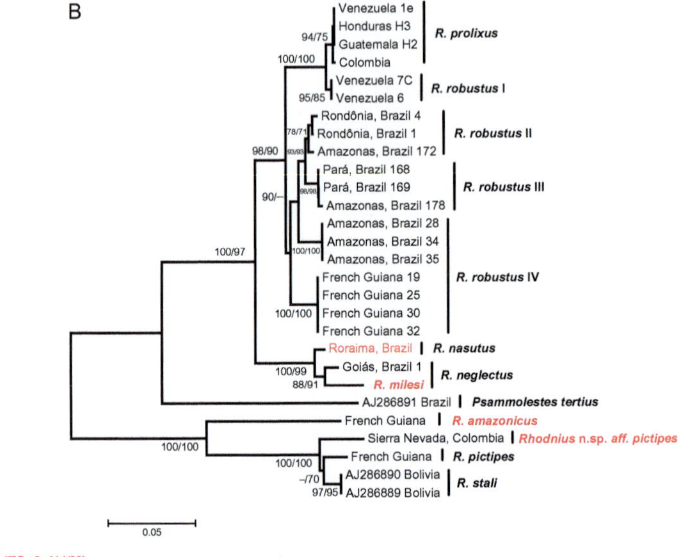

Fig. 2 Rhodniini mitochondrial cytochrome *b* (A) and nuclear ITS-2 (B) phylogenetic trees. *Numbers on branches* are bootstrap values above 70 (1000 pseudoreplicates) for the neighbour-joining (K2P) and maximum likelihood (HKY) methods, respectively. Important observations are highlighted in *red*: (1) the high genetic similarity between *R. milesi* and *R. neglectus*; (2) *R. montenegrensis* is part of the variability of *R. robustus* II; (3) a recently discovered form, sister to *R. nasutus*, from the Guianan savannahs of Roraima, Brazil; (4) a newly discovered form similar to *R. pictipes* from the Sierra Nevada de Santa Marta, Colombia; and (5) genetic confirmation of the distinctiveness of *R. amazonicus*.

and several events of taxon splitting and lumping—e.g., the description of three subspecies, *T. d. maculipennis*, *T. d. dimidiata*, and *T. d. capitata*, which Lent and Wygodzinsky (1979) synonymized back under *T. dimidiata* (see Dorn et al. (2007) for a detailed review).

Given the relevance of *T. dimidiata* as a disease vector, and the apparent variation of such relevance with geography, it is not surprising that this species or species complex has received substantial research attention. Recent examples include assessments of divergence based on genetics (Bargues et al., 2008; Dorn et al., 2007, 2009, 2016; Monteiro et al., 2013; Panzera et al., 2006), cuticular hydrocarbons (Calderón-Fernández et al., 2005, 2011), or ecological niche traits (Gómez-Palacio et al., 2015). Some studies were particularly helpful in clarifying the number of meaningful taxonomic units within *T. dimidiata*. The existence of cryptic species within the *T. dimidiata* complex was first reported by Marcilla et al. (2001) based on the large ITS-2 sequence divergence between bugs from Yucatán

(Mexico) and specimens from elsewhere in Mexico and from Central and South America. Subsequent studies based on cytogenetics and genome size (Panzera et al., 2006) and mtDNA (Dorn et al., 2009; Monteiro et al., 2013) also supported these results. Slightly smaller genetic differences among all other populations were interpreted as consistent with clinal variation along a north–south axis (Marcilla et al., 2001). This clinal-axis hypothesis was challenged by the comprehensive ITS-2 work of Bargues et al. (2008), who recovered four groups in two well-supported subclades—one comprising subgroups 1A and 1B plus group 2, and another comprising group 3. Bargues et al. (2008) suggested that their group 2 represents *T. d. maculipennis*, the darker northern form from Mexico and Guatemala, plus *Triatoma hegneri* from Cozumel Island; subgroup 1A represents *T. d. dimidiata* from Central America (plus introduced populations of coastal Ecuador); and subgroup 1B represents *T. d. capitata*, the southern forms from Panama, Colombia, and northern Venezuela. Finally, group 3 bugs, dubbed *T.* sp. *aff. dimidiata*, were identified with the cryptic species first discovered in Yucatán but that also occurs in Holbox and Cozumel islands, northern Chiapas, northern Honduras and Guatemala, and Belize (Bargues et al., 2008; Dorn et al., 2009).

Taking advantage of the better resolution of the mitochondrial markers (cyt *b* and ND4) and a comprehensive geographic sampling (126 specimens from 32 localities across *T. dimidiata*'s range), Monteiro et al. (2013) found five genetically well differentiated, monophyletic groups (named groups I–IV plus *T. hegneri*). The results showed that mtDNA groups I, II, and III match, respectively, ITS-2 groups 1 (including subgroups 1A and 1B), 2, and 3. Group IV was new and comprised of cave specimens from Belize. As observed in other studies, there was partial overlap in the geographic distribution of some groups (Bargues et al., 2008; Dorn et al., 2009; Gómez Palacio et al., 2015). Thus, groups I, II, and III co-occur in Petén, Guatemala (Monteiro et al., 2013), and ITS-2 groups 2 (including *T. hegneri*) and 3 seem to be sympatric in Cozumel Island (Bargues et al., 2008; Dorn et al., 2009). Dorn et al. (2016) corroborated the cyt *b* and ND4 findings of Monteiro et al. (2013) and added information on yet another mtDNA marker (177 bp of the COI gene). These authors, however, chose not to recognize group II and lumped it with group I into a single unit referred to as *T. dimidiata s.s.* (Dorn et al., 2016).

Taken as a whole, current evidence supports the view that *T. dimidiata* is a complex of four major independently evolving lineages plus *T. hegneri* (Table 3). Some of these lineages deserve specific status and likely play different roles as vectors of Chagas disease, from the apparently strictly sylvatic group IV populations in Belize to the heavily synanthropic groups 1/I and 2/II populations in Mesoamerica, Colombia, and Ecuador—and with the Yucatán clade (group 3/III) apparently presenting intermediate behaviour (Bargues et al., 2008; Dorn et al., 2009, 2016; Monteiro et al., 2013).

3.3.3 The T. sordida *Complex*

T. sordida shares parts of its geographic range with the phenotypically similar *Triatoma patagonica*, *T. guasayana*, and *Triatoma garciabesi*. Range overlap and similarity fuelled some debate about the taxonomy and evolutionary relations of these bugs (Gorla et al., 1993; Lent and Wygodzinsky, 1979; Usinger et al., 1966). *T. sordida s.l.* has a wide distribution across parts of Argentina, Bolivia, Brazil, Paraguay, and Uruguay. *T. patagonica* and *T. garciabesi* are found in Argentina, while *T. guasayana* occurs in Argentina, Bolivia, and Paraguay (see also Section 4.1.3.6). Although *T. sordida* is often found in and around dwellings, it seems to play a relatively minor role as a vector of human Chagas disease—at least relative to primary vectors such as *T. infestans* (Noireau et al., 1996, 1997a).

T. garciabesi was described by Carcavallo et al. (1967) after observations suggesting that some sylvatic bugs from Argentina were smaller and darker than peridomestic and domestic populations of *T. sordida* and *T. guasayana* from Argentina and Brazil, respectively (Ábalos and Wygodzinsky, 1951; Carcavallo et al., 1967). *T. garciabesi* and *T. sordida* were synonymized by Lent and Wygodzinsky (1979), who considered these differences insufficient to separate the two species. Since then, molecular, cytogenetic, morphometric, and ecological studies have shown that *T. sordida* constitutes a complex of species and helped to settle the systematics of this group (García et al., 2001; García and Powell, 1998; Noireau et al., 1998; Panzera et al., 1997, 2015). Allozyme electrophoresis (Jurberg et al., 1998; Panzera et al., 1997), cytogenetics (Panzera et al., 1997), morphometrics (Gurgel-Gonçalves et al., 2011), and mtDNA sequences (García and Powell, 1998) all support the hypothesis that *T. garciabesi* and *T. sordida* are, in fact, different species. Similarly, the status of *T. guasayana* and *T. patagonica* as separate species was supported by allozyme (Panzera et al., 1997) and mtDNA (12S, 16S, and COI) data (García et al., 2001; Hypša et al., 2002). Likewise, morphometric (Gorla et al., 1993), allozyme (García et al., 1995; Noireau et al., 1998), and DNA sequence data (García et al., 2001; Hypša et al., 2002) were consistent with *T. sordida* and *T. guasayana* being separate species. Using allozymes, Noireau and colleagues revealed the existence of two sympatric, cryptic species (named groups 1 and 2) within *T. sordida*, and went on to study the ecology and genetic relations of these two cryptic taxa and *T. guasayana* in the Bolivian Chaco (Noireau et al., 1998, 1999). More recently, Panzera et al. (2015) used chromosomal markers to analyse 139 specimens of *T. sordida*, *T. garciabesi*, *T. guasayana*, and *T. patagonica* sampled in Argentina, Bolivia, Brazil, and Paraguay. Cytogenetic analyses identified five chromosomal taxa that overall agreed well with clades recovered in an mtDNA COI phylogeny. They revealed that a cryptic *T. sordida* group is widely distributed in Argentina, and suggested that *T. sordida* group 2 (Noireau et al., 1998) and *T. garciabesi* are probably conspecific. Cytogenetics also detected a further putative cryptic taxon within *T. sordida* in the Andean highlands of La Paz, Bolivia (Panzera et al., 2015).

At the interspecific level, morphology-based assessments suggest that *T. sordida* and *T. guasayana* are very closely related and fairly close to *T. patagonica*, whereas *T. garciabesi* and *T. sordida* are more clearly distinct from each other. Phylogenetic analyses have revealed, however, closer evolutionary relations within the species pairs *T. guasayana* + *T. patagonica* and *T. sordida* (*s.l.*) + *T. garciabesi* (García et al., 2001; Hypša et al., 2002; Panzera et al., 1997, 2015). Moreover, the phenotypically distinct *T. rubrovaria*, *Triatoma circummaculata*, and *Triatoma klugi* form a clade with *T. guasayana* + *T. patagonica*, while *Triatoma matogrossensis* groups with *T. sordida* + *T. garciabesi* (García et al., 2001; Hypša et al., 2002; Panzera et al., 2015; Schofield and Galvão, 2009). This example, then, again demonstrates that morphology-based groupings within *Triatoma* may not reflect evolutionary history (see also Tables 1–3 and Section 3.3.5).

3.3.4 The T. brasiliensis *Complex*

T. brasiliensis is the most important vector of Chagas disease in the dry Caatinga of northeastern Brazil. Marked morphological and chromatic variability prompted its division into three subspecies—*T. b. brasiliensis*, *T. b. macromelasoma*, and *T. b. melanica*. These subspecies were later synonymized based on the observation that intermediate forms are found in nature (Lent and Wygodzinsky, 1979). The results of genetic (allozyme-based) and ecological studies contradicted this idea and again favoured the existence of the three *T. brasiliensis* subspecies. Yet another chromatic variant, initially referred to as the 'juazeiro form', was identified (Costa et al., 1997, 1998). These studies, however, used specimens that were only sampled from the type localities of the three previously named subspecies, and hence did not test whether these might represent the extremes of a morphological/

chromatic gradient with naturally occurring intermediate forms as suggested by Lent and Wygodzinsky (1979).

To address this issue, Monteiro et al. (2004) analysed mtDNA cyt *b* sequence variation among 136 bugs sampled from 16 populations across the species' range. Phylogenetic and nested clade analyses indicated that *T. brasiliensis* includes four genetically distinct clusters corresponding to the chromatic forms 'brasiliensis', 'macromelasoma', 'juazeiro', and 'melanica'. Genetic divergence was high (K2P 0.027–0.119), and the geographic distribution of haplotypes was compatible with the existence of a species complex. Specific status was suggested for the 'melanica' and 'juazeiro' forms, and Costa et al. (2006) and Costa and Felix (2007) formally described, respectively, *T. melanica* and *Triatoma juazeirensis*. Given the smaller genetic divergence (K2P 0.027), the 'macromelasoma' form was kept as a chromatic variant of *T. brasiliensis*. Later on, mtDNA sequence analyses revealed that *Triatoma sherlocki* and *Triatoma lenti* also belong in this complex (Mendonça et al., 2009, 2016). Note that the Brasiliensis subcomplex proposed by Schofield and Galvão (2009) based on morphological similarities and geography includes the distantly related *Triatoma melanocephala* and is therefore non-monophyletic (see Tables 2 and 3 and Section 4.1.3).

3.3.5 Other Triatoma *Complexes and Subcomplexes*

Molecular research is also helping to clarify the relations among species within other putative complexes and subcomplexes loosely defined based on morphology and, at times, distribution patterns (see Tables 1–3). For example, the Maculata subcomplex of Schofield and Galvão (2009) includes *T. pseudomaculata*, *Triatoma arthurneivai*, and *Triatoma wygodzinskyi* (Table 2), yet none of these is closely related to *T. maculata*; instead, they appear to be closer to *Triatoma deaneorum*, *Triatoma williami*, *Triatoma baratai*, *Triatoma guazu*, *Triatoma costalimai*, and *Triatoma jatai*—which were included, together with *T. matogrossensis*, *Triatoma jurbergi*, and *Triatoma vandae*, in Schofield and Galvão's (2009) Matogrossensis subcomplex (see Tables 2 and 3 and Section 4.1.3.6). These latter three species, in turn, appear to be closer to the fairly diverse group including *T. sordida* and related species (see Table 3 and Sections 3.3.3 and 4.1.3.6). Finally, *T. guasayana* and *T. patagonica* are only distantly related to *T. sordida*, and therefore their inclusion in the Sordida subcomplex (Schofield and Galvão, 2009) seems unwarranted; instead, they appear to be more closely to members of the group comprising *T. rubrovaria* and related species (see Tables 2 and 3 and Sections 3.3.3 and 4.1.3.6).

3.4 Taxonomy: Describing and Sorting Out Newly Uncovered Diversity

In this section, we present a quick overview of recently described triatomine species, including a few whose taxonomic status does not appear to be well established.

3.4.1 Mepraia parapatrica

M. parapatrica occupies a small coastal area between the ranges of *Mepraia gajardoi* and *Mepraia spinolai* in Chile (see also Section 4.1.2.1). It was described based on morphology and cytogenetics (Frías-Lasserre, 2010), with molecular comparisons based on mtDNA and nuclear rDNA sequences providing further evidence of differentiation (Calleros et al., 2010; Campos et al., 2013a; Campos-Soto et al., 2015).

3.4.2 New Triatoma *Species*

T. juazeirensis and *T. melanica* were described as distinct from *T. brasiliensis* after joint consideration of phenotypic, ecological, and genetic data (Costa and Felix, 2007; Costa et al., 1997, 2006; Monteiro et al., 2004) (see Section 3.3.4). Within the same species complex, *Triatoma bahiensis* was removed from synonymy with *T. lenti* by Mendonça et al. (2016) based on morphological, morphometric, and molecular data. Cyt *b* divergence between *T. bahiensis* and *T. lenti* was however low (K2P 0.025) and similar to that reported for *T. b. brasiliensis* and *T. b. macromelasoma* (K2P 0.027; Monteiro et al., 2004). This distance is much smaller than the cut-off value of 0.075 proposed by Monteiro et al. (2004) to validate *Triatoma* species based on cyt *b* sequence data. Of note, the phylogenetic tree of Mendonça et al. (2016, p. 248) lacks any sequence from true *T. b. macromelasoma*: the sequence used to represent this taxon was instead from *T. juazeirensis* (haplotype B; AY336526; Monteiro et al., 2004).

Triatoma boliviana was described from specimens collected in the Bolivian highlands and said to resemble *Triatoma nigromaculata*; its overall phenotype, however, reminds that of the probably parapatric *Triatoma carrioni* (Martínez Avendaño et al., 2007; see also Section 4.1.1).

T. jatai was discovered in the state of Tocantins, Brazil and described by Gonçalves et al. (2013) based on morphology. Analysis of two mtDNA gene fragments (16S and COI) later supported the status of *T. jatai* as a distinct species and revealed its sister taxon relation to the sympatric *T. costalimai* (Teves et al., 2016).

Triatoma pintodiasi was described as similar to *T. circummaculata* (with perhaps some small differences in haemolymph proteins) and *Triatoma carcavalloi* (Jurberg et al., 2013). *T. pintodiasi* appears to differ from these and other closely related species in sex-chromosome meiotic behaviour (Alevi et al., 2015), but genetic distances were reported as <0.012 for a fragment of the mitochondrial 16S rDNA gene (Tamura 3-parameter model, unspecified fragment length; the sequences were not available in GenBank as of July 7, 2017) (Alevi et al., 2017a).

3.4.3 Panstrongylus: *Two New Species and a Shifting Synonym*

Panstrongylus mitarakaensis from French Guiana and *Panstrongylus martinezorum* from the Venezuelan Amazon were described using morphological characters and do not appear to have been studied with molecular methods. The original descriptions mention similarities with, respectively, *Panstrongylus lignarius* (Bérenger and Blanchet, 2007) and *Panstrongylus lenti* (Ayala, 2009). At least superficially, however, both look quite similar to the phenotypically very variable, widespread *P. geniculatus* (Lent and Wygodzinsky, 1979; see also Patterson et al., 2009; Table 3 and Section 4.1.2.4). This is also the case for another Venezuelan species, *Panstrongylus turpiali*, which was first synonymized with *Panstrongylus chinai* but whose holotype was later shown to clearly match *P. geniculatus* at many phenotypic characters (Ayala, 2016; Lent, 1997; Valderrama et al., 1996).

3.4.4 Revalidated, New, and Dubious Rhodnius *Species*

Rhodnius amazonicus was described as closely related to *Rhodnius pictipes* based on a female specimen (Almeida et al., 1973). Lent and Wygodzinsky (1979) synonymized it with *R. pictipes*, arguing that this particular specimen was simply an atypical *R. pictipes*. Bérenger and Pluot-Sigwalt (2002) challenged this view based on their discovery of further specimens collected in French Guiana that matched the original description of *R. amazonicus* by Almeida et al. (1973). We have examined new mtDNA (cyt *b*) and rDNA (ITS-2) sequences from French Guianan bugs with *R. amazonicus* phenotypes (C.L. and F.A.M., unpublished), comparing them with sequences from *R. pictipes* and closely related species; the results confirmed the distinctiveness of *R. amazonicus* (Fig. 2).

Rhodnius barretti was discovered serendipitously during an investigation on mtDNA sequence variation in *Rhodnius* species and populations from Ecuador (Abad-Franch, 2003; Abad-Franch et al., 2009; Abad-Franch and Monteiro, 2005, 2007). Márquez et al. (2011) studied specimens from the same region using morphometric and molecular techniques and also suggested that this western Amazon population deserved species status. The new species was formally described as *R. barretti* based on the assessment of mtDNA sequence data, morphometrics, and qualitative phenotype traits (Abad-Franch et al., 2013).

Five species of *Rhodnius* have been described in the 21st century whose status does not appear to be well established and merits further scrutiny. *Rhodnius milesi* from southeastern Amazonia (Valente et al., 2001) appears to be genetically very close to *R. neglectus* (Fig. 2; see also Pita et al., 2013); in spite of some cytogenetic differences, this might also be the case for the recently described *Rhodnius taquarussuensis* from southern Brazil (Rosa et al., 2017). Although *Rhodnius zeledoni* was described as similar to *Rhodnius paraensis*, it looks very much like the sympatric *Rhodnius domesticus* (Jurberg et al., 2009; Lent and Wygodzinsky, 1979).

Rhodnius montenegrensis was described by Rosa et al. (2012) based on a *R. robustus*-like population from Montenegro, Rondônia state, Brazil. As discussed in Section 3.3.1, *R. robustus* is a paraphyletic assemblage including *R. robustus* II from southern and western Amazonia (see Monteiro et al., 2003; Pavan et al., 2013; Pavan and Monteiro, 2007; and Section 4.2.3). Cyt *b* sequences from Monteiro et al. (2003; *R. robustus* from Porto Velho, EF011720) and Maia da Silva et al. (2007; *R. robustus* from Montenegro, EF071583) differ at just 1–2 bp from that used to support *R. montenegrensis* as a distinct species by Rosa et al. (2012; KR072682) (see Fig. 2). Similarly, *Rhodnius marabaensis* was described by Souza et al. (2016) from *R. robustus*-like material collected in Marabá, state of Pará, Brazil—that is, the southeastern Amazon subregion where *R. robustus* III is widely distributed (see Monteiro et al., 2003; Pavan et al., 2013; Pavan and Monteiro, 2007; and Section 4.2.3). *R. marabaensis*, in fact, appears to be genetically very close to *R. robustus* III, but Souza et al. (2016) did not provide details (or sequence data) that could be used to further examine the issue.

In attempting to explain the phenotypic distinctiveness that led to the descriptions of *R. montenegrensis* and *R. marabaensis*, we noted that both were compared to putative *R. robustus* material from laboratory colonies founded in the early 1970s with bugs from Peru (see Rosa et al., 2012). This was also the case for the transcriptome-based comparisons between *R. montenegrensis* and putative *R. robustus* made by Carvalho (2016) and later published (without colony details) by Carvalho et al. (2017). Although rarely studied (and, hence, seldom recognized), the cross-contamination of *R. prolixus* and *R. robustus* colonies is fairly frequent (see Mesquita et al., 2015). It seems quite likely to us that the '*R. robustus*' colonies used by Rosa et al. (2012), Souza et al. (2016), and Carvalho et al. (2017) became at some time point contaminated with non-*R. robustus* material—almost certainly *R. prolixus*, which can interbreed with *R. robustus* from different origins including Peru (Barrett, 1996). Monteiro et al. (2000) showed that bugs from a colony labelled as '*R. robustus*' from 'Lima, Peru' have cyt *b* sequences nearly identical to those of *R. prolixus*. That *R. montenegrensis* (i.e. *R. robustus* II) and *R. marabaensis* (i.e. *R. robustus* III) are distinct from mixed *R. robustus*–*R. prolixus* material is unsurprising. In any case, assuming that the recently founded *R. montenegrensis* and *R. marabaensis* colonies have not themselves become contaminated, these reports provide detailed, and hence potentially valuable, descriptions of two of the cryptic *R. robustus* taxa discovered by Monteiro et al. (2003).

3.4.5 Strengthening Triatomine Taxonomy

One obvious implication of the cautionary remarks we have often made in this section is that colony material must always be genotyped and compared against established taxonomic standards (e.g. as in Mesquita et al., 2015), particularly when research involves taxa with known cryptic variation—i.e., most *Rhodnius* or the groups including *T. sordida* or *T. dimidiata*. More generally, we strongly advise that, from now on, reliable and openly available molecular data be required by journal editors and reviewers for publishing results on the taxonomy of the Triatominae.

4. BIOGEOGRAPHY OF THE TRIATOMINAE

Most triatomine taxa are currently found in the Americas. The exceptions are eight *Triatoma* and six *Linshcosteus* species from the Old World. In this section, we present a global overview of the major biogeographic patterns in the Triatominae, and explore (with some cautious speculation) the likely underlying processes. We pay special attention to the less studied groups, and briefly summarize current knowledge about those that have been investigated in more detail. We use a few, general references reviewing triatomine bug ecology and distribution patterns (Abad-Franch et al., 2009, 2015; Abad-Franch and Monteiro, 2007; Barrett, 1991; Carcavallo et al., 1985, 1999; de Paula et al., 2007; Galvão, 2014; Leite, 2013; Lent and Wygodzinsky, 1979; Patterson et al., 2009; Santana, 2014; Usinger, 1944), and refer to more specific studies in the text. Our descriptive ecoregional approach largely follows the updated version of Olson et al.

(2001) available at https://worldmap.harvard.edu/maps/4616. We propose a hierarchy of (1) major evolutionary 'lineages'; (2) 'clades' within lineages, defined by common ancestry and broad biogeographic correspondences; and (3) 'species groups' within clades, with some of these groups matching 'species complexes' as defined in Section 3.3. For the Triatomini, this arrangement is summarized in Table 3 and Fig. 3.

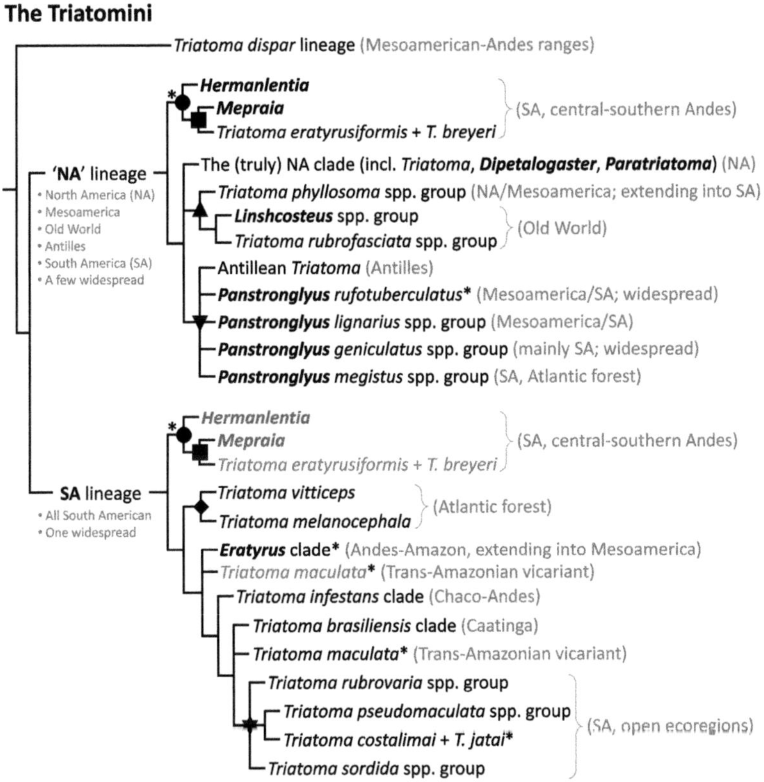

The Triatomini

Triatoma dispar lineage (Mesoamerican-Andes ranges)

'NA' lineage
• North America (NA)
• Mesoamerica
• Old World
• Antilles
• South America (SA)
• A few widespread

Hermanlentia
Mepraia
Triatoma eratyrusiformis + *T. breyeri*
(SA, central-southern Andes)
The (truly) NA clade (incl. *Triatoma*, ***Dipetalogaster***, ***Paratriatoma***) (NA)
Triatoma phyllosoma spp. group (NA/Mesoamerica; extending into SA)
Linshcosteus spp. group
Triatoma rubrofasciata spp. group
(Old World)
Antillean *Triatoma* (Antilles)
Panstrongylus *rufotuberculatus** (Mesoamerica/SA; widespread)
Panstrongylus *lignarius* spp. group (Mesoamerica/SA)
Panstrongylus *geniculatus* spp. group (mainly SA; widespread)
Panstrongylus *megistus* spp. group (SA, Atlantic forest)

SA lineage
• All South American
• One widespread

Hermanlentia
Mepraia
Triatoma eratyrusiformis + *T. breyeri*
(SA, central-southern Andes)
Triatoma vitticeps
Triatoma melanocephala
(Atlantic forest)
Eratyrus clade* (Andes-Amazon, extending into Mesoamerica)
*Triatoma maculata** (Trans-Amazonian vicariant)
Triatoma infestans clade (Chaco-Andes)
Triatoma brasiliensis clade (Caatinga)
*Triatoma maculata** (Trans-Amazonian vicariant)
Triatoma rubrovaria spp. group
Triatoma pseudomaculata spp. group
Triatoma costalimai + *T. jatai**
Triatoma sordida spp. group
(SA, open ecoregions)

Fig. 3 Schematic representation of the relations among lineages, clades, and species groups in the Triatomini. The order of these hierarchical levels follows the structure of Section 4—where many further details are provided (see also Table 3). Genera other than *Triatoma* are in **bold typeface** to highlight paraphyly. The symbols stand for: *circle* (●), *Hermanlentia*+*Mepraia* clade; *square* (■), *Mepraia* species group; *triangle* (▲), North-Mesoamerican+Old World clade; *inverted triangle* (▼), Antillean *Triatoma* +*Panstrongylus* clade; *diamond* (◆), Atlantic forest *Triatoma* clade; and *star* (✱), *Triatoma sordida* clade (see Table 3). *Asterisks* indicate that the positions of the *Hermanlentia* +*Mepraia* clade (including *T. eratyrusiformis*+*T. breyeri*), *P. rufotuberculatus*, *Eratyrus* spp., *T. maculata*, and *T. costalimai*+*T. jatai* remain unresolved; for the *Hermanlentia* +*Mepraia* clade and *T. maculata*, the position we deem more likely is in *darker font*. Some general biogeographic notes are presented in *light-grey font*.

4.1 The Tribe Triatomini

The monophyly of the tribe Triatomini seems largely uncontroversial. Within the tribe, however, the current, morphology-based systematic arrangement often breaks down, in full or in part, at every level from genera to populations. For example, molecular phylogenetic analyses suggest patterns of descent that challenge the status of basically every genus in the tribe. The widespread paraphyly of *Triatoma* with respect to *Panstrongylus*, *Dipetalogaster*, *Paratriatoma*, *Mepraia*, *Eratyrus*, and *Linshcosteus* illustrates well this point (e.g. Hwang and Weirauch, 2012; Hypša et al., 2002; Ibarra-Cerdeña et al., 2014; Justi et al., 2016; Marcilla et al., 2002) (Fig. 3). At lower levels, phylogeographic studies have shown that some species encompass several cryptic (or near-cryptic) taxa—and that these include, in some cases, other named species (e.g. Bargues et al., 2008; Monteiro et al., 2013). This means that to track the biogeography of the Triatomini we need the guidance of phylogenetic relations, even if our knowledge about them is still fragmentary—unfortunately, available studies suffer from limited taxon sampling, limited character sampling, or both.

The first cladogenesis within the Triatomini (~20 to 35 Mya; Hwang and Weirauch, 2012; Justi et al., 2016; Fig. 3) seems to have given rise to a little-known lineage of five extant species primarily distributed along the Andean axis but extending into Central America and into the lower Orinoco and Venezuelan coastal basins. This group is defined by the distinct morphology of its members, which suggests monophyly, and includes *Triatoma dispar*, *Triatoma venosa*, *T. nigromaculata*, *T. carrioni*, and *T. boliviana* (Lent and Wygodzinsky, 1979; Martínez Avendaño et al., 2007; see Table 3).

4.1.1 The T. dispar *Lineage*

T. dispar occurs from the Central American moist forests through the Colombian western Andes, the Cauca River valley, and the Chocó forests along the Pacific coast, and down to the montane forests of the western Ecuadorian Andes (Abad-Franch et al., 2001; Guhl et al., 2007; Pinto et al., 2015; Zeledón et al., 2001). The range of *T. dispar* largely overlaps that of *T. venosa*, but this latter species does not seem to occur in Central America—and in Colombia it extends into the lower-middle Magdalena valley and the Cordillera Occidental (Guhl et al., 2007; Zeledón et al., 2001). In Ecuador, *T. venosa* occurs in montane forests on both sides of the Andes (Abad-Franch et al., 2001). *T. nigromaculata* has been reported from several dry ecoregions of the lower Orinoco and Lake Maracaibo basins, and might also occur in the moister forests of the Venezuelan coastal ranges and the northeastern edge of the Guiana Shield (Abad-Franch and Monteiro, 2007). *T. carrioni* has been reported from the wet premontane forests of northwestern Ecuador (where the bugs are heavily melanized) through the drier southern Ecuadorian Andes and inter-Andean valleys (where the bugs are paler and a well-known domestic pest) and down to the Andes of northern Peru (Abad-Franch et al., 2001; Cuba Cuba et al., 2002). Two records of *T. nigromaculata* from, respectively, the upper Cauca valley montane forests of southwestern Colombia (Vásquez et al., 2005) and the eastern foothills of the Cordillera Oriental in northern Peru (Calderón and Monzón, 1995) may refer to pale forms of *T. carrioni* (Cuba Cuba et al., 2002). *T. boliviana* was described in 2007 based on specimens collected in the Central Andean wet Puna near Lake Titicaca in Bolivia. It was also said to bear a resemblance to *T. nigromaculata*, yet it looks very much like a pale *T. carrioni* (Martínez Avendaño et al., 2007). Although the genetic relation between *T. carrioni* and *T. boliviana* remains to be studied, we anticipate that these two species will turn out to be close relatives—much closer that any of them is to *T. nigromaculata*.

In sharp contrast to the species-poor and geographically restricted *T. dispar* lineage, the next diverging Triatomini lineage diversified extensively to yield over 100 known species in 8 named genera, spread throughout the Americas and the Antilles, and successfully colonized large parts of Asia and some spots in Oceania (see Figs 3 and 4). What follows is a brief approximation to the biogeography and evolution of this most successful triatomine lineage.

The second major cladogenetic event within the Triatomini (~17 to 30 Mya; Hwang and Weirauch, 2012; Justi et al., 2016) appears to have given rise to two major lineages—one primarily composed of currently North/Mesoamerican and Caribbean species (but including all Old World and some South American species and populations) and the other entirely composed of currently South American species (Fig. 3). Hereafter, we will refer to these two major clades as the 'North American' lineage (which includes non-North American taxa) and the South American lineage (whose members are all South American) (Fig. 3 and Table 3).

4.1.2 The 'North American' Lineage

The deeper splits within the 'North American' lineage are not particularly well resolved in published phylogenies; there are, however, some apparently consistent basic trends, and the shallower groupings are overall better

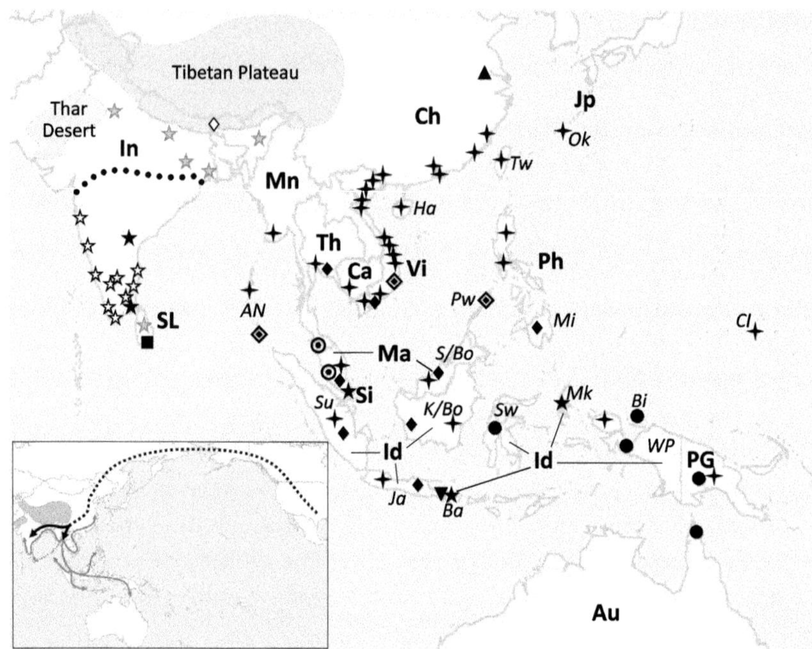

Fig. 4 *Triatoma* species in the Old World. *Stars* (★, ✦), *T. rubrofasciata* (*five-pointed stars*, records within the hypothesized native range of the species on the Deccan Peninsula (south of the *dotted line* in India); *five-pointed black*, wild *Triatoma* specimens likely belonging to *T. rubrofasciata*; *five-pointed grey*, sites where the species may or may not be native; *four-pointed stars*, populations most likely introduced through man-mediated passive dispersal); *diamonds* (◆), *T. migrans* (*white*, a dubious record from the Himalayas in Sikkim, India); *target diamonds* (◈), *T. bouvieri; targets* (◎), *T. cavernicola*; *circles* (●), *T. leopoldi*; *square* (■), *T. amicitiae*; *inverted triangle* (▼), *T. pugasi*; and *triangle* (▲), *T. sinica*. Note that most locations are approximate. The *inset* illustrates the 'out of America' hypothesis for the origin of the Old World triatomines, involving (i) old ($\sim$20 Mya) long-distance dispersal from North America through the Bering land bridge into Asia (*black broken line*), (ii) basal cladogenesis (perhaps $\sim$15 Mya) to yield the ancestors of *Linshcosteus* in continental India and the Asian *Triatoma* (*thick black arrows*), and (iii) a (likely Pliocene–Pleistocene) radiation of *Triatoma* throughout the region (*grey arrows*; the *broken arrows* pointing towards Sri Lanka represent our uncertainty about whether *T. amicitiae* arrived into the island from continental India or from Southeast Asia). The Tibetan Plateau and the Thar Desert between India and Pakistan are also highlighted. Countries (in **bold**): **In**, India; **SL**, Sri Lanka; **Mn**, Myanmar; **Th**, Thailand; **Ca**, Cambodia; **Si**, Singapore; **Vi**, Vietnam; **Ch**, China; **Jp**, Japan; **Ph**, the Philippines; **Ma**, Malaysia; **In**, Indonesia; **PG**, Papua New Guinea; **Au**, Australia. Islands mentioned in the text (in *italics*): *AN*, Andaman and Nicobar (India); *S/Bo*, Sarawak/Borneo (Malaysia); *Su*, Sumatra, *Ja*, Java, *K/Bo*, Kalimantan/Borneo, *Sw*, Sulawesi, *Ba*, Bali, *Mk*, Maluku, *WP*, West Papua and *Bi*, Biak (Indonesia); *Mi*, Mindanao and *Pw*, Palawan (Philippines); *Ha*, Hainan (China); *Tw*, Taiwan; *Ok*, Okinawa (Japan); *Cl*, Caroline Islands (Palau and Micronesia).

resolved. The common ancestor of the 'North American' lineage appears to have originated three major clades—two likely North American and one likely Caribbean. There is, though, a fourth clade of rather uncertain affinities that seems related to this lineage—even if its living descendants are all to be found in South America (Fig. 3 and Table 3).

4.1.2.1 The Hermanlentia + Mepraia Clade

The evolutionary relations of *Hermanlentia matsunoi* have not been adequately addressed in available phylogenetic studies. This monotypic genus is known only from cave habitats in the upper Marañón dry forests of the Peruvian Andes (Barrett, 1991; Cuba Cuba et al., 2002). Somewhat surprisingly, Justi et al. (2014) show *Hermanlentia* as sister to a cluster of *T. dimidiata* sequences, i.e., firmly nested within the 'North American' lineage. This result is, however, based on a 219-bp fragment of the mtDNA COII gene (KC249400) that is indeed 100% identical (for a 208-bp overlap) to a COII *T. dimidiata* sequence also obtained by Justi and colleagues (KC249430). This *H. matsunoi* sequence is labelled as 'UNVERIFIED' and 'similar to cytochrome oxidase subunit II' in GenBank (accessed July 6, 2017) and was excluded from phylogenies later published by Justi et al. (2016).

A second South American genus with a sub-Andean distribution is *Mepraia*. It is composed of three named species that occur along the narrow, semiarid to arid strip between the Andes and the Pacific in Chile. *M. gajardoi* is known from the Atacama Desert coastline in northern Chile, whereas *M. spinolai* is probably endemic to the Chilean Matorral; the recently described *M. parapatrica* seems to be restricted to a small coastal area in the transition between Atacama and the Matorral at approximately 24° to 28°S (Calleros et al., 2010; Campos et al., 2013a; Frías-Lasserre, 2010).

Two *Triatoma* species from the eastern side of the Andes south of the Tropic of Capricorn are closely related to *Mepraia* (Calleros et al., 2010; Campos et al., 2013a, b; Campos-Soto et al., 2015; Hypša et al., 2002; Justi et al., 2016). *Triatoma eratyrusiformis* occupies arid environments in the Argentinean Monte and the southwestern dry Chaco; a population formerly known as *Triatoma ninioi* seems restricted to more humid habitats along gallery forests, where it associates with the underground burrows of caviid rodents. *Triatoma breyeri* occurs in the transition between the dry high Monte on the eastern slope of the Andes and the lowland dry Chaco around the saline lake system of central-northern Argentina (Abrahan et al., 2011; Barrett, 1991; Carcavallo et al., 1985; Cécere et al., 2016; Lent and Wygodzinsky, 1979); bugs identified as *T. breyeri* have been found in the

Andean montane dry forest near Mataral, Bolivia, but their cyt *b* sequence (KC249242; Justi et al., 2014) appears to diverge by ~4% from that of Argentinean material collected in the Monte-Chaco transition (JN102361, KC236980; Campos et al., 2013a, b) (Table 3). Molecular phylogenies suggest that *Mepraia* spp., *T. eratyrusiformis*, and *T. breyeri* form a monophyletic clade whose relation with the main 'North' and South American Triatomini lineages is not well resolved (Campos et al., 2013b; García et al., 2001; Hypša et al., 2002; Justi et al., 2014; Sainz et al., 2004). However, recent results favour the 'North American' lineage hypothesis (Justi et al., 2016). *Mepraia* also share their chromosome complement (20A $+X_1X_2Y$) with North American rather than South American *Triatoma*, as well as with most *Panstrongylus*, whereas *T. eratyrusiformis* has a 20A $+X_1X_2X_3Y$ complement (Panzera et al., 2010).

Although cytogenetic or molecular data remain unavailable, some phenotype traits seem to suggest a close relation between *Mepraia* and *Hermanlentia*. First, they share some male genitalia traits apparently to the exclusion of other Triatomini (Jurberg and Galvão, 1997). Second, both genera (and *T. eratyrusiformis* and *T. breyeri*) have light-coloured intersegment connexivum sutures; this is typical of North American *Triatoma* species, whereas South American ones almost invariably have dark-coloured sutures (Lent and Wygodzinsky, 1979). Taken together, these data and observations suggest the hypothesis that the *Hermanlentia* + *Mepraia* + (*T. breyeri* + *T. eratyrusiformis*) clade derives from a relatively ancient offshoot of the 'North American' lineage that followed the Andes axis in a southward range expansion—i.e., largely paralleling what members of the *T. dispar* lineage did some time earlier (Fig. 3).

4.1.2.2 The (Truly) North American Clade

The first currently North American clade appears to include several *Triatoma* species plus the monotypic *Paratriatoma* and *Dipetalogaster* (Fig. 3 and Table 3). Although the relations among species (and genera) within this clade are, again, not well resolved, it overall seems to encompass 14 named taxa. *T. protracta*, *Triatoma barberi*, *Triatoma lecticularia*, *Paratriatoma hirsuta*, and *D. maxima* appear to be more closely related to each other than any of them is to a second subclade including *Triatoma nitida*, *Triatoma rubida*, *Triatoma ryckmani*, and *Triatoma bolivari*; how other species likely belonging to this clade, namely *Triatoma neotomae*, *Triatoma peninsularis*, *Triatoma incrassata*, *Triatoma sinaloensis*, and *Triatoma indictiva*, relate to these two groupings remains to be ascertained (Bargues et al., 2008; de la Rúa et al., 2014;

Espinoza et al., 2013; Hwang and Weirauch, 2012; Hypša et al., 2002; Ibarra-Cerdeña et al., 2014; Justi et al., 2014, 2016; Lyman et al., 1999; Marcilla et al., 2001; Pfeiler et al., 2006; Sainz et al., 2004; Weirauch and Munro, 2009) (see Table 3). In any case, a key trait these species share is that, with the exception of *T. nitida*, all seem to occur only north of the Isthmus of Tehuantepec—and this would define them as truly North American species. They occupy terrestrial habitats in dry ecoregions including deserts, often in association with ground-burrowing rodents (notably *Neotoma* woodrats) or, in the case of *D. maxima* in Baja California, rock-dwelling lizards (Barrett, 1991; Bern et al., 2011; Ibarra-Cerdeña et al., 2009; Kjos et al., 2009; Lent and Wygodzinsky, 1979; Peterson et al., 2002; Ramsey et al., 2015; Usinger, 1944).

4.1.2.3 The North-Mesoamerican + Old World Clade

The second clade includes the *T. dimidiata* complex (see Section 3.3.2) and its close relatives (*Triatoma bassolsae, Triatoma mazzottii, Triatoma longipennis, Triatoma pallidipennis, Triatoma phyllosoma, T. picturata*, and perhaps *Triatoma brailovskyi* and *Triatoma gomeznunezi*) plus *T. mexicana, Triatoma gerstaeckeri*, and probably also the North American *Triatoma sanguisuga* and *Triatoma recurva*. The Old World lineage, of which *T. rubrofasciata* and two *Linshcosteus* species have so far been included in molecular phylogenies, also seems to belong here (Fig. 3 and Table 3) (Bargues et al., 2008; Espinoza et al., 2013; Hwang and Weirauch, 2012; Hypša et al., 2002; Ibarra-Cerdeña et al., 2014; Justi et al., 2014, 2016; Lyman et al., 1999; Marcilla et al., 2001; Martínez et al., 1994; Patterson and Gaunt, 2010; Pfeiler et al., 2006; Sainz et al., 2004; Weirauch and Munro, 2009). The species within this clade are now mainly found north of the Isthmus of Tehuantepec, except for the diverse *T. dimidiata* including *T. hegneri* from Cozumel Island offshore eastern Yucatán (Ramsey et al., 2015). This suggests a North American origin for this clade, with an early northward dispersal of the Old World group ancestor and a much more recent southward dispersal (successfully overcoming Tehuantepec) of the *T. dimidiata* complex ancestor—whose descendants spread over Central America and crossed the humid Darién forests to reach northern Colombia and parts of the lower Orinoco and Venezuelan coastal basins (Abad-Franch and Monteiro, 2007; Bargues et al., 2008; Dorn et al., 2007; Gómez-Palacio et al., 2013; Gómez-Palacio and Triana, 2014; Monteiro et al., 2013; Parra-Henao et al., 2016). Although the southernmost Colombian *T. dimidiata* populations (from Huila, in the upper Magdalena River valley) appear geographically close to Ecuadorian conspecifics, both are separated by the Colombian Central Massif (Abad-Franch et al., 2001; Gómez-Palacio and Triana, 2014; Parra-Henao et al., 2016). This observation, together with the absence of *T. dimidiata* records from the coastal Chocó forests and archaeological evidence of intense sea trade between Mesoamerican and coastal Ecuadorian pre-Columbian peoples, led to the hypothesis that the strongly synanthropic Ecuadorian (and Peruvian) *T. dimidiata* populations were probably not native (Abad-Franch et al., 2001; Álvarez, 1984). This hypothesis has received firm support from molecular genetics (Bargues et al., 2008; Dorn et al., 2016; Marcilla et al., 2001; Monteiro et al., 2013; Wong et al., 2016).

All the Old World Triatominae probably descend from a common ancestor (Hypša et al., 2002; Justi et al., 2016; Patterson and Gaunt, 2010) that radiated after reaching Eurasia through the Bering land bridge ~20 Mya—i.e., coinciding with the onset of global warming that led to the mid-Miocene climatic optimum (Feakins et al., 2012; Justi et al., 2016; Zachos et al., 2001) (Figs 3 and 4). Neotropical/Asian disjunct distributions are common in many taxa; most of them seem to involve very old cladogenetic events that are best explained by vicariance (Sequeira and Farrell, 2001; Sharma and Giribet, 2012), but some disjunct groups are of about the same age as suggested for Old World triatomines—e.g., the Camelidae and the Tapiridae (Steiner and Ryder, 2011; Wu et al., 2014). The close association of many extant North American triatomines with cricetid rodents, which are widespread and highly diverse across Asia, suggests a simple mechanism for triatomine long-distance dispersal. The alternative hypothesis is that all Old World triatomine taxa derive from a single modern species, *T. rubrofasciata*, which was carried from the Americas to Asia on trade ships since the XVI–XVII centuries. This implies that the six *Linshcosteus* and seven *Triatoma* Old World species all evolved in about 300 years (Dujardin et al., 2015a, b; Gorla et al., 1997; Patterson et al., 2001; Schofield, 1988; Schofield and Galvão, 2009). This hypothesis is clearly at odds with what we know about the *tempo* of insect speciation and DNA evolution, and has indeed been refuted by every molecular study examining the issue (Hypša et al., 2002; Justi et al., 2014, 2016; Patterson and Gaunt, 2010). It now seems clear (i) that Old World triatomines are indeed of New World origin, yet the divergence of both lineages most likely dates back to the Miocene and (ii) that ship-mediated dispersal of *T. rubrofasciata* did indeed occur, yet most likely starting in Asia. Man-driven dispersal included the Americas, where introduced *T. rubrofasciata* met their older, far-off relatives (Barrett, 1991; Neiva, 1914; Neiva and Lent, 1936). In Asia, the Old World triatomine ancestor gave rise to two groups. One includes the morphologically divergent *Linshcosteus* and the other the Asian/Australasian *Triatoma* species (Table 3).

The genus *Linshcosteus* is composed of six named species, all restricted to continental India south of the Himalayas and southeast of the Thar Desert (see Fig. 4). *Linshcosteus carnifex* seems to occur only in northern India—possibly in the Upper Gangetic Plains moist forests near Kanpur, Uttar Pradesh. *Linshcosteus confumus* and *Linshcosteus costalis* are sympatric (and share rocky ecotopes) in the South Deccan Plateau dry deciduous forests, whereas *Linshcosteus kali* has only been reported from the Southwestern Ghats moist deciduous forests in Coimbatore (formerly Madras and now Tamil Nadu). *Linshcosteus karupus* occupies rocky ecotopes in the South Deccan Plateau dry deciduous forests of southern Tamil Nadu. Finally, *Linshcosteus chota* is only reported as occurring in 'south India' (Ambrose, 2006; Dujardin et al., 2015b; Galvão et al., 2002; Lent and Wygodzinsky, 1979).

T. rubrofasciata is the only *Triatoma* species recorded in continental India (Ambrose, 2006; see below for a record of *Triatoma migrans*). There, it has been often collected along the western coast—i.e., in the Malabar Coast moist forests of Maharashtra, Goa, Karnataka, and Kerala. It is unclear whether *T. rubrofasciata* occurs in the moist deciduous and montane rainforests on the Western Ghats range; it has been reported from Mysore and Bangalore, in the South Deccan Plateau dry deciduous forests of Karnataka. On the southeast coast, *T. rubrofasciata* has been found in Tamil Nadu (Malabar Coast moist forests, South Deccan Plateau dry deciduous forests, East Deccan dry-evergreen forests, and Deccan thorn scrub forests) and Pondicherry (East Deccan dry-evergreen forests) (Ambrose, 2006; Dujardin et al., 2015a, b; Lent and Wygodzinsky, 1979; Schofield, 1988) (Fig. 4). One record of wild '*T. sanguisuga*' from the Godavari Krishna mangroves in Muthupet, Tamil Nadu (Rahaman, 2002), most likely corresponds to *T. rubrofasciata*. There is a second record of apparently wild '*Triatoma* sp.' from the Nallamalai Hill Ranges in the Eastern Ghats of Andhra Pradesh (Rao et al., 2007) (Fig. 4). The Nallamala forests are part of the Central Deccan Plateau dry deciduous forests and are separated by about 150 km of mainly Deccan thorn scrub forests from the coastal Godavari–Krishna mangroves south of the Krishna River mouth.

The records of *T. rubrofasciata* in continental India discussed so far correspond to the Deccan Peninsula including the Deccan Plateau, the Ghats, and the lowland coastal ecoregions (Fig. 4). To the northeast, *T. rubrofasciata* has been collected in Calcutta, which lies on the boundary between the

West Bengal Lower Gangetic Plains moist deciduous forests and the Sundarbans freshwater swamp forests of the Ganges Delta. There are, in addition, a few records of the species within continental India north of the Deccan—one from Assam (likely in the Brahmaputra Valley semievergreen forests), one from the Chhota–Nagpur dry deciduous forests (Ranchi in Jharkhand), and two from the Upper Gangetic Plains moist deciduous forests (Delhi and Gorakhpur, Uttar Pradesh) (Claver and Yaqub, 2015) (Fig. 4).

These distribution patterns, including records of *Triatoma* sp. from wild habitats, suggest the hypothesis that *T. rubrofasciata* is native to the Deccan Peninsula, where it mainly associates with low-elevation forests on the Ghats and with coastal humid habitats (including mangroves on occasion) (Fig. 4). Remarkably, the Deccan Peninsula is also the region where a particular *Rattus rattus* lineage (with $2n=38$ chromosomes) likely originated to become a major human commensal and pest (Aplin et al., 2011). This is the black rat lineage that dispersed with people to some parts of Asia and into Oceania, the Middle East, Africa, Europe, and the Americas—i.e., including the 'ship rats' (Aplin et al., 2011). If, as widely believed, *T. rubrofasciata* developed an association with the Indian *R. rattus* lineage, this may have helped it disperse out of its natural habitats in the Deccan and then hitchhike on ships from coastal India to many ports throughout the tropics—and reaching also some inland sites. We now quickly review the records of *T. rubrofasciata* out of India; many refer to whole countries or large regions, and in those cases we roughly follow Schofield (1988, especially fig. 18.1 on p. 290), Schofield et al. (2009), and Dujardin et al. (2015a). We stress that many records are rather old, and note that a few introduced specimens may easily fail to found a viable local population. Thus, although *T. rubrofasciata* was reported once from the continental USA (Jacksonville, Florida) in the 1940s, it has apparently never been recorded since (Bern et al., 2011; Usinger, 1944).

We could not find any locality-specific record of *T. rubrofasciata* from Sri Lanka. The island has both drier (including some South Deccan thorn scrub) and moister forests (including montane and lowland rainforests to the southwest); with the data at hand, we cannot exclude the possibility that wild *T. rubrofasciata* occur in Sri Lanka. However, *R. rattus* sampled from this island belong to a clade found mainly in Indochina and with $2n=40$ chromosomes—i.e., these black rats are fairly distant relatives of the Deccan populations (Aplin et al., 2011). We consider the rest of records of *T. rubrofasciata* as most likely corresponding to artificially introduced populations; we therefore will not go into much ecoregional detail. Claver and Yaqub (2015) mention the occurrence of '*Triatoma*' in Karachi,

Pakistan, but this does not seem to have been confirmed. *T. rubrofasciata* has been found in rainy sites in the Andaman and Nicobar Islands, Myanmar, Thailand, Malaysia, Singapore, Indonesia (Sumatra, Kalimantan/Borneo, Java, Bali, Maluku, and West Papua), Papua New Guinea, and the Philippines. In Cambodia and Vietnam, where it has become a widespread urban pest (Dujardin et al., 2015a), *T. rubrofasciata* occurs in areas with moist, yet more seasonal climate. To the north, there are records from temperate China (Hainan Island, Guangzhou, Hong Kong, Xiamen, Fuzhou, and Taiwan) and from Okinawa (Japan) (Fig. 4). The species has been found on small islands of the Pacific (Caroline Islands, Tonga, and Hawaii) and Indian Oceans (Seychelles, Rodrigues in Mauritius, Réunion, and Comoros), as well as in both seasonally moist and dry areas of Madagascar. The driest-climate site with a record is Saudi Arabia (Jeddah on the Red Sea coast), while the only record in east Africa is from the seasonally rainy Zanzibar in Tanzania. On west Africa, *T. rubrofasciata* has been reported from Guinea, Sierra Leone, and the drier Angola; some sources (e.g. Dujardin et al., 2015a; Lent and Wygodzinsky, 1979) cite also Katanga (a province of the Democratic Republic of the Congo with no coast or major rivers) and the Central African Republic (whose capital city, Bangui, has a port on the Ubangi River). Finally, there are records from South Africa.

In the Atlantic, *T. rubrofasciata* was found in Santa Maria Island (Azores, Portugal) and in the Bahamas. In the Caribbean, records include Cuba, Haiti and the Dominican Republic, Jamaica, the Virgin Islands, Antigua, Guadeloupe and Martinique, St. Vincent, Granada, and Trinidad. In continental America, *T. rubrofasciata* has been recorded in Florida (United States), in El Salvador (apparently the only record from the Pacific side of the continent), in the Venezuelan coast, in Suriname and French Guiana on the moist Guiana Shield, and in Brazil—where coastal sites have produced the vast majority of the records (from the moist Pará and Maranhão through the drier northeast in Rio Grande do Norte, Paraíba, Pernambuco, Alagoas, Sergipe, and Bahia, and down to the humid Atlantic forests of Rio de Janeiro and São Paulo); some sources cite also the inland states of Mato Grosso do Sul and Minas Gerais (e.g. Borges-Pereira et al., 2001; Silveira et al., 1984). In the current century, records from Brazil seem restricted to coastal cities: Belém, São Luís, Salvador, Recife, Rio de Janeiro, and Santos (Galvão, 2014; R. Gurgel-Gonçalves, pers. comm.). *T. rubrofasciata* has been reported (apparently only once) from Buenos Aires, Argentina (cf. Galvão et al., 2003).

The other Old World *Triatoma* species have been collected much less often; we summarize the records in Fig. 4. *Triatoma sinica* is known only from

Nanjing (Changjiang Plain evergreen forests), and *Triatoma amicitiae* was collected just once in the dry-evergreen forests of southern Sri Lanka. The remaining species occur mainly in Southeast Asia. *Triatoma leopoldi*, however, occurs in the Cape York Peninsula of north Queensland and is the only triatomine known to be native to Australia. It was collected at the transition between the eastern Iron Range rainforests and the tropical savannah (Monteith, 1974) and occurs also in the rainforests of Sulawesi, West Papua including Biak (as *Triatoma novaeguineae*, a synonym of *T. leopoldi*), and Papua New Guinea. *Triatoma pugasi* has been reported from the East Java rainforests in Indonesia; Schofield et al. (2009) also mention Malaysia, but this does not seem to have been confirmed. *Triatoma cavernicola* inhabits limestone caves in rainforests of continental Malaysia. Bugs identified as *Triatoma bouvieri* have a wider distribution, with records from Nicobar rainforests, the Philippine Palawan rainforests, and Vietnam (Southern Vietnam lowland dry forests in Khánh Hòa and, more recently, the limits between the Northern Indochina subtropical forests and the South China-Vietnam subtropical evergreen forests in Vĩnh Phúc; VAST, 2016). Finally, the most widely spread species in this group is *T. migrans*. It has been reported as occurring in Sikkim, India (e.g. Lent and Wygodzinsky, 1979; Fig. 4), yet triatomines seem unlikely to occur in the Himalayan ecoregions of that state; a relatively recent checklist of Indian Reduviidae does not include this record (Ambrose, 2006). Other records seem more likely and in line with the overall distribution of *T. migrans*' close relatives—i.e., the lowland tropical rainforests of Thailand, Malaysia (peninsular and Sarawak/Borneo), Indonesia (Sumatra, Java, and Kalimantan/Borneo), and Mindanao in the Philippines. *T. migrans* was recently reported from Ho Chi Minh city, Vietnam (VAST, 2016). We note that *T. migrans* and *T. bouvieri* are very similar, possibly to the point of inducing identification errors.

In summary, our overview of the Old World triatomines suggests an early arrival into India (perhaps ~15 Mya; cf. Justi et al., 2016) to yield *Linshcosteus* and a later, Pliocene–Pleistocene radiation of *Triatoma*—mainly to currently moist-forest regions in mainland Southeast Asia and across Sundaland, the Philippines, and Wallacea (including one recent arrival into Australia), and with more limited northward dispersal into subtropical and temperate regions along the Vietnam–southern China coast (Fig. 4). This scenario implies that ancestral populations in northwestern America and northeastern Asia became extinct, most likely because of sharp global cooling as the mid-Miocene climate optimum ended ~15 to 14 Mya (cf. Feakins et al., 2012; Zachos et al., 2001); this led to the current disjunct distribution (Fig. 4).

4.1.2.4 The Antillean Triatoma + Panstrongylus Clade

The third clade within the 'North American' lineage includes the Antillean *Triatoma* species (*Triatoma flavida* and *T. bruneri* from Cuba, and probably also *Triatoma obscura* from Jamaica) plus *Panstrongylus* species including *Panstrongylus geniculatus*, *Panstrongylus tupynambai*, *Panstrongylus lutzi* (plus *Panstrongylus sherlocki*), *P. lignarius* (plus *P. herreri*), and *P. megistus*. Molecular phylogenies most often recover *P. megistus* as sister to *Triatoma tibiamaculata*. *P. chinai* and *Panstrongylus howardi* likely belong to this clade too, whereas the precise position of *Panstrongylus rufotuberculatus* within the 'North American' lineage remains less well defined (Hypša et al., 2002; Justi et al., 2014, 2016; Marcilla et al., 2002) (Fig. 3). All species in this clade, like most 'North American' lineage species, have a $20A+X_1X_2Y$ chromosome complement, with the exception of *P. megistus* ($18A+X_1X_2Y$, shared only by *T. nitida*) (Panzera et al., 2010). It is interesting to note that the only two triatomine fossils known to date are from Dominican Republic (i.e. Antillean) amber—one *Triatoma*-like nymph exuvia (named *Triatoma dominicana*), and, remarkably, one very well preserved, *Panstrongylus*-like adult recently described as *Panstrongylus hispaniolae* (Poinar, 2005, 2013). These fossils may be from 45 to 15 My old (Poinar, 2013), but a narrower range of 20–15 My (Iturralde-Vinent, 2001; Iturralde-Vinent and MacPhee, 1996) has been widely used in Dominican amber fossil studies (e.g. Ramírez et al., 2007; Sherratt et al., 2015).

These data and observations clearly suggest a Mesoamerican–Antillean origin for this clade, whose descendants either remained on the Greater Antilles or reached the continent to disperse and diversify widely. Leaving aside the uncertain affinities of the widespread *P. rufotuberculatus*, we may tentatively distinguish three contemporary groups within the Panstrongylus subclade this clade (see also Table 3):

(i) The phenotypically diverse *P. geniculatus* is about as widespread as *P. rufotuberculatus* in both humid forest and drier lowland ecoregions across most of Central and South America (Abad-Franch and Monteiro, 2007; Lent and Wygodzinsky, 1979; Patterson et al., 2009); in Colombia and Ecuador, but not further south, it also occurs along the Pacific coast west of the Andes (Abad-Franch et al., 2001; Guhl et al., 2007). *P. geniculatus*' likely closest relatives include *P. mitarakaensis* (humid Guiana Shield forests; Bérenger and Blanchet, 2007), *P. martinezorum* (transition between the Orinoco Llanos and the Guianan highlands moist forests; Ayala, 2009; Ayala et al., 2014), and a set of species occupying the drier ecoregions south of the Amazon—*P. lutzi* (Caatinga, Pernambuco interior forests, and possibly the moister Caatinga Enclaves) and its likely synonym *P. sherlocki* (Caatinga); *P. lenti* (Cerrado); the more widely spread *Panstrongylus diasi* (Caatinga-Cerrado and perhaps parts of the Pantanal and Chaco); *Panstrongylus guentheri* (dry Chaco and parts of the Argentinean Monte and Espinal); and *P. tupynambai* (Uruguayan savannahs) (Barrett, 1991; Jurberg et al., 2001; Lent and Wygodzinsky, 1979; Patterson et al., 2009). One further group possibly related to *P. geniculatus* is restricted to a small area including the Ecuadorian coastal dry forests, where the light-coloured *P. howardi* occurs (Villacís et al., 2015), and the Andes of southern Ecuador and neighbouring northern Peru (including the southernmost Eastern Cordillera Real montane forests and the lowland Tumbes/Piura dry forests), where the melanic *P. chinai* is common (Abad-Franch et al., 2001; Cuba Cuba et al., 2002; Patterson et al., 2009). One record of *P. chinai* from the Catatumbo moist forests by the Lake Maracaibo in Venezuela (Ayala, 2016; Carcavallo et al., 1994) could represent an undescribed species, a melanic form of an otherwise known *Panstrongylus* species, or a labelling error.

(ii) The second group is composed of two similar species that dwell on humid forest trees—*Panstrongylus humeralis* from Central America and *P. lignarius* from the greater Amazon including the Guiana Shield and the Orinoco (Abad-Franch and Monteiro, 2007; Barrett, 1991; Lent and Wygodzinsky, 1979; Patterson et al., 2009). A parapatric *P. lignarius* population formerly known as *P. herreri* occurs in the dry forests of the middle-upper Marañón River valley in northeastern Peru, where it transmits Chagas disease (Alroy et al., 2015; Cuba Cuba et al., 2002; Patterson et al., 2009). *P. lignarius* and *P. herreri* are phenotypically very similar, interfertile (with intermediate forms found in nature), and indistinguishable both cytogenetically and at one ribosomal locus (Abad-Franch et al., 2001; Abad-Franch and Monteiro, 2007; Barrett, 1991; Crossa et al., 2002; Lent and Wygodzinsky, 1979; Marcilla et al., 2002; Panzera et al., 2010; Patterson et al., 2009); they have therefore been synonymized (Galvão et al., 2003).

(iii) Finally, two representatives of this subclade likely descend from a common ancestor (Hypša et al., 2002; Justi et al., 2014, 2016) that colonized the moist Atlantic forests of eastern Brazil south of parallel 7S. *P. megistus* is widespread across the Atlantic forests but also occurs in gallery forests throughout the drier Cerrado and stretches into the semiarid Caatinga, the Chaco, and parts of the Pantanal and Uruguayan savannahs (Barbosa et al., 2006; Barrett, 1991; Cavassin et al., 2014; Forattini, 1980; Galvão, 2014; Gurgel-Gonçalves et al., 2012; Lent and Wygodzinsky, 1979; Patterson et al., 2009; Silveira et al., 1984). Southern *P. megistus* populations seem tightly associated with humid forests and rarely, if ever, colonize in man-made habitats, whereas populations in seasonally drier regions are important domestic vectors (Barbosa et al., 2003, 2006; Forattini, 1980; Forattini et al., 1978; Lent and Wygodzinsky, 1979). *T. tibiamaculata* is associated with palms and bromeliads along a narrow strip of coastal Brazil including the Pernambuco, Bahia, and Serra do Mar coastal moist forests (Barrett, 1991; Galvão, 2014; Gurgel-Gonçalves et al., 2012; Lent and Wygodzinsky, 1979). If, as seems likely, the sister species relation of *P. megistus* and *T. tibiamaculata* is confirmed, this case will provide yet another striking illustration of how misleading morphology-based triatomine taxonomy can be.

4.1.3 The South American Lineage

The South American lineage is composed of all Triatomini species not dealt with above—i.e., it excludes *T. dispar* and its allies, the *Panstrongylus* with *T. tibiamaculata*, and probably the *Hermanlentia* + *Mepraia* clade (Table 3). With the exception of *Eratyrus cuspidatus*, all species in this highly diverse lineage seem to occur only south of the Panamanian Isthmus. Moreover, the vast majority of these taxa have only been recorded south of the Amazon.

4.1.3.1 The Atlantic Forest Triatoma Clade

The first, likely early-Miocene, cladogenesis within the truly South American Triatomini (Fig. 3) seems to have given rise to a two-species clade (with chromosome complement $20A + X_1X_2X_3Y$) currently found associated with the Brazilian Atlantic region between parallel 5S and the Tropic of Capricorn (Alevi et al., 2012, 2017b; Justi et al., 2016; Panzera et al., 2010). *T. melanocephala* occurs along the southeastern fringe of the dry Caatinga and adjacent ecoregions including the Pernambuco coastal and interior forests, the Campos Rupestres montane savannahs, the Atlantic dry forests, and the Bahia interior forests. To the east, *T. melanocephala* has rarely been recorded in the more humid Bahia coastal forests; its distribution seems limited by the Borborema Plateau to the north and by the São Francisco valley to the west. To the south, *Triatoma vitticeps* replaces *T. melanocephala* in the transition between the Chapada Diamantina and the Serra do Espinhaço. *T. vitticeps* seems primarily associated with the Bahia interior forests on the hills and mountains of the Serra do Espinhaço and Serra da Mantiqueira; it is also common in the southern third of the Bahia coastal forests—i.e., on the Serra do Castelo and Serra do Caparaó in the state of Espírito Santo. Further south, it also occurs in the northernmost

Serra do Mar coastal forests and Alto Paraná Atlantic forests; to the west, *T. vitticeps* has been collected in the eastern Cerrado and the Campos Rupestres montane savannahs, and may also occur in southern parts of the Atlantic dry forests (see Galvão, 2014; Gurgel-Gonçalves et al., 2012; Leite et al., 2011; Silveira et al., 1984).

4.1.3.2 The Eratyrus Clade

This two-species genus has most often been recovered within the South American Triatomini clade (e.g. Justi et al., 2014; Patterson and Gaunt, 2010; but see Hypša et al., 2002; Sainz et al., 2004). *E. cuspidatus* occurs in mangroves, in humid and seasonally drier forests (lowland and montane), and in xeric shrub ecoregions from southern Mexico including Yucatán to northern-northwestern South America. In South America, *E. cuspidatus* records cover the area between the lower third of the greater Orinoco basin (including the Maracaibo and Venezuelan coastal basins) and northern-central Peru (Pacific slope of the Andes and coast) (Abad-Franch et al., 2001; Abad-Franch and Monteiro, 2007; Cuba Cuba et al., 2002; Guhl et al., 2007; Ramsey et al., 2015; Reyes-Novelo and Ruiz-Piña, 2012; Sandoval-Ruiz et al., 2012; Sousa and Johnson, 1973; Zeledón et al., 2001). *E. cuspidatus* distribution hence strikingly resembles that of the *T. dispar* lineage. *E. mucronatus* occurs primarily east of the Andes, including humid forests over the whole Amazon-Orinoco system and on the eastern foothills of the northern and central Andes. Records from northern Colombia (Magdalena and Maracaibo basins) fall east of the Cordillera Occidental; we consider it unlikely that *E. mucronatus* occurs in Mesoamerica or along the Pacific coastal strip west of the Andes (Abad-Franch et al., 2001; Abad-Franch and Monteiro, 2007; Cuba Cuba et al., 2002; Guhl et al., 2007). The association of *Eratyrus* with bats in live hollow trees and other arboreal ecotopes (Barrett, 1991; Monte et al., 2014) might have helped them successfully occupy very diverse ecoregions.

4.1.3.3 T. maculata

T. maculata is the only truly (evolutionarily) South American representative of the genus *Triatoma* that occurs north of the Amazon. It occupies open land in the northern lowlands of Colombia east of the Cordillera Occidental, the lower Orinoco basin and associated coastal basins and ecoregions, and the Guianan savannahs of Roraima in Brazil (Abad-Franch and Monteiro, 2007; Barrett, 1991; Guhl et al., 2007; Lent and Wygodzinsky, 1979). In part due to confusion about the origins and identification of specimens used for DNA sequencing in some studies (see dos Santos et al., 2007; Pita et al., 2016), and in part due to superficial phenotypic similarity with some congenerics (Lent and Wygodzinsky, 1979), the phylogenetic affinities of *T. maculata* remain uncertain (e.g. Díaz et al., 2016; García et al., 2001; Hypša et al., 2002; Justi et al., 2014, 2016; Sainz et al., 2004). It seems however clear that (i) *T. maculata* shares a most recent common ancestor with the South American Triatomini lineage, and not with the 'North American' lineage, (ii) it is not sister to *T. pseudomaculata* or any species within the *T. pseudomaculata* group (dos Santos et al., 2007; Hypša et al., 2002; Justi et al., 2014; Pita et al., 2016), and (iii), although it harbours moderate genetic and phenotypic diversity (dos Santos et al., 2007; Monsalve et al., 2016), *T. maculata* it is not part of a highly genetically or phenotypically diverse group of northern species (as would be expected had it been evolving in isolation over several million years). These facts argue in favour of a relatively recent isolation (perhaps Pliocene–Pleistocene) from its relatives south of the Amazon (Abad-Franch and Monteiro, 2007). Based on 16S and 28S sequence data, Justi et al. (2016) suggest, however, that *T. maculata* is basal to all South American *Triatoma* except the *T. vitticeps* + *T. melanocephala* clade, implying a much deeper (likely Miocene) divergence. This raises the question of why *T. maculata* did not diversify as profusely as did its putative sister clade, which would include the *T. infestans*, *T. brasiliensis*, and *T. sordida* clades we survey below (Fig. 3).

4.1.3.4 The T. infestans Clade

Wild *T. infestans* populations are widespread in the dry Chaco and in the montane dry forests of the eastern Bolivian Andes (Buitrago et al., 2010; Ceballos et al., 2009, 2011; Dujardin et al., 1987; Noireau, 2009; Noireau et al., 1997b, 2000a, b, 2005; Rojas Cortez et al., 2007; Rolón et al., 2011; Waleckx et al., 2012). A few records from stone piles and terrestrial bromeliads in anthropic landscapes of the Matorral in central Chile (Bacigalupo et al., 2006, 2010) might represent feral populations (Torres-Pérez et al., 2011), although introduced domestic triatomines seem unlikely to move back into new wild habitats.

T. infestans shares a relatively recent (Pleistocene) common ancestor with *Triatoma platensis* and *Triatoma delpontei*, with either the latter (Bargues et al., 2006; García et al., 2001; Ibarra-Cerdeña et al., 2014; Panzera et al., 1995; Pereira et al., 1996; Sainz et al., 2004) or the former (Justi et al., 2016) recovered as basal. Since *T. platensis* and *T. delpontei* are both typical of the Chaco (dry and humid) and adjacent savannah ecoregions (Barrett, 1991; Carcavallo et al., 1985; Lent and Wygodzinsky, 1979), this suggests that all three species evolved and diversified in the Gran Chaco lowlands, likely in association with birds in arboreal microhabitats (Barrett, 1991; Carcavallo et al., 1985; Ceballos et al., 2009; Lent and Wygodzinsky, 1979; Noireau et al., 2000b). More generally, the highly diverse *Triatoma* clades sister to *T. delpontei* + *T. platensis* + *T. infestans* are all typically lowland and, except for some *T. maculata*, cis-Andean (see Fig. 3 and below). East-Andean wild *T. infestans* populations, which exploit stony microhabitats and associate with rodents, would thus be relatively recent derivatives (Torres-Pérez et al., 2011), and not ancestral as often suggested (e.g. Bargues et al., 2006; Forattini, 1980; Panzera et al., 2004; Schofield, 1988). Once adapted to man-made habitats, these originally east-Andean populations spread back throughout southern South America, aided by human migration, to eventually become the most important vectors of human Chagas disease (Schofield, 1988). In the lowlands, wild *T. infestans* populations do not seem to have been recorded in the Chiquitania seasonally dry forests, the Pantanal, the humid Chaco, or, to the west and south, the dry Monte shrublands and arid southern Chaco. East-Andean wild *T. infestans* appear to be restricted to the Bolivian montane dry forests of the Andean valleys in La Paz, Cochabamba, Potosí, and Tarija, with a few records from the Southern Andean Yungas (Buitrago et al., 2010; Ceballos et al., 2009, 2011; Noireau, 2009; Noireau et al., 1997b, 2000b, 2005; Rojas Cortez et al., 2007; Waleckx et al., 2012). Wild *T. infestans* appear to be absent from the humid Bolivian Yungas and the Andean Puna—except perhaps for some parts of the Central Andean wet Puna, which connects pockets of dry montane forest in high-altitude valleys where wild *T. infestans* occur (e.g. Buitrago et al., 2010).

4.1.3.5 The T. brasiliensis Clade

The semiarid Caatinga is home to a group of fairly closely related *Triatoma* species that probably diversified from a late Miocene-early Pliocene common ancestor (Justi et al., 2014, 2016; Monteiro et al., 2004). *T. brasiliensis* (including *T. b. macromelasoma*) and *T. juazeirensis* likely represent a monophyletic subclade sister to a second one including *T. melanica*, *T. sherlocki*, *T. lenti*, and *T. bahiensis*; finally, *T. petrocchiae* has been recovered as basal to these two subclades (Justi et al., 2016; Mendonça et al., 2009, 2016; Monteiro et al., 2004; Oliveira et al., 2017).

T. petrocchiae appears to be restricted to the north of the lower São Francisco River, primarily occupying the Caatinga on the Borborema Plateau and the Chapada do Araripe; we regard the few records from southern Bahia (São Francisco valley and Serra do Espinhaço; Caranha et al., 2011; Galvão, 2014) as dubious. *T. bahiensis* is known from a relatively small area on the northern tip of the Serra do Espinhaço between the São Francisco and Paramirim river valleys (Mendonça et al., 2015, 2016); in this hilly region, the Caatinga is interspersed with Campos Rupestres montane savannahs and Atlantic dry forests. *T. lenti* seems to be more widespread over the Caatinga and Atlantic dry forests on the Chapada Diamantina; records from the Cerrado west of the São Francisco valley (Galvão, 2014) seem also dubious. *T. melanica* occupies the São Francisco River valley to the west of the Serra do Espinhaço in Minas Gerais. Although reported from areas of Caatinga and Atlantic dry forests on the northern tip of its distribution in southern Bahia, it spreads southwards well into the Cerrado along the São Francisco valley, reaching its upper stretches in the Alto Paraná Atlantic forests (Costa et al., 2014; Souza et al., 2015). *T. melanica* seems to be the only species in this clade that is not primarily associated with the Caatinga. *T. sherlocki* is only known from the northernmost tip of the Serra da Mangabeira on the southwestern Chapada Diamantina (Costa et al., 2014).

T. brasiliensis and its closest relatives occur across the entire Caatinga and associated dry ecoregions (Barrett, 1991; Costa et al., 2014; Gurgel-Gonçalves et al., 2012; Lent and Wygodzinsky, 1979). *T. b. brasiliensis* occupies the Caatinga northern lowlands and the northern half of the Borborema Plateau; to the west, it extends into the transitional Atlantic dry forests and then into the northeastern tip of the Cerrado. To the north–northwest, it does not seem to cross the Serra da Ibiapaba-Serra Grande range between the states of Ceará and Piauí (Costa et al., 2014; Gurgel-Gonçalves et al., 2012; Souza et al., 2015). To the southeast, *T. b. brasiliensis* is replaced by *T. b. macromelasoma* on the southern half of the Borborema Plateau; to the southwest, its distribution seems limited by the mountain ranges that bound the São Francisco valley (Costa et al., 2014). *T. brasiliensis* has therefore very rarely been reported from the Caatinga southern lowlands (Costa et al., 2014; Gurgel-Gonçalves et al., 2012). Instead, those lowlands and associated ecoregions (including the Atlantic dry forests on the Chapada Diamantina and perhaps the Bahia interior forests) are occupied by *T. juazeirensis*. Most *T. juazeirensis* records fall south and east of the São Francisco River (Costa et al., 2014).

4.1.3.6 The T. sordida Clade

Sister to the *T. brasiliensis* clade is a highly diversified suite of species that occupy a large area across the South American open lowlands of the Caatinga, Cerrado, Pantanal, Chaco, Pampas, and down to the Patagonian steppe, with some species occurring also in parts of the Atlantic forests, the dry Monte and Espinal, and a few Andean ecoregions. These diverse *Triatoma* are subdivided into three main clusters of closely related species that we will refer to as 'species groups' (Table 3; see Almeida et al., 2009; Carbajal de la Fuente et al., 2011; dos Santos et al., 2007; García et al., 2001; Hypša et al., 2002; Ibarra-Cerdeña et al., 2014; Justi et al., 2014, 2016; Noireau et al., 1998, 1999, 2002; Panzera et al., 2015; Pita et al., 2016; Sainz et al., 2004; Teves et al., 2016).

4.1.3.6.1 The T. rubrovaria Species Group This group seems to represent a southwestern radiation with early (likely late Pliocene) offshoots in the Chaco and a more derived, Pleistocene subclade that dispersed eastwards across the lower Rio de la Plata basin and then colonized the lowland savannahs east and south of the Uruguay River and up to the southern limit of the moist Atlantic forests (Almeida et al., 2009; García et al., 2001; Hypša et al., 2002; Ibarra-Cerdeña et al., 2014; Justi et al., 2014, 2016; Sainz et al., 2004). *T. patagonica* and *Triatoma limai* occur in the dry Chaco and Espinal, although *T. patagonica* has also been recorded from the humid Chaco, the low Monte, the humid Pampas, and, as one would expect, the Patagonian steppe. *T. guasayana* has a wide distribution across the dry Chaco but also occurs on neighbouring ecoregions on the eastern Andean foothills and valleys (high Monte, Southern Andean Yungas, and Bolivian montane dry forests) as well as in the Espinal and the humid Pampas; it is apparently absent from the humid Chaco. To the east, *T. rubrovaria* and *T. circummaculata* occupy the Uruguayan savannahs (where *T. carcavalloi*, *Triatoma oliveirai*, and *T. pintodiasi* have been recorded) and reach the southernmost limit of the Alto Paraná Atlantic forests (where *T. klugi* also occurs) (Barrett, 1991; Carcavallo et al., 1985, 1999; Galvão, 2014; Gurgel-Gonçalves et al., 2012; Lent and Wygodzinsky, 1979; Noireau et al., 1999; Pita et al., 2016). Overall, it seems that the Atlantic forests on the southern tip of the Serra Geral, including the Alto Paraná and Araucaria forests, represent unsuitable habitat for the species in this clade.

4.1.3.6.2 The T. pseudomaculata Species Group *T. pseudomaculata* is widespread across the Caatinga and the eastern half of the Cerrado, but also occurs in other ecoregions including the Atlantic dry forests, the

Pernambuco and Bahia interior forests, the northernmost Alto Paraná Atlantic forests, and perhaps the easternmost Mato Grosso seasonal forests in the Cerrado-Amazon transition; all records but a dubious one fall east of the Araguaia River (Carcavallo et al., 1999; Galvão, 2014; Gurgel-Gonçalves et al., 2012; Silveira et al., 1984). *T. arthurneivai* is known only from a small area of transition between the Cerrado and Bahia interior forests on the southern stretches of the Serra do Espinhaço, whereas the closely related *T. wygodzinskyi* occupies the eastern Alto Paraná Atlantic forests associated with the Serra da Mantiqueira complex (Carbajal de la Fuente et al., 2011; dos Santos et al., 2007; Galvão, 2014). *T. deaneorum* has only been collected in the central-western Cerrado, and *T. williami* and *T. baratai* occur both in the Cerrado and in the seasonally flooded Pantanal lowlands. The only two records of *T. guazu* are from the central-western Cerrado in Brazil and from the southwestern Alto Paraná Atlantic forests in Paraguay (Barrett, 1991; Galvão et al., 2001; Lent et al., 1996; Lent and Wygodzinsky, 1979; Obara et al., 2012). Although this will need to be confirmed, *T. costalimai* and the closely related *T. jatai* might also belong to this (probably Pliocene–Pleistocene) group (Justi et al., 2014, 2016; Pita et al., 2016; Teves et al., 2016) (see Table 3). *T. costalimai* occurs in the central-eastern Cerrado, and perhaps also the northernmost patches of Alto Paraná Atlantic forests, along the eastern–southern limits of the upper Tocantins basin and east of the Araguaia River. *T. jatai* has so far only been collected at one site in the eastern Cerrado (Brito et al., 2017; Gonçalves et al., 2013; Gurgel-Gonçalves et al., 2012).

4.1.3.6.3 The T. sordida Species Group *T. sordida*, as defined by morphological characters, encompasses at least three genetic subclades that share a (likely Pliocene–Pleistocene) common ancestor with three further, very closely related species (Ibarra-Cerdeña et al., 2014; Justi et al., 2016; Noireau et al., 1998, 1999; Panzera et al., 2015; Pita et al., 2016) (see also Section 3.3.3). The first *T. sordida* subclade occupies the core Cerrado; it also extends, however, into the Caatinga and the Bahia interior and Atlantic dry forests to the east, into the Alto Paraná Atlantic forests to the south (with a few records from the neighbouring humid Chaco), and into the northern dry Chaco, the Chiquitania dry forests (and perhaps parts of the Pantanal), and up into the Andean Bolivian Yungas of La Paz to the west. The second subclade occurs over the humid Chaco and the central-southern dry Chaco, reaching the Southern Andean Yungas and the extending northwards along the eastern slope of the Andes to the southern parts of the Bolivian dry montane forests. The third subclade includes specimens collected in the core dry Chaco and identified as either *T. sordida* or *T. garciabesi*; this subclade hence occurs in the area between the northern and central-southern dry Chaco occupied by the two other *T. sordida* subclades (Gurgel-Gonçalves et al., 2011; Panzera et al., 2015; Pita et al., 2016). Sister to the *T. sordida* + *T. garciabesi* clade is a group of central-western/southwestern Cerrado species including *T. jurbergi* plus the genetically and phenotypically very similar *T. matogrossensis* (which has also been recorded in the Pantanal) and *T. vandae* (García et al., 2001; Hypša et al., 2002; Justi et al., 2014, 2016; Noireau et al., 2002; Panzera et al., 2015; Pita et al., 2016; Sainz et al., 2004).

4.2 The Tribe Rhodniini

Our views about the biogeography and evolution of the clearly monophyletic Rhodniini were discussed in detail in a previous paper (Abad-Franch et al., 2009); therefore, here we will only summarize (and, to the extent possible, update) those views. The most relevant news regarding the deep evolutionary relations of the Rhodniini is that they seem to be more closely related to *Microtriatoma* (Bolboderini) and *Cavernicola* (Cavernicolini) than to the Triatomini (Hwang and Weirauch, 2012; Patterson and Gaunt, 2010). The second point of some interest concerns the early split of the two basal lineages within the tribe. One view suggests that the earliest divergence separated the trans-Andean lineage (*Rhodnius pallescens* and allies) from the cis-Andean lineage (*R. pictipes* and allies + *R. robustus* and allies including *Psammolestes*). An alternative possibility is that the basal cladogenesis yielded the 'pictipes lineage' including the trans-Andean species cluster and the 'robustus lineage' including *Psammolestes*. We have favoured this latter view based on some phenotypic traits, biogeographic–ecological patterns, allozyme relations, and taxon-rich mitochondrial gene phylogenies (Abad-Franch et al., 2009; see also Díaz et al., 2014; Monteiro et al., 2000, 2002, 2003; see also Fig. 2). Recently, Justi et al. (2016) boldly stated that their molecular data analyses refute this latter hypothesis because they recover *R. pallescens* and allies as a sister clade to all other Rhodniini; this conclusion, however, is based on a limited data subset including few taxa with only partial sequence overlap (e.g. a 434-bp 16S overlap with up to three segregating sites between *R. pallescens*, *Rhodnius colombiensis*, and *Rhodnius ecuadoriensis*, or a 405-bp 16S overlap between *R. pallescens* and *R. pictipes*). In addition, Justi et al. (2016) seem to have misread or misunderstood the parsimony analysis of endemicity (PAE) cladograms of

Abad-Franch et al. (2009), perhaps by overlooking the caption to Fig. 2, where it is noted that '... PAE cladograms depict mainly contemporary patterns of relatedness among areas—and not historical processes of diversification'. Finally, Justi et al. (2016) emphasize that their results match those of previous phylogenies—as they indeed should, because the same nucleotide sequences, retrieved from GenBank, were used (e.g. Hypša et al., 2002; Lyman et al., 1999; see also de Paula et al., 2007). We therefore conclude that judgement about the basal split in the Rhodniini must be cautiously suspended until further data become available. It seems uncontroversial, however, that the tribe Rhodniini is composed of three major clades.

4.2.1 *The* R. pallescens *Lineage: The Trans-Andean* Rhodnius

Three named, closely related species of *Rhodnius* occur on the western side of the Andes (or the Cordillera de Mérida at the northernmost tip). The basal split within this clade separated (i) a northern clade including two *R. pallescens* subclades (one extending into moister Central American ecoregions and the other from the drier Magdalena basin lowlands) and *R. colombiensis* (from the middle-upper Magdalena valley dry forests) and (ii) a southern clade including two young *R. ecuadoriensis* subclades—one from western and southwestern Ecuador and the other from northwestern Peru (see Abad-Franch et al., 2009; Díaz et al., 2014; Gómez-Palacio et al., 2012; Pita et al., 2013 for further details). *R. pallescens* (both subclades) and *R. colombiensis* are associated with *Attalea butyracea* palms, whereas wild *R. ecuadoriensis* occur in *Phytelephas aequatorialis* palms and squirrel or bird nests in Ecuador and in hollow trees and cacti in northern Peru (Abad-Franch et al., 2015; Barrett, 1991).

4.2.2 *The* R. pictipes *Lineage*

R. pictipes has been recorded from palms in six genera throughout the Amazon and Orinoco basins; it seems to be sister to *R. amazonicus* from the moist forests of the Guiana Shield. A second clade includes the closely related *Rhodnius brethesi* and *Rhodnius stali* (from, respectively, *Leopoldinia piassaba* palms in northwestern and *Attalea phalerata* palms in southwestern Amazonia) plus a recently discovered form, superficially similar to *R. pictipes*, collected in the montane forests on the Sierra Nevada de Santa Marta in northern Colombia (Fig. 2). We also note that atypical *R. pictipes* forms have been reported from northwestern Venezuela (Aldana et al., 2003). Finally, *R. paraensis*, known from the eastern–northeastern Amazon basin, is probably also a member of this lineage, but its genetic affinities have not, to the best of our knowledge, been investigated. Although *R. zeledoni* was described as resembling *R. paraensis*, the illustrations in the original paper (Jurberg et al., 2009) are quite clearly suggestive of close similarity with *R. domesticus*; in addition, the only record of *R. zeledoni* is from coastal Sergipe, Brazil, which is within the distribution range of *R. domesticus* (Galvão, 2014). See Abad-Franch and Monteiro (2007) and Abad-Franch et al. (2009, 2015) for further details.

4.2.3 The R. robustus *Lineage, Including the* Psammolestes

Although the basal cladogenetic events are not well resolved in available phylogenies (see Fig. 2), it seems likely that they yielded two extant species apparently restricted to Atlantic coastal sites (*Rhodnius neivai* and *R. domesticus*) and the *Psammolestes*. *R. neivai* occurs around the northern tip of the Andean Cordillera Oriental and the Maracaibo basin, and *R. domesticus* (presumably including *R. zeledoni*) in the Brazilian Atlantic moist forests. These two early offshoots now occur, respectively, north and south of the Amazon-Orinoco system. The *Psammolestes* adapted to exploit bird nest microhabitats and currently occur over the open ecoregions north and south of the moist Amazon forests—*Psammolestes arthuri* in the Orinoco and Venezuelan coastal basins, *Psammolestes tertius* primarily in the Cerrado-Caatinga, and *Ps. coreodes* primarily in the Chaco (Abad-Franch et al., 2009). Sister to these basal taxa is a diverse group of species and sublineages including *R. barretti* from the Napo moist forests (and perhaps *R. dalessandroi* from the Orinoco Llanos), *R. nasutus* from the Caatinga (and a newly discovered form, sister to *R. nasutus*, from the Guianan savannahs of Roraima in northern Brazil; see Fig. 2), and the *R. robustus* + *R. neglectus* + *R. prolixus* subclade (Abad-Franch et al., 2009, 2013). This latter subclade (Fig. 2) seems to represent a late Pliocene–Pleistocene radiation that yielded (i) a northern cluster from the Orinoco basin (*R. prolixus* and *R. robustus* I) and the central-northern Amazon (*R. robustus* V); (ii) a southern cluster represented by *R. neglectus* (primarily from the Cerrado, but also including *R. milesi* from eastern Amazonia (Fig. 2) and possibly *R. taquarussuensis* from the transition between the southeastern edge of the Cerrado and the Alto Paraná Atlantic forests); and (iii) a core-Amazon cluster including *R. robustus* II (to the south and west; described as *R. montenegrensis*), *R. robustus* III (to the southeast; recently described as *R. marabaensis*), and *R. robustus* IV (northeastern Amazon including the Guianan highlands) (see Fig. 2 and Section 3.3.1). While these cyt *b*-based results are largely consistent across markers, the subclade including *R. neglectus* and *R. milesi* is recovered as sister to *R. nasutus* in phylogenies based on ITS-2 sequences (Fig. 2) and in allozyme-based comparisons (Chávez et al., 1999; Monteiro et al., 2002). This highlights the challenges of disentangling the evolutionary history of relatively young yet highly diversified triatomine groups. (Further details on *R. robustus* and related species can be found in Abad-Franch et al., 2009, 2013, 2015; Abad-Franch and Monteiro, 2007; Conn and Mirabello, 2007; Fitzpatrick et al., 2008; Justi et al., 2016; Lyman et al., 1999; Monteiro et al., 2000, 2002, 2003; Pavan and Monteiro, 2007; Rosa et al., 2012, 2017; Souza et al., 2016.)

4.3 Other Tribes

4.3.1 The Cavernicolini

Cavernicola pilosa is very widely spread from Panama to southeastern Brazil; the close association with bats (in caves or hollow trees) may have contributed to its dispersal across a large and ecologically diverse area including humid forests (Isthmian-Atlantic, Amazonian *terra firme* and *várzea*, Bahia interior, and Alto Paraná Atlantic forests), montane forests (Venezuelan Andes, Eastern Cordillera Real), open savannahs (Orinoco Llanos and Cerrado), and drier forests and shrubland (in, for example, the Magdalena valley or along the Venezuelan coast). In sharp contrast, *C. lenti* is only known from the type locality in the Uatumã-Trombetas moist forests—i. e., between the left banks of the Branco-Negro-Amazon rivers and the southern limit of the Guiana Shield (Abad-Franch and Monteiro, 2007; Barrett, 1991; Lent and Wygodzinsky, 1979; Oliveira et al., 2007).

4.3.2 The Bolboderini

The tribe Bolboderini includes 4 genera and 13 named species whose phylogenetic relations remain uncharted. A genus-level cladogram based on 18 morphological characters suggests *Bolbodera* (a genus with a single known species, *Bolbodera scabrosa*) as the more highly divergent taxon and *Belminus* as sister to the very similar *Microtriatoma* and *Parabelminus* (Lent and Wygodzinsky, 1979, p. 440). These small, inconspicuous, mostly arboreal triatomines are only rarely collected, and their distribution is therefore only partially known (Santana, 2014).

B. scabrosa is only known from Cuba, where it seems to occur in both dry and moist forests (Lent and Wygodzinsky, 1979). *Belminus costaricensis* occurs in the Costa Rican seasonal and Isthmian-Atlantic moist forests. *Belminus* specimens collected from bromeliads in the Sierra de los Tuxtlas moist forests, just north of the Isthmus of Tehuantepec, Mexico, closely resemble *B. costaricensis* (Lent and Wygodzinsky, 1979, p. 445). This suggests that intermediate *Belminus* populations likely occur along the Mesoamerican moist forests—perhaps including the Yucatán peninsula, which would suggest a possible link with Cuban *Bolbodera*. To the south, *B. herreri* has been reported from Panama (Chocó-Darién moist forests) and from the montane forests on the western slope of the Cordillera Oriental in northern Colombia, where *Belminus ferroae* and *Belminus corredori* also occur. *Belminus rugulosus* has been collected on both sides of the northern Andes in Colombia and in montane forests along the Cordillera de la Costa in Venezuela—where it is sympatric with *Belminus pittieri*. *B. laportei* has only been reported from moist forests close to the mouth of the Amazon River—the eastern Uatumã-Trombetas, the northern Tocantins/Pindaré, and the Marajó *várzea*. To the west, the genus *Belminus* is represented by *Belminus peruvianus* of the northeastern Peruvian Andes, where it has been collected in the southernmost tip of the Eastern Cordillera Real montane forests and in the Marañón valley dry forests (Abad-Franch and Monteiro, 2007; Barrett, 1991; Carcavallo et al., 1999; Galvão, 2014; Galvão and Angulo, 2006; Lent et al., 1995; Lent and Wygodzinsky, 1979; Osuna and Ayala, 1993; Sandoval et al., 2004, 2007, 2010). *Microtriatoma trinidadensis* is very widely distributed, with records from the Isthmian-Atlantic moist forests of Central America, the Orinoco basin Llanos and coastal moist and swamp forests, the Amazonian moist and freshwater swamp forests, the Mato Grosso seasonal forests, the Peruvian and Bolivian Yungas, and the Chiquitania dry forests of eastern Bolivia—and, of course, also from the Trinidad and Tobago moist forests. The finding of a *Microtriatoma* nymph in Costa Rica implies that the genus extends further north into Mesoamerica. *Microtriatoma borbai* seems to occur primarily in the Brazilian Serra do Mar and Araucaria moist Atlantic forests, but has also been found in the Cerrado. Finally, the two known species of *Parabelminus* are also associated with the Brazilian Atlantic ecoregions. *Parabelminus carioca* is known only from one site in the Serra do Mar coastal forests, and *Parabelminus yurupucu* from a small area encompassing the northern tips of the Bahia interior and coastal forests (Abad-Franch and Monteiro, 2007; Carcavallo et al., 1999; Galvão, 2014; Lent and Wygodzinsky, 1979).

4.3.3 The Alberproseniini

Alberprosenia goyovargasi, the smallest known triatomine, has only been collected in the Maracaibo basin dry forests, whereas *Alberprosenia malheiroi* was found in hollow palm trunks of the Tocantins/Pindaré moist forests of eastern Amazonia (Abad-Franch et al., 2015; Carcavallo et al., 1995, 1999; Galvão, 2014).

5. CLOSING THOUGHTS AND CONCLUSIONS

A full understanding of the evolution, systematics, and biogeography of the Triatominae will require better phylogenies—with denser taxon and character sampling—and, for the many little-known species, more and better-documented field records. Thoughtful, hypothesis-driven selection of nontriatomine reduviid taxa and a more extensive use of explicit biogeographic procedures will also help further that understanding. The quick development of highly efficient technologies to generate and analyse DNA sequence data suggests that improved phylogenies will soon become available. To be genuinely useful, however, such technologies must be coupled with, first, careful identification and vouchering of (preferably field-collected) material, and, second, rigorous processing of output data including strict sequence quality checks. Sloppy ascertainment or documentation (e.g. mislabelling) of the identity and places of capture of specimens used in molecular studies, together with doubts about the genetic integrity of many laboratory colonies, yield results that, when published, only add confusion to an already complicated field of study. Similarly, substandard-quality sequences can mislead not only those who generated them: once made available in public databases, their misleadingness is bound to swiftly propagate across the literature. The information generated and published up to the present, with its problems and all, has however allowed us to perceive certain general trends and patterns; we summarize them below.

From an evolutionary perspective, currently recognized tribes almost certainly represent monophyletic assemblages. Major rearrangements will nonetheless be required if genus-level systematics are to match actual patterns of ancestry and descent. Subgeneric classifications including species groups, complexes, and subcomplexes have traditionally been based on superficial phenotype similarities, and are often confusing and biologically unsound. We suggest that any such classification, if one is really needed, should reflect phylogenetic affinities and should perhaps use terms like 'lineage', 'clade', 'species group', and (occasionally) 'subclade' along the lines followed in this review. The term 'species complex' would have a more restrictive meaning involving closely related species whose relations are difficult to establish based on morphology and were usually resolved using molecular approaches (see Section 3.3 and Table 3).

The Mesoamerican-Caribbean region seems key to the origin of the Triatomini, with several waves (some ancient, some recent) of southward and northward dispersal—one of which reached Asia. In the Old World, the *Linshcosteus* seem to have evolved on the Indian subcontinent, whereas *Triatoma* spp. likely evolved in Southeast Asia with a later insular radiation over Sundaland, the Philippines, Wallacea, and northern Australasia. Northern South America, in turn, seems key to the origin of the Rhodniini, which appear to have first evolved in what is now the orographically complex northern tip of the Andes; later, they colonized the western side of the northern Andes and the Guianan and Brazilian shields—and then dispersed to moist forests on the Isthmus of Panama, on the Atlantic coastal ranges, and across the Amazon-Orinoco lowlands. Extensive taxon diversity in the northern Andes suggests that the Bolboderini might have evolved in a quite similar way. The Cavernicolini seem to be Amazonian in origin, whereas the evolutionary origins and affinities of Alberproseniini remain unknown. We also note that there seems to be a general, recurrent pattern among South American triatomines in which basal lineages are associated with the Andes and the Atlantic coastal ranges north and south of the Amazon basin, whereas more derived taxa tend to occur in lowland areas—occasionally with recolonization of Andean valleys including a few high-altitude spots (as in the case, for instance, of east-Andean *T. infestans* populations).

Triatomine bugs have colonized almost every terrestrial ecoregion available to them in the Americas. Different species occur across the current extreme range of humidity in the continent, from the hyperarid Atacama Desert to the hypermoist Napo forests of western Amazonia. However, with only a few exceptions, no species seems to occur in areas where average annual temperatures are below ~10°C. This applies, for example, to the high-altitude Andean grasslands (Páramo and Puna) or the southern Valdivian temperate and Magellanic subpolar forests. Exceptions to this '10°C rule' are (i) the northernmost records of *T. protracta* from the Colorado Plateau shrublands and the Great Basin shrub steppe (northwestern Colorado and northern Utah, USA), (ii) *T. boliviana* and *T. infestans* from the Bolivian Central Andean wet Puna, and (iii) the southernmost record of *T. patagonica* from the Patagonian steppe (southwestern Chubut, Argentina). Old World *Triatoma* appear to be primarily associated with lowland rainforests, whereas the *Linshcosteus* seem to have evolved in drier, stony habitats.

From a public health standpoint, we finally note that the capacity of any given triatomine species to stably infest houses (and, hence, to become a major vector of human Chagas disease) does not seem to map consistently onto the phylogeny of any lineage, group, clade, or subclade (see also Abad-Franch, 2016). For example, none of the species within the subclades sister to *T. infestans* (*T. platensis* and *T. delpontei*), *R. prolixus* (*R. robustus* I and V), *T. dimidiata* (Yucatán genotypes, *T. hegneri*), *T. b. brasiliensis* (*T. b. macromelasoma* and *T. juazeirensis*), or *P. megistus* (*T. tibiamaculata*) are important domestic vectors of *T. cruzi*. At higher systematic levels, these key 'domestic' vectors are scattered across clades and genera within the Triatomini and Rhodniini—with, respectively, four and one major vector species in three and one genera (as currently recognized). This suggests that the set of traits conferring 'domiciliation capacity' arose several times in a few terminal branches (sometimes even population-level) of the many Triatomini and Rhodniini lineages. In very recent times, humans began building bug-friendly habitats (adequate physical substrate, plenty of vertebrate blood) in what we now call the Americas. On occasion, this happened within the ranges of bug species or populations that, by historical accident, expressed the combination of traits that would allow them to successfully colonize such man-made habitats—microclimate-stress tolerance, adaptability to new blood sources, painless bites, high reproductive potential, and so forth. Some of those bugs started to exploit the man-built environment and were, in addition, prone to get infected by *T. cruzi*—and good at transmitting it. And thus began human Chagas disease.

ACKNOWLEDGEMENTS

F.A.M. received support from the Conselho Nacional de Desenvolvimento Científico e Tecnológico (CNPq) of Brazil. F.A.-F. acknowledges support by the Instituto René Rachou (IRR) and the Vice-Presidência de Pesquisa e Coleções Biológicas (VPPCB), both at Fiocruz.

REFERENCES

Abad-Franch, F., 2003. The ecology and genetics of Chagas disease vectors in Ecuador, with emphasis on *Rhodnius ecuadoriensis* (Triatominae). PhD thesis, London School of Hygiene and Tropical Medicine, University of London, UK.

Abad-Franch, F., 2016. A simple, biologically sound, and potentially useful working classification of Chagas disease vectors. Mem. Inst. Oswaldo Cruz 111, 649–651.

Abad-Franch, F., Monteiro, F.A., 2005. Molecular research and the control of Chagas disease vectors. An. Acad. Bras. Cienc. 77, 437–454.

Abad-Franch, F., Monteiro, F.A., 2007. Biogeography and evolution of Amazonian triatomines (Heteroptera: Reduviidae): implications for Chagas disease surveillance in humid forest eco-regions. Mem. Inst. Oswaldo Cruz 102, 57–69.

Abad-Franch, F., Paucar, C.A., Carpio, C.C., Cuba, C.A.C., Aguilar, H.M., Miles, M.A., 2001. Biogeography of Triatominae (Hemiptera: Reduviidae) in

Ecuador: implications for the design of control strategies. Mem. Inst. Oswaldo Cruz 96, 611–620.

Abad-Franch, F., Monteiro, F.A., Jaramillo, O.N., Gurgel-Gonçalves, R., Dias, F.B.S., Diotaiuti, L., 2009. Ecology, evolution, and the long-term surveillance of vector-borne Chagas disease: a multi-scale appraisal of the tribe Rhodniini (Triatominae). Acta Trop. 110, 159–177.

Abad-Franch, F., Pavan, M.G., Jaramillo, O.N., Palomeque, F.S., Dale, C., Chaverra, D., Monteiro, F.A., 2013. *Rhodnius barretti*, a new species of Triatominae (Hemiptera: Reduviidae) from western Amazonia. Mem. Inst. Oswaldo Cruz 108, 92–99.

Abad-Franch, F., Lima, M.M., Sarquis, O., Gurgel-Gonçalves, R., Sánchez-Martín, M., Calzada, J., Saldaña, A., Monteiro, F.A., Palomeque, F.S., Santos, W.S., Angulo, V.M., Esteban, L., Dias, F.B.S., Diotaiuti, L., Bar, M.E., Gottdenker, N.L., 2015. On palms, bugs, and Chagas disease in the Americas. Acta Trop. 151, 126–141.

Ábalos, J.W., Wygodzinsky, P., 1951. Las Triatominae argentinas (Reduviidae, Hemiptera). Inst. Med. Reg. Tucumán 601, 1–179.

Abrahan, L.B., Gorla, D.E., Catalá, S.S., 2011. Dispersal of *Triatoma infestans* and other Triatominae species in the arid Chaco of Argentina—flying, walking or passive carriage? The importance of walking females. Mem. Inst. Oswaldo Cruz 106, 232–239.

Adams, R.R., Ryckman, R.E., 1969. A comparative electrophoresis study of the *Triatoma rubida* Complex (Hemiptera: Reduviidae: Triatominae). J. Med. Entomol. 6, 1–18.

Aldana, E., Lizano, E., Valderrama, A., 2003. Revisión del estatus taxonómico de *Rhodnius pictipes* Stål, 1872 (Hemiptera, Reduviidae, Triatominae). Bol. Malariol. Salud Amb. 43, 31–38.

Alevi, K.C.C., Mendonça, P.P., Pereira, N.P., Rosa, J.A., Azeredo-Oliveira, M.T.V., 2012. Karyotype of *Triatoma melanocephala* Neiva and pinto (1923). Does this species fit in the Brasiliensis subcomplex? Infect. Genet. Evol. 12, 1652–1653.

Alevi, K.C.C., Moreira, F.F.F., Jurberg, J., Azeredo-Oliveira, M.T.V., 2015. Description of the diploid chromosome set of *Triatoma pintodiasi* (Hemiptera, Triatominae). Genet. Mol. Res. 15, gmr.15026343.

Alevi, K.C.C., Guerra, A.L., Imperador, C.H., Jurberg, J., Moreira, F.F.F., Azeredo-Oliveira, M.T.V., 2017a. Mitochondrial gene confirms the specific status of *Triatoma pintodiasi* Jurberg, Cunha, and Rocha, 2013 (Hemiptera, Triatominae), an endemic species in Brazil. Am. J. Trop. Med. Hyg. 96, 200–201.

Alevi, K.C.C., Oliveira, J., Azeredo-Oliveira, M.T.V., Rosa, J.A., 2017b. *Triatoma vitticeps* subcomplex (Hemiptera, Reduviidae, Triatominae): a new grouping of Chagas disease vectors from South America. Parasit. Vectors 10, 180.

Almeida, F.B., Santos, E.I., Sposina, G., 1973. Triatomíneos da Amazônia III. Acta Amaz. 3, 43–66.

Almeida, C.E., Marcet, P.L., Gumiel, M., Takiya, D.M., Cardozo-de-Almeida, M., Pacheco, R.S., Lopes, C.M., Dotson, E.M., Costa, J., 2009. Phylogenetic and phenotypic relationships among *Triatoma carcavalloi* (Hemiptera: Reduviidae: Triatominae) and related species collected in domiciles in Rio Grande do Sul state, Brazil. J. Vector Ecol. 34, 164–173.

Alroy, K.A., Huang, C., Gilman, R.H., Quispe-Machaca, V.R., Marks, M.A., Ancca-Juárez, J., Hillyard, M., Verastegui, M., Sánchez, G., Cabrera, L., Vidal, E., Billig, E.M.W., Cama, V.A., Náquira, C., Bern, C., Levy, M.Z., Working Group on Chagas Disease in Peru, 2015. Prevalence and transmission of *Trypanosoma cruzi* in people of rural communities of the high jungle of northern Peru. PLoS Negl. Trop. Dis. 9, e0003779.

Álvarez, C.J., 1984. Historia de la medicina tropical ecuatoriana III. Enfermedad de Chagas en el Ecuador, Ed. Arquidiocesana Justicia y Paz, Guayaquil, Ecuador.

Ambrose, D.P., 2006. A checklist of Indian assassin bugs (Insecta: Hemiptera: Reduviidae) with taxonomic status, distribution and diagnostic morphological characteristics. Zoos Print J. 21, 2388–2406.

Anderson, J.M., Lai, J.E., Dotson, E.M., Cordón-Rosales, C., Ponce, C., Norris, D.E., Beard, C.B., 2002. Identification and characterization of microsatellite markers in the Chagas disease vector *Triatoma dimidiata*. Infect. Genet. Evol. 1, 243–248.

Aplin, K.P., Suzuki, H., Chinen, A.A., Chesser, R.T., Ten Have, J., Donnellan, S.C., Austin, J., Frost, A., Gonzalez, J.P., Herbreteau, V., Catzeflis, F., Soubrier, J., Fang, Y.P., Robins, J., Matisoo-Smith, E., Bastos, A.D., Maryanto, I., Sinaga, M.H., Denys, C., Van Den Bussche, R.A., Conroy, C., Rowe, K., Cooper, A., 2011. Multiple geographic origins of commensalism and complex dispersal history of black rats. PLoS One 6, e26357.

Araujo, R.N., Santos, A., Pinto, F.S., Gontijo, N.F., Lehane, M.J., Pereira, M.H., 2006. RNA interference of the salivary gland nitrophorin 2 in the triatomine bug *Rhodnius prolixus* (Hemiptera: Reduviidae) by dsRNA ingestion or injection. Insect Biochem. Mol. Biol. 36, 683–693.

Araujo, R.N., Soares, A.C., Paim, R.M., Gontijo, N.F., Gontijo, A.F., Lehane, M.J., Pereira, M.H., 2009. The role of salivary nitrophorins in the ingestion of blood by the triatomine bug *Rhodnius prolixus* (Reduviidae: Triatominae). Insect Biochem. Mol. Biol. 39, 83–89.

Ayala, L.J.M., 2009. Una nueva especie de *Panstrongylus* Berg de Venezuela (Hemiptera: Reduviidae, Triatominae). Entomotropica 24, 105–109.

Ayala, L.J.M., 2016. Presencia de *Panstrongylus chinai* (Del Ponte, 1929) en Venezuela, con notas aclaratorias sobre su sinonimia con *Panstrongylus turpiali* (Heteroptera: Reduviidae, Triatominae). Bol. Soc. Entomol. Aragon. 59, 233–236.

Ayala, L.J.M., Mattei, R., Mattei, R., 2014. Descripción de la hembra de *Panstrongylus martinezorum* Ayala, 2009 (Hemiptera: Reduviidae, Triatominae) con comentarios sobre la distribución geográfica de la especie en el Estado Amazonas, Venezuela. Bol. Soc. Entomol. Aragon. 54, 383–389.

Bacigalupo, A., Segura, J.A., García, A., Hidalgo, J., Galuppo, S., Cattan, P.E., 2006. Primer hallazgo de vectores de la enfermedad de Chagas asociados a matorrales silvestres en la Región Metropolitana, Chile. Rev. Med. Chil. 134, 1230–1236.

Bacigalupo, A., Torres-Pérez, F., Segovia, V., García, A., Correa, J.P., Moreno, L., Arroyo, P., Cattan, P.E., 2010. Sylvatic foci of the Chagas disease vector *Triatoma infestans* in Chile: description of a new focus and challenges for control programs. Mem. Inst. Oswaldo Cruz 105, 633–641.

Barbosa, S.E., Dujardin, J.P., Soares, R.P., Pires, H.H., Margonari, C., Romanha, A.J., Panzera, F., Linardi, P.M., Duque-de-Melo, M., Pimenta, P.F., Pereira, M.H., Diotaiuti, L., 2003. Interpopulation variability among *Panstrongylus megistus* (Hemiptera: Reduviidae) from Brazil. J. Med. Entomol. 40, 411–420.

Barbosa, S.E., Belisário, C.J., Souza, R.C., Paula, A.S., Linardi, P.M., Romanha, A.J., Diotaiuti, L., 2006. Biogeography of Brazilian populations of *Panstrongylus megistus* (Hemiptera, Reduviidae, Triatominae) based on molecular marker and paleo-vegetational data. Acta Trop. 99, 144–154.

Bargues, M.D., Klisiowicz, D.R., Panzera, F., Noireau, F., Marcilla, A., Pérez, R., Rojas, M.G., O'Connor, J.E., González-Candelas, F., Galvão, C., Jurberg, J., Carcavallo, R.U., Dujardin, J.P., Mas-Coma, S., 2006. Origin and phylogeography of the Chagas disease main vector *Triatoma infestans* based on nuclear rDNA sequences and genome size. Infect. Genet. Evol. 6, 46–62.

Bargues, M.D., Klisiowicz, D.R., González-Candelas, F., Ramsey, J.M., Monroy, C., Ponce, C., Salazar-Schettino, P.M., Panzera, F., Abad-Franch, F., Sousa, O.E., Schofield, C.J., Dujardin, J.P., Guhl, F., Mas-Coma, S., 2008. Phylogeography and genetic variation of *Triatoma dimidiata*, the main Chagas disease vector in Central America, and its position within the genus *Triatoma*. PLoS Negl. Trop. Dis. 2, e233.

Bargues, M.D., Schofield, C.J., Dujardin, J.P., 2010. Classification and phylogeny of the Triatominae. In: Telleria, J., Tibayrenc, M. (Eds.), American Trypanosomiasis Chagas Disease One Hundred Years of Research. Elsevier, London, UK, pp. 117–147.

Barrett, T.V., 1991. Advances in triatomine bug ecology in relation to Chagas disease. In: Harris, K.H. (Ed.), Advances in Disease Vector Research, vol. 8. Springer-Verlag, New York, pp. 143–176.

Barrett, T.V., 1996. Species interfertility and crossing experiments in triatomine systematics. In: Schofield, C.J., Dujardin, J.P., Jurberg, J. (Eds.), Proceedings of the International Workshop on Population Genetics and Control of Triatominae, Santo Domingo de los Colorados, Ecuador, Sept. 1995. Mexico City, Indre, pp. 72–77.

Benoit, P.L.G., van Sande, M., 1959. Etude des protéines de l'hemolymphe de *Triatoma infestans* et *Rhodnius prolixus* par la ultra-microelectrophorese en gel de gélose. Ann. Soc. Belg. Med. Trop. 39, 135–143.

Bérenger, J.M., Blanchet, D., 2007. A new species of the genus *Panstrongylus* from French Guiana (Heteroptera; Reduviidae; Triatominae). Mem. Inst. Oswaldo Cruz 102, 733–736.

Bérenger, J.M., Pluot-Sigwalt, D., 2002. *Rhodnius amazonicus* Almeida, Santos and Sposina, 1973, bona species, close to *R. pictipes* Stål, 1872 (Heteroptera: Reduviidae: Triatominae). Mem. Inst. Oswaldo Cruz 97, 73–77.

Bern, C., Kjos, S., Yabsley, M.J., Montgomery, S.P., 2011. *Trypanosoma cruzi* and Chagas' disease in the United States. Clin. Microbiol. Rev. 24, 655–681.

Bininda-Emonds, O.R.P., Cardillo, M., Jones, K.E., MacPhee, R.D.E., Beck, R.M.D., Grenyer, R., Price, S.A., Vos, R.A., Gittleman, J.L., Purvis, A., 2007. The delayed rise of present-day mammals. Nature 446, 507–512.

Borges-Pereira, J., Zauza, P.L., Galhardo, M.C., Nogueira, J.S., Pereira, G.R., Cunha, R.V., 2001. Doença de Chagas na população urbana do distrito sanitário de Rio Verde, Mato Grosso do Sul, Brasil. Rev. Soc. Bras. Med. Trop. 34, 459–466.

Brito, R.N., Diotaiuti, L., Gomes, A.C.F., Souza, R.C.M., Abad-Franch, F., 2017. *Triatoma costalimai* Verano & Galvão in and around houses of Tocantins state, Brazil, 2005–2014. J. Med. Entomol. 54, 1771–1774.

Brodie, H.D., Ryckman, R.E., 1967. Molecular taxonomy of Triatominae (Hemiptera: Reduviidae). J. Med. Entomol. 4, 497–517.

Buitrago, R., Waleckx, E., Bosseno, M.F., Zoveda, F., Vidaurre, P., Salas, R., Mamani, E., Noireau, F., Brenière, S.F., 2010. First report of widespread wild populations of *Triatoma infestans* (Reduviidae, Triatominae) in the valleys of La Paz, Bolivia. Am. J. Trop. Med. Hyg. 82, 574–579.

Bustamante, D.M., Monroy, C., Menes, M., Rodas, A., Salazar-Schettino, P.M., Rojas, G., Pinto, N., Guhl, F., Dujardin, J.P., 2004. Metric variation among geographic populations of the Chagas vector *Triatoma dimidiata* (Hemiptera: Reduviidae: Triatominae) and related species. J. Med. Entomol. 41, 296–301.

Calderón, F.G.E., Monzón, L., 1995. Nota científica: primer hallazgo de *Triatoma nigromaculata* (Stål, 1872) en el Perú. Rev. Peruana Entomol. 37, 124.

Calderón-Fernández, G., Juárez, M.P., Ramsey, J., Salazar-Schettino, P.M., Monroy, M.C., Ordóñez, R., Cabrera, M., 2005. Cuticular hydrocarbon variability among *Triatoma dimidiata* (Hemiptera: Reduviidae) populations from Mexico and Guatemala. J. Med. Entomol. 42, 780–788.

Calderón-Fernández, G.M., Girotti, J.R., Juárez, M.P., 2011. Cuticular hydrocarbons of *Triatoma dimidiata* (Hemiptera: Reduviidae): intraspecific variation and chemotaxonomy. J. Med. Entomol. 48, 262–271.

Calleros, L., Panzera, F., Bargues, M.D., Monteiro, F.A., Klisiowicz, D.R., Zuriaga, M.A., Mas-Coma, S., Pérez, R., 2010. Systematics of *Mepraia* (Hemiptera-Reduviidae): cytogenetic and molecular variation. Infect. Genet. Evol. 10, 221–228.

Campos, R., Torres-Pérez, F., Botto-Mahan, C., Coronado, X., Solari, A., 2013a. High phylogeographic structure in sylvatic vectors of Chagas disease of the genus *Mepraia* (Hemiptera: Reduviidae). Infect. Genet. Evol. 19, 280–286.

Campos, R., Botto-Mahan, C., Coronado, X., Catalá, S.S., Solari, A., 2013b. Phylogenetic relationships of the Spinolai complex and other Triatomini based on mitochondrial DNA sequences (Hemiptera: Reduviidae). Vector Borne Zoonotic Dis. 13, 73–76.

Campos-Soto, R., Torres-Pérez, F., Solari, A., 2015. Phylogenetic incongruence inferred with two mitochondrial genes in *Mepraia* spp. and *Triatoma eratyrusiformis* (Hemiptera, Reduviidae). Genet. Mol. Biol. 38, 390–395.

Caranha, L., Gurgel-Gonçalves, R., Ramalho, R.D., Galvão, C., 2011. New records and geographic distribution map of *Triatoma petrocchiae* Pinto and Barreto, 1925 (Hemiptera: Reduviidae: Triatominae). Check List 7, 508–509.

Carbajal de la Fuente, A.L., Jaramillo, N., Barata, J.M., Noireau, F., Diotaiuti, L., 2011. Misidentification of two Brazilian triatomes, *Triatoma arthurneivai* and *Triatoma wygodzinskyi*, revealed by geometric morphometrics. Med. Vet. Entomol. 25, 178–183.

Carcavallo, R.U., Cichero, J.A., Martínez, A., Prosen, A.F., Ronderos, R., 1967. Una nueva especie del género *Triatoma* Laporte (Hemiptera, Reduviidae, Triatominae). Segundas Jornadas Entomoepidemiológicas Argentinas 2, 43–48.

Carcavallo, R.U., Rabinovich, J.E., Tonn, R.J. (Eds.), 1985. Factores biológicos y ecológicos en la enfermedad de Chagas, vols I and II. Organización Panamericana de la Salud/Ministerio de Salud y Acción Social (República Argentina), Buenos Aires, Argentina.

Carcavallo, R.U., Galíndez, I., Martínez, A., 1994. *Panstrongylus chinai* (Del Ponte 1929), nueva especie para la entomofauna de Venezuela (Hemiptera, Reduviidae, Triatominae). Triatominae. Entomol. Vectores 1, 195–199.

Carcavallo, R.U., Barata, J.M.S., Costa, A.I.P., Serra, O.P., 1995. *Alberprosenia malheiroi* Serra, Atzingen & Serra, 1987 (Hemiptera, Reduviidae), redescription and bionomics. Rev. Saude Publica 29, 488–495.

Carcavallo, R.U., Curto de Casas, S.I., Sherlock, I.A., Galíndez Girón, I., Jurberg, J., Galvão, C., Mena Segura, C.A., Noireau, F., 1999. Geographical distribution and alti-latitudinal dispersion. In: Carcavallo, R.U., Galíndez Girón, I., Jurberg, J., Lent, H. (Eds.), Atlas of Chagas Disease Vectors in the Americas. Vol. III. Editora Fiocruz, Rio de Janeiro, Brazil, pp. 747–792.

Carcavallo, R.U., Jurberg, J., Lent, H., Noireau, F., Galvão, C., 2000. Phylogeny of the Triatominae (Hemiptera: Reduviidae). Proposals for taxonomic arrangements. Entomol. Vectores 7, 1–99.

Carvalho, D.B., 2016. Estudo de transcriptomas por RNAseq em tecidos de cabeça e glândula salivar de *Rhodnius montenegrensis* e *Rhodnius robustus* (Hemiptera, Reduviidade, Triatominae). PhD thesis, Universidade Estadual Paulista Júlio de Mesquita Filho, Araraquara, Brazil.

Carvalho, D.B., Congrains, C., Chahad-Ehlers, S., Pinotti, H., Brito, R.A., Rosa, J.A., 2017. Differential transcriptome analysis supports *Rhodnius montenegrensis* and *Rhodnius robustus* (Hemiptera, Reduviidae, Triatominae) as distinct species. PLoS One 12, e0174997.

Cavassin, F.B., Kuehn, C.C., Kopp, R.L., Thomaz-Soccol, V., Rosa, J.A., Luz, E., Mas-Coma, S., Bargues, M.D., 2014. Genetic variability and geographical diversity of the main Chagas' disease vector *Panstrongylus megistus* (Hemiptera: Triatominae) in Brazil based on ribosomal DNA intergenic sequences. J. Med. Entomol. 51, 616–628.

Ceballos, L.A., Piccinali, R.V., Berkunsky, I., Kitron, U., Gürtler, R.E., 2009. First finding of melanic sylvatic *Triatoma infestans* (Hemiptera: Reduviidae) colonies in the Argentine Chaco. J. Med. Entomol. 46, 1195–1202.

Ceballos, L.A., Piccinali, R.V., Marcet, P.L., Vázquez-Prokopec, G.M., Cardinal, M.V., Schachter-Broide, J., Dujardin, J.P., Dotson, E.M., Kitron, U., Gürtler, R.E., 2011. Hidden sylvatic foci of the main vector of Chagas disease *Triatoma infestans*: threats to the vector elimination campaign? PLoS Negl. Trop. Dis. 5, e1365.

Cécere, M.C., Leporace, M., Fernández, M.P., Zárate, J.E., Moreno, C., Gürtler, R.E., Cardinal, M.V., 2016. Host-feeding sources and infection with *Trypanosoma cruzi* of *Triatoma infestans* and *Triatoma eratyrusiformis* (Hemiptera: Reduviidae) from the Calchaqui valleys in northwestern Argentina. J. Med. Entomol. 53, 666–673.
Chagas, C., 1909. Nova Trypanozomiaze humana. Estudos sobre a morfologia e o ciclo evolutivo do *Schizotrypanum cruzi* n. gen., n. sp., ajente etiologico de nova entidade morbida no homem. Mem. Inst. Oswaldo Cruz 1, 159–218.
Chávez, T., Moreno, J., Dujardin, J.P., 1999. Isoenzyme electrophoresis of *Rhodnius* species: a phenetic approach to relationships within the genus. Ann. Trop. Med. Parasitol. 93, 229–307.
Claver, M.A., Yaqub, A., 2015. Morphometric analysis of tropicopolitan bug *Triatoma rubrofasciata* (De Geer) in two different parts of India. Int. J. Res. Stud. Biosci. 3, 130–138.
Clayton, R.A., 1990. A phylogenetic analysis of the Reduviidae (Hemiptera: Heteroptera) with redescription of the subfamilies and tribes. MSc dissertation, The George Washington University, Washington, DC.
Cobben, R.H., 1978. Evolutionary trends in Heteroptera. Part II. Mouth-part structures and feeding strategies. Meded. Landbouwhogeschool Wageningen 78, 1–407.
Conn, J.E., Mirabello, L., 2007. The biogeography and population genetics of neotropical vector species. Heredity 99, 245–256.
Costa, J., Felix, M., 2007. *Triatoma juazeirensis* sp. nov. from the state of Bahia, northeastern Brazil (Hemiptera: Reduviidae: Triatominae). Mem. Inst. Oswaldo Cruz 102, 87–90.
Costa, J., Freitas Sibajev, M.G.R., Marcon Silva, V., Pires, M.Q., Pacheco, R.S., 1997. Isoenzymes detect variation in populations of *Triatoma brasiliensis* (Hemiptera: Reduviidae: Triatominae). Mem. Inst. Oswaldo Cruz 92, 459–464.
Costa, J., Almeida, J.R., Britto, C., Duarte, R., Marchon-Silva, V., Pacheco, R.S., 1998. Ecotopes, natural infection and trophic resources of *Triatoma brasiliensis* (Hemiptera, Reduviidae, Triatominae). Mem. Inst. Oswaldo Cruz 93, 7–13.
Costa, J., Argolo, A.M., Felix, M., 2006. Redescription of *Triatoma melanica* Neiva & Lent, 1941, new status (Hemiptera: Reduviidae: Triatominae). Zootaxa 1385, 47–52.
Costa, J., Dornak, L.L., Almeida, C.E., Peterson, A.T., 2014. Distributional potential of the *Triatoma brasiliensis* species complex at present and under scenarios of future climate conditions. Parasit. Vectors 7, 238.
Cracraft, J., 1989. Speciation and its ontology: the empirical consequences of alternative species concepts for understanding patterns and processes of differentiation. In: Otte, D., Endler, J.A. (Eds.), Speciation and Its Consequences. Sinauer Associates, Inc., Sunderland, MA, pp. 28–59
Crossa, R.P., Hernández, M., Caraccio, M.N., Rose, V., Valente, S.A., Valente, V.C., Mejía, J.M., Angulo, V.M., Ramírez, C.M., Roldán, J., Vargas, F., Wolff, M., Panzera, F., 2002. Chromosomal evolution trends of the genus *Panstrongylus* (Hemiptera, Reduviidae), vectors of Chagas disease. Infect. Genet. Evol. 2, 47–56.
Cuba Cuba, C.A., Abad-Franch, F., Roldán, R.J., Vargas, F.J., Pollack, V.L., Miles, M.A., 2002. The triatomines of northern Peru, with emphasis on the ecology and infection by trypanosomes of *Rhodnius ecuadoriensis* (Hemiptera: Reduviidae: Triatominae). Mem. Inst. Oswaldo Cruz 97, 175–183.
Darwin, C., 1839. Voyages of the Adventure and Beagle. vol. III—Journal and Remarks, 1832–1836. Henry Colburn, London.
Davis, N.T., 1961. Morphology and phylogeny of the Reduvioidea (Hemiptera: Heteroptera). Part II. Wing venation. Ann. Entomol. Soc. Am. 54, 340–354.
Dayrat, B., 2005. Towards integrative taxonomy. Biol. J. Linn. Soc. 85, 407–415.
De Geer, C., 1773. Mémoires pour servir à l'historie des insectes. V. Stockholm, 448 pp. +16 pl.

de la Rúa, N.M., Bustamante, D.M., Menes, M., Stevens, L., Monroy, C., Kilpatrick, C.W., Rizzo, D., Klotz, S.A., Schmidt, J., Axen, H.J., Dorn, P.L., 2014. Towards a phylogenetic approach to the composition of species complexes in the North and Central American *Triatoma*, vectors of Chagas disease. Infect. Genet. Evol. 24, 157–166.

de Paula, A.S., Diotaiuti, L., Schofield, C.J., 2005. Testing the sister-group relationship of the Rhodniini and Triatomini (Insecta: Hemiptera: Reduviidae: Triatominae). Mol. Phylogenet. Evol. 35, 712–718.

de Paula, A.S., Diotaiuti, L., Galvão, C., 2007. Systematics and biogeography of Rhodniini (Heteroptera: Reduviidae: Triatominae) based on 16S mitochondrial rDNA sequences. J. Biogeogr. 34, 699–712.

de Queiroz, K., 2007. Species concepts and species delimitation. Syst. Biol. 56, 879–886.

Díaz, S., Panzera, F., Jaramillo-O, N., Pérez, R., Fernández, R., Vallejo, G., Saldaña, A., Calzada, J.E., Triana, O., Gómez-Palacio, A., 2014. Genetic, cytogenetic and morphological trends in the evolution of the *Rhodnius* (Triatominae: Rhodniini) trans-Andean group. PLoS One 9, e87493.

Díaz, S., Triana-Chávez, O., Gómez-Palacio, A., 2016. The nuclear elongation factor-1α gene: a promising marker for phylogenetic studies of Triatominae (Hemiptera: Reduviidae). Infect. Genet. Evol. 43, 274–280.

Dorn, P.L., Monroy, C., Curtis, A., 2007. *Triatoma dimidiata* (Latreille, 1811): a review of its diversity across its geographic range and the relationship among populations. Infect. Genet. Evol. 7, 343–352.

Dorn, P.L., Calderón, C., Melgar, S., Moguel, B., Solórzano, E., Dumonteil, E., Rodas, A., de la Rúa, N., Garnica, R., Monroy, C., 2009. Two distinct *Triatoma dimidiata* (Latreille, 1811) taxa are found in sympatry in Guatemala and Mexico. PLoS Negl. Trop. Dis. 3, e393.

Dorn, P.L., de la Rúa, N.M., Axen, H., Smith, N., Richards, B.R., Charabati, J., Suárez, J., Woods, A., Pessoa, R., Monroy, C., Kilpatrick, C.W., Stevens, L., 2016. Hypothesis testing clarifies the systematics of the main Central American Chagas disease vector, *Triatoma dimidiata* (Latreille, 1811), across its geographic range. Infect. Genet. Evol. 44, 431–443.

dos Santos, S.M., Lopes, C.M., Dujardin, J.P., Panzera, F., Pérez, R., Carbajal de la Fuente, A.L., Pacheco, R.S., Noireau, F., 2007. Evolutionary relationships based on genetic and phenetic characters between *Triatoma maculata*, *Triatoma pseudomaculata* and morphologically related species (Reduviidae: Triatominae). Infect. Genet. Evol. 7, 469–475.

Dotson, E.M., Beard, C.B., 2001. Sequence and organization of the mitochondrial genome of the Chagas disease vector, *Triatoma dimidiata*. Insect Mol. Biol. 10, 205–225.

Dujardin, J.P., Tibayrenc, M., 1985a. Etude de 11 enzymes et données de génétique formelle pour 19 loci enzymatiques chez *Triatoma infestans* (Hemiptera: Reduviidae). Ann. Soc. Belg. Med. Trop. 65, 271–280.

Dujardin, J.P., Tibayrenc, M., 1985b. Etudes isoenzymatiques du vecteur principal de la maladie de Chagas: *Triatoma infestans* (Hemiptera: Reduviidae). Ann. Soc. Belg. Med. Trop. 65 (Suppl. 1), 165–169.

Dujardin, J.P., Tibayrenc, M., Venegas, E., Maldonado, L., Desjeux, P., Ayala, F.J., 1987. Isoenzyme evidence of lack of speciation between wild and domestic *Triatoma infestans* (Heteroptera: Reduviidae) in Bolivia. J. Med. Entomol. 24, 40–45.

Dujardin, J.P., García-Zapata, M.T., Jurberg, J., Poelants, P., Cardozo, L., Panzera, F., Dias, J.C.P., Schofield, C.J., 1991. Which species of *Rhodnius* is invading houses in Brazil? Trans. R. Soc. Trop. Med. Hyg. 85, 679–680.

Dujardin, J.P., Panzera, F., Schofield, C.J., 1999a. Triatominae as a model of morphological plasticity under ecological pressure. Mem. Inst. Oswaldo Cruz 94 (Suppl. 1), 223–228.

Dujardin, J.P., Chávez, T., Moreno, J.M., Machane, M., Noireau, F., Schofield, C.J., 1999b. Comparison of isoenzyme electrophoresis and morphometric analysis for phylogenetic reconstruction of the Rhodniini (Hemiptera: Reduviidae: Triatominae). J. Med. Entomol. 36, 653–659.

Dujardin, J.P., Costa, J., Bustamante, D., Jaramillo, N., Catalá, S., 2009. Deciphering morphology in Triatominae: the evolutionary signals. Acta Trop. 110, 101–111.

Dujardin, J.P., Lam, T.X., Khoa, P.T., Schofield, C.J., 2015a. The rising importance of *Triatoma rubrofasciata*. Mem. Inst. Oswaldo Cruz 110, 319–323.

Dujardin, J.P., Pham Thi, K., Truong Xuan, L., Panzera, F., Pita, S., Schofield, C.J., 2015b. Epidemiological status of kissing-bugs in South East Asia: a preliminary assessment. Acta Trop. 151, 142–149.

Dumonteil, E., Tripet, F., Ramírez-Sierra, M.J., Payet, V., Lanzaro, G., Menu, F., 2007. Assessment of *Triatoma dimidiata* dispersal in the Yucatán peninsula of Mexico by morphometry and microsatellite markers. Am. J. Trop. Med. Hyg. 76, 930–937.

Espinoza, B., Martínez-Ibarra, J.A., Villalobos, G., de la Torre, P., Laclette, J.P., Martínez-Hernández, F., 2013. Genetic variation of North American triatomines (Insecta: Hemiptera: Reduviidae): initial divergence between species and populations of Chagas disease vector. Am. J. Trop. Med. Hyg. 88, 275–284.

Feakins, S.J., Warny, S., Lee, J.E., 2012. Hydrologic cycling over Antarctica during the middle Miocene warming. Nature Geosci. 5, 557–560.

Fitzpatrick, S., Feliciangeli, M.D., Sánchez-Martín, M.J., Monteiro, F.A., Miles, M.A., 2008. Molecular genetics reveal that silvatic *Rhodnius prolixus* do colonise rural houses. PLoS Negl. Trop. Dis. 2, e210.

Flores, A., Magallón-Gastélum, E., Bosseno, M.F., Ordóñez, R., Lozano-Kasten, F., Espinoza, B., Ramsey, J., Breniére, S.F., 2001. Isoenzyme variability of five principal triatomine vector species of Chagas disease in Mexico. Infect. Genet. Evol. 1, 21–28.

Forattini, O.P., 1980. Biogeografia, origem e distribuição da domiciliação de triatomíneos no Brasil. Rev. Saude Publica 14, 265–299.

Forattini, O.P., Ferreira, O.A., Rocha e Silva, E.O., Rabello, E.X., 1978. Aspectos ecológicos da Trypanosomíase Americana. XII—Variação regional da tendência de *Panstrongylus megistus* à domiciliação. Rev. Saude Publica 12, 209–233.

Forthman, M., Weirauch, C., 2012. Toxic associations: a review of the predatory behaviors of millipede assassin bugs (Hemiptera: Reduviidae: Ectrichodiinae). Eur. J. Entomol. 109, 147–153.

Frías-Lasserre, D., 2010. A new species and karyotype variation in the bordering distribution of *Mepraia spinolai* (Porter) and *Mepraia gajardoi* Frías *et al* (Hemiptera: Reduviidae: Triatominae) in Chile and its parapatric model of speciation. Neotrop. Entomol. 39, 572–583.

Galvão, C., 2014. Vetores da doença de chagas no brasil. Sociedade Brasileira de Zoologia, Curitiba, Brazil.

Galvão, C., Angulo, V.M., 2006. *Belminus corredori*, a new species of Bolboderini (Hemiptera: Reduviidae: Triatominae) from Santander, Colombia. Zootaxa 1241, 61–68.

Galvão, C., de Paula, A.S., 2014. Sistemática e evolução dos vetores. In: Galvão, C. (Org.), Vetores da doença de Chagas no Brasil. Sociedade Brasileira de Zoologia. Curitiba, Brazil, pp. 25–32.

Galvão, C., Rocha, D.S., Jurberg, J., Carcavallo, R.U., 2001. Ampliação da distribuição geográfica de *Triatoma deaneorum* Galvão, Souza & Lima 1967, nova denominação para *Triatoma deanei* (Hemiptera, Reduviidae). Rev. Soc. Bras. Med. Trop. 34, 587–589.

Galvão, C., Patterson, J.S., Rocha, D.S., Jurberg, J., Carcavallo, R., Rajen, K., Ambrose, D.P., Miles, M.A., 2002. A new species of Triatominae from Tamil Nadu, India. Med. Vet. Entomol. 16, 75–82.

Galvão, C., Carcavallo, R.U., Rocha, D.S., Jurberg, J., 2003. A checklist of the current valid species of the subfamily Triatominae Jeannel, 1919 (Hemiptera, Reduviidae) and their geographical distribution, with nomenclatural and taxonomic notes. Zootaxa 202, 1–36.

García, B.A., Powell, J.R., 1998. Phylogeny of species of *Triatoma* (Hemiptera: Reduviidae) based on mitochondrial DNA sequences. J. Med. Entomol. 35, 232–238.
García, B.A., Canale, D.M., Blanco, A., 1995. Genetic structure of four species of *Triatoma* (Hemiptera: Reduviidae) from Argentina. J. Med. Entomol. 32, 134–137.
García, B.A., Moriyama, E.N., Powell, J.R., 2001. Mitochondrial DNA sequences of triatomines (Hemiptera: Reduviidae): phylogenetic relationships. J. Med. Entomol. 38, 675–683.
García, B.A., Manfredi, C., Fichera, L., Segura, E.L., 2003. Short report: variation in mitochondrial 12S and 16S ribosomal DNA sequences in natural populations of *Triatoma infestans* (Hemiptera: Reduviidae). Am. J. Trop. Med. Hyg. 68, 692–694.
García, B.A., Zheng, L., Pérez de Rosas, A.R., Segura, E.L., 2004. Isolation and characterization of polymorphic microsatellite loci in the Chagas' disease vector *Triatoma infestans* (Hemiptera.Reduviidae). Mol. Ecol. Notes 4, 568–571.
Garcia, M.H.H.M., Souza, L., Souza, R.C.M., de Paula, A.S., Borges, E.C., Barbosa, S.E., Schofield, C.J., Diotaiuti, L., 2005. Occurrence and variability of *Panstrongylus lutzi* in the state of Ceará, Brazil. Rev. Soc. Bras. Med. Trop. 38, 410–415.
Gaunt, M., Miles, M., 2000. The ecotopes and evolution of triatomine bugs (Triatominae) and their associated trypanosomes. Mem. Inst. Oswaldo Cruz 95, 557–565.
Godfray, H.C.J., 2002. Challenges for taxonomy. The discipline will have to reinvent itself if it is to survive and flourish. Nature 417, 17–19.
Gómez-Palacio, A., Triana, O., 2014. Molecular evidence of demographic expansion of the Chagas disease vector *Triatoma dimidiata* (Hemiptera, Reduviidae, Triatominae) in Colombia. PLoS Negl. Trop. Dis. 8, e2734.
Gómez-Palacio, A., Jaramillo-O, N., Caro-Riaño, H., Diaz, S., Monteiro, F.A., Pérez, R., Panzera, F., Triana, O., 2012. Morphometric and molecular evidence of intraspecific biogeographical differentiation of *Rhodnius pallescens* (Hemiptera: Reduviidae: Rhodniini) from Colombia and Panama. Infect. Genet. Evol. 12, 1975–1983.
Gómez-Palacio, A., Triana, O., Jaramillo-O, N., Dotson, E.M., Marcet, P.L., 2013. Ecogeographical differentiation among Colombian populations of the Chagas disease vector *Triatoma dimidiata* (Hemiptera: Reduviidae). Infect. Genet. Evol. 20, 352–361.
Gómez-Palacio, A., Arboleda, S., Dumonteil, E., Peterson, A.T., 2015. Ecological niche and geographic distribution of the Chagas disease vector, *Triatoma dimidiata* (Reduviidae: Triatominae): evidence for niche differentiation among cryptic species. Infect. Genet. Evol. 36, 15–22.
Gonçalves, T.C.M., Teves-Neves, S.C., Santos-Mallet, J.R., Carbajal-de-la-Fuente, A.L., Lopes, C.M., 2013. *Triatoma jatai* sp. nov. in the state of Tocantins, Brazil (Hemiptera: Reduviidae: Triatominae). Mem. Inst. Oswaldo Cruz 108, 429–437.
Gonçalves, L.O., Oliveira, L.M., D'Ávila Pessoa, G.C., Rosa, A.C.L., Bustamante, M.G., Belisário, C.J., Resende, D.M., Diotaiuti, L.G., Ruiz, J.C., 2017. Insights from tissue-specific transcriptome sequencing analysis of *Triatoma infestans*. Mem. Inst. Oswaldo Cruz 112, 456–457.
Gordon, E.R., Weirauch, C., 2016. Efficient capture of natural history data reveals prey conservatism of cryptic termite predators. Mol. Phylogenet. Evol. 94, 65–73.
Gorla, D.E., Jurberg, J., Catalá, S.S., Schofield, C.J., 1993. Systematics of *Triatoma sordida*, *T. guasayana* and *T. patagonica* (Hemiptera, Reduviidae). Mem. Inst. Oswaldo Cruz 88, 379–385.
Gorla, D.E., Dujardin, J.P., Schofield, C.J., 1997. Biosystematics of Old World Triatominae. Acta Trop. 63, 127–140.
Graham, A., 2011. The age and diversification of terrestrial New World ecosystems through Cretaceous and Cenozoic time. Am. J. Bot. 98, 336–351.
Guhl, F., Aguilera, G., Pinto, N., Vergara, D., 2007. Actualización de la distribución geográfica y ecoepidemiología de la fauna de triatominos (Reduviidae: Triatominae) en Colombia. Biomedica 27 (Suppl. 1), 143–162.

Gurgel-Gonçalves, R., Ferreira, J.B., Rosa, A.F., Bar, M.E., Galvão, C., 2011. Geometric morphometrics and ecological niche modelling for delimitation of near-sibling triatomine species. Med. Vet. Entomol. 25, 84–93.
Gurgel-Gonçalves, R., Galvão, C., Costa, J., Peterson, A.T., 2012. Geographic distribution of Chagas disease vectors in Brazil based on ecological niche modeling. J. Trop. Med. 2012, 705326.
Haridass, E.T., Ananthakrishnan, T.N., 1981. Functional morphology of the salivary system in some Reduviidae (Insecta: Heteroptera). Proc. Indian Acad. Sci. B 90, 145–160.
Harry, M., 1993. Isozymic data question the specific status of some blood-sucking bugs of the genus *Rhodnius*, vectors of Chagas disease. Trans. R. Soc. Trop. Med. Hyg. 87, 492–493.
Harry, M., 1994. Morphometric variability in the Chagas disease vector *Rhodnius prolixus*. Jpn. J. Genet. 69, 233–250.
Harry, M., Galíndez, I., Cariou, M.L., 1992. Isozyme variability and differentiation between *Rhodnius prolixus*, *R. robustus* and *R. pictipes*, vectors of Chagas disease in Venezuela. Med. Vet. Entomol. 6, 37–43.
Harry, M., Poyet, G., Romaña, C.A., Solignac, M., 1998. Isolation and characterization of microsatellite markers in the bloodsucking bug *Rhodnius pallescens* (Heteroptera, Reduviidae). Mol. Ecol. 7, 1784–1786.
Harry, M., Roose, C.L., Vautrin, D., Noireau, F., Romaña, C.A., Solignac, M., 2008. Microsatellite markers from the Chagas disease vector, *Rhodnius prolixus* (Hemiptera, Reduviidae), and their applicability to *Rhodnius* species. Infect. Genet. Evol. 8, 381–385.
Hennig, W., 1966. Phylogenetic Systematics. University of Illinois Press, Urbana.
Hwang, W.S., Weirauch, C., 2012. Evolutionary history of assassin bugs (Insecta: Hemiptera: Reduviidae): insights from divergence dating and ancestral state reconstruction. PLoS One 7, e45523.
Hypša, V., Tietz, D.F., Zrzavý, J., Rego, R.O., Galvão, C., Jurberg, J., 2002. Phylogeny and biogeography of Triatominae (Hemiptera: Reduviidae): molecular evidence of a New World origin of the Asiatic clade. Mol. Phylogenet. Evol. 23, 447–457.
Ibarra-Cerdeña, C.N., Sánchez-Cordero, V., Peterson, A.T., Ramsey, J.M., 2009. Ecology of North American Triatominae. Acta Trop. 110, 178–186.
Ibarra-Cerdeña, C.N., Zaldívar-Riverón, A., Peterson, A.T., Sánchez-Cordero, V., Ramsey, J.M., 2014. Phylogeny and niche conservatism in North and Central American triatomine bugs (Hemiptera: Reduviidae: Triatominae), vectors of Chagas' disease. PLoS Negl. Trop. Dis. 8, e3266.
Iturralde-Vinent, M.A., 2001. Geology of the amber-bearing deposits of the greater Antilles. Caribb. J. Sci. 17, 141–167.
Iturralde-Vinent, M.A., MacPhee, R.D.E., 1996. Age and paleogeographical origin of Dominican amber. Science 273, 1850–1852.
Jeannel, R., 1919. Insectes Hémiptères, iii. Henicocephalidae et Reduviidae. In: Voyage de Ch. Alluaud et R. Jeannel en Afrique Orientale (1911–1912). A. Schulz, Paris, pp. 133–313.
Jörger, K.M., Schrödl, M., 2013. How to describe a cryptic species? Practical challenges of molecular taxonomy. Front. Zool. 10, 59.
Jurberg, J., Galvão, C., 1997. *Hermanlentia* n. gen. da tribo Triatomini, com um rol de espécies de Triatominae (Hemiptera, Reduviidae). Mem. Inst. Oswaldo Cruz 92, 181–185.
Jurberg, J., Galvão, C., Lent, H., Monteiro, F., Lopes, C.M., Panzera, F., Pérez, R., 1998. Revalidação de *Triatoma garciabesi* Carcavallo, Cichero, Martínez, Prosen & Ronderos, 1967 (Hemiptera: Reduviidae). Entomol. Vectores 5, 107–122.
Jurberg, J., Carcavallo, R.U., Lent, H., 2001. *Panstrongylus sherlocki* sp. n. do estado da Bahia, Brasil (Hemiptera, Reduviidae, Triatominae). Entomol. Vectores 8, 261–274.
Jurberg, J., Rocha, D.S., Galvão, C., 2009. *Rhodnius zeledoni* sp. nov. afim de *Rhodnius paraensis* Sherlock, Guitton & Miles, 1977 (Hemiptera, Reduviidae, Triatominae). Biota Neotrop. 9, 123–128.

Jurberg, J., Cunha, V., Cailleaux, S., Raigorodschi, R., Lima, M.S., Rocha, D.S., Moreira, F.F.F., 2013. *Triatoma pintodiasi* sp. nov. do subcomplexo *T. rubrovaria* (Hemiptera, Reduviidae, Triatominae). Rev. Pan-Amaz. Saude 4, 43–56.

Justi, S.A., Russo, C.A., Mallet, J.R., Obara, M.T., Galvão, C., 2014. Molecular phylogeny of Triatomini (Hemiptera: Reduviidae: Triatominae). Parasit. Vectors 7, 149.

Justi, S.A., Galvão, C., Schrago, C.G., 2016. Geological changes of the Americas and their influence on the diversification of the Neotropical kissing bugs (Hemiptera: Reduviidae: Triatominae). PLoS Negl. Trop. Dis. 10, e0004527.

Kato, H., Jochim, R.C., Gómez, E.A., Sakoda, R., Iwata, H., Valenzuela, J.G., Hashiguchi, Y., 2010. A repertoire of the dominant transcripts from the salivary glands of the blood-sucking bug, *Triatoma dimidiata*, a vector of Chagas disease. Infect. Genet. Evol. 10, 184–191.

Kjos, S.A., Snowden, K.F., Olson, J.K., 2009. Biogeography and *Trypanosoma cruzi* infection prevalence of Chagas disease vectors in Texas, USA. Vector Borne Zoonotic Dis. 9, 41–49.

Kuntner, M., Agnarsson, I., 2006. Are the Linnean and phylogenetic nomenclatural systems combinable? Recommendations for biological nomenclature. Syst. Biol. 55, 774–784.

Lacombe, D., 1999. Anatomy and histology of salivary glands of triatomine bugs. Mem. Inst. Oswaldo Cruz 94, 557–564.

Laporte, F.L.C., 1832–1833. Essai d'une classification systématique de l'ordre des Hémiptères (Hémiptères Hétéroptères, Latr.). Magasin Zool. 2, 1–88.

Leite, G.R., 2013. Biogeografia do gênero *Triatoma* Laporte, 1832 (Hemiptera, Reduviidae, Triatominae): distribuição, padrões de riqueza, endemismo, e diversificação. PhD thesis, Universidade Federal do Espírito Santo, Vitória, Brazil.

Leite, G.R., dos Santos, C.B., Falqueto, A., 2011. Influence of the landscape on dispersal of sylvatic triatomines to anthropic habitats in the Atlantic Forest. J. Biogeogr. 38, 651–663.

Lent, H., 1997. Novos sinônimos de duas espécies de Triatominae da Venezuela (Hemiptera, Reduviidae). Entomol. Vectores 4, 67–70.

Lent, H., Jurberg, J., 1985. Sobre a variaçao intra-específica em *Triatoma dimidiata* (Latreille) e *Triatoma infestans* (Klug) (Hemiptera: Reduviidae). Mem. Inst. Oswaldo Cruz 80, 285–299.

Lent, H., Wygodzinsky, P., 1956. Situação atual do gênero *Opisthacidius* Berg, 1879 (Hemiptera, Reduviidae). Rev. Bras. Biol. 16, 327–334.

Lent, H., Wygodzinsky, P., 1979. Revision of the Triatominae (Hemiptera, Reduviidae) and their significance as vectors of Chagas disease. Bull. Am. Mus. Nat. Hist. 163, 125–520.

Lent, H., Jurberg, J., Carcavallo, R.U., 1995. *Belminus laportei* sp. n. da região Amazônica. (Hemiptera: Reduviidae: Triatominae). Mem. Inst. Oswaldo Cruz 90, 33–39.

Lent, H., Jurberg, J., Galvão, C., 1996. Descrição do alótipo (macho) de *Triatoma guazu* Lent & Wygodzinsky, 1979 proveniente do estado do Mato Grosso, Brasil (Hemiptera, Reduviidae). Mem. Inst. Oswaldo Cruz 91, 313–315.

Linnaeus, C., 1753. Species plantarum. L. Salvii, Stockholm.

Linnaeus, C., 1758. Systema naturae. vol. 1. L. Salvii, Stockholm.

López, G., Moreno, M.J., 1995. Genetic variability and differentiation between populations of *Rhodnius prolixus* and *Rhodnius pallescens*, vectors of Chagas disease in Colombia. Mem. Inst. Oswaldo Cruz 90, 353–357.

Louis, D., Kumar, R., 1973. Morphology of the alimentary and reproductive organs in the Reduviidae (Hemiptera: Heteroptera) with comments on interrelationships within the family. Ann. Entomol. Soc. Am. 66, 635–639.

Lyman, D.F., Monteiro, F.A., Escalante, A.A., Cordón-Rosales, C., Wesson, D.M., Dujardin, J.P., Beard, C.B., 1999. Mitochondrial DNA sequence variation among triatomine vectors of Chagas disease. Am. J. Trop. Med. Hyg. 60, 377–386.

Maia da Silva, F., Junqueira, A.C., Campaner, M., Rodrigues, A.C., Crisante, G., Ramírez, L.E., Caballero, Z.C., Monteiro, F.A., Coura, J.R., Añez, N., Teixeira, M.M., 2007. Comparative phylogeography of *Trypanosoma rangeli* and

Rhodnius (Hemiptera: Reduviidae) supports a long coexistence of parasite lineages and their sympatric vectors. Mol. Ecol. 16, 3361–3373.

Maldonado Capriles, J., 1990. Systematic catalogue of the Reduviidae of the world (Insecta: Heteroptera). Caribb. J. Sci, 1–694 (special ed.).

Marchant, A., Mougel, F., Almeida, C., Jacquin-Joly, E., Costa, J., Harry, M., 2015. De novo transcriptome assembly for a non-model species, the blood-sucking bug *Triatoma brasiliensis*, a vector of Chagas disease. Genetica 143, 225–239.

Marcilla, A., Bargues, M.D., Ramsey, J.M., Magallón-Gastélum, E., Salazar-Schettino, P.M., Abad-Franch, F., Dujardin, J.P., Schofield, C.J., Mas-Coma, S., 2001. The ITS-2 of the nuclear rDNA as a molecular marker for populations, species, and phylogenetic relationships in Triatominae (Hemiptera: Reduviidae), vectors of Chagas disease. Mol. Phylogenet. Evol. 18, 136–142.

Marcilla, A., Bargues, M.D., Abad-Franch, F., Panzera, F., Noireau, F., Galvão, C., Jurberg, J., Miles, M.A., Dujardin, J.P., Mas-Coma, S., 2002. Nuclear rDNA ITS-2 sequences reveal polyphyly of *Panstrongylus* species (Hemiptera: Reduviidae: Triatominae), vectors of *Trypanosoma cruzi*. Infect. Genet. Evol. 1, 225–235.

Márquez, E., Jaramillo-O, N., Gómez-Palacio, A., Dujardin, J.P., 2011. Morphometric and molecular differentiation of *a Rhodnius robustus*-like form from *R. robustus* Larousse, 1927 and *R. prolixus* Stål, 1859 (Hemiptera, Reduviidae). Acta Trop. 120, 103–109.

Martínez, A., Carcavallo, R.U., Jurberg, J., 1994. *Triatoma gomeznunezi* a new species of Triatomini from Mexico (Hemiptera, Reduviidae, Triatominae). Entomol. Vectores 1, 15–19.

Martínez Avendaño, E., Chávez Espada, T., Sossa Gil, D., Aranda Asturizaga, R., Vargas Mamani, B., Vidaurre Prieto, P., 2007. *Triatoma boliviana* sp. n. (Hemiptera: Reduviidae: Triatominae) de los valles subandinos de La Paz—Bolivia, similar a *Triatoma nigromaculata* Stål, 1859. Cuadernos Hosp. Clinicas 52, 9–16.

Mayr, E., 1963. Animal Species and Evolution. The Belknap Press of Harvard University Press, Cambridge.

Mendonça, V.J., Silva, M.T., Araújo, R.F., Martins Jr., J., Bacci Jr., M., Almeida, C.E., Costa, J., Graminha, M.A., Cicarelli, R.M., Rosa, J.A., 2009. Phylogeny of *Triatoma sherlocki* (Hemiptera: Reduviidae: Triatominae) inferred from two mitochondrial genes suggests its location within the *Triatoma brasiliensis* Complex. Am. J. Trop. Med. Hyg. 81, 858–864.

Mendonça, V.J., Oliveira, J., Rimoldi, A., Ferreira Filho, J.C.R., Araújo, R.F., Rosa, J.A., 2015. Triatominae survey (Hemiptera: Reduviidae: Triatominae) in the south-central region of the state of Bahia, Brazil between 2008 and 2013. Am. J. Trop. Med. Hyg. 92, 1076–1080.

Mendonça, V.J., Alevi, K.C.C., Pinotti, H., Gurgel-Gonçalves, R., Pita, S., Guerra, A.L., Panzera, F., Araújo, R.F., Azeredo-Oliveira, M.T.V., Rosa, J.A., 2016. Revalidation of *Triatoma bahiensis* Sherlock & Serafim, 1967 (Hemiptera: Reduviidae) and phylogeny of the *T. brasiliensis* species complex. Zootaxa 4107, 239–254.

Mesquita, R.D., Vionette-Amaral, R.J., Lowenberger, C., Rivera-Pomar, R., Monteiro, F.A., Minx, P., Spieth, J., Carvalho, A.B., Panzera, F., Lawson, D., Torres, A.Q., Ribeiro, J.M., Sorgine, M.H., Waterhouse, R.M., Montague, M.J., Abad-Franch, F., Alves-Bezerra, M., Amaral, L.R., Araujo, H.M., Araujo, R.N., Aravind, L., Atella, G.C., Azambuja, P., Berni, M., Bittencourt-Cunha, P.R., Braz, G.R., Calderón-Fernández, G., Carareto, C.M., Christensen, M.B., Costa, I.R., Costa, S.G., Dansa, M., Daumas-Filho, C.R., de Paula, I.F., Dias, F.A., Dimopoulos, G., Emrich, S.J., Esponda-Behrens, N., Fampa, P., Fernández-Medina, R.D., Fonseca, R.N., Fontenele, M., Fronick, C., Fulton, L.A., Gandara, A.C., Garcia, E.S., Genta, F.A., Giraldo-Calderón, G.I., Gomes, B., Gondim, K.C., Granzotto, A., Guarneri, A.A., Guigó, R., Harry, M., Hughes, D.S., Jablonka, W.,

Jacquin-Joly, E., Juárez, M.P., Koerich, L.B., Lange, A.B., Latorre-Estivalis, J.M., Lavore, A., Lawrence, G.G., Lazoski, C., Lazzari, C.R., Lopes, R.R., Lorenzo, M.G., Lugon, M.D., Majerowicz, D., Marcet, P.L., Mariotti, M., Masuda, H., Megy, K., Melo, A.C., Missirlis, F., Mota, T., Noriega, F.G., Nouzova, M., Nunes, R.D., Oliveira, R.L., Oliveira-Silveira, G., Ons, S., Orchard, I., Pagola, L., Paiva-Silva, G.O., Pascual, A., Pavan, M.G., Pedrini, N., Peixoto, A.A., Pereira, M.H., Pike, A., Polycarpo, C., Prosdocimi, F., Ribeiro-Rodrigues, R., Robertson, H.M., Salerno, A.P., Salmon, D., Santesmasses, D., Schama, R., Seabra-Junior, E.S., Silva-Cardoso, L., Silva-Neto, M.A., Souza-Gomes, M., Sterkel, M., Taracena, M.L., Tojo, M., Tu, Z.J., Tubio, J.M., Ursic-Bedoya, R., Venancio, T.M., Walter-Nuno, A.B., Wilson, D., Warren, W.C., Wilson, R.K., Huebner, E., Dotson, E.M., Oliveira, P.L., 2015. Genome of *Rhodnius prolixus*, an insect vector of Chagas disease, reveals unique adaptations to hematophagy and parasite infection. Proc. Natl. Acad. Sci. U. S. A. 112, 14936–14941.

Monsalve, Y., Panzera, F., Herrera, L., Triana-Chávez, O., Gómez-Palacio, A., 2016. Population differentiation of the Chagas disease vector *Triatoma maculata* (Erichson, 1848) from Colombia and Venezuela. J. Vector Ecol. 41, 72–79.

Monte, G.L., Tadei, W.P., Farias, T.M., 2014. Ecoepidemiology and biology of *Eratyrus mucronatus* Stål, 1859 (Hemiptera: Reduviidae: Triatominae), a sylvatic vector of Chagas disease in the Brazilian Amazon. Rev. Soc. Bras. Med. Trop. 47, 723–727.

Monteiro, F.A., Costa, J., Solé-Cava, A.M., 1998. Genetic confirmation of the specific status of *Triatoma petrochii* (Hemiptera: Reduviidae: Triatominae). Ann. Trop. Med. Parasitol. 92, 897–900.

Monteiro, F.A., Wesson, D.M., Dotson, E.M., Schofield, C.J., Beard, C.B., 2000. Phylogeny and molecular taxonomy of the Rhodniini derived from mitochondrial and nuclear DNA sequences. Am. J. Trop. Med. Hyg. 62, 460–465.

Monteiro, F.A., Escalante, A.A., Beard, C.B., 2001. Molecular tools and triatomine systematics: a public health perspective. Trends Parasitol. 17, 344–347.

Monteiro, F.A., Lazoski, C., Noireau, F., Solé-Cava, A.M., 2002. Allozyme relationships among ten species of Rhodniini, showing paraphyly of *Rhodnius* including *Psammolestes*. Med. Vet. Entomol. 16, 83–90.

Monteiro, F.A., Barrett, T.V., Fitzpatrick, S., Cordón-Rosales, C., Feliciangeli, D., Beard, C.B., 2003. Molecular phylogeography of the Amazonian Chagas disease vectors *Rhodnius prolixus* and *R. robustus*. Mol. Ecol. 12, 997–1006.

Monteiro, F.A., Donnelly, M.J., Beard, C.B., Costa, J., 2004. Nested clade and phylogeographic analyses of the Chagas disease vector *Triatoma brasiliensis* in northeast Brazil. Mol. Phylogenet. Evol. 32, 46–56.

Monteiro, F.A., Jurberg, J., Lazoski, C., 2009. Very low levels of genetic variation in natural peridomestic populations of the Chagas disease vector *Triatoma sordida* (Hemiptera: Reduviidae) in southeastern Brazil. Am. J. Trop. Med. Hyg. 81, 223–227.

Monteiro, F.A., Peretolchina, T., Lazoski, C., Harris, K., Dotson, E.M., Abad-Franch, F., Tamayo, E., Pennington, P.M., Monroy, C., Cordón-Rosales, C., Salazar-Schettino, P.M., Gómez-Palacio, A.M., Grijalva, M.J., Beard, C.B., Marcet, P.L., 2013. Phylogeographic pattern and extensive mitochondrial DNA divergence disclose a species complex within the Chagas disease vector *Triatoma dimidiata*. PLoS One 8, e70974.

Monteith, G.B., 1974. Confirmation of the presence of Triatominae (Hemiptera: Reduviidae) in Australia, with notes on Indo-Pacific species. J. Australian Entomol. Soc. 13, 89–94.

Nattero, J., Malerba, R., Rodríguez, C.S., Crocco, L., 2013. Phenotypic plasticity in response to food source in *Triatoma infestans* (Klug, 1834) (Hemiptera, Reduviidae: Triatominae). Infect. Genet. Evol. 19, 38–44.

Neiva, A., 1914. Revisão do gênero *Triatoma* Lap. PhD thesis, Universidade do Rio de Janeiro, Rio de Janeiro, Brazil.

Neiva, A., Lent, H., 1936. Notas e commentarios sobre triatomideos. Lista de especies e sua distribuição geographica. Rev. Entomol. 6, 153–190.

Noireau, F., 2009. Wild *Triatoma infestans*, a potential threat that needs to be monitored. Mem. Inst. Oswaldo Cruz 104 (Suppl. I), 60–64.

Noireau, F., Brenière, F., Cardozo, L., Bosseno, M.F., Vargas, F., Peredo, C., Medinacelli, M., 1996. Current spread of *Triatoma infestans* at the expense of *Triatoma sordida* in Bolivia. Mem. Inst. Oswaldo Cruz 91, 271–272.

Noireau, F., Brenière, F., Ordóñez, J., Cardozo, L., Morochi, W., Gutiérrez, T., Bosseno, M.F., García, S., Vargas, F., Yaksic, N., Dujardin, J.P., Peredo, C., Wisnivesky-Colli, C., 1997a. Low probability of transmission of *Trypanosoma cruzi* to humans by domiciliary *Triatoma sordida* in Bolivia. Trans. R. Soc. Trop. Med. Hyg. 91, 653–656.

Noireau, F., Flores, R., Gutiérrez, T., Dujardin, J.P., 1997b. Detection of wild dark morphs of *Triatoma infestans* in the Bolivian Chaco. Mem. Inst. Oswaldo Cruz 92, 583–584.

Noireau, F., Gutiérrez, T., Zegarra, M., Flores, R., Brenière, S.F., Cardozo, L., Dujardin, J.P., 1998. Cryptic speciation in *Triatoma sordida* (Hemiptera: Reduviidae) from the Bolivian Chaco. Trop. Med. Int. Health 3, 364–372.

Noireau, F., Gutiérrez, T., Flores, R., Brenière, F., Bosseno, M.F., Wisnivesky-Colli, C., 1999. Ecogenetics of *Triatoma sordida* and *Triatoma guasayana* (Hemiptera: Reduviidae) in the Bolivian Chaco. Mem. Inst. Oswaldo Cruz 94, 451–457.

Noireau, F., Bastrenta, B., Catalá, S.S., Dujardin, J.P., Panzera, F., Torres, M., Pérez, R., Jurberg, J., Galvão, C., 2000a. Sylvatic population of *Triatoma infestans* from the Bolivian Chaco: from field collection to characterization. Mem. Inst. Oswaldo Cruz 95, 119–122.

Noireau, F., Flores, R., Gutiérrez, T., Abad-Franch, F., Flores, E., Vargas, F., 2000b. Natural ecotopes of *Triatoma infestans* dark morph and other sylvatic triatomines in the Bolivian Chaco. Trans. R. Soc. Trop. Med. Hyg. 94, 23–27.

Noireau, F., dos Santos, S.M., Gumiel, M., Dujardin, J.P., Soares, M.S., Carcavallo, R.U., Galvão, C., Jurberg, J., 2002. Phylogenetic relationships within the *oliveirai* complex (Hemiptera: Reduviidae: Triatominae). Infect. Genet. Evol. 2, 11–17.

Noireau, F., Rojas Cortez, M.G., Monteiro, F.A., Jansen, A.M., Torrico, F., 2005. Can wild *Triatoma infestans* foci in Bolivia jeopardize Chagas disease control efforts? Trends Parasitol. 21, 7–10.

Nouvellet, P., Ramírez-Sierra, M.J., Dumonteil, E., Gourbière, S., 2011. Effects of genetic factors and infection status on wing morphology of *Triatoma dimidiata* species complex in the Yucatán peninsula, Mexico. Infect. Genet. Evol. 11, 1243–1249.

Obara, M.T., Barata, J.M.S., Rosa, J.A., Ceretti Jr., W., Almeida, P.S., Gonçalves, G.A., Dale, C., Gurgel-Gonçalves, R., 2012. Description of the female and new records of *Triatoma baratai* Carcavallo & Jurberg, 2000 (Hemiptera: Heteroptera: Reduviidae: Triatominae) from Mato Grosso do Sul, Brazil, with a key to the species of the *Triatoma matogrossensis* subcomplex. Zootaxa 3151, 63–68.

Oliveira, M.A., Souza, R.C.M., Diotaiuti, L., 2007. Redescription of the genus *Cavernicola* and the tribe Cavernicolini (Hemiptera: Reduviidae: Triatominae), with morphological and morphometric parameters. Zootaxa 1457, 57–68.

Oliveira, J., Marcet, P.L., Takiya, D.M., Mendonça, V.J., Belintani, T., Bargues, M.D., Mateo, L., Chagas, V., Folly-Ramos, E., Cordeiro-Estrela, P., Gurgel-Gonçalves, R., Costa, J., Rosa, J.A., Almeida, C.E., 2017. Combined phylogenetic and morphometric information to delimit and unify the *Triatoma brasiliensis* species complex and the Brasiliensis subcomplex. Acta Trop. 170, 140–148.

Olson, D.M., Dinerstein, E., Wikramanayake, E.D., Burgess, N.D., Powell, G.V.N., Underwood, E.C., D'Amico, J.A., Itoua, I., Strand, H.E., Morrison, J.C.,

Loucks, C.J., Allnutt, T.F., Ricketts, T.H., Kura, Y., Lamoreux, J.F., Wettengel, W.W., Hedao, P., Kassem, K.R., 2001. Terrestrial ecoregions of the world: a new map of life on earth. Bioscience 51, 933–938.

Osuna, E., Ayala, J.M., 1993. *Belminus pittieri*, nueva especie de Bolboderini (Triatominae: Reduviidae: Heteroptera). Bol. Entomol. Venez. 8, 147–150.

Paim, R.M., Araujo, R.N., Lehane, M.J., Gontijo, N.F., Pereira, M.H., 2013. Application of RNA interference in triatomine (Hemiptera: Reduviidae) studies. Insect Sci. 20, 40–52.

Panzera, F., Pérez, R., Panzera, Y., Álvarez, F., Scvortzoff, E., Salvatella, R., 1995. Karyotype evolution in holocentric chromosomes of three related species of triatomines (Hemiptera-Reduviidae). Chromosome Res. 3, 143–150.

Panzera, F., Hornos, S., Pereira, J., Cestau, R., Canale, D., Diotaiuti, L., Dujardin, J.P., Pérez, R., 1997. Genetic variability and geographic differentiation among three species of triatomine bugs (Hemiptera-Reduviidae). Am. J. Trop. Med. Hyg. 57, 732–739.

Panzera, F., Dujardin, J.P., Nicolini, P., Caraccio, M.N., Rose, V., Téllez, T., Bermúdez, H., Bargues, M.D., Mas-Coma, S., O'Connor, J.E., Pérez, R., 2004. Genomic changes of Chagas disease vector, South America. Emerg. Infect. Dis. 10, 438–446.

Panzera, F., Ferrandis, I., Ramsey, J.M., Ordóñez, R., Salazar-Schettino, P.M., Cabrera, M., Monroy, M.C., Bargues, M.D., Mas-Coma, S., O'Connor, J.E., Angulo, V.M., Jaramillo, N., Cordón-Rosales, C., Gómez, D., Pérez, R., 2006. Chromosomal variation and genome size support existence of cryptic species of *Triatoma dimidiata* with different epidemiological importance as Chagas disease vectors. Trop. Med. Int. Health 11, 1092–1103.

Panzera, F., Pérez, R., Panzera, Y., Ferrandis, I., Ferreiro, M.J., Calleros, L., 2010. Cytogenetics and genome evolution in the subfamily Triatominae (Hemiptera, Reduviidae). Cytogenet. Genome Res. 128, 77–87.

Panzera, F., Pita, S., Nattero, J., Panzera, Y., Galvão, C., Chávez, T., Rojas de Arias, A., Téllez, L.C., Noireau, F., 2015. Cryptic speciation in the *Triatoma sordida* subcomplex (Hemiptera, Reduviidae) revealed by chromosomal markers. Parasit. Vectors 8, 495.

Parra-Henao, G., Angulo, V.M., Osorio, L., Jaramillo-O, N., 2016. Geographic distribution and ecology of *Triatoma dimidiata* (Hemiptera: Reduviidae) in Colombia. J. Med. Entomol. 53, 122–129.

Patterson, J.S., 2007. Comparative morphometric and molecular genetic analyses of Triatominae (Hemiptera: Reduviidae). PhD thesis, London School of Hygiene and Tropical Medicine, University of London, London, UK.

Patterson, J.S., Barbosa, S.E., Feliciangeli, M.D., 2009. On the genus *Panstrongylus* Berg 1879: evolution, ecology and epidemiological significance. Acta Trop. 110, 187–199.

Patterson, J.S., Gaunt, M.W., 2010. Phylogenetic multi-locus codon models and molecular clocks reveal the monophyly of haematophagous reduviid bugs and their evolution at the formation of South America. Mol. Phylogenet. Evol. 56, 608–621.

Patterson, J.S., Schofield, C.J., Dujardin, J.P., Miles, M.A., 2001. Population morphometric analysis of the tropicopolitan bug *Triatoma rubrofasciata* and relationships with Old World species of *Triatoma*: evidence of New World ancestry. Med. Vet. Entomol. 15, 443–451.

Pavan, M.G., Monteiro, F.A., 2007. A multiplex PCR assay that separates *Rhodnius prolixus* from members of the *Rhodnius robustus* cryptic species complex (Hemiptera: Reduviidae). Trop. Med. Int. Health 12, 751–758.

Pavan, M.G., Mesquita, R.D., Lawrence, G.G., Lazoski, C., Dotson, E.M., Abubucker, S., Mitreva, M., Randall-Maher, J., Monteiro, F.A., 2013. A nuclear single-nucleotide polymorphism (SNP) potentially useful for the separation of *Rhodnius prolixus* from members of the *Rhodnius robustus* cryptic species complex (Hemiptera: Reduviidae). Infect. Genet. Evol. 14, 426–433.

Pavan, M.G., Corrêa-Antônio, J., Peixoto, A.A., Monteiro, F.A., Rivas, G.B.S., 2016. *Rhodnius prolixus* and *R. robustus* (Hemiptera: Reduviidae) nymphs show different locomotor patterns on an automated recording system. Parasit. Vectors 9, 239.

Pereira, J., Dujardin, J.P., Salvatella, R., Tibayrenc, M., 1996. Enzymatic variability and phylogenetic relatedness among *Triatoma infestans*, *T. platensis*, *T. delpontei* and *T. rubrovaria*. Heredity 77, 47–54.

Pérez de Rosas, A.R., Segura, E.L., García, B.A., 2007. Microsatellite analysis of genetic structure in natural *Triatoma infestans* (Hemiptera: Reduviidae) populations from Argentina: its implication in assessing the effectiveness of Chagas disease vector control programmes. Mol. Ecol. 16, 1401–1412.

Peterson, A.T., Sánchez-Cordero, V., Beard, C.B., Ramsey, J.M., 2002. Ecologic niche modeling and potential reservoirs for Chagas disease, Mexico. Emerg. Infect. Dis. 8, 662–667.

Pfeiler, E., Bitler, B.G., Ramsey, J.M., Palacios-Cardiel, C., Markow, T.A., 2006. Genetic variation, population structure, and phylogenetic relationships of *Triatoma rubida* and *T. recurva* (Hemiptera: Reduviidae: Triatominae) from the Sonoran Desert, insect vectors of the Chagas' disease parasite *Trypanosoma cruzi*. Mol. Phylogenet. Evol. 41, 209–221.

Pinto, C., 1926. Classificação dos Triatomideos (Hemiptera-Heteroptera hematofagos). Sciencia Medica 9, 485–490.

Pinto, C.M., Ocaña-Mayorga, S., Tapia, E.E., Lobos, S.E., Zurita, A.P., Aguirre-Villacís, F., MacDonald, A., Villacís, A.G., Lima, L., Teixeira, M.M., Grijalva, M.J., Perkins, S.L., 2015. Bats, trypanosomes, and triatomines in Ecuador: new insights into the diversity, transmission, and origins of *Trypanosoma cruzi* and Chagas disease. PLoS One 10, e0139999.

Pita, S., Panzera, F., Ferrandis, I., Galvão, C., Gómez-Palacio, A., Panzera, Y., 2013. Chromosomal divergence and evolutionary inferences in Rhodniini based on the chromosomal location of ribosomal genes. Mem. Inst. Oswaldo Cruz 108, 376–382.

Pita, S., Lorite, P., Nattero, J., Galvão, C., Alevi, K.C., Teves, S.C., Azeredo-Oliveira, M.T., Panzera, F., 2016. New arrangements on several species subcomplexes of *Triatoma* genus based on the chromosomal position of ribosomal genes (Hemiptera–Triatominae). Infect. Genet. Evol. 43, 225–231.

Pita, S., Panzera, F., Vela, J., Mora, P., Palomeque, T., Lorite, P., 2017. Complete mitochondrial genome of *Triatoma infestans* (Hemiptera, Reduviidae, Triatominae), main vector of Chagas disease. Infect. Genet. Evol. 54, 158–163.

Poinar Jr., G.O., 2005. *Triatoma dominicana* sp. n. (Hemiptera: Reduviidae: Triatominae), and *Trypanosoma antiquus* sp. n. (Stercoraria: Trypanosomatidae), the first fossil evidence of a triatomine-trypanosomatid vector association. Vector Borne Zoonotic Dis. 5, 72–81.

Poinar Jr., G.O., 2013. *Panstrongylus hispaniolae* sp. n. (Hemiptera: Reduviidae: Triatominae), a new fossil triatomine in Dominican amber, with evidence of gut flagellates. Palaeodiversity 6, 1–8.

Putshkov, V.G., Putshkov, P.V., 1985. A catalogue of assassin-bug genera of the world (Heteroptera, Reduviidae). Published by the authors, deposited in VINITI [All-Union Institute of Science & Technology Information], No. 4738-B 85. [Russian preface, English main text], Kiev.

Rahaman, A.A., 2002. Mangrove insect fauna of Muthupet, Tamil Nadu. In: National Seminar on Conservation of Eastern Ghats, Tirupati, Andhra Pradesh, India, pp. 327–338.

Ramírez, S.R., Gravendeel, B., Singer, R.B., Marshall, C.R., Pierce, N.E., 2007. Dating the origin of the Orchidaceae from a fossil orchid with its pollinator. Nature 448, 1042–1045.

Ramsey, J.M., Peterson, A.T., Carmona-Castro, O., Moo-Llanes, D.A., Nakazawa, Y., Butrick, M., Tun-Ku, E., de la Cruz-Félix, K., Ibarra-Cerdeña, C.N., 2015. Atlas of Mexican Triatominae (Reduviidae: Hemiptera) and vector transmission of Chagas disease. Mem. Inst. Oswaldo Cruz 110, 339–352.

Rao, K.T., Krishna, I.S.R., Reddy, C.S., 2007. Biodiversity of Nallamalai Hill Ranges, Eastern Ghats, India. In: Proceedings of the National Seminar on Conservation of Eastern Ghats: December 28–29, 2007, Chennai, Tamil Nadu, India, pp. 216–223.

Rassi Jr., A., Rassi, A., Marin-Neto, J.A., 2010. Chagas disease. Lancet 375, 1388–1402.

Reyes-Novelo, E., Ruiz-Piña, H.A., 2012. New finding of *Eratyrus cuspidatus* Stål (Hemiptera: Reduviidae) in Yucatán. Dugesiana 18, 143–145.

Ribeiro, J.M.C., Assumpção, T.C.F., Francischetti, I.M.B., 2012. An insight into the sialomes of bloodsucking Heteroptera. Psyche 2012, e470436.

Ribeiro, J.M., Genta, F.A., Sorgine, M.H., Logullo, R., Mesquita, R.D., Paiva-Silva, G.O., Majerowicz, D., Medeiros, M., Koerich, L., Terra, W.R., Ferreira, C., Pimentel, A.C., Bisch, P.M., Leite, D.C., Diniz, M.M., Junior, J.L., Silva, M.L., Araujo, R.N., Gandara, A.C., Brosson, S., Salmon, D., Bousbata, S., González-Caballero, N., Silber, A.M., Alves-Bezerra, M., Gondim, K.C., Silva-Neto, M.A., Atella, G.C., Araujo, H., Dias, F.A., Polycarpo, C., Vionette-Amaral, R.J., Fampa, P., Melo, A.C., Tanaka, A.S., Balczun, C., Oliveira, J.H., Gonçalves, R.L., Lazoski, C., Rivera-Pomar, R., Diambra, L., Schaub, G.A., Garcia, E.S., Azambuja, P., Braz, G.R., Oliveira, P.L., 2014. An insight into the transcriptome of the digestive tract of the bloodsucking bug, *Rhodnius prolixus*. PLoS Negl. Trop. Dis. 8, e2594.

Rojas Cortez, M., Emperaire, L., Piccinali, R.V., Gürtler, R.E., Torrico, F., Jansen, A.M., Noireau, F., 2007. Sylvatic *Triatoma infestans* (Reduviidae, Triatominae) in the Andean valleys of Bolivia. Acta Trop. 102, 47–54.

Rolón, M., Vega, M.C., Román, F., Gómez, A., Rojas de Arias, A., 2011. First report of colonies of sylvatic *Triatoma infestans* (Hemiptera: Reduviidae) in the Paraguayan Chaco, using a trained dog. PLoS Negl. Trop. Dis. 5, e1026.

Rosa, J.A., Rocha, C.S., Gardim, S., Pinto, M.C., Mendonça, V.J., Ferreira Filho, J.C.R., Carvalho, E.O.C., Camargo, L.M.A., Oliveira, J., Nascimento, J.D., Cilense, M., Almeida, C.E., 2012. Description of *Rhodnius montenegrensis* sp. nov. (Hemiptera: Reduviidae: Triatominae) from the state of Rondônia, Brazil. Zootaxa 3478, 62–76.

Rosa, J.A., Justino, H.H.G., Nascimento, J.D., Mendonça, V.J., Rocha, C.S., Carvalho, D.B., Falcone, R., Azeredo-Oliveira, M.T.V., Alevi, K.C.C., Oliveira, J., 2017. A new species of *Rhodnius* from Brazil (Hemiptera, Reduviidae, Triatominae). ZooKeys 675, 1–25.

Ryckman, R.E., 1962. Biosystematics and hosts of the *Triatoma protracta* complex in North America (Hemiptera: Reduviidae) (Rodentia: Cricetidae). Univ. Calif. Publ. Entomol. 27, 93–240.

Sainz, A.C., Mauro, L.V., Moriyama, E.N., García, B.A., 2004. Phylogeny of triatomine vectors of *Trypanosoma cruzi* suggested by mitochondrial DNA sequences. Genetica 121, 229–240.

Sandoval, C.M., Joya, M.I., Gutierez, R., Angulo, V.M., 2000. Cleptohaematophagy of the triatomine bug *Belminus herreri*. Med. Vet. Entomol. 14, 100–101.

Sandoval, C.M., Duarte, R., Gutiérrez, R., Rocha, D.S., Angulo, V.M., Esteban, L., Reyes, M., Jurberg, J., Galvão, C., 2004. Feeding sources and natural infection of *Belminus herreri* (Hemiptera, Reduviidae, Triatominae) from dwellings in Cesar, Colombia. Mem. Inst. Oswaldo Cruz 99, 137–140.

Sandoval, C.M., Pabón, E., Jurberg, J., Galvão, C., 2007. *Belminus ferroae* n. sp. from the Colombian north-east, with a key to the species of the genus (Hemiptera: Reduviidae: Triatominae). Zootaxa 1443, 55–64.

Sandoval, C.M., Ortiz, N., Jaimes, D., Lorosa, E., Galvão, C., Rodríguez, O., Scorza, J.V., Gutiérrez, R., 2010. Feeding behaviour of *Belminus ferroae* (Hemiptera: Reduviidae), a predaceous Triatominae colonizing rural houses in Norte de Santander, Colombia. Med. Vet. Entomol. 24, 124–131.

Sandoval-Ruiz, C.A., Cervantes-Peredo, L., Mendoza-Palmero, F.S., Ibáñez-Bernal, S., 2012. The Triatominae (Hemiptera: Heteroptera: Reduviidae) of Veracruz, Mexico: geographic distribution, taxonomic redescriptions, and a key. Zootaxa 3487, 1–23.

Santana, H.R.G., 2014. Estudo taxonômico da tribo Bolboderini (Hemiptera-Heteroptera, Reduviidae, Triatominae), com análise cladística. PhD thesis, Fundação, Oswaldo Cruz, Rio de Janeiro, Brazil.

Santos, A., Ribeiro, J.M., Lehane, M.J., Gontijo, N.F., Veloso, A.B., Sant'Anna, M.R., Araujo, R.N., Grisard, E.C., Pereira, M.H., 2007. The sialotranscriptome of the blood-sucking bug *Triatoma brasiliensis* (Hemiptera, Triatominae). Insect Biochem. Mol. Biol. 37, 702–712.

Schaefer, C.W., 2003. Triatominae (Hemiptera: Reduviidae): systematic questions and some others. Neotrop. Entomol. 32, 1–10.

Schofield, C.J., 1988. Biosystematics of the Triatominae. In: Sevice, M.W. (Ed.), Biosystematics of Haematophagous Insects, Systematics Association, Special vol. 37. Clarendon Press, Oxford, pp. 284–312.

Schofield, C.J., 2000. *Trypanosoma cruzi*—the vector-parasite paradox. Mem. Inst. Oswaldo Cruz 95, 535–544.

Schofield, C.J., Dujardin, J.P., 1999. Theories on the evolution of *Rhodnius*. Actual. Biol. 21, 183–197.

Schofield, C.J., Galvão, C., 2009. Classification, evolution and species groups within the Triatominae. Acta Trop. 110, 88–100.

Schofield, C.J., Grijalva, M.J., Diotaiuti, L., 2009. Distribución de los vectores de la enfermedad de Chagas en los países "no endémicos": la posibilidad de transmisión vectorial fuera de América Latina. Enferm. Emerg. 11 (Suppl. 1), 20–27.

Schuh, R.T., Slater, J.A., 1995. True bugs of the world (Hemiptera: Heteroptera): classification and natural history. In: True Bugs of the World. Cornell University Press, New York (xii 336 pp).

Sequeira, A.S., Farrell, B.D., 2001. Evolutionary origins of Gondwanan interactions: how old are araucaria beetle herbivores? Biol. J. Linnean Soc. 74, 459–474.

Sharma, P.P., Giribet, G., 2012. Out of the Neotropics: Late Cretaceous colonization of Australasia by American arthropods. Proc. Biol. Sci. 279, 3501–3509.

Sherratt, E., Castañeda, M.R., Garwood, R.J., Mahler, D.L., Sangere, T.J., Herrel, A., de Queiroz, K., Losos, J.B., 2015. Amber fossils demonstrate deep-time stability of Caribbean lizard communities. Proc. Natl. Acad. Sci. U. S. A. 112, 9961–9966.

Silveira, A.C., Feitosa, V.R., Borges, R., 1984. Distribuição de triatomíneos capturados no ambiente domiciliar, no período 1975/83. Brasil. Rev. Bras. Malariol. D. Trop. 36, 15–312.

Soares, R.P.P., Barbosa, S.E., Dujardin, J.P., Schofield, C.J., Siqueira, A.M., Diotaiuti, L., 1999. Characterization of *Rhodnius neglectus* from two regions of Brazil using isoenzymes, genitalia morphology and morphometry. Mem. Inst. Oswaldo Cruz 94, 161–166.

Solano, P., Dujardin, J.P., Schofield, C.J., Romaña, C., Tibayrenc, M., 1996. Isoenzymes as a tool for the identification of *Rhodnius* species. Res. Rev. Parasitol. 56, 41–47.

Sousa, O.E., Johnson, C.M., 1973. Prevalence of *Trypanosoma cruzi* and *Trypanosoma rangeli* in triatomines (Hemiptera: Reduviidae) collected in the Republic of Panama. Am. J. Trop. Med. Hyg. 22, 18–23.

Souza, R.C.M., Campolina-Silva, G.H., Bezerra, C.M., Diotaiuti, L., Gorla, D.E., 2015. Does *Triatoma brasiliensis* occupy the same environmental niche space as *Triatoma melanica*? Parasit. Vectors 8, 361.

Souza, E.S., von Atzingen, N.C.B., Furtado, M.B., Oliveira, J., Nascimento, J.D., Vendrami, D.P., Gardim, S., Rosa, J.A., 2016. Description of *Rhodnius marabaensis* sp. n. (Hemiptera, Reduviidade, Triatominae) from Pará state, Brazil. ZooKeys 621, 45–62.

Steiner, C.C., Ryder, O.A., 2011. Molecular phylogeny and evolution of the Perissodactyla. Zool. J. Linn. Soc. 163, 1289–1303.

Stothard, J.R., Yamamoto, Y., Cherchi, A., García, A.L., Valente, S.A.S., Schofield, C.J., Miles, M.A., 1998. A preliminary survey of mitochondrial sequence variation within triatomine bugs (Hemiptera: Reduviidae) using polymerase chain reaction-based single strand conformational polymorphism (SSCP) analysis and direct sequencing. Bull Entomol. Res. 88, 553–560.

Taracena, M.L., Oliveira, P.L., Almendares, O., Umaña, C., Lowenberger, C., Dotson, E.M., Paiva-Silva, G.O., Pennington, P.M., 2015. Genetically modifying the insect gut microbiota to control Chagas disease vectors through systemic RNAi. PLoS Negl. Trop. Dis. 9, e0003358.

Teves, S.C., Gardim, S., Carbajal de la Fuente, A.L., Lopes, C.M., Gonçalves, T.C., dos Santos-Mallet, J.R., Rosa, J.A., Almeida, C.E., 2016. Mitochondrial genes reveal *Triatoma jatai* as a sister species to *Triatoma costalimai* (Reduviidae: Triatominae). Am. J. Trop. Med. Hyg. 94, 686–688.

Tibayrenc, M., 1980. Note préliminaire sur les isoenzymes de *Triatoma infestans* (Hemiptera, Reduviidae), vecteur majeur de la maladie de Chagas en Amérique latine. Cah. ORSTOM sér. Ent. Méd. Parasitol XVIII, 71–73.

Torres-Pérez, F., Acuña-Retamar, M., Cook, J.A., Bacigalupo, A., García, A., Cattan, P.E., 2011. Statistical phylogeography of Chagas disease vector *Triatoma infestans*: testing biogeographic hypotheses of dispersal. Infect. Genet. Evol. 11, 167–174.

Traverso, L., Sierra, I., Sterkel, M., Francini, F., Ons, S., 2016. Neuropeptidomics in *Triatoma infestans*. Comparative transcriptomic analysis among triatomines. J. Physiol. Paris 3, 83–98.

Usinger, R.L., 1943. A revised classification of the Reduvioidea with a new subfamily from South America (Hemiptera). Ann. Entomol. Soc. Am. 36, 602–618.

Usinger, R.L., 1944. The Triatominae of North and Central America and the West Indies and their public health significance. U.S. Publ. Health Bull. 288, 1–83.

Usinger, R.L., Wygodzinsky, P., Ryckman, R.E., 1966. The biosystematics of Triatominae. Annu. Rev. Entomol. 11, 309–330.

Valderrama, A., Lizano, E., Cabello, D., Valera, M., 1996. *Panstrongylus turpiali*, n. sp. (Hemiptera: Reduviidae: Triatominae) from Venezuela. Caribb. J. Sci. 32, 142–144.

Valente, V.C., Valente, S.A.S., Carcavallo, R.U., Rocha, D.S., Galvão, C., Jurberg, J., 2001. Considerações sobre uma nova espécie do gênero *Rhodnius* Stål, do Estado de Pará, Brasil (Hemiptera: Reduviidae: Triatominae). Entomol. Vectores 8, 65–80.

van Sande, M., Karcher, D., 1960. Species differentiation of insects by hemolymph electrophoresis. Science 131, 1103–1104.

Vásquez, L.R., Galvão, C., Pinto, N.A., Granados, H., 2005. Primer registro de *Triatoma nigromaculata* (Stål, 1859) (Hemiptera, Reduviidae, Triatominae) para Colombia. Biomedica 25, 417–421.

VAST—Vietnam Academy of Science and Technology, 2016. Research on biology, ecology of blood sucking bugs belong to sbbfamily Triatominae (Heteroptera: Reduviidae) and their distribution characteristic in Vietnam [sic]. Available at http://vast.ac.vn/en/science-and-technology-research-projects?view=categories&start=180?option=com_detai&view=detai&id=817. (accessed June 6, 2017).

Villacís, A.G., Ocaña-Mayorga, S., Lascano, M.S., Yumiseva, C.A., Baus, E.G., Grijalva, M.J., 2015. Abundance, natural infection with trypanosomes, and food source of an endemic species of triatomine, *Panstrongylus howardi* (Neiva 1911), on the Ecuadorian central coast. Am. J. Trop. Med. Hyg. 92, 187–192.

Waleckx, E., Depickère, S., Salas, R., Aliaga, C., Monje, M., Calle, H., Buitrago, R., Noireau, F., Brenière, S.F., 2012. New discoveries of sylvatic *Triatoma infestans*

(Hemiptera: Reduviidae) throughout the Bolivian Chaco. Am. J. Trop. Med. Hyg. 86, 455–458.

Weirauch, C., 2007. Hairy attachment structures in Reduviidae (Cimicomorpha, Heteroptera), with observations on the fossula spongiosa in some other Cimicomorpha. Zool. Anz. 246, 155–175.

Weirauch, C., 2008. Cladistic analysis of Reduviidae (Heteroptera: Cimicomorpha) based on morphological characters. Syst. Entomol. 33, 229–274.

Weirauch, C., Munro, J.B., 2009. Molecular phylogeny of the assassin bugs (Hemiptera: Reduviidae), based on mitochondrial and nuclear ribosomal genes. Mol. Phylogenet. Evol. 53, 287–299.

Weirauch, C., Bérenger, J.-M., Berniker, L., Forero, D., Forthman, M., Frankenberg, S., Freedman, A., Gordon, E., Hoey-Chamberlain, R., Hwang, W.S., Marshall, S.A., Michael, A., Paiero, S.M., Udah, O., Watson, C., Yeo, M., Zhang, G., Zhang, J., 2014. An illustrated identification key to assassin bug subfamilies and tribes (Hemiptera: Reduviidae). Can. J. Arthropod Identific. 26, 1–115.

Wong, Y.Y., Sornosa Macias, K.J., Guale Martínez, D., Solórzano, L.F., Ramírez-Sierra, M.J., Herrera, C., Dumonteil, E., 2016. Molecular epidemiology of *Trypanosoma cruzi* and *Triatoma dimidiata* in costal Ecuador. Infect. Genet. Evol. 41, 207–212.

Wu, H., Guang, X., Al-Fageeh, M.B., Cao, J., Pan, S., Zhou, H., Zhang, L., Abutarboush, M.H., Xing, Y., Xie, Z., Alshanqeeti, A.S., Zhang, Y., Yao, Q., Al-Shomrani, B.M., Zhang, D., Li, J., Manee, M.M., Yang, Z., Yang, L., Liu, Y., Zhang, J., Altammami, M.A., Wang, S., Yu, L., Zhang, W., Liu, S., Ba, L., Liu, C., Yang, X., Meng, F., Wang, S., Li, L., Li, E., Li, X., Wu, K., Zhang, S., Wang, J., Yin, Y., Yang, H., Al-Swailem, A.M., Wang, J., 2014. Camelid genomes reveal evolution and adaptation to desert environments. Nat. Commun. 5, 5188.

Zachos, J., Pagani, M., Sloan, L., Thomas, E., Billups, K., 2001. Trends, rhythms, and aberrations in global climate 65 Ma to present. Science 292, 686–693.

Zeledón, R., Ugalde, J.A., Paniagua, L.A., 2001. Entomological and ecological aspects of six sylvatic species of triatomines (Hemiptera, Reduviidae) from the collection of the National Biodiversity Institute of Costa Rica, Central America. Mem. Inst. Oswaldo Cruz 96, 757–764.

Zhang, J., Weirauch, C., Zhang, G., Forero, D., 2016a. Molecular phylogeny of Harpactorinae and Bactrodinae uncovers complex evolution of sticky trap predation in assassin bugs (Heteroptera: Reduviidae). Cladistics 32, 538–554.

Zhang, J., Gordon, E., Forthman, M., Hwang, W.S., Walden, K.K., Swanson, D.R., Johnson, K., Meier, R., Weirauch, C., 2016b. Evolution of the assassin's arms: insights from a phylogeny of combined transcriptomic and ribosomal DNA data (Heteroptera: Reduvioidea). Sci. Rep. 6, 22177.

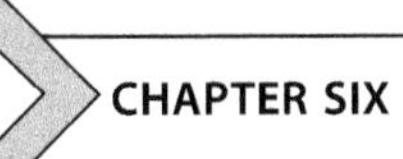

CHAPTER SIX

Expanding the Vector Control Toolbox for Malaria Elimination: A Systematic Review of the Evidence

Yasmin A. Williams*[,1,2], Lucy S. Tusting[†,2], Sophia Hocini*, Patricia M. Graves[‡], Gerry F. Killeen[§,¶], Immo Kleinschmidt[||,#,], Fredros O. Okumu[§], Richard G.A. Feachem*, Allison Tatarsky*, Roly D. Gosling***

*Malaria Elimination Initiative, Global Health Group, University of California, San Francisco, San Francisco, CA, United States
[†]Big Data Institute, University of Oxford, Oxford, United Kingdom
[‡]College of Public Health, Medical and Veterinary Sciences and Australian Institute of Tropical Health and Medicine, James Cook University, Cairns, QLD, Australia
[§]Ifakara Health Institute, Ifakara, Tanzania
[¶]Liverpool School of Tropical Medicine, Liverpool, United Kingdom
[||]MRC Tropical Epidemiology Group, London School of Hygiene and Tropical Medicine, London, United Kingdom
[#]School of Pathology, Faculty of Health Sciences, University of Witwatersrand, Johannesburg, South Africa
**Elimination 8, Windhoek, Namibia
[1]Corresponding author: e-mail address: yasmin.williams@ucsf.edu

Contents

1. Introduction 346
2. Methods 348
   2.1 Eligibility Criteria 348
   2.2 Search Strategy and Selection Criteria 348
   2.3 Data Abstraction 354
   2.4 Quality of Systematic Reviews and Risk of Bias in Phase III Studies 354
3. Results 354
   3.1 VCTs With a Recent Systematic Review 355
   3.2 Other VCTs With a Phase III Evaluation 368
   3.3 VCTs With No Phase III Evaluation 370
4. Discussion 370
Acknowledgements 375

[2] These authors are joint first authors and contributed equally.

Advances in Parasitology, Volume 99
ISSN 0065-308X
https://doi.org/10.1016/bs.apar.2018.01.003

Contributors 375
Conflict of Interest 375
References 375

Abstract

Background: Additional vector control tools (VCTs) are needed to supplement insecticide-treated nets (ITNs) and indoor residual spraying (IRS) to achieve malaria elimination in many settings. To identify options for expanding the malaria vector control toolbox, we conducted a systematic review of the availability and quality of the evidence for 21 malaria VCTs, excluding ITNs and IRS.

Methods: Six electronic databases and grey literature sources were searched from January 1, 1980 to September 28, 2015 to identify systematic reviews, Phase I–IV studies, and observational studies that measured the effect of malaria VCTs on epidemiological or entomological outcomes across any age groups in all malaria-endemic settings. Eligible studies were summarized qualitatively, with quality and risk of bias assessments undertaken where possible. Of 17,912 studies screened, 155 were eligible for inclusion and were included in a qualitative synthesis.

Results: Across the 21 VCTs, we found considerable heterogeneity in the volume and quality of evidence, with 7 VCTs currently supported by at least one Phase III community-level evaluation measuring parasitologically confirmed malaria incidence or infection prevalence (insecticide-treated clothing and blankets, insecticide-treated hammocks, insecticide-treated livestock, larval source management (LSM), mosquito-proofed housing, spatial repellents, and topical repellents). The remaining VCTs were supported by one or more Phase II ($n = 13$) or Phase I evaluation ($n = 1$). Overall the quality of the evidence base remains greatest for LSM and topical repellents, relative to the other VCTs evaluated, although existing evidence indicates that topical repellents are unlikely to provide effective population-level protection against malaria.

Conclusions: Despite substantial gaps in the supporting evidence, several VCTs may be promising supplements to ITNs and IRS in appropriate settings. Strengthening operational capacity and research to implement underutilized VCTs, such as LSM and mosquito-proofed housing, using an adaptive, learning-by-doing approach, while expanding the evidence base for promising supplementary VCTs that are locally tailored, should be considered central to global malaria elimination efforts.

1. INTRODUCTION

Great advances have been made in malaria control and elimination, with a 37% global decline in malaria incidence during 2000–2015 (Global Malaria Programme, 2015). New targets include the elimination of malaria from at least 35 countries by 2030 (Global Malaria Programme, 2017), with renewed calls for eradication within a generation (Gates and Chambers, 2015). In sub-Saharan Africa (SSA), vector control with insecticide-treated nets (ITNs) and indoor residual spraying (IRS) has averted an estimated 524 million malaria cases since 2000 (Global Malaria Programme, 2015). However, after an extraordinary period of success in global malaria control, progress has stalled with 216 million malaria cases in 2016, up 5 million cases from 2015 (Global Malaria Programme, 2017). There remain important obstacles to achieving and sustaining progress towards elimination, including operational inefficiencies that lead to low effective coverage (Bhatt et al., 2015), insecticide resistance (Ranson and Lissenden, 2016), and residual transmission mediated by mosquito behaviours such as outdoor biting and resting, feeding upon animals, and early exit from houses immediately after entering, which are not effectively targeted by ITNs and IRS (Govella and Ferguson, 2012; Killeen, 2014).

To achieve malaria elimination goals in the face of such challenges, what evidence-based vector control tools (VCTs) can national malaria control and elimination programmes access today or within the next decade to supplement ITNs and IRS? To date, ITNs and IRS are the only VCTs to have been recommended for wide-scale implementation by the World Health Organization (WHO), while larval source management (LSM) and personal protection measures against mosquitoes are recommended in some settings (World Health Organization, 2015). Recognizing the need for additional VCTs, WHO recently established mechanisms for expedited vector control recommendations, including new technical expert panels (Malaria Policy Advisory Committee, 2015; WHO Vector Control Advisory Group, 2013) and the Innovation to Impact (I2I) initiative to support VCT development and access (Innovation to Impact (I2I), 2016). Recent calls for novel vector control interventions with proven effectiveness elevated the global demand for new VCTs (malERA Refresh Consultative Panel on Tools for Malaria Elimination, 2017; World Health Organization, 2017). Here, to guide the identification of promising VCTs to expand the vector control toolbox for malaria elimination, we conducted a systematic review to collate published and unpublished evidence on the effect of selected VCTs on confirmed clinical malaria and malaria infection in people of any ages and on *Anopheles*-specific entomological outcomes in malaria-endemic regions. This is the first study to collate systematically the evidence across the spectrum of malaria vector control, excluding ITNs and IRS. Innovations in ITN and IRS technologies are also important contributions to the vector control toolbox (e.g. new active ingredients, insecticide combinations, and application technologies, among others) with significant product development and evaluation efforts underway but are outside the scope of this review (Innovative Vector Control Consortium, 2016; Wagman et al., 2018).

2. METHODS

We conducted a systematic review of the literature to summarize the availability and quality of the evidence for 21 malaria VCTs, excluding ITNs and IRS (Table 1). We followed guidelines of the Preferred Reporting Items for Systematic Reviews and Meta-analyses (PRISMA) (Additional File 1 in the online version at https://doi.org/10.1016/bs.apar.2018.01.003) (Wilson et al., 2015). The candidate VCTs for evaluation were selected through consultation with experts (including a meeting held on June 1–3, 2015 in San Francisco, USA) and the review of policy documents (WHO Vector Control Advisory Group, 2013, 2014).

2.1 Eligibility Criteria

Studies were included that evaluated any VCT targeting *Anopheles* mosquitoes in Table 1 and that met the eligibility criteria described in Table 2. Eligible study designs were categorized as observational, Phase I, Phase II, or Phase III studies. Observational studies included those with case–control, cohort, or cross-sectional designs. Phase I studies were defined as laboratory assays to determine the mode of action. Phase II was defined as semifield, experimental hut, and small-scale field studies, generally with entomological outcomes. Finally, Phase III studies were defined as trials measuring the efficacy of the VCT against epidemiological outcomes under optimal conditions (Wilson et al., 2015). Categories based on level of evidence were used since level of evidence is the basis for WHO policy recommendation.

2.2 Search Strategy and Selection Criteria

PubMed; EMBASE; LILACS; the Cochrane Infectious Diseases Group Specialized Register; Cochrane Central Register of Controlled Trials (CENTRAL), published in The Cochrane Library; and the Meta-Register of Controlled Trials were searched for studies published in English from January 1, 1980 to September 28, 2015 with the search terms described in Additional File 2 in the online version at https://doi.org/10.1016/bs.apar.2018.01.003. Search dates were restricted because systematic reviews included in this review captured the historical evidence on older VCTs, including LSM. Additionally, we searched reference lists of identified studies and contacted authors and field experts for unpublished data. To identify studies in progress, we searched the ClinicalTrials.gov registry.

Table 1 Description of Malaria Vector Control Tools (VCTs) Included in the Review

VCT[a]	Description	Primary Mode(s) of Action Against Malaria Vectors
Interventions targeting immature mosquitoes		
Larval source management (LSM)	Management of potential larval habitats to prevent the development of immature mosquitoes into adults; includes habitat modification and manipulation; biological control with natural enemies of mosquitoes; aerial and ground-based larviciding	Reduced adult emergence and density
Interventions targeting adult mosquitoes		
Adult sterilization by contamination	Sterilization of adult mosquitoes through contact with pyriproxyfen, using delivery mechanisms other than ITNs	Reduced adult reproduction and density
Other attract-and-kill mechanisms	Traps and targets that attract blood-seeking mosquitoes using a combination of odours from humans and other mammals (e.g. carbon dioxide, L-lactic acid, ammonia, and short-chain fatty acids), some of which are treated with chemical or biological insecticides (e.g. pyrethroids organophosphates, entomopathogenic fungi)	Increased adult mortality
Attractive toxic sugar baits (ATSBs)	Lethal traps that exploit sugar-feeding behaviour to attract mosquitoes using sugar and that contain insecticides (e.g. boric acid)	Reduced adult survival and density
Biological control of adult vector capacity/longevity	Infection of adult mosquitoes with bacteria (e.g. *Wolbachia* spp.) or entomopathogenic fungi to reduce longevity and/or upregulate immune genes	Reduced adult survival and infection rates
Eave tubes and eave baffles	A variety of different eave (space between the roof and walls of a house or structure) modifications that kill mosquitoes with traps or insecticides when they try to enter or exit from those houses	Reduced adult survival and density
Endectocide administration in humans	Mass administration to humans of a systemic insecticide, sometimes described as an endectocide (e.g. ivermectin)	Reduced adult survival and density

Continued

Table 1 Description of Malaria Vector Control Tools (VCTs) Included in the Review—cont'd

VCT	Description	Primary Mode(s) of Action Against Malaria Vectors
Endectocide administration in livestock	Mass administration to livestock of an endectocide (e.g. ivermectin, fipronil, eprinomectin) to kill zoophagic *Anopheles*	Reduced adult survival and density
Genetic modification	Mass release of mosquitoes, which are genetically modified (e.g. homing endonuclease genes (HEG) and RNA interference (RNAi); radiation- or chemo-sterilized males (sterile insect technique, SIT))	Reduced adult reproduction and density and/or reduced competence as the primary host for malaria parasites
Insecticide-treated clothing and blankets	Clothing and/or blankets treated with an insecticide (e.g. permethrin)	Reduced adult survival and density, as well as human exposure to biting
Insecticide-treated durable wall linings	Thin, durable sheets of insecticide-treated cloths that cover interior wall surfaces; insecticides remain efficacious for a period of 3–4 years	Reduced adult survival and density
Insecticide-treated fencing	Insecticide-treated netting used as fencing around livestock enclosures	Reduced adult survival and density
Insecticide-treated hammocks	Hammocks treated with an insecticide (e.g. permethrin)	Reduced adult survival and density, as well as human exposure to biting
Insecticide-treated livestock	Application of topical insecticide (e.g. pyrethroids) or entomopathogenic fungus to livestock to kill zoophilic mosquitoes	Reduced adult survival and density
Mosquito-proofed housing	Houses with features that reduce mosquito house entry (e.g. use of modern wall, floor and roof materials, use of insecticide-treated or untreated door and window screens, presence of a ceiling)	Reduced human exposure to biting mosquitoes
Push–pull systems	The simultaneous use of attractive and repellent volatiles (e.g. baited trap near home with insecticide-treated fabric in eaves)	Reduced adult survival and density, as well as human exposure to biting

Space spraying (ground application)	Liquid insecticide (e.g. pyrethroids, malathion) dispersed as fine droplets in the air (either thermal or cold fog) using hand-held or vehicle-mounted devices; can be used indoors or outdoors. Includes targeted spraying of male-mating swarms	Reduced adult survival and density
Spatial repellents	Products that release chemical active ingredients into the air as vapours, which repel, incapacitate, or kill adult mosquitoes (e.g. mosquito coils and emanators to release pyrethroids)	Reduced human biting, increased adult mortality
Topical repellents	Insect repellent (e.g. DEET, *Citronella*, picaridin, lemon eucalyptus) applied to the skin to provide personal protection from biting	Reduced human biting
Zooprophylaxis	Presence of animals/livestock to divert vector biting away from humans (which if applied at the individual level may also result in increased individual human risk, known as zoopotentiation)	Reduced exposure of humans to infectious adult mosquitoes and mosquitoes to infectious human beings
Interventions targeting immature mosquitoes via adults		
Larvicide application by autodissemination	Delivery of larvicide (e.g. pyriproxyfen) to larval habitats by adult female mosquitoes that are exposed to contaminated artificial resting sites	Reduced adult density

[a]VCTs excluded from the study: adult mosquito traps with no kill mechanism, aerial application of larvicide or adulticide, electronic mosquito repellents, indoor residual spraying, insecticide-treated curtains and nets, insecticide-treated paint, insecticide-treated plastic sheeting in tents or in temporary shelters, insecticide-treated tents, live plants as spatial repellents, nanoparticles for larviciding. Additionally, studies of the insecticidal properties of compounds and formulations were excluded.

Table 2 Criteria for Inclusion or Exclusion of Studies

	Inclusion Criteria	Exclusion Criteria
Study design	**Systematic reviews of experimental studies**	Review articles Opinion papers Modelling studies
	Phase III studies: randomized controlled (RCT), controlled before-and-after (CBA)[a], crossover[b], interrupted time-series[c]	
	Phase II studies[d]: small-scale, semifield, experimental hut	
	Phase I studies: laboratory	
	Observational studies: case–control, cohort, cross-sectional	
Intervention	Any malaria vector control tool (VCT) targeting *Anopheles* mosquitoes described in Table 1	Adult mosquito traps with no kill mechanism, electronic mosquito repellents, indoor residual spraying (IRS), insecticide-treated curtains and nets, insecticide-treated paint, insecticide-treated plastic sheeting in tents or in temporary shelters, insecticide-treated nets (ITNs), insecticide-treated tents, live plants as spatial repellents, studies of the insecticidal properties of compounds and formulations
Primary epidemiological outcomes	Malaria incidence and infection prevalence in any age group, diagnostically confirmed by microscopy or rapid diagnostic test	Malaria incidence and infection prevalence not diagnostically confirmed by rapid diagnostic test or microscopy

Primary entomological outcomes	Entomological inoculation rate (EIR)[e]	
	Human biting rate (HBR)[f]	
	Adult mosquito density metrics other than HBR[g]	
Secondary entomological outcomes[h]	Additional entomological outcomes appropriate to the intervention including adult mosquito fecundity, adult mosquito fitness, adult emergence rates, knockdown postexposure, blood-feeding inhibition	
Dates	Studies published from January 1, 1980 to September 28, 2015	Studies published before January 1, 1980 and after September 28, 2015

[a]Controlled before-and-after studies: if arms were comparable at baseline, there were at least two units per arm, follow-up periods were the same for the intervention and control arms, and baseline characteristics were comparable between arms.

[b]Crossover studies: if there was adequate allowance for washout (time between two intervention periods to allow the effect of the first intervention to be washed out).

[c]Interrupted time-series studies: if data were collected during at least three time points pre- and postfollow-up, if no cointerventions were introduced after baseline data collection, and if the intervention was implemented for a clearly defined period.

[d]Phase III studies were differentiated from Phase II studies in being conducted in real-life settings (not semifield or experimental hut systems) and having a minimum intervention period of one transmission season or year.

[e]Entomological inoculation rate (EIR): the number of bites by sporozoite-infected mosquitoes per person per unit time.

[f]Human biting rate (HBR): the number of host-seeking mosquitoes attempting to attack humans per person or house per time period.

[g]Secondary entomological outcomes, such as adult mosquito fecundity, adult mosquito fitness, adult emergence rates, knockdown postexposure, blood-feeding inhibition, included were reported in Phase I and II studies.

[h]Density measures other than HBR (e.g. number of mosquitoes per person, house, or catch), measured directly using human landing catches or indirectly using light traps, knockdown catches or other methods of biting rate determination.

Y.A.W. and S.H. independently screened titles and abstracts, followed by full-text screening of relevant studies for eligibility using a standard form in Qualtrics (Qualtrics, Provo, UT). Disagreements were resolved by L.S.T.

2.3 Data Abstraction

Study characteristics (including participants, intervention, control group, outcomes, and sample size, as applicable) and findings were double-entered into a standard form in Microsoft Excel by Y.A.W. and verified by L.S.T. Since we aimed to assess evidence availability, not VCT efficacy, we did not combine studies in a meta-analysis. Instead, for each VCT we summarized the current evidence by the number and type of completed studies and, where possible, stratified this information by outcome. We presented in tables all eligible studies for every VCT, except for VCTs with a recent ($\leq$5 years old) high-quality systematic review (Measurement Tool to Assess Systematic Reviews (AMSTAR) (Shea et al., 2007) score $\geq$50%; see later), for which we presented only the systematic review (Wilson et al., 2015).

2.4 Quality of Systematic Reviews and Risk of Bias in Phase III Studies

The quality of systematic reviews was assessed using the AMSTAR tool (Shea et al., 2007). Risk of bias for randomized controlled trials (RCTs), controlled before-and-after studies (CBA), crossover studies, and interrupted time-series studies was assessed using the Effective Practice and Organisation of Care (EPOC) tool (Effective Practice and Organisation of Care (EPOC), 2015). Risk of bias was not assessed for Phase I, Phase II, or observational studies due to wide heterogeneity in study designs. We did not perform a statistical test for publication bias because we did not conduct any meta-analyses.

3. RESULTS

The search results yielded 17,912 unique studies after removing duplicates (Fig. 1). A total of 155 studies met the eligibility criteria and were included in the qualitative synthesis; these were of the following designs: systematic reviews (n=7); Phase III (n=7), Phase II (n=76), and Phase I (n=54) experimental studies; and cross-sectional (n=7), case–control (n=3), and cohort (n=1) observational studies (Fig. 2; Additional File 3 in the online version at https://doi.org/10.1016/bs.apar.2018.01.003).

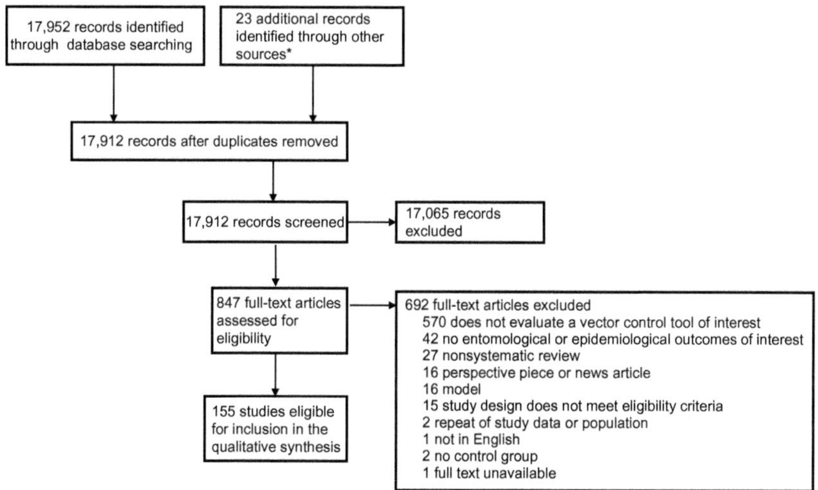

Fig. 1 Study flow for a systematic review of the evidence for 21 malaria vector control tools. **Other sources: reference lists of included studies.*

Methodological quality was variable across the seven eligible systematic reviews, with AMSTAR scores ranging from 18% to 100% (Additional File 4 in the online version at https://doi.org/10.1016/bs.apar.2018.01.003A). The systematic reviews of LSM ($n=2$), mosquito-proofed housing ($n=1$), and topical repellents ($n=1$) were determined to be of the highest quality (AMSTAR scores $\geq$50%), while those of spatial repellents ($n=2$) and zooprophylaxis ($n=1$) were judged to be of lower quality. Of the 21 VCTs evaluated, we identified 7 with one or more completed Phase III study, including some that were included in systematic reviews: LSM, insecticide-treated clothing and blankets, insecticide-treated hammocks, insecticide-treated livestock, mosquito-proofed housing, spatial repellents, and topical repellents, with recent, high-quality systematic reviews available for LSM, mosquito-proofed housing, and topical repellents (Table 3).

3.1 VCTs With a Recent Systematic Review

3.1.1 Larval Source Management

A 2013 Cochrane review compared biological control with larvivorous fish to biological control without larvivorous fish (Walshe et al., 2013). No eligible studies included in this review measured malaria incidence, entomological inoculation rate, or adult vector density (Table 3). Nine quasi-experimental studies measured larval mosquito density, with variable effects. A second 2013 Cochrane review compared LSM (excluding

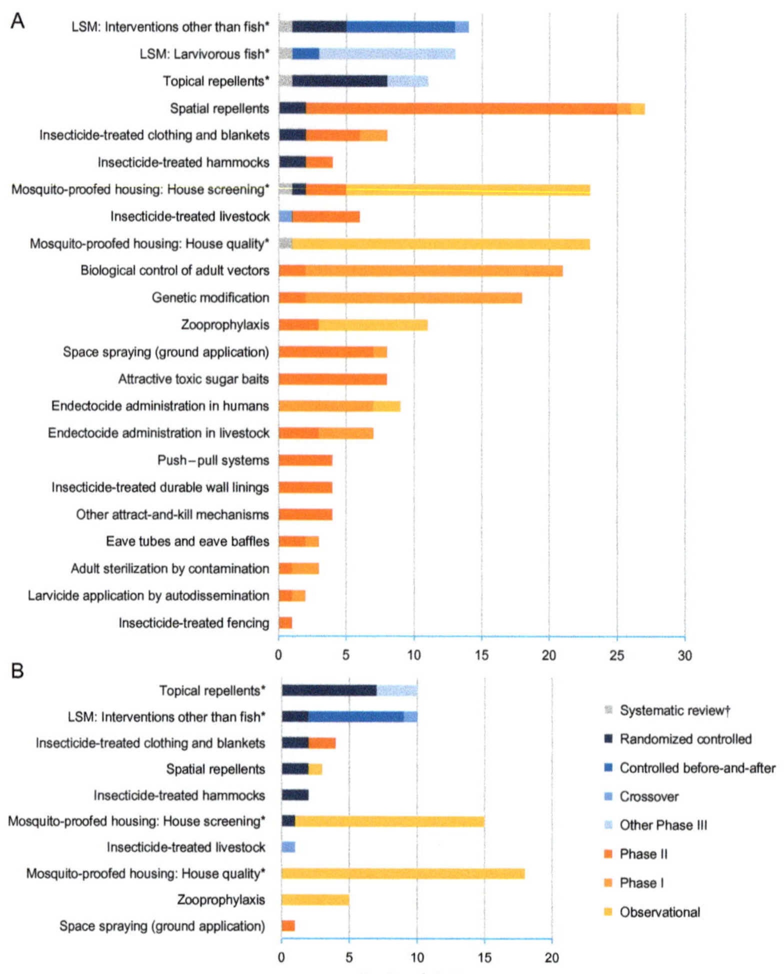

Fig. 2 Frequency of eligible studies of 21 malaria vector control tools (VCTs), stratified by study design. (A) Studies with any outcome of interest; (B) studies with diagnostically confirmed malaria incidence or prevalence. †Only systematic reviews with AMSTAR (A Measurement Tool to Assess Systematic Reviews 14) scores of ≥50% are included. *For topical repellents, larval source management, and mosquito-proofed housing, the frequency of studies represents all eligible studies within the referenced systematic review. For all other VCTs, the frequency of studies represents all eligible studies within the present review.

biological control with larvivorous fish) with no LSM (Tusting et al., 2013). Compared to the control, LSM reduced malaria incidence by 74% in two cluster RCTs, but there was no consistent effect on malaria incidence in three CBA studies. GRADE quality (Atkins et al., 2004) of evidence

Table 3 Summary of Malaria Vector Control Tools (VCTs), Excluding ITNs and IRS, With At Least One Phase III Evaluation, Stratified by Outcome

VCT or Study Reference	Outcome	Study Design	Number of Participants (*N* Studies)	Measure of Effect	Relative Effect (95% CI)[a]	Quality of Evidence (GRADE)[b]	Risk of Bias (EPOC)[c]	Comments
VCTs with a recent systematic review[d]								
Larval source management (LSM)								
Biological control with larvivorous fish								
Reference: Walshe et al. (2013) **Study type:** Cochrane review (AMSTAR score 91%) **Participants:** All age groups in malaria-endemic settings **Countries:** Kenya, Korea, India, Indonesia, Sri Lanka, Sudan	Malaria incidence; EIR; density of adult mosquitoes	—	0 studies	—	—	—	—	No studies eligible
	Density of mosquito larvae	Quasi-experimental	9 studies	Not pooled	—	Very low	—	Variable effects reported

Continued

Table 3 Summary of Malaria Vector Control Tools (VCTs), Excluding ITNs and IRS, With At Least One Phase III Evaluation, Stratified by Outcome—cont'd

VCT or Study Reference	Outcome	Study Design	Number of Participants (*N* Studies)	Measure of Effect	Relative Effect (95% CI)	Quality of Evidence (GRADE)	Risk of Bias (EPOC)	Comments
Larval source management, excluding biological control with larvivorous fish								
Reference: Tusting et al. (2013) **Study type:** Cochrane review (AMSTAR score 100%) **Participants:** All age groups in malaria-endemic settings **Countries:** Eritrea, The Gambia, Greece, Kenya, India, Mali, Philippines, Sri Lanka, Tanzania	Malaria incidence	Cluster RCT	20,124 (2 studies)	Rate ratio	0.26 (0.22, 0.31)	Moderate	—	95% CI may be falsely narrow because studies did not adjust for cluster design
		CBA	98,233 (3 studies)	Rate ratio	0.51 (0.18, 1.44)	Very low	—	—
	Parasite prevalence	Cluster RCT	2963 (1 study)	Risk ratio	0.11 (0.05, 0.22)	Moderate	—	95% CI may be falsely narrow because studies did not adjust for cluster design
		CBA	8041 (5 studies)	Risk ratio	0.32 (0.19, 0.55)	Moderate	—	—
	EIR	Cluster RCT	1 study	Percent reduction	84.6 (year 2 data); not estimable (year 1 data)	—	—	95% CI not available
		CBA	3 studies	Percent reduction	21.3 (−42.3, 56.4) to 73.0 (22.0, 90.7)	—	—	—

	HBR	Cluster RCT	1 study	Percent reduction	45.8 (year 2) to 49.0 (year 1)	—	—	95% CI not available
		CBA	2 studies	Percent reduction	31.3 (−59.2, 70.4) to 73.1 (20.3, 90.9)	—	—	—
Mosquito-proofed housing								
Screened vs unscreened housing								
Reference: Tusting et al. (2015) **Study type:** Systematic review (AMSTAR score 91%) **Participants:** All age groups in malaria-endemic settings **Countries:** Benin, Equatorial Guinea, Ethiopia, The Gambia, Ghana, Kenya, Nigeria, Peru, São Tomé and	Clinical malaria	Case–control	1 study	Crude odds ratio	1.16 (0.82, 1.64)	Not assessed	—	—
		Cohort	3 studies	Adjusted rate ratio	0.56 (0.46, 0.67)	Not assessed	—	—
	Malaria infection	RCT	1 study	Adjusted odds ratio	0.95 (0.63, 1.43)	Not assessed	—	—
		Case–control, cross-sectional, cohort	2 studies	Adjusted odds ratio	0.93 (0.82, 1.05)	Not assessed	—	—
	Anaemia in children aged 0–11 years	RCT	1 study	Adjusted odds ratio	0.52 (0.34, 0.80)	Not assessed	—	—
		Case–control	1 study	Adjusted odds ratio	0.56 (0.24, 1.27)	Not assessed	—	—

Continued

Table 3 Summary of Malaria Vector Control Tools (VCTs), Excluding ITNs and IRS, With At Least One Phase III Evaluation, Stratified by Outcome—cont'd

VCT or Study Reference	Outcome	Study Design	Number of Participants (*N* Studies)	Measure of Effect	Relative Effect (95% CI)	Quality of Evidence (GRADE)	Risk of Bias (EPOC)	Comments
Príncipe, Sudan, Tanzania, United States	EIR	RCT	1 study	Abundance ratio	0.34 (0.21, 0.54) (year 1); 0.31 (0.16, 0.59) (year 2)	Not assessed	—	—
	HBR	RCT	1 study	Ratio of means	0.46 (0.34, 0.63)	Not assessed	—	—
		Phase II	2 studies	Rate ratio	0.50 (0.38, 0.67) to 0.61 (0.44, 0.83)	Not assessed	—	—
		Cohort; cross-sectional	3 studies	—	—	Not assessed	—	Variable associations reported
Modern vs traditional housing								
Reference: Tusting et al. (2015) **Study type:** Systematic review (AMSTAR score 91%) **Participants:** All age groups in malaria-endemic	Clinical malaria	Case–control	357 (1 study)	Adjusted odds ratio	0.35 (0.20, 0.62)	Very low	—	—
		Cohort	2237 (3 studies)	Adjusted rate ratio	0.55 (0.36, 0.84)	Very low	—	—
	Malaria infection	Case–control; cross-sectional; cohort	3949 (5 studies)	Adjusted odds ratio	0.53 (0.42, 0.67)	Very low	—	—

settings **Countries:** East Timor, Egypt, Ethiopia, Greece, Malawi, Mexico, Sri Lanka, Tanzania, Thailand, Uganda, Yemen	HBR	Cohort	2 studies	Adjusted odds ratio	0.48 (0.37, 0.64) to 0.57 (0.37, 0.87)	Not assessed	—	—
Topical repellents								
Reference: Wilson et al. (2014) **Study type**: Systematic review (AMSTAR score: 64%) **Participants:** All age groups in malaria-endemic settings **Countries**: Bolivia, Ecuador, Ethiopia, Ghana, India, Lao People's Democratic Republic (PDR), Pakistan, Peru, Tanzania, Thailand	*P. falciparum* malaria or infection	RCT; CBA	7 studies	Crude risk ratio	0.82 (0.62, 1.08)	Not assessed	Low/ unclear	—
	P. vivax malaria or infection	RCT; CBA	6 studies	Crude risk ratio	0.80 (0.47, 1.37)	Not assessed	Low/ unclear	—

Continued

Table 3 Summary of Malaria Vector Control Tools (VCTs), Excluding ITNs and IRS, With At Least One Phase III Evaluation, Stratified by Outcome—cont'd

VCT or Study Reference	Outcome	Study Design	Number of Participants (*N* Studies)	Measure of Effect	Relative Effect (95% CI)	Quality of Evidence (GRADE)	Risk of Bias (EPOC)	Comments
Other VCTs with a Phase III evaluation[e]								
Insecticide-treated clothing and blankets								
Reference: Macintyre et al. (2003) **Study type:** Phase III **Participants:** All age groups **Country:** Kenya	Malaria incidence (aged >5 years)	RCT	375	Crude rate ratio	0.19 (0.05, 0.77)	—	Moderate	—
	Malaria incidence (aged ≤5 years)	RCT	97	Crude rate ratio	1.87 (0.31, 11.30)	—	Moderate	—
Reference: Rowland et al. (1999) **Study type:** Phase III **Participants:** All age groups; refugees **Country:** Pakistan	*P. falciparum* incidence	RCT	825	Adjusted odds ratio	0.51 (0.30, 0.86)	—	Low	—
	P. vivax incidence	RCT	825	Adjusted odds ratio	0.70 (0.43, 1.13)	—	Low	—

Insecticide-treated hammocks								
Reference: Magris et al. (2007) **Study type:** Phase III **Participants:** All age groups **Country:** Venezuela	Malaria incidence	RCT	924	Adjusted rate ratio	0.44 (0.41, 0.48)	—	Low	—
	Malaria prevalence (aged ≤6 months)	RCT	924	Adjusted risk ratio	0.17 (0.00, 0.53)	—	Low	—
Reference: Thang et al. (2009) **Study type:** Phase III **Participants:** All age groups **Country:** Vietnam	Malaria incidence	RCT	18,646	—	—	—	Low	See footnote f
	Malaria prevalence	RCT	18,646	—	—	—	Low	
Insecticide-treated livestock								
Reference: Rowland et al. (2001) **Study type:** Phase III	Malaria incidence (*P. falciparum*)	Crossover	56,329	Adjusted rate ratio	0.44 (0.22, 0.86)	—	Moderate	—
	Malaria incidence (*P. vivax*)	Crossover	56,329	Adjusted rate ratio	0.69 (0.50, 0.95)	—	Moderate	—

Continued

Table 3 Summary of Malaria Vector Control Tools (VCTs), Excluding ITNs and IRS, With At Least One Phase III Evaluation, Stratified by Outcome—cont'd

VCT or Study Reference	Outcome	Study Design	Number of Participants (*N* Studies)	Measure of Effect	Relative Effect (95% CI)	Quality of Evidence (GRADE)	Risk of Bias (EPOC)	Comments
Participants: All age groups; refugees **Country:** Pakistan	Malaria prevalence (*P. falciparum*)	Crossover	56,329	Adjusted rate ratio	0.46 (0.31, 0.70)	—	Moderate	—
	Malaria prevalence (*P. vivax*)	Crossover	56,329	Adjusted rate ratio	0.60 (0.33, 1.08)	—	Moderate	—
	Density of adult *A. stephensi*	Crossover	15 sentinel rooms/village	Density ratio	0.53 (0.32, 0.88)	—	Moderate	—
	Density of adult *A. subpictus*	Crossover	15 sentinel rooms/village	Density ratio	0.67 (0.25, 1.85)	—	Moderate	—
Spatial repellents								
Reference: Hill et al. (2014) **Study type:** Phase III	Malaria prevalence (*P. falciparum*)	RCT	7413	Adjusted odds ratio	0.23 (0.11, 0.50)	—	Moderate	—

Participants: All age groups **Country:** China	Malaria prevalence (*P. vivax*)	RCT	7413	Adjusted odds ratio	0.20 (0.09, 0.44)	—	Moderate	—
	HBR	RCT	Four sentinel houses per arm	Percent reduction	88%	—	Moderate	95% CI not reported
Reference: Lawrence and Croft (2004) **Study type:** Systematic review (AMSTAR score: 18%) **Participants:** Not reported **Countries:** Not reported	Biting or feeding inhibition; mosquito mortality, knockdown	Laboratory, Phase II	15 studies	No meta-analysis reported	—	Not assessed	—	No included studies measured the effect of mosquito coils on the incidence of clinical malaria. Mosquito coils inhibited nuisance biting in 13 of 15 included studies (though the effect was not always significant)

Continued

Table 3 Summary of Malaria Vector Control Tools (VCTs), Excluding ITNs and IRS, With At Least One Phase III Evaluation, Stratified by Outcome—cont'd

VCT or Study Reference	Outcome	Study Design	Number of Participants (*N* Studies)	Measure of Effect	Relative Effect (95% CI)	Quality of Evidence (GRADE)	Risk of Bias (EPOC)	Comments
Reference: Ogoma et al. (2012) **Study type:** Systematic review (AMSTAR score: 18%) **Participants:** n/a **Countries:** Not reported	Adult mosquito mortality; knockdown time postexposure; deterrence; human feeding	Laboratory, Phase II	17 studies	No meta-analysis reported	—	Not assessed	—	There was evidence that coils and emanators increased mosquito mortality and deterrence and reduced human feeding
Reference: Syafruddin et al. (2014) **Study type**: Phase III **Participants:** Men aged 18–60 years **Country:** Indonesia	Incidence of new malaria infections	RCT	170	Relative risk	0.48 (0.31, 0.75)		Low	—
	HBR	RCT	Five sentinel houses	Percent reduction	32.9%		Low	95% CI not reported

Zooprophylaxis							
Reference: Donnelly et al. (2015) **Study type:** Systematic review (AMSTAR score: 18%) **Participants:** All age groups **Countries:** Bolivia, Burkina Faso, Ethiopia, The Gambia, Ghana, Guinea Bissau, Kenya, Mozambique, Pakistan, Lao PDR, Zambia	Malaria prevalence, human blood index, HBR	Not reported	20 studies	n/a	No meta-analysis	n/a	Variable effects reported

[a] *CI*, confidence interval.
[b] GRADE: GRADE Working Group (Atkins et al., 2004) grades of evidence for each outcome, as evaluated by the authors of the cited review. Grades range from high quality (further research is very unlikely to change our confidence in the estimate of effect) to moderate quality (further research is likely to have an important impact on our confidence in the estimate of effect and may change the estimate), low quality (further research is very likely to have an important impact on our confidence in the estimate of effect and is likely to change the estimate), and very low quality (we are very uncertain about the estimate).
[c] For VCTs with a systematic review with an AMSTAR score of ≥50%, individual Phase III studies are not presented.
[d] For VCTs without a systematic review with an AMSTAR score of ≥50%, both systematic reviews and individual Phase III studies are presented.
[e] After 24 months' follow-up, malaria incidence decreased in both control (IRR 0.48; 95% CI 0.28, 0.82) and intervention (IRR 0.23, 95% CI 0.14, 0.38) arms, compared to the baseline, and infection prevalence decreased in both control (OR 0.26, 95% CI 0.20, 0.35) and intervention (OR 0.15, 95% CI 0.09, 0.26) arms. Malaria incidence and infection prevalence in the intervention group decreased significantly more in the intervention arm than in the control.
[f] EPOC (2015) risk of bias scores for Phase III studies, as evaluated by the authors of the present review.

ranged from very low to moderate. Parasite prevalence was reduced by 89% in another cluster RCT and by an average of 68% in five CBA studies. GRADE quality of evidence was assessed to be moderate for both subgroups.

3.1.2 Mosquito-Proofed Housing

A 2015 systematic review included one Phase III RCT and four observational studies in a meta-analysis comparing screened with unscreened housing, in which findings on the effect on clinical malaria, malaria infection, and anaemia in children were inconsistent (Table 3) (Tusting et al., 2015). A further 15 observational studies were included in a meta-analysis comparing 'modern' housing (e.g. brick or cement walls and metal roofs) with 'traditional' housing (e.g. mud walls, thatched roofs, open eaves, and no screening) (Tusting et al., 2015). Modern housing was associated with a 45%–65% lower odds of clinical malaria and 47% lower odds of malaria infection, compared to traditional housing, although the GRADE quality of evidence was assessed to be very low.

3.1.3 Topical Repellents

In a systematic review of experimental studies comparing topical repellents with no repellent or placebo repellents (Wilson et al., 2014), the risk of *Plasmodium falciparum* malaria or infection was reduced by 18% in six RCTs and one CBA. *Plasmodium vivax* malaria or infection was reduced by 20% in five RCTs and one CBA, compared to the control, but neither reduction was statistically significant. EPOC risk of bias in the included studies ranged from low to unclear (Table 3).

3.2 Other VCTs With a Phase III Evaluation

3.2.1 Insecticide-Treated Clothing and Blankets

Malaria incidence was measured in two RCTs with low to moderate risk of bias, where the effect of insecticide-treated clothing and blankets ranged from an 81% decrease to no effect, compared to the control (Table 3) (Macintyre et al., 2003; Rowland et al., 1999). Outcomes assessed by the four Phase II studies included parasite prevalence ($n=2$) and adult mosquito mortality ($n=2$) (Additional File 3 in the online version at https://doi.org/10.1016/bs.apar.2018.01.003B).

3.2.2 Insecticide-Treated Hammocks

Malaria incidence and parasite prevalence were measured in two Phase III RCTs, with EPOC risk of bias for both studies assessed to be low (Table 3). In Venezuela, insecticide-treated hammocks reduced malaria incidence by 56% and parasite prevalence by 83%, compared to the control (Magris et al., 2007), and in Vietnam a greater reduction in malaria incidence and parasite prevalence was observed in the intervention arm than in the control (footnote to Table 3) (Thang et al., 2009). One Phase II study measured adult *Anopheles gambiae* mortality, hut entry, and blood-feeding inhibition (Additional File 3 in the online version at https://doi.org/10.1016/bs.apar.2018.01.003C).

3.2.3 Insecticide-Treated Livestock

Malaria incidence and parasite prevalence were measured in one Phase III crossover study, with EPOC risk of bias assessed to be moderate, in which insecticide-treated livestock reduced malaria incidence by 31%–56% and parasite prevalence by 40%–54% compared to the control, though the effect was not consistently significant (Table 3) (Rowland et al., 2001). Entomological outcomes measured in five Phase II studies included adult mosquito mortality and blood-feeding preference (Additional File 3 in the online version at https://doi.org/10.1016/bs.apar.2018.01.003C).

3.2.4 Spatial Repellents

Two systematic reviews included laboratory and Phase II field studies only, with no meta-analyses (Table 3) (Lawrence and Croft, 2004; Ogoma et al., 2012). No eligible studies measured the effect of spatial repellents on malaria incidence. Parasite prevalence was measured in two RCTs, with the EPOC risk of bias assessed to be low for both studies, and in one cross-sectional study. In the RCTs, transfluthrin coils reduced parasite prevalence by 77% compared to long-lasting insecticide-treated nets (LLINs) alone and by 94% when combined with LLINs, compared to no intervention in China (Hill et al., 2014); metofluthrin mosquito coils reduced parasite prevalence by 52% compared to a placebo in Indonesia (Syafruddin et al., 2014). Entomological outcomes measured in 23 Phase II studies and 1 Phase I study included human biting rate (HBR), adult mosquito mortality, and repellency (Additional File 3 in the online version at https://doi.org/10.1016/bs.apar.2018.01.003C).

3.3 VCTs With No Phase III Evaluation

Fourteen VCTs had Phase I, II, and/or observational evidence only: adult sterilization by contamination, attractive toxic sugar baits, other attract-and-kill mechanisms, biological control of adult vectors, eave tubes and eave baffles, endectocide administration in humans, endectocide administration in livestock, genetic modification, insecticide-treated durable wall linings, insecticide-treated fencing, larvicide application by autodissemination, push–pull systems, space spraying (ground application), and zooprophylaxis (Fig. 2; Additional File 3 in the online version at https://doi.org/10.1016/bs.apar.2018.01.003C and D). For these VCTs we included a total of 103 studies, comprising 42 Phase II, 51 Phase I, and 10 observational studies. All VCTs had at least one eligible Phase II study, except endectocide administration in humans. Three VCTs had at least one eligible observational study: endectocide administration in humans, spatial repellents, and zooprophylaxis. For zooprophylaxis, we also identified one systematic review (AMSTAR score 18%), which reported no meta-analysis (Donnelly et al., 2015). Entomological outcomes were measured for all VCTs, while epide tsavemiological outcomes were measured for two VCTs only (space spraying and zooprophylaxis).

4. DISCUSSION

To address the challenges of insecticide resistance and residual transmission, strengthen malaria vector control, and maintain progress towards elimination, additional malaria VCTs are needed. In this systematic review assessing the availability and quality of evidence for 21 supplementary VCTs, we included 155 studies dating from January 1, 1980 to September 28, 2015. This is the first study to collate evidence systematically across the malaria vector control toolbox beyond ITNs and IRS. Our study highlights the expanding pipeline of research into supplementary VCTs, while identifying substantial heterogeneity in the availability and quality of the evidence required by WHO to provide normative guidance on implementation (i.e. standardized epidemiological data from Phase III trials in multiple settings) (Malaria Policy Advisory Committee, 2012; WHO Vector Control Advisory Group, 2013).

For each VCT, we summarized the current evidence by the number and quality of studies and stratified this information by outcome where possible since this information forms the basis of WHO policy considerations.

Within this framework, the evidence base was the most extensive for LSM and topical repellents, which both have multiple published Phase III evaluations and recent systematic reviews assessed to be of high methodological quality. While the evidence for LSM was assessed to be of very low to moderate quality (Tusting et al., 2013), combinations of larviciding and environmental management have been effective in reducing malaria transmission in certain eco-epidemiological settings in Africa and Asia and larviciding has been recommended by WHO as a supplementary intervention in SSA since 2013 (Global Malaria Programme, 2015). This recommendation is limited to discrete settings where habitats are relatively 'few, fixed, and findable', far narrower than settings in high-income countries where larviciding is used routinely and successfully for mosquito and disease control (Global Malaria Programme, 2015). In contrast, the evidence for topical repellents is of relatively high quality (Wilson et al., 2014) but indicates that topical repellents are unsuitable as a large-scale public health intervention, although they can provide individual protection against mosquitoes (Wilson et al., 2014). We identified five further VCTs with at least one Phase III evaluation with epidemiological outcomes: insecticide-treated clothing and blankets, insecticide-treated hammocks, insecticide-treated livestock, mosquito-proofed housing, and spatial repellents. These VCTs offer additional options for supplementing ITNs and IRS, often with complementary modes of action. Further Phase III community-level trials will help to clarify their roles in malaria vector control in different epidemiological settings (Killeen, 2014; Lobo et al., 2014; Pinder et al., 2016).

Our assessment of evidence was based on study design and outcomes, but in the future it may be necessary to consider evidence complementary to standard epidemiological assessments (Vontas et al., 2014). First, making recommendations across diverse transmission settings and local vector ecologies is difficult; what works in one or two settings may not work in all settings. Growing understanding of the genetic diversity among *Anopheles* further contributes to this complexity (The *Anopheles gambiae* 1000 Genomes Consortium, 2017). Trends in malaria transmission and performance of VCTs are also confounded by longer-term changes in environmental and infrastructural landscapes and climate (Snow et al., 2017). Although Cochrane reviews remain the gold standard in evidence-based policy, it is often inappropriate to combine findings from studies across different eco-epidemiological settings when VCT efficacy is tied to local transmission ecology (Tusting et al., 2013; Walshe et al., 2013). Second, some emerging VCTs remain years away from accumulating a full dossier of epidemiological evidence, and although further Phase III studies are planned (Thomas and Knols, 2015), nearing completion (Mtove et al., 2016), or recently concluded (Homan et al., 2016), we identified 14 VCTs for which no Phase III epidemiological data were available within the search dates. Demonstrating protection against disease and/or infection is critical before any VCTs can be recommended for large-scale deployment (Wilson et al., 2015). However, in some circumstances, evidence of effect might be built by adopting underutilized VCTs as supplementary interventions within a 'learning-by-doing' framework. This iterative, adaptive approach involves the incorporation of rigorous monitoring and evaluation of epidemiological and entomological outcomes in control and intervention areas to support the gradual scale-up of additional VCTs within existing programme infrastructure, such as through adaptable Phase IV effectiveness studies (Global Malaria Programme, 2014; Killeen, 2014; Wilson et al., 2015). For example, while only one RCT of house screening for malaria control has been completed (Kirby et al., 2009), a large body of observational evidence suggests that screened housing is associated with reduced malaria risk and national malaria control programmes are encouraged to explore opportunities to build 'healthier' housing (Roll Back Malaria, 2015). This approach would also allow for a more rapid expansion of the evidence base across a wider diversity of eco-epidemiological settings to inform locally tailored solutions as well as iteration over time as the transmission landscape changes.

Direct transition to Phase IV 'learning-by-doing' approaches is controversial and inappropriate for VCTs with a poor or absent evidence base (Wilson et al., 2015). The history of ITNs and IRS demonstrates varying routes to establishing effectiveness against malaria disease or infection; ITNs underwent rigorous evaluation through Phase III RCTs (Darriet et al., 1984), while IRS effectiveness was established decades before evaluation in RCTs (Sadasivaia et al., 2007). Given adequate funding, promising new VCTs should reach approval far faster than ITNs, but depending on the entomological mode of action, efficacy of a VCT in one ecological setting is not always guaranteed elsewhere. Recent examples illustrate the importance of demonstrating efficacy against epidemiological as well as entomological outcomes. Topical repellents reduce vector biting, but it took a cluster RCT with epidemiological outcomes to show their unsuitability as a generalizable public health intervention due to the high user compliance required (Messenger et al., 2012). Conversely, odour-baited traps have recently been shown to reduce malaria infection prevalence in a rigorous RCT, but entomological data from that study suggest caution before deploying this VCT at scale in different settings since the traps were largely effective against *Anopheles funestus* only (Homan et al., 2016). Such information may be obtainable through 'learning-by-doing' evaluations, as long as evaluations of outcomes are of high quality. Research institutions will need to support control programmes in design, technical capacity, and analysis to ensure meaningful findings are obtained from Phase IV effectiveness evaluations. A recent call for more adaptive strategies responding to shifting transmission also highlights the need for optimizing combinations of interventions to maximize impact and mitigate the risk of insecticide resistance (malERA Refresh Consultative Panel on Tools for Malaria Elimination, 2017).

Despite limited evidence on their efficacy against malaria, the 14 VCTs with no complete Phase III evaluation offer diverse modes of action to complement those of ITNs and IRS within a comprehensive intervention package. Some may only be suitable for niche application, for example, insecticide-treated clothing may be effective for individuals working outdoors at night, but not as a general public health intervention. Others such as insecticide-treated durable wall linings (which are impregnable with alternative insecticides to those used for IRS) might reduce reliance on the main classes of insecticides currently available for ITNs and IRS; a multicountry Phase III evaluation is currently underway (Messenger et al., 2012). Similarly, administration of endectocides such as ivermectin to people or livestock could circumvent insecticide resistance and target zoophagic behaviours in vectors, although epidemiological effect remains to be demonstrated (Chaccour et al., 2015; Foy et al., 2011). Some emerging VCTs might reduce transmission by vectors biting outdoors, including larvicide application by autodissemination using pyriproxyfen, which targets immature mosquitoes regardless of adult biting and resting behaviour (Mbare et al., 2014). Some emerging VCTs exploit vulnerability in alternative vector life stages to those targeted by ITNs and IRS. ATSBs, which target sugar feeding, consistently reduced adult mosquito density and HBR in Phase II studies in Israel, Mali, and the United States. However, Phase III trials of ATSBs with epidemiological outcomes are certainly needed. Genetic modification of mosquitoes aims to suppress populations thereby reducing vectorial competence (Alphey and Alphey, 2014), but our review highlights how such approaches have yet to progress fully beyond laboratory evaluations.

Overall the expansion of research on supplementary VCTs is encouraging, but arguably the first step to strengthening vector control for malaria elimination is to improve operational capacity to deliver and sustain existing interventions effectively (Brady et al., 2016). For example, major inefficiencies persist within LLIN delivery systems across SSA, limiting population access (Bhatt and Gething, 2014). There are also opportunities to explore new or improved delivery mechanisms for existing supplementary interventions, such as aerial application of larvicides (Knapp et al., 2015). Some VCTs may not be highly effective individually, but could potentially be highly effective when used in combinations. The malERA-updated research agenda highlights this need for optimizing combinations of interventions to maximize impact and mitigate the risk of insecticide resistance (malERA Refresh Consultative Panel on Tools for Malaria Elimination, 2017). Use of mathematical models could help to address such questions, where no epidemiological evidence is available (Kiware et al., 2017). Critical to improving vector control is the development of strong local entomological capacity (Mnzava et al., 2014), together with a much more significant focus on community engagement and effective integration of control across vector-borne diseases and government sectors (Brady et al., 2016; World Health Organization, 2009, 2017).

Our study has several limitations. First, our VCTs of interest were selected a priori through expert consultation and are not an exhaustive list. Second, our search was restricted to English language papers only, potentially excluding experiences from some regions. Third, we did not combine data across studies in a meta-analysis, precluding evaluation of effect on entomological and epidemiological outcomes and statistical tests for publication bias. Fourth, for studies with entomological outcomes there was no mechanism to standardize outcomes and assess how heterogeneity in the choice of control affected study findings. Fifth, this review focused on individual interventions and did not consider the potential benefits of combining two or more of the new VCTs in communities already using ITNs and/or IRS. Finally, we did not assess methodological quality and risk of bias in Phase I and II studies due to heterogeneity in study design.

In conclusion, our review highlights the expanding pipeline of research into new and underutilized approaches to malaria vector control and the critical need to prioritize and fund robust evaluation of supplementary VCTs. Despite substantial gaps in the supporting evidence, several VCTs are promising supplements to ITNs and IRS. Strengthening operational capacity to implement and evaluate underutilized VCTs, such as LSM and mosquito-proofed housing, while expanding the evidence base for newer VCTs through strategic assessment of existing evidence and rigorous epidemiological evaluation, should be central to global malaria control and elimination efforts. A practical, programme-oriented research agenda to evaluate where, when, and in what combination to use these supplemental VCTs should be developed and prioritized for funding and implementation in the near term. Future research should also assess the cost, cost-effectiveness, scalability, and availability of supplemental VCTs to inform vector control strategies and intervention selection as countries and regions accelerate towards elimination.

Supplementary data to this article can be found online at https://doi.org/10.1016/bs.apar.2018.01.003.

ACKNOWLEDGEMENTS

This work was supported by the University of California, Group Health Group Malaria Elimination Initiative through funding from The Parker Foundation (www.parker.org). L.S.T. is a Skills Development Fellow (#N011570) jointly funded by the UK Medical Research Council (MRC) and the UK Department for International Development (DFID) under the MRC/DFID Concordat agreement (http://www.mrc.ac.uk/). F.O.O. is also supported by a Wellcome Trust Intermediate Research Fellowship (#WT102350/Z/13/Z). We thank Dr William Hawley for his review of the manuscript, Dr Jimee Hwang for input on the study protocol, and Nicolas Simon for his help with study screening.

CONTRIBUTORS

R.D.G., A.T., and G.F.K. conceived of the study. Y.A.W., L.S.T., R.D.G., G.F.K., and A.T. developed the study design. Y.A.W., L.S.T., and S.H. searched the literature. Y.A.W. and L.S.T. extracted the data and prepared the manuscript. P.M.G. advised on the systematic review. All authors had access to study data and reviewed the final manuscript. All authors read and approved the final manuscript.

CONFLICT OF INTEREST

The authors declare that they have no conflict of interests. The study sponsors had no role in study design, in the collection, analysis and interpretation of data, in writing the report, and in the decision to submit for publication.

REFERENCES

Alphey, L., Alphey, N., 2014. Five things to know about genetically modified (GM) insects for vector control. PLoS Pathog. 10 (3), e1003909.

Atkins, D., Best, D., Briss, P.A., Eccles, M., Falck-Ytter, Y., Flottorp, S., Guyatt, G.H., Harbour, R.T., Haugh, M.C., Henry, D., Hill, S., Jaeschke, R., Leng, G., Liberati, A., Magrini, N., Mason, J., Middleton, P., Mrukowicz, J., O'Connell, D., Oxman, A.D., Phillips, B., Schünemann, H.J., Edejer, T., Varonen, H., Vist, G.E.,

Williams Jr., J.W., Zaza, S., GRADE Working Group, 2004. Grading quality of evidence and strength of recommendations. BMJ 328 (7454), 1490.

Bhatt, S., Gething, P.W., 2014. Insecticide-treated nets (ITNs) in Africa 2000-2016: coverage, system efficiency and future needs for achieving international targets. Malar. J. 13, O29.

Bhatt, S., Weiss, D.J., Mappin, B., Dalrymple, U., Cameron, E., Bisanzio, D., Smith, D.L., Moyes, C.L., Tatem, A.J., Lynch, M., Fergus, C.A., Yukich, J., Bennett, A., Eisele, T.P., Kolaczinski, J., Cibulskis, R.E., Hay, S.I., Gething, P.W., 2015. Coverage and system efficiencies of insecticide-treated nets in Africa from 2000 to 2017. eLife 4, 1–37.

Brady, O.J., Godfray, H.C.J., Tatem, A.J., Gething, P.W., Cohen, J.M., Mckenzie, F.E., Perkins, T.A., Reiner, R.C., Tusting, L.S., Sinka, M.E., Moyes, C.L., Eckhoff, P.A., Scott, T.W., Lindsay, S.W., Hay, S.I., Smith, D.L., 2016. Vectorial capacity and vector control: reconsidering sensitivity to parameters for malaria elimination. Trans. R. Soc. Trop. Med. Hyg. 110 (2), 107–117.

Chaccour, C.J., Rabinovich, N.R., Slater, H., Canavati, S.E., Bousema, T., Lacerda, M., Ter Kuile, F., Drakeley, C., Bassat, Q., Foy, B.D., Kobylinski, K., 2015. Establishment of the Ivermectin Research for Malaria Elimination Network: updating the research agenda. Malar. J. 14, 243.

Darriet, F.D.R., Robert, V., Vien, N.T., Carnevale, P., 1984. Evaluation of the Efficacy of Permethrin-Impregnated Intact and Perforated Mosquito Nets Against Vectors of Malaria. World Health Organization, Geneva.

Donnelly, B., Berrang-Ford, L., Ross, N.A., Michel, P., 2015. A systematic, realist review of zooprophylaxis for malaria control. Malar. J. 14, 313.

Effective Practice and Organisation of Care (EPOC). Suggested risk of bias criteria for EPOC reviews. EPOC Resources for review authors. Oslo: Norwegian Knowledge Centre for the Health Services, 2015.

Foy, B.D., Kobylinski, K.C., Da Silva, I.M., Rasgon, J.L., Sylla, M., 2011. Endectocides for malaria control. Trends Parasitol. 27, 423–428.

Gates, B., & Chambers, R. (2015). "From Aspiration to Action: What Will It Take to End Malaria?". http://endmalaria2040.org/ (accessed Feb 18, 2016).

Global Malaria Programme, 2014. Control of Residual Malaria Parasite Transmission, Guidance Note—September. World Health Organization, Geneva.

Global Malaria Programme, 2015. World Malaria Report 2015. World Health Organization, Geneva.

Global Malaria Programme, 2017. World Malaria Report 2017. World Health Organization, Geneva.

Govella, N.J., Ferguson, H., 2012. Why use of interventions targeting outdoor biting mosquitoes will be necessary to achieve malaria elimination. Front. Physiol. 3, 199.

Hill, N., Zhou, H.N., Wang, P., Guo, X., Carneiro, I., Moore, S.J., 2014. A household randomized, controlled trial of the efficacy of 0.03% transfluthrin coils alone and in combination with long-lasting insecticidal nets on the incidence of *Plasmodium falciparum* and *Plasmodium vivax* malaria in Western Yunnan Province, China. Malar. J. 13, 208.

Homan, T., Hiscox, A., Mweresa, C.K., Masiga, D., Mukabana, W.R., Oria, P., Maire, N., Pasquale, A.D., Silkey, M., Alaii, J., Bousema, T., Leeuwis, C., Smith, T.A., Takken, W., 2016. The effect of mass mosquito trapping on malaria transmission and disease burden (SolarMal): a stepped-wedge cluster-randomised trial. Lancet 388 (10050), 1193–1201.

Innovation to Impact (I2I), 2016. http://innovationtoimpact.org/. Accessed 20 August 2016.

Innovative Vector Control Consortium, 2016. IVCC Annual Report 2015-16. IVCC, Liverpool.

Killeen, G.F., 2014. Characterizing, controlling and eliminating residual malaria transmission. Malar. J. 13, 330.

Kirby, M.J., Ameh, D., Bottomley, C., Green, C., Jawara, M., Milligan, P.J., Snell, P.C., Conway, D.J., Lindsay, S.W., 2009. Effect of two different house screening interventions on exposure to malaria vectors and on anaemia in children in The Gambia: a randomised controlled trial. Lancet 374 (9694), 998–1009.

Kiware, S.S., Chitnis, N., Tatarsky, A., Wu, S., Castellanos, H.M.S., Gosling, R., Smith, D., Marshall, J.M., 2017. Attacking the mosquito on multiple fronts: insights from the Vector Control Optimization Model (VCOM) for malaria elimination. PLoS One 12 (2), e0187680.

Knapp, J., Macdonald, M., Malone, D., Hamon, N., Richardson, J., 2015. Disruptive technology for vector control: the Innovative Vector Control Consortium and the US Military join forces to explore transformative insecticide application technology for mosquito control programmes. Malar. J. 14, 371.

Lawrence, C.E., Croft, A.M., 2004. Do mosquito coils prevent malaria? A systematic review of trials. J. Travel Med. 11, 92–96.

Lobo, N.F., Achee, N.L., Syafruddin, D., 2014. Spatial Repellent Products for Control of Vector-Borne Diseases—Malaria—Indonesia (SR-M-IDR). University of Notre Dame. https://clinicaltrials.gov/show/NCT02294188. Accessed 10 June 2016.

Macintyre, K., Sosler, S., Letipila, F., Lochigan, M., Hassig, S., Omar, S.A., Githure, J., 2003. A new tool for malaria prevention?: results of a trial of permethrin-impregnated bedsheets (shukas) in an area of unstable transmission. Int. J. Epidemiol. 32, 157–160.

Magris, M., Rubio-Palis, Y., Alexander, N., Ruiz, B., Galván, N., Frias, D., Blanco, M., Lines, J., 2007. Community-randomized trial of lambdacyhalothrin-treated hammock nets for malaria control in Yanomami communities in the Amazon region of Venezuela. Trop. Med. Int. Health 12, 392–403.

Malaria Policy Advisory Committee, WHO, 2012. Technical Expert Group (TEG) on Malaria Vector Control: Terms of Reference. World Health Organization, Geneva.

Malaria Policy Advisory Committee, WHO, 2015. Innovation to Impact—WHO Change Plan for Strengthening Innovation, Quality and Use of Vector Control Tools. World Health Organization, Geneva.

malERA Refresh Consultative Panel on Tools for Malaria Elimination, 2017. malERA: an updated research agenda for diagnostics, drugs, vaccines, and vector control in malaria elimination and eradication. PLoS Med. 14 (11), e1002455.

Mbare, O., Lindsay, S., Fillinger, U., 2014. Pyriproxyfen for mosquito control: female sterilization or horizontal transfer to oviposition substrates by *Anopheles gambiae sensu stricto* and *Culex quinquefasciatus*. Parasit. Vectors 7 (1), 280.

Messenger, L.A., Matias, A., Manana, A.N., Stiles-Ocran, J.B., Knowles, S., Boakye, D.A., Coulibaly, M.B., Larsen, M.-L., Traoré, A.S., Diallo, B., Konaté, M., Guindo, A., Traoré, S.F., Mulder, C.E., Le, H., Kleinschmidt, I., Rowland, M., 2012. Multicentre studies of insecticide-treated durable wall lining in Africa and South-East Asia: entomological efficacy and household acceptability during one year of field use. Malar. J. 11 (1), 1–13.

Mnzava, A.P., Macdonald, M.B., Knox, T.B., Temu, E.A., Shiff, C.J., 2014. Malaria vector control at a crossroads: public health entomology and the drive to elimination. Trans. R. Soc. Trop. Med. Hyg. 108 (9), 550–554.

Mtove, G., Mugasa, J.P., Messenger, L.A., Malima, R.C., Mangesho, P., Magogo, F., Plucinski, M., Hashimu, R., Matowo, J., Shepard, D., Batengana, B., Cook, J., Emidi, B., Halasa, Y., Kaaya, R., Kihombo, A., Lindblade, K.A., Makenga, G., Mpangala, R., Mwambuli, A., Mzava, R., Mziray, A., Olang, G., Oxborough, R.M., Seif, M., Sambu, E., Samuels, A., Sudi, W., Thomas, J., Weston, S., Alilio, M., Binkin, N., Gimnig, J., Kleinschmidt, I., McElroy, P., Moulton, L.H., Norris, L., Ruebush, T., Venkatesan, M., Rowland, M., Mosha, F.W., Kisinza, W.N., 2016. The effectiveness of non-pyrethroid insecticide-treated durable wall lining to control malaria in rural Tanzania: study protocol for a two-armed cluster randomized trial. BMC Public Health 16 (1), 633.

Ogoma, S.B., Moore, S.J., Maia, M.F., 2012. A systematic review of mosquito coils and passive emanators: defining recommendations for spatial repellency testing methodologies. Parasit. Vectors 5, 287.

Pinder, M., Conteh, L., Jeffries, D., Jones, C., Knudsen, J., Kandeh, B., Jawara, M., Sicuri, E., D'Alessandro, U., Lindsay, S.W., 2016. The RooPfs study to assess whether improved housing provides additional protection against clinical malaria over current best practice in The Gambia: study protocol for a randomized controlled study and ancillary studies. Trials 17 (1), 275.

Ranson, H., Lissenden, N., 2016. Insecticide resistance in African *Anopheles* mosquitoes: a worsening situation that needs urgent action to maintain malaria control. Trends Parasitol. 32, 187–196.

Roll Back Malaria, 2015. Draft Consensus Statement on Housing and Malaria. World Health Organization, Geneva.

Rowland, M., Durrani, N., Hewitt, S., Mohammed, N., Bourna, M., Carneiro, I., 1999. Permethrin-treated protection against chaddars and top-sheets: appropriate technology for malaria in Afghanistan and other complex emergencies. Trans. R. Soc. Trop. Med. Hyg. 93, 465–472.

Rowland, M., Durrani, N., Kenward, M., Mohammed, N., Urahman, H., Hewitt, S., 2001. Control of malaria in Pakistan by applying deltamethrin insecticide to cattle: a community-randomised trial. Lancet 357, 1837–1841.

Sadasivaia, S., Tozan, Y., Breman, J., 2007. Dichlorodiphenyltrichloroethane (DDT) for indoor residual spraying in Africa: how can it be used for malaria control? Am. J. Trop. Med. Hyg. 77, 249–263.

Shea, B.J., Grimshaw, J.M., Wells, G.A., et al., 2007. Development of AMSTAR: a measurement tool to assess the methodological quality of systematic reviews. BMC Med. Res. Methodol. 7, 10.

Snow, R.W., Sartorius, B., Kyalo, D., Maina, J., Amratia, P., Mundia, C.W., Bejon, P., Noor, A.M., 2017. The prevalence of *Plasmodium falciparum* in sub-Saharan Africa since 1900. Nature 550 (4677), 515–518.

Syafruddin, D., Bangs, M.J., Sidik, D., Elyazar, I., Asih, P.B., Chan, K., Nurleila, S., Nixon, C., Hendarto, J., Wahid, I., Ishak, H., Bogh, C., Grieco, J.P., Achee, N.L., Baird, J.K., 2014. Impact of a spatial repellent on malaria incidence in two villages in Sumba, Indonesia. Am. J. Trop. Med. Hyg. 91, 1079–1087.

Thang, N.D., Erhart, A., Speybroeck, N., Xa, N.X., Thanh, N.N., Van Ky, P., Hung, L.X., Thuan, L.K., Coosemans, M., D'alessandro, U., 2009. Long-lasting insecticidal hammocks for controlling forest malaria: a community-based trial in a rural area of Central Vietnam. PLoS One 4, e7369.

The Anopheles gambiae 1000 Genomes Consortium, 2017. Genetic diversity of the African malaria vector *Anopheles gambiae*. Nature 552 (1476–4687 (Electronic)), 96–100.

Thomas, M., Knols, B., N'guessan, R., 2015. Transition of Eave Tubes from Concept to Implementation. Pennsylvania State University.

Tusting, L.S., Thwing, J., Sinclair, D., Fillinger, U., Gimnig, J., Bonner, K.E., Bottomley, C., Lindsay, S.W., 2013. Mosquito larval source management for controlling malaria. Cochrane Database Syst. Rev. 8, CD008923.

Tusting, L.S., Ippolito, M.M., Willey, B.A., Kleinschmidt, I., Dorsey, G., Gosling, R.D., Lindsay, S.W., 2015. The evidence for improving housing to reduce malaria: a systematic review and meta-analysis. Malar. J. 14, 209.

Vontas, J., Moore, S., Kleinschmidt, I., Ranson, H., Lindsay, S., Lengeler, C., Hamon, N., Mclean, T., Hemingway, J., 2014. Framework for rapid assessment and adoption of new vector control tools. Trends Parasitol. 30, 191–204.

Wagman, J., Gogue, C., Tynuv, K., Mihigo, J., Bankineza, E., Bah, M., Diallo, D:, Saibu, A., Richardson, J.H., Kone, D., Fomba, S., Bernson, J., Steketee, R., Slutsker, L., Robertson, M., 2018. An observational analysis of the impact of indoor residual spraying with non-pyrethroid insecticides on the incidence of malaria in Ségou Region, Mali: 2012–2015. Malar. J. 17, 19.

Walshe, D.P., Garner, P., Abdel-Hameed Adeel, A.A., Pyke, G.H., Burkot, T., 2013. Larvivorous fish for preventing malaria transmission. Cochrane Database Syst. Rev. 12, CD008090.

WHO Vector Control Advisory Group, 2013. Report on the First Meeting of the WHO Vector Control Advisory Group. World Health Organization Vector Control Advisory Group, Geneva.

WHO Vector Control Advisory Group, 2014. Report on the Second Meeting of the WHO Vector Control Advisory Group. World Health Organization Vector Control Advisory Group, Geneva.

Wilson, A.L., Chen-Hussey, V., Logan, J.G., Lindsay, S.W., 2014. Are topical insect repellents effective against malaria in endemic populations? A systematic review and meta-analysis. Malar. J. 13, 446.

Wilson, A.L., Boelaert, M., Kleinschmidt, I., Pinder, M., Scott, T.W., Tusting, L.S., Lindsay, S.W., 2015. Evidence-based vector control? Improving the quality of vector control trials. Trends Parasitol. (8), 380–390.

World Health Organization, 2009. Development of a Global Action Plan for Integrated Vector Management (IVM). World Health Organization, Geneva.

World Health Organization, 2015. Global Technical Strategy for Malaria 2016–2030. World Health Organization, Geneva. http://www.who.int/malaria/publications/atoz/9789241564991/en/ (accessed May 27, 2016).

World Health Organization, 2017. Global Vector Control Response 2017–2030. World Health Organization, Geneva. http://apps.who.int/iris/bitstream/10665/259205/1/9789241512978-eng.pdf?ua=1 (accessed January 10, 2018).

CPI Antony Rowe
Chippenham, UK
2018-05-24 21:27